Feedstock-based Bioethanol Fuels. II. Waste Feedstocks

This book provides an overview of research on the production of bioethanol fuels from waste feedstocks such as second-generation residual sugar and starch feedstocks, food waste, industrial waste, urban waste, forestry waste, and lignocellulosic biomass at large with 17 chapters. In this context, there are eight sections where the first two chapters cover the production of bioethanol fuels from waste feedstocks at large.

This book is the fourth volume in the ***Handbook of Bioethanol Fuels*** (Six-Volume Set). It shows that pretreatments and hydrolysis of the waste feedstocks, fermentation of hydrolysates, and separation and distillation of bioethanol fuels are the fundamental processes for bioethanol fuel production from these waste feedstocks.

This book is a valuable resource for stakeholders primarily in research fields of energy and fuels, chemical engineering, environmental science and engineering, biotechnology, microbiology, chemistry, physics, mechanical engineering, agricultural sciences, food science and engineering, materials science, biochemistry, genetics, molecular biology, plant sciences, water resources, economics, business and management, transportation science and technology, ecology, public, environmental and occupational health, social sciences, toxicology, multi-disciplinary sciences, and humanities among others.

Feedstock-based Bioethanol Fuels. II. Waste Feedstocks

Agricultural, Food, Industrial, Urban, Forestry, and Lignocellulosic Waste-based Bioethanol Fuels

Edited by
Ozcan Konur

CRC Press
Taylor & Francis Group
Boca Raton London New York

CRC Press is an imprint of the
Taylor & Francis Group, an **informa** business

Designed cover image: © Shutterstock

First edition published 2024
by CRC Press
2385 NW Executive Center Drive, Suite 320, Boca Raton FL 33431

and by CRC Press
4 Park Square, Milton Park, Abingdon, Oxon, OX14 4RN

CRC Press is an imprint of Taylor & Francis Group, LLC

ISBN: 978-1-032-12754-5 (hbk)
ISBN: 978-1-032-12867-2 (pbk)
ISBN: 978-1-003-22655-0 (ebk)

DOI: 10.1201/9781003226550

Typeset in Times
by codeMantra

Contents

PART 18 Second Generation Bioethanol Fuels from Residual Starch Feedstocks

Preface

Recent supply shocks caused first by the COVID-19 pandemic and later by the Ukrainian war have shown that biofuels such as bioethanol, biohydrogen, biogas, biosyngas, and biodiesel fuels could play a vital role to maintain the energy security and indirectly food security at the global scale. These shocks have also resulted in the need for further setup of incentive structures for the production and consumption of bioethanol fuels in blends with crude oil-based gasoline, petrodiesel, or liquefied natural gas (LNG) in gasoline and diesel engines, for their direct utilization in direct ethanol fuel cells (DEFCs), and for the production of biohydrogen fuels for fuel cells and valuable biochemicals from bioethanol fuels.

Thus, it is essential to assess the research on the production, evaluation, and utilization of bioethanol fuels from a wide range of biomass including first generation starch and sugar feedstocks, wood, and grass; and second generation lignocellulosic biomass including waste biomass and agricultural residues such as starch feedstock residues and sugar feedstock residues; and third generation algal biomass.

Thus, this six-volume *Handbook of Bioethanol Fuels* assesses research on the production, evaluation, and utilization of bioethanol fuels and presents a representative sample of this interdisciplinary research population with a collection of 110 chapters (Table 1.1).

The first two volumes provide an overview of research on fundamental processes for bioethanol fuel production with a collection of 39 chapters: Pretreatments of the biomass, hydrolysis of the pretreated biomass, microbial fermentation of the hydrolysates with yeasts, and separation and distillation of bioethanol fuels from the fermentation broth. They also provide an overview of the research on bioethanol fuels and production processes for bioethanol fuels (Tables 1.2 and 1.3).

The third and fourth volumes provide an overview of research on the production of bioethanol fuels from non-waste and waste biomass, respectively, with a collection of 36 chapters. In this context, the third volume covers the production of bioethanol fuels from first generation starch feedstocks and sugar feedstocks, grass biomass, wood biomass, cellulose, biosyngas, and third generation algae (Table 1.4) while the fourth volume covers the production of second generation bioethanol fuels from residual sugar feedstocks, residual starch feedstocks, food waste, industrial waste, urban waste, forestry waste, and lignocellulosic biomass at large (Table 1.5). They also provide an overview of research on feedstock-based bioethanol fuels, non-waste feedstock-based bioethanol fuels, and second generation waste biomass-based bioethanol fuels (Tables 1.4 and 1.5).

Finally, the fifth and sixth volumes provide an overview of research on the evaluation and utilization of bioethanol fuels with a collection of 37 chapters. In this context, the fifth volume covers the evaluation and utilization of bioethanol fuels in general, gasoline fuels, nanotechnology applications in bioethanol fuels, utilization of bioethanol fuels in transport engines, evaluation of bioethanol fuels, utilization of bioethanol fuels, and development and utilization of bioethanol fuel sensors (Table 1.6). Further, the sixth volume of this handbook provides an overview of research on the country-based experience of bioethanol fuels at large, Chinese, US, and European experience of bioethanol fuels, production of bioethanol fuel-based biohydrogen fuels for fuel cells, bioethanol fuel cells, and bioethanol fuel-based biochemicals with a collection of 19 chapters (Table 1.7).

Thus, the fourth volume of this handbook provides an overview of research on the production of bioethanol fuels from waste feedstocks such as second generation residual sugar and starch feedstocks, food waste, industrial waste, urban waste, forestry waste, and lignocellulosic biomass at large with 17 chapters (Table 1.5). In this context, there are eight sections where the first two chapters cover the production of bioethanol fuels from waste feedstocks at large.

Hence, the fourth volume indicates that research on the production of bioethanol fuels from these waste feedstocks has intensified in recent years to become a major part of the bioenergy and biofuels research together primarily with biodiesel, biohydrogen, and biogas research as a sustainable alternative to crude oil-based gasoline and petrodiesel fuels as well as LNG.

The fourth volume also indicates that a wide range of pretreatments alone or in combination with each other fractionate the biomass to its constituents of cellulose, lignin, and hemicellulose and improve both sugar and bioethanol fuel yields, making bioethanol fuels from these waste feedstocks more competitive in relation to crude oil and natural gas-based fuels.

The fourth volume also indicates that hydrolysis of the biomass, microbial hydrolysate fermentation, and separation and distillation of bioethanol fuels from fermentation broth together with biomass pretreatments are the fundamental production processes for bioethanol fuel production from these waste feedstocks, making bioethanol fuels more competitive in relation to crude oil- and natural gas-based fuels.

The fourth volume also indicates that a small number of documents, authors, institutions, publication years, source titles, countries, Scopus subject categories, Scopus keywords, and research fronts have shaped the research on bioethanol fuels from these waste feedstocks.

The fourth volume also indicates that the level of funding for research on both bioethanol fuels from these waste feedstocks has not been sufficient with the resulting loss of momentum in the research output in recent years. Thus, there is a crucial need to improve the incentive structures for major stakeholders such as researchers and their institutions as well source titles and academic databases to improve the volume and quality of the research output in these fields. There is a crucial need to maintain the energy security and indirectly food security at a global scale in light of the recent supply shocks caused by the COVID-19 pandemic and the Ukrainian war.

The fourth volume also indicates that the contribution of social sciences and humanities to the research in these fields has been minimal, due in part to the restrictive editorial policies of the source titles in these fields toward social science- and humanities-based interdisciplinary studies. Thus, there is ample room to improve incentive structures for the inclusion of social sciences and humanities into these fields.

The fourth volume also indicates that China, Europe as a whole, and the USA have been major producers of research in these research fields, and there has been heavy competition among them in terms of both volume and citation impact of the research output. The USA and Europe as a whole have had a higher citation impact in relation to China, benefiting from their first-mover advantage starting their research in these fields in the 1970s. China as a late mover have had more intensive research funding initiatives in relation to the USA and Europe, improving its both research output and citation impact through the provision of efficient incentive structures for its major stakeholders in the last two decades. In this way, China might also overtake both the USA and Europe in terms of citation impact of the research output in addition to the volume of the research output in the future.

This handbook at large and fourth volume are a valuable resource for stakeholders primarily in research fields of energy and fuels, chemical engineering, environmental science and engineering, biotechnology, microbiology, chemistry, physics, mechanical engineering, agricultural sciences, food science and engineering, materials science, biochemistry, genetics, molecular biology, plant sciences, water resources, economics, business and management, transportation science and technology, ecology, public, environmental and occupational health, social sciences, toxicology, multidisciplinary sciences, and humanities among others.

Ozcan Konur

Acknowledgments

This handbook has been a multi-stakeholder project from its conception to its publication. CRC Press and Taylor & Francis Group have been the major stakeholders in financing and executing it. Marc Gutierrez has been the executive director of the project. A large number of teams from the publisher contributed immensely to the production of the handbook. Only a limited number of authors have participated in this project due to the low level of incentives, compared to journals. A small number of highly cited scholars have shaped the research on bioethanol fuels. The contribution of all these and other stakeholders to this handbook has been greatly acknowledged.

Editor

Ozcan Konur has interdisciplinary research interests and has published primarily in areas of bioenergy and biofuels, algal bioenergy and biofuels, nanoenergy and nanofuels, nanobiomedicine, algal biomedicine, disability studies, higher education, biodiesel fuels, algal biomass, lignocellulosic biomass, scientometrics, and bioethanol fuels. He has edited a book titled *Bioenergy and Biofuels* (CRC Press, 2018), a handbook titled *Handbook of Algal Science, Technology, and Medicine* (Elsevier, 2020), and a handbook titled *Handbook of Biodiesel and Petrodiesel Fuels: Science, Technology, Health, and Environment* (CRC Press, 2021) in three volumes.

Contributors

María C. Area
IMAM, UNaM, CONICET, FCEQYN, Pulp and Paper Program (PROCyP)
National University of Misiones
Posadas, Argentina

Ozcan Konur
(Formerly) Department of Materials Engineering
Ankara Yildirim beyazit University
Ankara, Turkey

Julia Kruyeniski
IMAM, UNaM, CONICET, FCEQYN, Pulp and Paper Program (PROCyP)
National University of Misiones
Posadas, Argentina

Carolina M. Mendieta
IMAM, UNaM, CONICET, FCEQYN, Pulp and Paper Program (PROCyP)
National University of Misiones
Posadas, Argentina

María E. Vallejos
IMAM, UNaM, CONICET, FCEQYN, Pulp and Paper Program (PROCyP)
National University of Misiones
Posadas, Argentina

Part 16

Introduction to Second Generation Waste Biomass-based Bioethanol Fuels

59 Second Generation Waste Biomass-based Bioethanol Fuels

Scientometric Study

Ozcan Konur
(Formerly) Ankara Yildirim Beyazit University

59.1 INTRODUCTION

Crude oil-based gasoline fuels (Ma et al., 2002; Newman and Kenworthy, 1989) have been widely used in the transportation sector since the 1920s. However, there have been great public concerns over the adverse environmental and human impact of these fuels (Hill et al., 2006, 2009). Hence, biomass-based bioethanol fuels (Hill et al., 2006; Konur, 2012e, 2015, 2019, 2020a) have increasingly been used in blending gasoline fuels (Hsieh et al., 2002; Najafi et al., 2009), in fuel cells (Antolini, 2007, 2009), and in biochemical production (Angelici et al., 2013; Morschbacker, 2009) in a biorefinery context (Fernando et al., 2006; Huang et al., 2008).

Bioethanol fuels also play a critical role in maintaining energy security (Kruyt et al., 2009; Winzer, 2012) in supply shocks (Kilian, 2008, 2009) related to oil price shocks (Hamilton, 1983, 2003), the COVID-19 pandemic (Fauci et al., 2020; Li et al., 2020), or wars (Hamilton, 1983; Jones, 2012) in the aftermath of the Russian invasion of Ukraine (Reeves, 2014).

However, it is necessary to pretreat the biomass (Taherzadeh and Karimi, 2008; Yang and Wyman, 2008) to enhance the yield of the bioethanol (Hahn-Hagerdal et al., 2006; Sanchez and Cardona, 2008) prior to bioethanol production through hydrolysis (Sun and Cheng, 2002; Taherzadeh and Karimi, 2007) and fermentation (Lin and Tanaka, 2006; Olsson and Hahn-Hagerdal, 1996) of biomass and hydrolysates, respectively.

One of the most-studied feedstocks for bioethanol fuels has been waste biomass. The research in the field of second generation waste biomass-based bioethanol fuels has intensified in this context in the key research fronts of the pretreatment (Hendriks and Zeeman, 2009; Mosier et al., 2005) and hydrolysis (Jorgensen et al., 2007; Sun and Cheng, 2002) of waste biomass, fermentation (Palmqvist and Hahn-Hagerdal, 2000a,b) of waste biomass-based hydrolysates, and production (Limayem and Ricke, 2012; Zaldivar et al., 2001) and evaluation (Hamelinck et al., 2005; Sassner et al., 2008) of second generation waste biomass-based bioethanol fuels.

The research in this field has also intensified for feedstocks of lignocellulosic biomass at large (Mosier et al., 2005; Sun and Cheng, 2002), agricultural residues (Prasad et al., 2007; Saini et al., 2015), industrial wastes (Kadar et al., 2004; Koutinas et al., 2014), forestry wastes (Brandt et al., 2010; Duff and Murray, 1996), urban wastes (Guimaraes et al., 2010; Lenihan et al., 2010), and food wastes (Lenihan et al., 2010; Ravindran and Jaiswal, 2016). Thus, it emerges as a distinctive research field, complementing the primary research on first generation bioethanol fuels from agricultural feedstocks and wood among others.

However, it is essential to develop efficient incentive structures (North, 1991) for the primary stakeholders to enhance research in this field (Konur, 2000, 2002a–c, 2006a, b, 2007a, b).

DOI: 10.1201/9781003226550-79

Scientometric analysis has been used in this context to inform the primary stakeholders about the current state of research in this research field (Garfield, 1955; Konur, 2011, 2012a–i, 2015, 2018b, 2019, 2020a).

As there have been no published scientometric studies in this field, this chapter presents a scientometric study of the research in second generation waste biomass-based bioethanol fuels. It examines the scientometric characteristics of both the sample and population data presenting scientometric characteristics of both these datasets in the order of documents, authors, publication years, institutions, funding bodies, source titles, countries, Scopus subject categories, Scopus keywords, and research fronts.

59.2 MATERIALS AND METHODS

The search for this study was carried out using the Scopus database (Burnham, 2006) in October 2022.

As the first step for the search of the relevant literature, keywords were selected using the 300 most-cited population papers for each waste biomass. The selected keyword list was then optimized to obtain a representative sample of papers from this research field. These keyword lists were then integrated to obtain the keyword list for this research field (Konur, 2023a–g).

As the second step, two sets of data were used in this study. First, a population sample of 13,728 papers was used to examine the scientometric characteristics of the population data. Second, a sample of 275 most-cited papers, corresponding to 2% of the population papers, was used to examine the scientometric characteristics of these citation classics.

The scientometric characteristics of both these sample and population datasets were presented in the order of documents, authors, publication years, institutions, funding bodies, source titles, countries, Scopus subject categories, Scopus keywords, and research fronts.

Lastly, the key scientometric findings for both datasets were discussed to highlight the research landscape for second generation waste biomass-based bioethanol fuels. Additionally, a number of brief conclusions were drawn and a number of relevant recommendations were made to enhance the future research landscape.

59.3 RESULTS

59.3.1 The Most Prolific Documents on Second Generation Waste Biomass-based Bioethanol Fuels

Information on the types of documents for both datasets is given in Table 59.1. Articles and conference papers, published in journals, dominate both the sample (69%) and population (92%) papers with a 23% deficit. Further, review papers and short surveys are a 26% surplus as they are over-represented in the sample papers as they constitute 31% and 5% of the sample and population papers, respectively.

It is further notable that 97%, 2%, and 1% of the population papers were published in journals, books, and book series, respectively. Similarly, 98% and 2% of the sample papers were published in journals and book series, respectively.

59.3.2 The Most Prolific Authors on Second Generation Waste Biomass-based Bioethanol Fuels

Information about the 32 most prolific authors with at least 1.6% of sample papers each is given in Table 59.2. The most prolific author is Bruce E. Dale with 6.6% of the sample papers, followed by John N. Saddler and Charles E. Wyman with 4.9% of the sample papers each. Other prolific

TABLE 59.1
Documents in the Second Generation Waste Biomass-based Bioethanol Fuels

Documents	Sample Dataset (%)	Population Dataset (%)	Surplus (%)
Article	65.1	89.8	–24.7
Review	29.8	4.6	25.2
Conference paper	3.6	2.5	1.1
Short survey	1.5	0.1	1.4
Book Chapter	0.0	2.4	–2.4
Editorial	0.0	0.1	–0.1
Letter	0.0	0.2	–0.2
Book	0.0	0.1	–0.1
Note	0.0	0.2	–0.2
Sample size	275	13,728	

Sample dataset: The number of papers (%) in the set of 275 highly cited papers.
Population dataset: The number of papers (%) in the set of 13,728 population papers.

authors are Yoon Y. Lee, Barbel Hahn-Hagerdal, Mark T. Holtzapple, and Michael R. Ladisch with 3.8%–4.4% of the sample papers each.

On the other hand, the most influential author is Bruce E. Dale with a 4.4% surplus, followed by Yoon Y. Lee, Charles E. Wyman, and John N. Saddler with a 3.7%, 3.5%, and 3.4% surplus, respectively. Other influential authors are Barbel Hahn-Hagerdal and Mark T. Holtzapple with 3.2% and 3.1% surplus, respectively.

The most prolific institution for the sample dataset is the National Renewable Energy Laboratory (NREL) with four authors, followed by Imperial College and Lund University with three authors each. Other prolific institutions are the Center for Energy, Environmental, and Technological Research (CIEMAT), Purdue University, and the University of Copenhagen with two authors each. On the other hand, the most prolific country for the sample dataset is the USA with 15 authors, followed by Sweden with six authors, respectively. Other prolific countries are Denmark, the UK, and Spain with three, three, and two authors, respectively. In total, only eight countries house these top authors.

The most prolific research front for these top authors is pretreatments of waste biomass with 29 authors followed by hydrolysis of waste biomass with 24 authors. Other prolific research fronts are the fermentation of waste biomass-based hydrolysates and bioethanol production with 16 and 17 authors, respectively.

On the other hand, there is a significant gender deficit (Beaudry and Lariviere, 2016) for the sample dataset as surprisingly only six of these top researchers are female with a representation rate of 19%.

Additionally, there are other authors with a relatively low citation impact and with 0.5%–1.5% of the population papers each: Je Bao, Zhenhong Yuan, Qiang Yong, Ashok Pandey, Anuj K. Chandel, Verawat Champreda, Navadol Laosiripojana, Blake A. Simmons, Xuebing Zhao, Hongyan Chen, Akihiko Kondo, Carlos Martin, Johann F. Gorgens, Dehua Liu, Liangcai Peng, Zhenhong Yuan, Eulogio Castro, Solange I. Musatto, Wen Wang, Carlos R. Soccol, Hasan Jameel, Kyoung Heon Kim, Paloma Manzanares, Adriane M. F. Milagres, Anne S. Meyer, Jose M. Oliva, Wei Qi, Jinghuang Hu, Sarita C. Rabelo, Yan Xu, Zhanrong Zhang, Birgitte K. Ahring, Yongcan Jin, Luiz P. Ramos, S. Singh, Jibin Zhang, Bruce S. Dien, Yinbo Qu, Ying-Jin Yuan, Parameswaran Binod, Bing-Zhi Li, Hector A. Ruiz, Badal C. Saha, Jose A. Teixeria, Qian Wang, and Jing Zhao.

TABLE 59.2
Most Prolific Authors in Second Generation Waste Biomass-based Bioethanol Fuels

No.	Author Name	Author Code	Sample Papers (%)	Population Papers (%)	Surplus (%)	Institution	Country	HI	N	Res. Front
1	Dale, Bruce E.	7201511969	6.6	2.2	4.4	Michigan State Univ.	USA	92	430	P, H, F, R
2	Wyman, Charles E.	7004396809	4.9	1.4	3.5	Univ. Calf. Riverside	USA	80	287	P, H, F, R
3	Saddler, John N.	7005297559	4.9	1.5	3.4	Univ. British Columbia	Canada	99	420	P, H, F, R
4	Lee, Yoon Y	8948274900	4.4	0.7	3.7	Auburn Univ.	USA	45	102	P, H
5	Hahn-Hagerdal, Barbel*	7005389381	3.8	0.6	3.2	Lund Univ.	Sweden	76	258	P, H, F, R
6	Holtzapple, Mark T.	7004167004	3.8	0.7	3.1	Texas A&M Univ.	USA	47	199	P
7	Ladisch, Michael R.	7005670397	3.8	0.9	2.9	Purdue Univ.	USA	75	334	P, R
8	Mosier, Nathan S.	6602426392	3.3	0.4	2.9	Purdue Univ.	USA	43	117	P, H, F, R
9	Himmel, Michael E.	7007125552	3.3	0.7	2.6	Natl. Renew. Ener. Lab.	USA	73	423	P, H, R
10	Jonsson, Leif J.	7102349315	3.3	0.9	2.4	Umea Univ.	Sweden	41	148	F
11	Kim, Tae Hyun	57210847338	3.3	1.0	2.3	Hanyang Univ.	S. Korea	23	88	P, H, F, R
12	Zacchi, Guido	7006727748	3.3	1.1	2.2	Lund Univ.	Sweden	68	204	P, H, F, R
13	Taherzadeh, Mohammad J.	6701407496	3.3	1.6	1.7	Univ. Boras	Sweden	66	419	P, H, F, R
14	Chundawat, Shishir P. S.	12803763300	2.7	0.6	2.1	Rutgers	USA	32	90	PH
15	Jorgensen, Henning	7202554496	2.7	0.7	2.0	Univ. Copenhagen	Denmark	43	96	P, H
16	Thomsen, Anne B.*	7102150211	2.7	0.8	1.9	Tech. Univ. Denmark	Denmark	38	62	P, H, F, R
17	Ragauskas, Arthur J.	7006265204	2.7	1.2	1.5	Univ. Tennessee	USA	93	762	P, H
18	Galbe, Mats	7003788758	2.7	1.3	1.4	Lund Univ.	Sweden	51	131	P, H, F, R
19	Balan, Venkatesh	15757087100	2.7	1.8	0.9	Univ. Houston	USA	56	213	P, H
20	Karimi, Keikhosro	10046195700	2.7	1.8	0.9	Vrije Univ.	Netherlands	56	222	P, H, F, R

(*Continued*)

TABLE 59.2 (*Continued*)
Most Prolific Authors in Second Generation Waste Biomass-based Bioethanol Fuels

No.	Author Name	Author Code	Sample Papers (%)	Population Papers (%)	Surplus (%)	Institution	Country	HI	N	Res. Front
21	Welton, Tom	7003503272	2.2	0.3	1.9	Imperial Coll.	UK	74	191	P
22	Brandt, Agnieszka*	35785816800	2.2	0.3	1.9	Imperial Coll.	UK	23	41	P
23	Aden, Andy	35324090200	2.2	0.4	1.8	Natl. Renew. Ener. Lab.	USA	USA19	36	E
24	McMillan, James D.	7102040863	2.2	0.4	1.8	Natl. Renew. Natl. lab.	USA	30	63	P, H
25	Hallett, Jason P.	7102331746	2.2	0.4	1.8	Imperial Coll.	UK	47	181	P
26	Elander, Richard T.	6603931116	2.2	0.4	1.8	Natl. Renew. Ener. Lab.	USA	31	61	P, H
27	Bura, Renata*	6602335655	2.2	0.4	1.8	Univ. Washington	USA	23	51	P, H
28	Felby, Claus	6603368580	2.2	0.7	1.5	Univ. Copenhagen	Denmark	39	103	P, H
29	Cotta, Michael A.	7006656876	2.2	0.8	1.4	USDA Agr. Res. Serv	USA	53	186	P, H, F, R
30	Olsson, Lisbeth*	7203077540	2.2	0.9	1.3	Chalmers Univ. Technol.	Sweden	60	243	F, R
31	Ballesteros, Ignacio	6602732963	2.2	1.0	1.2	CIEMAT	Spain	38	70	P, H, F, R
32	Ballesteros, Mercedes*	7006110611	2.2	1.6	0.6	CIEMAT	Spain	51	135	P, H, F, R

Author code: the unique code given by Scopus to the authors. Sample papers: the number of papers authored in the sample dataset. Population papers: the number of papers authored in the population dataset.
*, Female; P, Pretreatment of the waste biomass; H, Hydrolysis of the waste biomass; F, Fermentation of the waste biomass-based hydrolysates; R, Bioethanol fuel production; E, Bioethanol fuel evaluation.

59.3.3 The Most Prolific Research Output by Years on Second Generation Waste Biomass-based Bioethanol Fuels

Information about papers published between 1970 and 2022 is given in Figure 59.1. This figure clearly shows that the bulk of the research papers in the population dataset were published primarily in the 2010s and the early 2020s with 58% and 30% of the population dataset, respectively. Similarly, the publication rates for the 2000s, 1990s, 1980s, and 1970s were 10%, 5%, 4%, and 1% respectively. Further, the rate for the pre-1970s was 1%.

Similarly, the bulk of the research papers in the sample dataset were published in the 2010s and 2000s with 51% and 33% of the sample dataset, respectively. Similarly, the publication rates for the early 2020s, 1990s, 1980s, and 1970s were 1%, 5%, 2%, and 0% of the sample papers, respectively.

The most prolific publication years for the population dataset were 2021, 2020, and 2022 with 8.0%, 7.5%, and 7.3% of the dataset, respectively, while 83% of the population papers were published between 2009 and 2022. Similarly, 92% of the sample papers were published between 2000 and 2018 while the most prolific publication years were 2009, 2010, and 2007 with 11.3%, 9.5%, and 9.1% of the sample papers, respectively.

59.3.4 The Most Prolific Institutions on Second Generation Waste Biomass-based Bioethanol Fuels

Information about the 26 most prolific institutions publishing papers on second generation waste biomass-based bioethanol fuels with at least 1.5% of the sample papers each is given in Table 59.3.

The most prolific institution is the NREL with 7.6% of the sample papers, followed by Lund University, Technical University of Denmark, and Michigan State University with 6.5%, 5.5%, and 5.1% of the sample papers, respectively. Other prolific institutions are the University of British Columbia, Dartmouth College, Purdue University, and Auburn University with 2.9%–3.6% of the sample papers each. Similarly, the top country for these most prolific institutions is the USA with 12 institutions. Other prolific countries are Sweden, China, and Denmark with three, three, and two institutions, respectively. In total, ten countries house these top institutions.

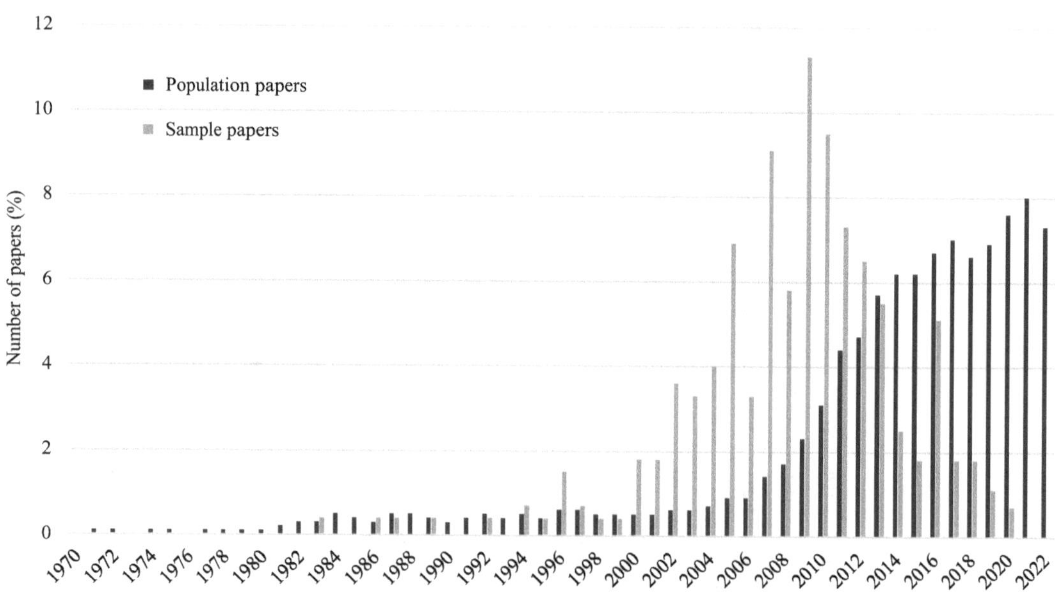

FIGURE 59.1 The research output by years regarding the second generation waste biomass-based bioethanol fuels.

TABLE 59.3
The Most Prolific Institutions in the Second Generation Waste Biomass-based Bioethanol Fuels

No.	Institutions	Country	Sample Papers (%)	Population Papers (%)	Surplus (%)
1	Natl. Renew. Ener. Lab.	USA	7.6	1.2	6.4
2	Lund Univ.	Sweden	6.5	1.0	5.5
3	Tech. Univ. Denmark	Denmark	5.5	1.2	4.3
4	Michigan State Univ.	USA	5.1	1.0	4.1
5	Univ. British Columbia	Canada	3.6	0.8	2.8
6	Dartmouth Coll.	USA	3.6	0.3	3.3
7	Purdue Univ.	USA	2.9	0.6	2.3
8	Auburn Univ.	USA	2.9	0.3	2.6
9	Univ. Sao Paulo	Brazil	2.5	2.5	0.0
10	USDA Agr. Res. Serv.	USA	2.5	0.9	1.6
11	Oak Ridge Natl. lab.	USA	2.5	0.5	2.0
12	Chinese Acad. Sci.	China	2.2	2.7	−0.5
14	Univ. Boras	Sweden	2.2	0.5	1.7
15	Texas A&M Univ.	USA	2.2	0.4	1.8
16	Imperial Coll.	UK	2.2	0.3	1.9
17	Univ. Wisconsin Madison	USA	1.8	0.5	1.3
18	Isfahan Univ. Technol.	Iran	1.8	0.5	1.3
19	Univ. Copenhagen	Denmark	1.8	0.5	1.3
20	USDA Forest Serv.	USA	1.8	0.5	1.3
21	S. China Univ. Technol.	China	1.5	1.6	−0.1
22	CIEMAT	Spain	1.5	0.6	0.9
23	Univ. Tennessee	USA	1.5	0.4	1.1
24	Umea Univ.	Sweden	1.5	0.4	1.1
25	Univ. Ill. U. C.	USA	1.5	0.3	1.2
26	KU Leuven	Belgium	1.5	0.3	1.2

On the other hand, the institutions with the highest impact are the NREL and Lund University with 6.4% and 5.5% surplus, respectively. Other influential institutions are Technical University of Denmark, Michigan State University, Dartmouth College, the University of British Columbia, and Auburn University with a 2.6%–4.3% surplus each.

Additionally, there are other institutions with relatively low citation impacts and with 0.5%–1.5% of the population papers each: State University of Campinas, Nanjing Forestry University, State Key Laboratory of Pulp and Paper Engineering, State University of Paulista, Tsinghua University, National Biorenewables Laboratory, Beijing Forestry University, Tianjin University, Ministry of Agriculture of China, East China University of Science and Technology, Korea University, King Mongkut's University of Technology, Federal University of Parana, China Agricultural University, Federal University of Rio de Janeiro, University of Putra Malaysia, DOE Bioenergy Research Centers, NC State University, Tianjin University of Science & Technology, Jiangnan University, Beijing University of Chemical Technology, Federal University of Sao Carlos, State Key Laboratory of Bioreactor Engineering, Scientific Research National Center (CNRS), National Technical University of Athens, Chalmers University of Technology, Stellenbosch University, Nanjing Tech University, Guangxi University, Kyoto University, Huazhong Agricultural University, Iowa State University, Thailand National Center for Genetic Engineering and Biotechnology, Wageningen University & Research, and Qilu University of Technology.

59.3.5 The Most Prolific Funding Bodies on Second Generation Waste Biomass-based Bioethanol Fuels

Information about the 23 most prolific funding bodies funding at least 1.1% of the sample papers each is given in Table 59.4. Further, only 39% and 51% of the sample and population papers were funded, respectively.

The most prolific funding body is the U.S. Department of Energy (US DOE) with 6.9% of the sample papers, followed by the European Commission, the National Natural Science Foundation of China, and Research Support Foundation of the State of Sao Paulo with 4.4%, 2.9%, and 2.5% of the sample papers, respectively. Other funding bodies are National Council for Scientific and

TABLE 59.4
The Most Prolific Funding Bodies in the Second Generation Waste Biomass-based Bioethanol Fuels

No.	Funding Bodies	Country	Sample Paper No. (%)	Population Paper No. (%)	Surplus (%)
1	U.S. Department of Energy	USA	6.9	2.1	4.8
2	European Commission	EU	4.4	1.4	3.0
3	National Natural Science Foundation of China	China	2.9	10.3	–7.4
4	Research Support Foundation of the State of São Paulo	Brazil	2.5	2.5	0.0
5	National Council for Scientific and Technological Development	Brazil	2.2	4.1	–1.9
6	National Renewable Energy Laboratory	USA	2.2	0.4	1.8
7	U.S. Department of Agriculture	USA	1.8	0.8	1.0
8	National Nuclear Security Administration	USA	1.8	0.3	1.5
9	Higher Education Personnel Improvement Coordination	Brazil	1.5	3.3	–1.8
10	National Science Foundation	USA	1.5	0.9	0.6
11	Office of Science	USA	1.5	0.9	0.6
12	Natural Sciences and Engineering Research Council of Canada	Canada	1.5	0.8	0.7
13	Seventh Framework Program	EU	1.5	0.7	0.8
14	Ministry of Science and Technology of China	China	1.1	2.5	–1.4
15	Ministry of Education of China	China	1.1	1.3	–0.2
16	Government of Canada	Canada	1.1	0.4	0.7
17	Engineering and Physical Sciences Research Council	UK	1.1	0.3	0.8
18	Swedish Research Council	Sweden	1.1	0.2	0.9
19	Knut and Alice Wallenberg Foundation	Sweden	1.1	0.1	1.0
20	National Institute of Standards and Technology	USA	1.1	0.1	1.0
21	Natural Resources Canada	Canada	1.1	0.1	1.0
22	Swedish National Board for Industrial and Technical Development	Sweden	1.1	0.1	1.0
23	University of California Riverside	USA	1.1	0.1	1.0

Technological Development, NREL, the U.S. Department of Agriculture, and National Nuclear Security Administration with 1.8%–2.2% of the sample papers each. It is notable that the National Natural Science Foundation of China is the largest funder of the population papers with a 10.3% funding rate.

On the other hand, the most prolific country among these top funding bodies is the USA with seven funding bodies, followed by Brazil, Canada, China, and Sweden with three funding bodies each. Additionally, the EU has two funding bodies. In total, only six countries and the EU house these top funding bodies.

The funding bodies with the highest citation impact are the US DOE and European Commission with 4.8% and 3.0% surplus, respectively. Other influential funding bodies are the NREL and the National Nuclear Security Administration with 1.8% and 1.5% surplus, respectively. Further, the funding body with the lowest citation impact is the National Natural Science Foundation of China with a 7.4% deficit, followed by National Council for Scientific and Technological Development, Higher Education Personnel Improvement Coordination, and Ministry of Science and Technology of China with 1.9%, 1.8%, and 1.4% deficit, respectively.

The other funding bodies with a relatively low citation impact and with 0.5%–2.4% of the population papers each are the National Key Research and Development Program of China, Fundamental Research Funds for the Central Universities, Ministry of Science, Technology and Innovation, European Regional Development Fund, National Research Foundation of Korea, Ministry of Education, Culture, Sports, Science and Technology, Chinese Academy of Sciences, Japan Society for the Promotion of Science, National Basic Research Program of China (973 Program), Priority Academic Program Development of Jiangsu Higher Education Institutions, Ministry of Science and Technology India, China Postdoctoral Science Foundation, Natural Science Foundation of Jiangsu Province, National Council of Science and Technology, Ministry of Finance, National High-tech Research and Development Program, Thailand Research Fund, Biotechnology and Biological Sciences Research Council, Energy Agency, Foundation for Science and Technology, Ministry of Education, Science and Technology, and Ministry of Higher Education Malaysia.

59.3.6 The Most Prolific Source Titles on Second Generation Waste Biomass-based Bioethanol Fuels

Information about the 15 most prolific source titles publishing at least 1.5% of the sample papers each in second generation waste biomass-based bioethanol fuels is given in Table 59.5.

The most prolific source title is Bioresource Technology with 28.4% of the sample papers, followed by Biotechnology and Bioengineering, Green Chemistry, and Biomass and Bioenergy with 6.5%, 4.7%, and 4.4% of the sample papers, respectively. Other prolific titles are Biotechnology for Biofuels and Enzyme and Microbial Technology with 3.6% of the sample papers each.

On the other hand, the source title with the highest citation impact is Bioresource Technology with a 16.4% surplus, followed by Biotechnology and Bioengineering and Green Chemistry with a 5.2% and 4.4% surplus, respectively. Other influential titles are the Enzyme and Microbial Technology and Biotechnology Advances with 2.7% and 2.4% surplus, respectively.

The other source titles with a relatively low citation impact with 0.5%–2.6% of the population papers each are Bioresources, Applied Biochemistry and Biotechnology, Biomass Conversion and Biorefinery, Renewable Energy, Waste and Biomass Valorization, Bioenergy Research, Fuel, ACS Sustainable Chemistry and Engineering, Journal of Cleaner Production, Journal of Chemical Technology and Biotechnology, Chemical Engineering Transactions, Biotechnology Letters, Cellulose, RSC Advances, Bioprocess and Biosystems Engineering, Biochemical Engineering Journal, Cellulose Chemistry and Technology, Industrial and Engineering Chemistry Research, Energies, Applied Energy, Energy and Fuels, Energy, and Chemical Engineering Journal.

TABLE 59.5
The Most Prolific Source Titles in the Second Generation Waste Biomass-based Bioethanol Fuels

No.	Source Titles	Sample Papers (%)	Population Papers (%)	Surplus (%)
1	Bioresource Technology	28.4	12.0	16.4
2	Biotechnology and Bioengineering	6.5	1.3	5.2
3	Green Chemistry	4.7	0.7	4.0
4	Biomass and Bioenergy	4.4	2.5	1.9
5	Biotechnology for Biofuels	3.6	2.4	1.2
6	Enzyme and Microbial Technology	3.6	0.9	2.7
7	Applied Biochemistry and Biotechnology Part A	2.5	0.8	1.7
8	Biotechnology Advances	2.5	0.1	2.4
9	Industrial Crops and Products	2.2	3.1	−0.9
10	Biofuels Bioproducts and Biorefining	1.8	0.4	1.4
11	Biotechnology Progress	1.8	0.5	1.3
12	Renewable and Sustainable Energy Reviews	1.8	0.4	1.4
13	Applied Microbiology and Biotechnology	1.5	1.1	0.4
14	Current Opinion on Biotechnology	1.5	0.1	1.4
15	Process Biochemistry	1.5	1.2	0.3

59.3.7 The Most Prolific Countries on Second Generation Waste Biomass-based Bioethanol Fuels

Information about the 19 most prolific countries publishing at least 1.5% of sample papers each in the second generation waste biomass-based bioethanol fuels is given in Table 59.6.

The most prolific country is the USA with 35% of the sample papers, followed by Sweden, Denmark, and China with 12%, 9%, and 9% of the sample papers, respectively. Other prolific countries are India, Canada, and Brazil with 7%, 6%, and 5% of the sample papers, respectively. It is notable that China is the largest producer of the population papers with a 21.7% publication rate. Additionally, seven European countries listed in Table 59.6 produce 38% and 16% of the sample and population papers, respectively, with a 22% surplus.

On the other hand, the country with the highest citation impact is the USA with a 20% surplus, followed by Sweden and Denmark with a 9% and 7% surplus, respectively. Other influential countries are Canada, the Netherlands, and the UK with 3%, 2%, and 2% surplus, respectively. Similarly, the country with the lowest citation impact is China with a 13% deficit, followed by Brazil, India, Malaysia, and Japan with a 2%–5% deficit each.

Additionally, there are other countries with relatively low citation impact and with 0.5%–2.7% of the sample papers each: Thailand, Germany, Indonesia, Italy, Turkey, Finland, Taiwan, Portugal, Pakistan, Egypt, Nigeria, Poland, Greece, Colombia, Austria, Russian Federation, Argentina, Cuba, Vietnam, and Saudi Arabia.

59.3.8 The Most Prolific Scopus Subject Categories on Second Generation Waste Biomass-based Bioethanol Fuels

Information about the nine most prolific Scopus subject categories indexing at least 2.5% of the sample papers each is given in Table 59.7.

The most prolific Scopus subject category in second generation waste biomass-based bioethanol fuels is Chemical Engineering with 63% of the sample papers, followed by Environmental Science, Energy, Biochemistry, Genetics and Molecular Biology, and Immunology and Microbiology with

TABLE 59.6
The Most Prolific Countries in the Second Generation Waste Biomass-based Bioethanol Fuels

No.	Countries	Sample Papers (%)	Population Papers (%)	Surplus (%)
1	USA	34.5	14.7	19.8
2	Sweden	12.0	3.0	9.0
3	Denmark	9.1	2.1	7.0
4	China	9.1	21.7	−12.6
5	India	7.3	10.2	−2.9
6	Canada	6.2	3.3	2.9
7	Brazil	4.7	9.2	−4.5
8	UK	4.4	2.9	1.5
9	Spain	4.0	3.7	0.3
10	Netherlands	3.3	1.2	2.1
11	S. Korea	3.3	3.4	−0.1
12	France	2.9	2.1	0.8
13	Japan	2.9	4.5	−1.6
14	Iran	2.2	1.8	0.4
15	Mexico	2.2	1.8	0.4
16	Belgium	1.8	0.9	0.9
17	Australia	1.5	1.6	−0.1
18	Malaysia	1.5	3.2	−1.7
19	S. Africa	1.5	1.4	0.1

TABLE 59.7
The Most Prolific Scopus Subject Categories in the Second Generation Waste Biomass-based Bioethanol Fuels

No.	Scopus Subject Categories	Sample Papers (%)	Population Papers (%)	Surplus (%)
1	Chemical Engineering	63.3	43.3	20.0
2	Environmental Science	51.6	36.8	14.8
3	Energy	46.5	35.3	11.2
4	Biochemistry, Genetics and Molecular Biology	38.2	28.5	9.7
5	Immunology and Microbiology	29.5	19.7	9.8
6	Agricultural and Biological Sciences	11.3	18.9	−7.6
7	Chemistry	8.4	15.2	−6.8
8	Engineering	5.8	11.7	−5.9
9	Materials Science	2.5	6.1	−3.6

52%, 47%, 38%, and 30% of the sample papers, respectively. It is notable that Social Sciences including Economics and Business accounts for 1% and 3% of the sample and population studies, respectively.

On the other hand, the Scopus subject categories with the highest citation impact are Chemical Engineering and Environmental Science with 20% and 15% surplus, respectively. Other influential subject areas are Energy, Immunology and Microbiology, and Biochemistry, Genetics and Molecular Biology with 11%, 10%, and 10% surplus, respectively. Similarly, the least influential subject categories are Agricultural and Biological Sciences, Chemistry, Engineering, and Materials Science with a 4%–7% deficit each.

59.3.9 The Most Prolific Keywords in Second Generation Waste Biomass-based Bioethanol Fuels

Information about the Scopus keywords used with at least 6.5% or 3.8% of the sample or population papers, respectively, is given in Table 59.8. For this purpose, keywords related to the keyword set given in the appendix of (Konur, 2023a–g) are selected from a list of the most prolific keyword set provided by the Scopus database.

These keywords are grouped under five headings: Waste biomass, pretreatments, fermentation, hydrolysis and hydrolysates, and products.

The most prolific keyword related to biomass and biomass constituents is lignin with 69% of the sample papers, followed by cellulose, biomass, and lignocellulose with 68%, 52%, and 47% of the sample papers, respectively. Other prolific keywords are hemicellulose, lignocellulosic biomass, zea, and triticum with 16%–28% of the sample papers each.

Further, the most prolific keyword related to pretreatments is pretreatment with 25% of the sample papers, followed by pre-treatment and enzymes with 24% and 22% of the sample papers, respectively. Other prolific keywords are temperature, ionic liquids, and ammonia with 9%–14% of the sample papers each.

The most prolific keyword related to fermentation is fermentation with 41% of the sample papers. Other prolific keywords are saccharomyces, fungi, bioreactors, yeasts, and furfural with 8%–14% of the sample papers each.

Further, the most prolific keyword related to hydrolysis and hydrolysates is hydrolysis with 63% of the sample papers. Other prolific keywords are sugar, enzymatic hydrolysis, cellulases, enzyme activity, glucose, and saccharification with 20%–38% of the sample papers each.

Finally, the most prolific keyword related to the products is ethanol with 46% of the sample papers, followed by biofuels with 30% of the sample papers. Other prolific keywords are bioethanol and ethanol production with 15% and 11% of the sample papers, respectively.

On the other hand, the most prolific keywords across all of the research fronts are lignin, cellulose, hydrolysis, biomass, lignocellulose, ethanol, fermentation, sugar, enzymatic hydrolysis, cellulases, biofuels, and enzyme activity with 30%–69% of the sample papers each. Other prolific keywords are hemicellulose, pretreatment, pre-treatment, lignocellulosic biomass, enzymes, zea, glucose, and saccharification with 20%–28% of the sample papers each.

Similarly, the most influential keywords are lignin, cellulose, hydrolysis, lignocellulose, biomass, sugar, and hemicellulose with a 20%–36% surplus each. Other influential keywords are cellulases, ethanol, enzyme activity, pretreatment, biofuels, zea, enzymes, and fermentation with a 12%–17% surplus each.

59.3.10 The Most Prolific Research Fronts in Second Generation Waste Biomass-based Bioethanol Fuels

Information about the research fronts for the sample papers in the waste biomass-based bioethanol fuels is given in Table 59.9. This table shows the most prolific research front for this field is the lignocellulosic biomass at large such as corn stover and wheat straw with 55% of the HCPs, followed by residual starch feedstocks, industrial wastes, and residual sugar feedstocks with 37%, 18%, and 10% of these HCPs, respectively. Other feedstocks are urban wastes, food wastes, and forestry wastes with 2% of the HCPs each.

Further, the most influential research front is the lignocellulosic biomass at large with a 28% surplus while the industrial waste biomass is the lowest influential feedstock with a 19% deficit, followed by urban wastes, food wastes, and residual sugar feedstocks with 10%, 8%, and 5% deficit, respectively.

Information about the thematic research fronts for the sample papers in second generation waste biomass-based bioethanol fuels is given in Table 59.10. As this table shows, the most prolific research

TABLE 59.8
The Most Prolific Keywords in the Second Generation Waste Biomass-based Bioethanol Fuels

No.	Keywords	Sample Papers (%)	Population Papers (%)	Surplus (%)
1.	**Waste Biomass**			
	Lignin	68.7	32.3	36.4
	Cellulose	68.4	32.7	35.7
	Biomass	51.6	25.3	26.3
	Lignocellulose	46.5	18.7	27.8
	Hemicellulose	27.6	8.1	19.5
	Lignocellulosic biomass	24.4	13.0	11.4
	Zea	20.7	8.4	12.3
	Triticum	16.4	7.2	9.2
	Corn stover	12.4	4.4	8.0
	Straw	11.3	9.1	2.2
	Corn	11.3	2.3	9.0
	Bagasse	8.0	11.5	–3.5
	Xylan	8.0	3.2	4.8
	Carbohydrate	7.3	8.9	–1.6
	Wheat	7.3	4.2	3.1
	Lignocellulosic materials	6.5	2.9	3.6
	Glucan	6.5	0.0	6.5
	Wheat straw	6.2	7.0	–0.8
	Sugarcane	5.8	4.8	1.0
	Maize	5.1	5.7	–0.6
	Rice straw	4.4	6.4	–2.0
	Sugarcane	4.0	4.8	–0.8
	Sugarcane bagasse		5.2	–5.2
	Sugarcane bagasse		4.5	–4.5
	Crop residues		3.8	–3.8
2.	**Pretreatments**			
	Pretreatment	25.1	11.7	13.4
	Pre-treatment	24.4	13.7	10.7
	Enzymes	21.8	9.8	12.0
	Temperature	13.5	7.1	6.4
	Ionic liquids	9.1	6.9	2.2
	Ammonia	9.1	2.6	6.5
	Delignification	8.7	6.3	2.4
	Sulfuric acid	8.7	4.5	4.2
	Water	6.5	4.0	2.5
	pH	6.2	6.7	–0.5
	Sodium hydroxide		4.2	–4.2
3.	**Fermentation**			
	Fermentation	41.1	29.2	11.9
	Saccharomyces	13.5	7.9	5.6
	Fungi	12.4	9.0	3.4

(Continued)

TABLE 59.8 (*Continued*)
The Most Prolific Keywords in the Second Generation Waste Biomass-based Bioethanol Fuels

No.	Keywords	Sample Papers (%)	Population Papers (%)	Surplus (%)
	Bioreactors	10.2	6.8	3.4
	Yeast	8.0	9.7	−1.7
	Furfural	8.0	3.6	4.4
	Detoxification	7.3	2.7	4.6
	Acetic acids	5.8	4.5	1.3
4.	**Hydrolysis and Hydrolysates**			
	Hydrolysis	63.3	31.2	32.1
	Sugar	37.8	18.2	19.6
	Enzymatic hydrolysis	30.5	20.5	10.0
	Cellulases	30.5	13.2	17.3
	Enzyme activity	30.2	16.3	13.9
	Glucose	20.0	15.7	4.3
	Saccharification	19.6	15.9	3.7
	Xylose	14.9	7.2	7.7
	Enzymolysis	9.5	4.5	5.0
	Fermentable sugars	6.5	3.0	3.5
	Enzymatic saccharification	3.6	4.2	−0.6
5.	**Products**			
	Ethanol	46.2	30.4	15.8
	Biofuels	30.2	16.9	13.3
	Bioethanol	14.5	15.1	−0.6
	Ethanol production	11.3	6.3	5.0
	Biofuel production	7.3	3.0	4.3
	Cellulosic ethanol	5.5	4.6	0.9
	Bio-ethanol production	3.6	6.0	−2.4

TABLE 59.9
The Most Prolific Thematic Research Fronts for the Waste Biomass-based Bioethanol Fuels

No.	Research Fronts	N Paper % Sample	N Paper Population (%)	Surplus (%)
1	Lignocellulosic biomass	54.9	26.5	28.4
2	Residual starch feedstocks	37.1	29.8	7.3
3	Industrial wastes	17.5	36.2	−18.7
4	Residual sugar feedstocks	9.5	14.9	−5.4
5	Urban wastes	2.2	11.9	−9.7
6	Food wastes	2.2	10.4	−8.2
7	Forestry wastes	2.2	3.9	−1.7
8	Sample size	275	13,728	

N paper (%) sample: The number of papers in the population sample of 275 papers. N paper population: The number of papers in the population sample of 13,728 papers.

TABLE 59.10
The Most Prolific Thematic Research Fronts for the Second Generation Waste Biomass-based Bioethanol Fuels

No.	Research Fronts	N Paper % Sample
1	Biomass pretreatments	75.6
2	Biomass hydrolysis	50.9
3	Bioethanol production	25.5
4	Hydrolysate fermentation	22.9
5	Bioethanol fuel evaluation	6.5

N paper (%) review: The number of papers in the sample of 25 reviewed papers. N paper (%) sample: The number of papers in the population sample of 275 papers.

front is pretreatments of waste feedstocks with 77% of the sample papers, followed by hydrolysis of waste feedstocks with 51% of the sample papers. Other prolific research fronts are bioethanol production and hydrolysate fermentation with 26% and 23% of the sample papers, respectively. Further, bioethanol fuel evaluation relates to 6.5% of the sample papers.

59.4 DISCUSSION

59.4.1 Introduction

Crude oil-based gasoline fuels have been widely used in the transportation sector since the 1920s. However, there have been great public concerns over the adverse environmental and human impact of these fuels. Hence, biomass-based bioethanol fuels have increasingly been used in blending gasoline fuels, in fuel cells, and in biochemical production in a biorefinery context.

However, it is necessary to pretreat the biomass to enhance the yield of the bioethanol prior to bioethanol production through hydrolysis and fermentation. One of the most-studied feedstocks for bioethanol fuels has been waste biomass at large. The research in the field of second generation waste biomass-based bioethanol fuels has intensified in this context in the key research fronts of pretreatment and hydrolysis of waste biomass, fermentation of waste biomass-based hydrolysates, and production and evaluation of second generation waste biomass-based bioethanol fuels.

The research in this field has also intensified for the feedstocks of lignocellulosic biomass at large, agricultural residues, industrial wastes, forestry wastes, urban wastes, and food wastes. Thus, it emerges as a distinctive research field, complementing the primary research on first generation bioethanol fuels from agricultural feedstocks and wood among others.

However, it is essential to develop efficient incentive structures for the primary stakeholders to enhance research in this field. This is especially important to maintain energy security in the cases of supply shocks such as oil price shocks, war-related shocks as in the case of the Russian invasion of Ukraine, or COVID-19 shocks.

Scientometric analysis has been used in this context to inform the primary stakeholders about the current state of research in this research field. As there has been no scientometric study in this field, this chapter presents a scientometric study of the research in second generation waste biomass-based bioethanol fuels. It examines scientometric characteristics of both the sample and population data presenting scientometric characteristics of these both datasets in the order of documents, authors, publication years, institutions, funding bodies, source titles, countries, Scopus subject categories, Scopus keywords, and research fronts.

As the first step in the search for the relevant literature, the keywords were selected using the 300 most-cited population papers for each research front. The selected keyword list was then optimized to obtain a representative sample of papers from this research field. These keyword lists were then integrated to obtain the keyword list for this research field (Konur, 2023a–g).

As the second step, two sets of data were used in this study. First, a population sample of 13,728 papers was used to examine the scientometric characteristics of the population data. Second, a sample of 275 most-cited papers, corresponding to 2% of the population papers, was used to examine the scientometric characteristics of these citation classics.

Scientometric characteristics of these sample and population datasets were presented in the order of documents, authors, publication years, institutions, funding bodies, source titles, countries, Scopus subject categories, Scopus keywords, and research fronts.

Lastly, key scientometric findings for both datasets were discussed to highlight the research landscape for second generation waste biomass-based bioethanol fuels. Additionally, a number of brief conclusions were drawn and a number of relevant recommendations were made to enhance the future research landscape.

59.4.2 The Most Prolific Documents on Second Generation Waste Biomass-based Bioethanol Fuels

Articles (together with conference papers) dominate both the sample (69%) and population (92%) papers with a 23% deficit (Table 59.1). Further, review papers have a surplus (26%) and the representation of reviews in the sample papers is quite extraordinary (31%).

Scopus differs from the Web of Science database in differentiating and showing articles (65%) and conference papers (4%) published in journals separately. However, it should be noted that these conference papers are also published in journals as articles, compared to those published only in the conference proceedings. Hence, the total number of articles and review papers in the sample dataset are 69% and 31%, respectively.

It is observed during the search process that there has been inconsistency in the classification of documents in Scopus as well as in other databases such as the Web of Science. This is especially relevant for the classification of papers as reviews or articles as papers not involving a literature review may be erroneously classified as a review paper. There is also a case of review papers being classified as articles. For example, the total number of reviews in the sample data set was manually found as nearly 37% compared to 31% as indexed by Scopus, decreasing the number of articles and conference papers to 63% for the sample dataset.

In this context, it would be helpful to provide a classification note for the published papers in books and journals at the first instance. It would also be helpful to use the document types listed in Table 59.1 for this purpose. Book chapters may also be classified as articles or reviews as an additional classification to differentiate review chapters from experimental chapters as is done by the Web of Science. It would be further helpful to additionally classify the conference papers as articles or review papers as well as is done in the Web of Science database.

59.4.3 The Most Prolific Authors on Second Generation Waste Biomass-based Bioethanol Fuels

There have been 32 most prolific authors with at least 1.6% of the sample papers each as given in Table 59.2. These authors have shaped the development of research in this field.

The most prolific authors are Bruce E. Dale, John N. Saddler, Charles E. Wyman, and to a lesser extent Yoon Y. Lee, Barbel Hahn-Hagerdal, Mark T. Holtzapple, and Michael R. Ladisch. Further, the most influential authors are Bruce E. Dale, Yoon Y. Lee, Charles E. Wyman, John N. Saddler, and to a lesser extent Barbel Hahn-Hagerdal and Mark T. Holtzapple.

It is important to note the inconsistencies in the indexing of author names in Scopus and other databases. It is especially an issue for names with more than two components such as 'Blake Sam de Hyun Ladisch'. The probable outcomes are 'Ladisch, B.S.D.H.', 'de Hyun Ladisch, B.S.', or 'Hyun Ladisch, B.S.D.'. The first choice is the gold standard of the publishing sector as the last word in the name is taken as the last name. In most of the academic databases such as PUBMED and EBSCO databases, this version is used predominantly. The second choice is a strong alternative while the last choice is an undesired outcome as the two last words are taken as the last name. It is good practice to combine the words of the last name with a hyphen: 'Hyun-Ladisch, B.S.D.'. It is notable that inconsistent indexing of the author names may cause substantial inefficiencies in the search process for papers as well as allocating credit to the authors as there are different author entries for each outcome in the databases.

There are also inconsistencies in the shortening of Chinese names. For example. 'YangYing Lee' is often shortened as 'Lee, Y.', 'Lee, Y.-Y.', and 'Lee, Y.Y.' as it is done in the Web of Science database as well. However, the gold standard, in this case is 'Lee, Y', where the last word is taken as the last name and the first word is taken as a single forename. In most of the academic databases such as PUBMED and EBSCO, this first version is used predominantly. However, it makes sense to use the third option to differentiate Chinese names efficiently: 'Lee, Y.Y.'. Therefore, there have been difficulties in locating papers for Chinese authors. In such cases, the use of the unique author codes provided for each author by the Scopus database has been helpful.

There is also a difficulty in allowing credit for authors, especially for the authors with common names such as 'Lee, X.' in conducting scientometric studies. These difficulties strongly influence the efficiency of scientometric studies as well as allocating credit to authors as there are the same author entries for different authors with the same name. For example, 'Lee, X.' in the databases.

In this context, the coding of authors in the Scopus database is a welcome innovation compared to other databases such as the Web of Science. In this process, Scopus allocates a unique number to each author in the database (Aman, 2018). However, there might still be substantial inefficiencies in this coding system, especially for common names. For example, some of the papers for a certain author may be allocated to another researcher with a different author code. It is possible that Scopus uses a number of software programs to differentiate the author names and the program may not be false-proof (Kim, 2018).

In this context, it does not help that author names are not given in full in some journals and books. This makes it difficult to differentiate authors with common names and makes the scientometric studies further difficult in the author domain. Therefore, the author names should be given in all books and journals at the first instance. There is also a cultural issue where some authors do not use their full names in their papers. Instead, they use initials for their forenames: 'Holtzapple, H.J.', 'Holtzapple, H.', or 'Holtzapple, J.' instead of 'Holtzapple, Hyun Jae'.

There are also inconsistencies in the naming of the authors with more than two components by the authors themselves in journal papers and book chapters. For example. 'Holtzapple, A.P.C.' might be given as 'Holtzapple, A.' or 'Holtzapple, A.C.' or 'Holtzapple, A.P.' or 'Holtzapple, C' in the journals and books. This also makes the scientometric studies difficult in the author domain. Hence, contributing authors should use their name consistently in their publications.

Another critical issue regarding the author names is the inconsistencies in the spelling of the author names in the national spellings (e.g., Özğümüş, Şençiğ) rather than in the English spellings (e.g., Ozgumus, Sencig) in the Scopus database. Scopus differs from the Web of Science database and many other databases in this respect where the author names are given only in English spellings. It is observed that national spellings of the author names do not help much in conducting scientometric studies as well as in allocating credits to the authors as sometimes there are different author entries for the English and national spellings in the Scopus database.

The most prolific institutions for the sample dataset are NREL, Imperial College, Lund University, and to a lesser extent CIEMAT, Purdue University, and the University of Copenhagen. Further, the most prolific countries for the sample dataset are the USA and to a lesser extent Sweden, Denmark, the UK, and Spain. These findings confirm the dominance of the USA and Europe in this field. On the other hand, pretreatments and hydrolysis of the waste biomass and to a lesser extent the fermentation of the waste biomass-based hydrolysates and the bioethanol fuels are the key research fronts studied by these top authors.

It is also notable that there is a significant gender deficit for the sample dataset surprisingly with a representation rate of 19%. This finding is the most thought-provoking with strong public policy implications. Hence, institutions, funding bodies, and policy makers should take efficient measures to reduce the gender deficit in this field as well as other scientific fields with strong gender deficits. In this context, it is worth noting the level of representation of the researchers from minority groups in science on the basis of race, sexuality, age, and disability, besides gender (Blankenship, 1993; Dirth and Branscombe, 2017; Konur, 2000, 2002a–c, 2006a, b, 2007a, b).

59.4.4 The Most Prolific Research Output by Years on Second Generation Waste Biomass-based Bioethanol Fuels

The research output observed between 1970 and 2022 is illustrated in Figure 59.1. This figure clearly shows that the bulk of the research papers in the population dataset were published primarily in the 2010s and early 2020s. Similarly, the bulk of the research papers in the sample dataset were published in the 2010s and 2000s.

These findings suggest that the most prolific sample and population papers were primarily published in the 2010s. Further, a significant portion of the sample and population papers was published in the early 2020s and 2000s, respectively.

These are thought-provoking findings as there has been a significant research boom since 2010 and 2007 for population and sample papers, respectively. In this context, the increasing public concerns about climate change (Change, 2007), greenhouse gas emissions (Carlson et al., 2017), and global warming (Kerr, 2007) have certainly been behind the boom in research in this field since 2007. Furthermore, the recent supply shocks experienced due to the COVID-19 pandemic and the Ukrainian war might also be behind the research boom in this field since 2019. Indeed, there has been a rise in the research output for the population papers in 2020 and 2021.

Based on these findings, the size of the population papers is likely to more than double in the current decade, provided that the public concerns about climate change, greenhouse gas emissions, and global warming, as well as supply shocks, are translated efficiently to the research funding in this field.

59.4.5 The Most Prolific Institutions on Second Generation Waste Biomass-based Bioethanol Fuels

The 26 most prolific institutions publishing papers on second generation waste biomass-based bioethanol fuels with at least 1.5% of the sample papers each given in Table 59.3 have shaped the development of the research in this field.

The most prolific institutions are the NREL, Lund University, Technical University of Denmark, Michigan State University, and to a lesser extent the University of British Columbia, Dartmouth College, Purdue University, and Auburn University. Similarly, the top countries for these most prolific institutions are the USA, and to a lesser extent Sweden, China, and Denmark. In total, 10 countries house these top institutions.

On the other hand, the institutions with the highest impact are the NREL, Lund University, and to a lesser extent Technical University of Denmark, Michigan State University, Dartmouth College, the University of British Columbia, and Auburn University. These findings confirm the dominance of the institutions from the USA, Europe, and to a lesser extent Canada.

59.4.6 The Most Prolific Funding Bodies on Second Generation Waste Biomass-based Bioethanol Fuels

The 23 most prolific funding bodies funding at least 1.1% of the sample papers each are given in Table 59.4. It is notable that only 39% and 51% of the sample and population papers were funded, respectively.

The most prolific funding bodies are the US DOE, the European Commission, and to a lesser extent National Natural Science Foundation of China, Research Support Foundation of the State of Sao Paulo, National Council for Scientific and Technological Development, NREL, U.S. Department of Agriculture, and National Nuclear Security Administration. On the other hand, the most prolific countries for these top funding bodies are the USA, and to a lesser extent Brazil, Canada, China, Sweden, and the EU. In total, only six countries and the EU house these top funding bodies.

The funding bodies with the highest citation impact are the European Commission, the US DOE, and to a lesser extent NREL and the National Nuclear Security Administration. Further, the funding bodies with the lowest citation impact are the National Natural Science Foundation of China and to a lesser extent National Council for Scientific and Technological Development, Higher Education Personnel Improvement Coordination, and Ministry of Science and Technology of China.

These findings on the funding of research in this field suggest that the level of funding, mostly since 2010, is highly intensive and it has been largely instrumental in enhancing the research in this field (Ebadi and Schiffauerova, 2016) in light of North's institutional framework (North, 1991). It is also notable that the funding rate in this field is relatively modest compared to those in other research fronts of bioethanol fuels such as algal bioethanol fuels. Further, it is expected that this high funding rate would continue in light of the recent supply shocks. Further, it emerges that the USA, and to a lesser extent Brazil, Canada, China, Sweden, and the EU have heavily funded the research on waste biomass-based bioethanol fuels.

59.4.7 The Most Prolific Source Titles on Second Generation Waste Biomass-based Bioethanol Fuels

The 15 most prolific source titles publishing at least 1.5% of the sample papers each in second generation waste biomass-based bioethanol fuels have shaped the development of the research in this field (Table 59.5).

The most prolific source titles are Bioresource Technology and to a lesser extent Biotechnology and Bioengineering, Green Chemistry, Biomass and Bioenergy, Biotechnology for Biofuels, and Enzyme and Microbial Technology. On the other hand, the source titles with the highest citation impact are Bioresource Technology and to a lesser extent by Biotechnology and Bioengineering, Green Chemistry, Enzyme and Microbial Technology, and Biotechnology Advances.

It is notable that these top source titles are primarily related to bioresources, biotechnology, and to a lesser extent energy and microbial technology. This finding suggests that Bioresource Technology and other prolific journals in these fields have significantly shaped the development of research in this field as they focus primarily on second generation waste biomass-based bioethanol fuels with a high yield. In this context, the influence of top journals is quite extraordinary.

59.4.8 The Most Prolific Countries on Second Generation Waste Biomass-based Bioethanol Fuels

The 19 most prolific countries publishing at least 1.5% of the sample papers each have significantly shaped the development of the research in this field (Table 59.6).

The most prolific countries are the USA and to a lesser extent Sweden, Denmark, China, India, Canada, and Brazil. It is notable that China is the largest producer of population papers with a 21.7%

publication rate. Additionally, seven European countries listed in Table 59.6 produce 38% and 16% of the sample and population papers, respectively, with a 22% surplus. Thus, Europe is the largest and second-largest producer of sample and population papers, respectively.

On the other hand, the countries with the highest citation impact are the USA and to a lesser extent Sweden, Denmark, Canada, the Netherlands, and the UK. Similarly, the countries with the lowest impact are China and to a lesser extent Brazil, India, Malaysia, and Japan.

A close examination of these findings suggests that the USA, Europe, and to a lesser extent China, Canada, India, and Brazil are the major producers of research in this field. It is notable that all these top countries have large access to wastes.

It is a fact that the USA has been a major player in science (Leydesdorff and Wagner, 2009). The USA has further developed a strong research infrastructure to support its corn and grass-based bioethanol industry (Gillon, 2010).

However, China has been a rising megastar in scientific research in competition with the USA and Europe (Leydesdorff and Zhou, 2005). China is also a major player in this field as a major producer of bioethanol (Fang et al., 2010).

Next, Europe has been a persistent player in scientific research in competition with both the USA and China (Leydesdorff, 2000). Europe has also been a persistent producer of bioethanol along with the USA and Brazil (Gnansounou, 2010).

Further, Canada (Tahmooresnejad et al., 2015), India (Basu and Kumar, 2000), and Brazil (Glanzel et al., 2006) are other countries with substantial research activities in bioethanol fuels.

59.4.9 The Most Prolific Scopus Subject Categories on Second Generation Waste Biomass-based Bioethanol Fuels

The nine most prolific Scopus subject categories indexing at least 2.5% of the sample papers each, given in Table 59.7 have shaped the development of the research in this field.

The most prolific Scopus subject categories on second generation waste biomass-based bioethanol fuels are Chemical Engineering, Environmental Science, and to a lesser extent Energy, Biochemistry, Genetics and Molecular Biology, and Immunology and Microbiology. It is also notable that Social Sciences including Economics and Business has a minimal presence in both sample and population studies.

On the other hand, the Scopus subject categories with the highest citation impact are Chemical Engineering, Environmental Science, and to a lesser extent Energy, Immunology and Microbiology, and Biochemistry, Genetics and Molecular Biology. Similarly, the least influential subject categories are Agricultural and Biological Sciences, Chemistry, Engineering, and Materials Science.

These findings are thought-provoking suggesting that the primary subject categories are related to chemical engineering, environmental sciences, and to a lesser extent energy, genetics, and microbiology as the core of research in this field concerns the production and utilization of second generation waste biomass-based bioethanol fuels. Another finding is that social sciences are not well represented in both the sample and population papers as in line with the highest fields in bioethanol fuels. Social, environmental, and economic studies account for the field of social sciences.

59.4.10 The Most Prolific Keywords on Second Generation Waste Biomass-based Bioethanol Fuels

A limited number of keywords have shaped the development of research in this field as shown in Table 59.8. These keywords are grouped under five headings: waste biomass, pretreatments, fermentation, hydrolysis and hydrolysates, and products.

The most prolific keywords across all research fronts are lignin, cellulose, hydrolysis, biomass, lignocellulose, ethanol, fermentation, sugar, enzymatic hydrolysis, cellulases, biofuels, enzyme activity, and to a lesser extent hemicellulose, pretreatment, pre-treatment, lignocellulosic biomass, enzymes, zea, glucose, and saccharification.

Similarly, the most influential keywords are lignin, cellulose, hydrolysis, lignocellulose, biomass, sugar, hemicellulose, and to a lesser extent cellulases, ethanol, enzyme activity, pretreatment, biofuels, zea, enzymes, and fermentation.

These findings suggest that it is necessary to determine the keyword set carefully to locate the relevant research in each of these research fronts. Additionally, the size of the samples for each keyword highlights the intensity of research in the relevant research areas for both sample and population datasets. These findings also highlight different spellings of some strategic keywords such as pretreatment v. pre-treatment and ethanol v. bio-ethanol, etc. However, there is a tendency toward the use of the connected keywords without using a hyphen. Another relevant issue is that it is necessary to include Latin keywords for wastes in the keyword set as they are often used by the authors in addition to English ones.

59.4.11 The Most Prolific Research Fronts on Second Generation Waste Biomass-based Bioethanol Fuels

Information about the research fronts for sample papers on second generation waste biomass-based bioethanol fuels is given in Table 59.9. As this table shows, the most prolific research front for this field is lignocellulosic biomass at large such as corn stover and wheat straw with 55% of the HCPs, followed by residual starch feedstocks such as corn stover, industrial wastes such as glycerol or sugarcane bagasse, and residual sugar feedstocks such as sugarcane bagasse. Other feedstocks are urban wastes such as waste paper, food wastes such as orange peels or spent coffee grounds, and forestry wastes such as tree branches and leaves.

Information about the thematic research fronts for sample papers in second generation waste biomass-based bioethanol fuels is given in Table 59.10. As this table shows, the most prolific research front is pretreatments of waste feedstocks with 77% of the sample papers, followed by the hydrolysis of the waste biomass with 51% of the sample papers. Other prolific research fronts are bioethanol production and hydrolysate fermentation with 26% and 23% of the sample papers, respectively. Further, bioethanol fuel evaluation relates to 6.5% of the sample papers.

These findings are thought-provoking in seeking ways to increase waste biomass feedstock-based bioethanol yield at the global scale. It is clear that all of these research fronts have public importance and merit substantial funding and other incentives. Further, it is notable that second generation waste biomass-based bioethanol fuels have become a core unit of bioethanol research to make it more competitive with crude oil-based gasoline and petrodiesel fuels, especially for the USA, Europe, and China. It also solves the tremendous waste treatment (Morrissey and Brown, 2004; Wilson, 2007) problem of huge amounts of waste biomass to avoid ecological disasters.

In comparison to other feedstock-based research fronts, it is notable that the pretreatment and hydrolysis of waste biomass emerge as primary research fronts for this field. However, the research fronts of fermentation of waste biomass-based hydrolysates and bioethanol production from waste biomass-based hydrolysates are also important.

Further, the field of evaluation and utilization of bioethanol fuels is a neglected area. This suggests that the primary stakeholders have been primarily interested in these key processes of bioethanol production. It is also notable that evaluation of second generation waste biomass-based bioethanol fuels such as technoeconomics, life cycle, economics, social science, land use, labor, and environment impact-related studies emerges as a case study for bioethanol fuels. Similarly, the utilization of these biofuels in gasoline or diesel engines and fuel cells is also an important research field from a societal perspective. In this context, the USA and Brazil have been global leaders in the production and use of corn- and sugarcane-based bioethanol fuels since the 1970s in the aftermath of the global crude oil crisis in the early 1970s.

It is further notable that the research on waste biomass-based bioethanol fuels complements the research on first generation bioethanol fuel research from sugar and starch feedstocks and other non-waste biomass such as wood, grass, algae, and syngas among others, extracting further value from these primary feedstocks.

In the end, these most-cited papers in this field hint that the production of second generation waste biomass-based bioethanol fuels could be optimized using the structure, processing, and property relationships of these waste biomass in the fronts of feedstock pretreatment and hydrolysis, and hydrolysate fermentation (Formela et al., 2016; Konur, 2018a, 2020b, 2021a–d; Konur and Matthews, 1989).

59.5 CONCLUSION AND FUTURE RESEARCH

The research on second generation waste biomass-based bioethanol fuels has been mapped through a scientometric study of both sample (275 papers) and population (13,728 papers) datasets.

The critical issue in this study has been to obtain a representative sample of the research as in any other scientometric study. Therefore, the keyword set has been carefully devised and optimized after a number of runs in the Scopus database. It is a representative sample of the wider population studies. This keyword set was provided in the Appendix of related studies and the relevant keywords are presented in Table 59.8. However, it should be noted that it has been very difficult to compile a representative keyword set since this research field has been connected closely with many other fields. Therefore, it has been necessary to compile a keyword list to exclude papers concerned with other research fields.

Another issue has been the selection of a multidisciplinary database to carry out the scientometric study of research in this field. For this purpose, the Scopus database has been selected. The journal coverage of this database has been notably wider than that of the Web of Science and other multi-subject databases.

The key scientometric properties of research in this field have been determined and discussed in this chapter. It is evident that a limited number of documents, authors, publication years, institutions, funding bodies, source titles, countries, Scopus subject categories, Scopus keywords, and research fronts have shaped the development of the research in this field.

There is ample scope to increase the efficiency of scientometric studies in this field in the author and document domains by developing consistent policies and practices in both domains across all academic databases. In this respect, it seems that authors, journals, and academic databases have a lot to do. Furthermore, the significant gender deficit as in most scientific fields emerges as a public policy issue. The potential deficits on the basis of age, race, disability, and sexuality need also to be explored in this field as in other scientific fields.

The research in this field has boomed since 2010 and 2007 for population and sample papers, respectively, possibly promoted by public concerns about global warming, greenhouse gas emissions, and climate change. Furthermore, the recent COVID-19 pandemic and the Russian invasion of Ukraine have resulted in a global supply shock shifting the recent focus of the stakeholders from crude oil-based fuels to biomass-based fuels such as bioethanol fuels.

It is expected that there would be further incentives for key stakeholders to carry out research for second generation waste biomass-based bioethanol fuels to increase the ethanol yield and to make it more competitive with crude oil-based gasoline and petrodiesel fuels. This might be truer for the crude oil- and foreign exchange-deficient countries to maintain energy and food security in the face of global supply shocks. It also solves the tremendous waste treatment problem of huge amounts of waste biomass and avoids ecological disasters.

The relatively modest funding rate of 39% and 51% for the sample and population papers, respectively, suggests that funding in this field significantly enhanced the research in this field primarily since 2010, possibly more than doubling in the current decade. However, it is evident that there is ample room for more funding and other incentives to enhance the research in this field further, especially in light of the recent supply shocks.

The institutions from the USA, and to a lesser extent Sweden, China, and Denmark have mostly shaped the research in this field. Further, the USA, Europe, and to a lesser extent China, India, Canada, and Brazil have been the major producers of research in this field as major producers and users of bioethanol fuels. It is evident that these countries have well-developed research infrastructure in bioethanol fuels and their derivatives and have large access to waste biomass.

It emerges that ethanol is more popular than bioethanol as a keyword with strong implications for the search strategy. In other words, the search strategy using only the bioethanol keyword would not be very helpful. On the other hand, the Scopus keywords are grouped under five headings: biomass, pretreatments, fermentation, hydrolysis and hydrolysates, and products. It is also important to include Latin keywords for waste biomass.

The most prolific research fronts for this field with respect to waste biomass are lignocellulosic biomass at large and to a lesser extent residual starch feedstocks, industrial wastes and residual sugar feedstocks, urban wastes, food wastes, and forestry wastes.

Further, the most prolific research fronts are the pretreatments of lignocellulosic feedstocks, hydrolysis of waste biomass, and to a lesser extent bioethanol production, hydrolysate fermentation, and bioethanol fuel evaluation.

The first four research fronts dominate the research in this field while the field of the utilization and evaluation of waste biomass-based bioethanol fuels is relatively a neglected research field. In this context, it is notable that there is ample room for the improvement of research on social science- and humanities- aspects of the research on the bioethanol fuels from the waste biomass such as scientometric, economic, policy, and user studies.

These findings are thought-provoking in seeking ways to increase waste biomass feedstock-based bioethanol yield at the global scale. It is clear that all of these research fronts have public importance and merit substantial funding and other incentives. Further, it is notable that second generation waste biomass-based bioethanol fuels have become a core unit of bioethanol research to make it more competitive with crude oil-based gasoline and petrodiesel fuels, especially for the USA, Europe, Brazil, and China. It is further notable that the research on waste biomass-based bioethanol emerges as a distinctive research field, complementing the research on first generation bioethanol fuels from the starch and sugar feedstocks as well as grass, algae, and wood among others.

Thus, scientometric analysis has a great potential to gain valuable insights into the evolution of the research in this field as in other scientific fields especially in the aftermath of the significant global supply shocks such as the COVID-19 pandemic and the Russian invasion of Ukraine.

It is recommended that further scientometric studies are carried out for the primary research fronts. It is further recommended that reviews of the most-cited papers are carried out for each primary research front to complement these scientometric studies. Next, the scientometric studies of the hot papers in these primary fields are carried out.

ACKNOWLEDGMENTS

The contribution of the highly cited researchers in the field of second generation waste biomass-based bioethanol fuels has been gratefully acknowledged.

REFERENCES

Aman, V. 2018. Does the scopus author ID suffice to track scientific international mobility? A case study based on Leibniz laureates. *Scientometrics* 117:705–720.

Angelici, C., B. M. Weckhuysen and P. C. A. Bruijnincx. 2013. Chemocatalytic conversion of ethanol into butadiene and other bulk chemicals. *ChemSusChem* 6:1595–1614.

Antolini, E. 2007. Catalysts for direct ethanol fuel cells. *Journal of Power Sources* 170:1–12.

Antolini, E. 2009. Palladium in fuel cell catalysis. *Energy and Environmental Science* 2:915–931.

Basu, A. and B. V. Kumar. 2000. International collaboration in Indian scientific papers. *Scientometrics* 48:381–402.

Beaudry, C. and V. Lariviere. 2016. Which gender gap? Factors affecting researchers' scientific impact in science and medicine. *Research Policy* 45:1790–1817.

Blankenship, K. M. 1993. Bringing gender and race in: US employment discrimination policy. *Gender & Society* 7:204–226.

Brandt, A., J. P. Hallett, D. J. Leak, R. J. Murphy and T. Welton. 2010. The effect of the ionic liquid anion in the pretreatment of pine wood chips. *Green Chemistry* 12:672–679.

Burnham, J. F. 2006. Scopus database: A review. *Biomedical Digital Libraries* 3:1–8.

Carlson, K. M., J. S. Gerber and D. Mueller, et al. 2017. Greenhouse gas emissions intensity of global croplands. *Nature Climate Change* 7:63–68.

Change, C. 2007. Climate change impacts, adaptation and vulnerability. *Science of the Total Environment* 326:95–112.

Dirth, T. P. and N. R. Branscombe. 2017. Disability models affect disability policy support through awareness of structural discrimination. *Journal of Social Issues* 73:413–442.

Duff, S. J. B. and W. D. Murray. 1996. Bioconversion of forest products industry waste cellulosics to fuel ethanol: A review. *Bioresource Technology* 55:1–33.

Ebadi, A. and A. Schiffauerova. 2016. How to boost scientific production? A statistical analysis of research funding and other influencing factors. *Scientometrics* 106:1093–1116.

Fang, X., Y. Shen, J. Zhao, X. Bao and Y. Qu. 2010. Status and prospect of lignocellulosic bioethanol production in China. *Bioresource Technology* 101:4814–4819.

Fauci, A. S., H. C. Lane and R. R. Redfield. 2020. Covid-19-navigating the uncharted. *New England Journal of Medicine* 382:1268–1269.

Fernando, S., S. Adhikari, C. Chandrapal and M. Murali. 2006. Biorefineries: Current status, challenges, and future direction. *Energy & Fuels* 20:1727–1737.

Formela, K., A. Hejna, L. Piszczyk, M. R. Saeb and X. Colom. 2016. Processing and structure-property relationships of natural rubber/wheat bran biocomposites. *Cellulose* 23:3157–3175.

Garfield, E. 1955. Citation indexes for science. *Science* 122:108–111.

Gillon, S. 2010. Fields of dreams: Negotiating an ethanol agenda in the Midwest United States. *Journal of Peasant Studies* 37:723–748.

Glanzel, W., J. Leta and B. Thijs. 2006. Science in Brazil. Part 1: A macro-level comparative study. *Scientometrics* 67:67–86.

Gnansounou, E. 2010. Production and use of lignocellulosic bioethanol in Europe: Current situation and perspectives. *Bioresource Technology* 101:4842–4850.

Guimaraes, P. M. R., J. A. Teixeira and L. Domingues. 2010. Fermentation of lactose to bio-ethanol by yeasts as part of integrated solutions for the valorisation of cheese whey. *Biotechnology Advances* 28:375–384.

Hahn-Hagerdal, B., M. Galbe, M. F. Gorwa-Grauslund, G. Liden and G. Zacchi. 2006. Bio-ethanol: The fuel of tomorrow from the residues of today. *Trends in Biotechnology* 24:549–556.

Hamelinck, C. N., G. van Hooijdonk and A. P. C. Faaij. 2005. Ethanol from lignocellulosic biomass: Techno-economic performance in short-, middle- and long-term. *Biomass and Bioenergy* 28:384–410.

Hamilton, J. D. 1983. Oil and the macroeconomy since World War II. *Journal of Political Economy* 91:228–248.

Hamilton, J. D. 2003. What is an oil shock? *Journal of Econometrics* 113:363–398.

Hendriks, A. T. W. M. and G. Zeeman. 2009. Pretreatments to enhance the digestibility of lignocellulosic biomass. *Bioresource Technology* 100:10–18.

Hill, J., E. Nelson, D. Tilman, S. Polasky and D. Tiffany. 2006. Environmental, economic, and energetic costs and benefits of biodiesel and ethanol biofuels. *Proceedings of the National Academy of Sciences of the United States of America* 103:11206–11210.

Hill, J., S. Polasky and E. Nelson, et al. 2009. Climate change and health costs of air emissions from biofuels and gasoline. *Proceedings of the National Academy of Sciences of the United States of America* 106:2077–2082.

Hsieh, W. D., R. H. Chen, T. L. Wu and T. H. Lin. 2002. Engine performance and pollutant emission of an SI engine using ethanol-gasoline blended fuels. *Atmospheric Environment* 36:403–410.

Huang, H. J., S. Ramaswamy, U. W. Tschirner and B. V. Ramarao. 2008. A review of separation technologies in current and future biorefineries. *Separation and Purification Technology* 62:1–21.

Jones, T. C. 2012. America, oil, and war in the Middle East. *Journal of American History* 99:208–218.

Jorgensen, H., J. B. Kristensen and C. Felby. 2007. Enzymatic conversion of lignocellulose into fermentable sugars: Challenges and opportunities. *Biofuels, Bioproducts and Biorefining* 1:119–134.

Kadar, Z., Z. Szengyel and K. Reczey. 2004. Simultaneous saccharification and fermentation (SSF) of industrial wastes for the production of ethanol. *Industrial Crops and Products* 20:103–110.

Kerr, R. A. 2007. Global warming is changing the world. *Science* 316:188–190.

Kilian, L. 2008. Exogenous oil supply shocks: How big are they and how much do they matter for the US economy? *Review of Economics and Statistics* 90:216–240.

Kilian, L. 2009. Not all oil price shocks are alike: Disentangling demand and supply shocks in the crude oil market. *American Economic Review* 99:1053–1069.

Kim, J. 2018. Evaluating author name disambiguation for digital libraries: A case of DBLP. *Scientometrics* 116:1867–1886.

Konur, O. 2000. Creating enforceable civil rights for disabled students in higher education: An institutional theory perspective. *Disability & Society* 15:1041–1063.

Konur, O. 2002a. Access to nursing education by disabled students: Rights and duties of nursing programs. *Nurse Education Today* 22:364–374.

Konur, O. 2002b. Assessment of disabled students in higher education: Current public policy issues. *Assessment and Evaluation in Higher Education* 27:131–152.

Konur, O. 2002c. Access to employment by disabled people in the UK: Is the Disability Discrimination Act working? *International Journal of Discrimination and the Law* 5:247–279.

Konur, O. 2006a. Participation of children with dyslexia in compulsory education: Current public policy issues. *Dyslexia* 12:51–67.

Konur, O. 2006b. Teaching disabled students in Higher Education. *Teaching in Higher Education* 11:351–363.

Konur, O. 2007a. A judicial outcome analysis of the Disability Discrimination Act: A windfall for the employers? *Disability & Society* 22:187–204.

Konur, O. 2007b. Computer-assisted teaching and assessment of disabled students in higher education: The interface between academic standards and disability rights. *Journal of Computer Assisted Learning* 23:207–219.

Konur, O. 2011. The scientometric evaluation of the research on the algae and bio-energy. *Applied Energy* 88:3532–3540.

Konur, O. 2012a. The evaluation of the biogas research: A scientometric approach. *Energy Education Science and Technology Part A: Energy Science and Research* 29:1277–1292.

Konur, O. 2012b. The evaluation of the educational research: A scientometric approach. *Energy Education Science and Technology Part B: Social and Educational Studies* 4:1935–1948.

Konur, O. 2012c. The evaluation of the global energy and fuels research: A scientometric approach. *Energy Education Science and Technology Part A: Energy Science and Research* 30:613–628.

Konur, O. 2012d. The evaluation of the research on the biodiesel: A scientometric approach. *Energy Education Science and Technology Part A: Energy Science and Research* 28:1003–1014.

Konur, O. 2012e. The evaluation of the research on the bioethanol: A scientometric approach. *Energy Education Science and Technology Part A: Energy Science and Research* 28:1051–1064.

Konur, O. 2012f. The evaluation of the research on the biofuels: A scientometric approach. *Energy Education Science and Technology Part A: Energy Science and Research* 28:903–916.

Konur, O. 2012g. The evaluation of the research on the biohydrogen: A scientometric approach. *Energy Education Science and Technology Part A: Energy Science and Research* 29:323–338.

Konur, O. 2012h. The evaluation of the research on the microbial fuel cells: A scientometric approach. *Energy Education Science and Technology Part A: Energy Science and Research* 29:309–322.

Konur, O. 2012i. The scientometric evaluation of the research on the production of bioenergy from biomass. *Biomass and Bioenergy* 47:504–515.

Konur, O. 2015. Current state of research on algal bioethanol. In *Marine Bioenergy: Trends and Developments*, Eds. S. K. Kim and C. G. Lee, pp. 217–244. Boca Raton, FL: CRC Press.

Konur, O., Ed. 2018a. *Bioenergy and Biofuels*. Boca Raton, FL: CRC Press, Boca Raton, FL.

Konur, O. 2018b. Bioenergy and biofuels science and technology: Scientometric overview and citation classics. In *Bioenergy and Biofuels*, Ed. O. Konur, pp. 3–63. Boca Raton, FL: CRC Press.

Konur, O. 2019. Cyanobacterial bioenergy and biofuels science and technology: A scientometric overview. In *Cyanobacteria: From Basic Science to Applications*, Eds. A. K. Mishra, D. N. Tiwari and A. N. Rai, pp. 419–442. Amsterdam: Elsevier.

Konur, O. 2020a. The scientometric analysis of the research on the bioethanol production from green macroalgae. In *Handbook of Algal Science, Technology and Medicine,* Ed. O. Konur, pp. 385–401. London: Academic Press.

Konur, O., Ed. 2020b. *Handbook of Algal Science, Technology and Medicine.* London: Academic Press.

Konur, O., Ed. 2021a. *Handbook of Biodiesel and Petrodiesel Fuels: Science, Technology, Health, and Environment.* Boca Raton, FL: CRC Press.

Konur, O., Ed. 2021b. *Handbook of Biodiesel and Petrodiesel Fuels: Science, Technology, Health, and Environment.* Volume 1. Biodiesel Fuels: Science, Technology, Health, and Environment. Boca Raton, FL: CRC Press.

Konur, O., Ed. 2021c. *Handbook of Biodiesel and Petrodiesel Fuels: Science, Technology, Health, and Environment.* Volume 2. Biodiesel Fuels based on the Edible and Nonedible Feedstocks, Wastes, and Algae: Science, Technology, Health, and Environment. Boca Raton, FL: CRC Press.

Konur, O., Ed. 2021d. *Handbook of Biodiesel and Petrodiesel Fuels: Science, Technology, Health, and Environment.* Volume 3. Petrodiesel Fuels: Science, Technology, Health, and Environment. Boca Raton, FL: CRC Press.

Konur, O. 2023a. Second generation bioethanol fuels from residual sugar feedstocks: Scientometric study. In *Feedstock-based Bioethanol Fuels. II. Waste Feedstocks: Agricultural, Food, Industrial, Urban, Forestry, and Lignocellulosic Waste-based Bioethanol Fuels. Handbook of Bioethanol Fuels Volume 4*, Ed. O. Konur, pp. 53–78. Boca Raton, FL: CRC Press.

Konur, O. 2023b. Second generation bioethanol fuels from residual starch feedstocks: Scientometric study. In *Feedstock-based Bioethanol Fuels. II. Waste Feedstocks: Agricultural, Food, Industrial, Urban, Forestry, and Lignocellulosic Waste-based Bioethanol Fuels. Handbook of Bioethanol Fuels Volume 4*, Ed. O. Konur, pp. 79–98. Boca Raton, FL: CRC Press.

Konur, O. 2023c. Second generation food waste-based bioethanol fuels: Scientometric study. In *Feedstock-based Bioethanol Fuels. II. Waste Feedstocks: Agricultural, Food, Industrial, Urban, Forestry, and Lignocellulosic Waste-based Bioethanol Fuels. Handbook of Bioethanol Fuels Volume 4*, Ed. O. Konur, pp. 147–172. Boca Raton, FL: CRC Press.

Konur, O. 2023d. Second generation industrial waste-based bioethanol fuels: Scientometric study. In *Feedstock-based Bioethanol Fuels. II. Waste Feedstocks: Agricultural, Food, Industrial, Urban, Forestry, and Lignocellulosic Waste-based Bioethanol Fuels. Handbook of Bioethanol Fuels Volume 4*, Ed. O. Konur, pp. 191–216. Boca Raton, FL: CRC Press.

Konur, O. 2023e. Second generation urban waste-based bioethanol fuels: Scientometric study. In *Feedstock-based Bioethanol Fuels. II. Waste Feedstocks: Agricultural, Food, Industrial, Urban, Forestry, and Lignocellulosic Waste-based Bioethanol Fuels. Handbook of Bioethanol Fuels Volume 4*, Ed. O. Konur, pp. 237–261. Boca Raton, FL: CRC Press.

Konur, O. 2023f. Second generation forestry waste-based bioethanol fuels: Scientometric study. In *Feedstock-based Bioethanol Fuels. II. Waste Feedstocks: Agricultural, Food, Industrial, Urban, Forestry, and Lignocellulosic Waste-based Bioethanol Fuels. Handbook of Bioethanol Fuels Volume 4*, Ed. O. Konur, pp. 281–304. Boca Raton, FL: CRC Press.

Konur, O. 2023g. Lignocellulosic biomass-based bioethanol fuels: Scientometric study. In *Feedstock-based Bioethanol Fuels. II. Waste Feedstocks: Agricultural, Food, Industrial, Urban, Forestry, and Lignocellulosic Waste-based Bioethanol Fuels. Handbook of Bioethanol Fuels Volume 4*, Ed. O. Konur, pp. 325–349. Boca Raton, FL: CRC Press.

Konur, O. and F. L. Matthews. 1989. Effect of the properties of the constituents on the fatigue performance of composites: A review. *Composites* 20:317–328.

Koutinas, A. A., A. Vlysidis and D. Pleissner, et al. 2014. Valorization of industrial waste and by-product streams via fermentation for the production of chemicals and biopolymers. *Chemical Society Reviews* 43:2587–2627.

Kruyt, B., D. P. van Vuuren, H. J. de Vries and H. Groenenberg. 2009. Indicators for energy security. *Energy Policy* 37:2166–2181.

Lenihan, P., A. Orozco and E. O'Neill, et al. 2010. Dilute acid hydrolysis of lignocellulosic biomass. *Chemical Engineering Journal* 156:395–403.

Leydesdorff, L. 2000. Is the European Union becoming a single publication system? *Scientometrics* 47:265–280.

Leydesdorff, L. and C. Wagner. 2009. Is the United States losing ground in science? A global perspective on the world science system. *Scientometrics* 78:23–36.

Leydesdorff, L. and P. Zhou. 2005. Are the contributions of China and Korea upsetting the world system of science? *Scientometrics* 63:617–630.

Li, H., S. M. Liu, X. H. Yu, S. L. Tang and C. K. Tang. 2020. Coronavirus disease 2019 (COVID-19): Current status and future perspectives. *International Journal of Antimicrobial Agents* 55:105951.

Limayem, A. and S. C. Ricke. 2012. Lignocellulosic biomass for bioethanol production: Current perspectives, potential issues and future prospects. *Progress in Energy and Combustion Science* 38:449–467.

Lin, Y. and S. Tanaka. 2006. Ethanol fermentation from biomass resources: Current state and prospects. *Applied Microbiology and Biotechnology* 69:627–642.

Ma, X., L. Sun and C. Song. 2002. A new approach to deep desulfurization of gasoline, diesel fuel and jet fuel by selective adsorption for ultra-clean fuels and for fuel cell applications. *Catalysis Today* 77:107–116.

Morrissey, A. J. and J. Browne. 2004. Waste management models and their application to sustainable waste management. *Waste Management* 24:297–308.

Morschbacker, A. 2009. Bio-ethanol based ethylene. *Polymer Reviews* 49:79–84.

Mosier, N., C. Wyman and B. Dale, et al. 2005. Features of promising technologies for pretreatment of lignocellulosic biomass. *Bioresource Technology* 96:673–686.

Najafi, G., B. Ghobadian and T. Tavakoli, et al. 2009. Performance and exhaust emissions of a gasoline engine with ethanol blended gasoline fuels using artificial neural network. *Applied Energy* 86:630–639.

Newman, P. W. G. and J. R. Kenworthy. 1989. Gasoline consumption and cities: A comparison of U.S. cities with a global survey. *Journal of the American Planning Association* 55:24–37.

North, D. C. 1991. Institutions. *Journal of Economic Perspectives* 5:97–112.

Olsson, L. and B. Hahn-Hagerdal. 1996. Fermentation of lignocellulosic hydrolysates for ethanol production. *Enzyme and Microbial Technology* 18:312–331.

Palmqvist, E. and B. Hahn-Hagerdal. 2000a. Fermentation of lignocellulosic hydrolysates. I: Inhibition and detoxification. *Bioresource Technology* 74:17–24.

Palmqvist, E. and B. Hahn-Hagerdal. 2000b. Fermentation of lignocellulosic hydrolysates. II: Inhibitors and mechanisms of inhibition. *Bioresource Technology* 74:25–33.

Prasad, S., A. Singhand and H. C. Joshi. 2007. Ethanol as an alternative fuel from agricultural, industrial and urban residues. *Resources, Conservation and Recycling* 50:1–39.

Ravindran, R. and A. K. Jaiswal. 2016. A comprehensive review on pre-treatment strategy for lignocellulosic food industry waste: Challenges and opportunities. *Bioresource Technology* 199:92–102.

Reeves, S. 2014. To Russia with love: How moral arguments for a humanitarian intervention in Syria opened the door for an invasion of the Ukraine. *Michigan State University International Law Review* 23:199.

Saini, J. K., R. Saini and L. Tewari. 2015. Lignocellulosic agriculture wastes as biomass feedstocks for second-generation bioethanol production: Concepts and recent developments. *3 Biotech* 5:337–353.

Sanchez, O. J. and C. A. Cardona. 2008. Trends in biotechnological production of fuel ethanol from different feedstocks. *Bioresource Technology* 99:5270–5295.

Sassner, P., M. Galbe and G. Zacchi. 2008. Techno-economic evaluation of bioethanol production from three different lignocellulosic materials. *Biomass and Bioenergy* 32:422–430.

Sun, Y. and J. Cheng. 2002. Hydrolysis of lignocellulosic materials for ethanol production: A review. *Bioresource Technology* 83:1–11.

Taherzadeh, M. J. and K. Karimi. 2007. Enzyme-based hydrolysis processes for ethanol from lignocellulosic materials: A review. *Bioresources* 2:707–738.

Taherzadeh, M. J. and K. Karimi. 2008. Pretreatment of lignocellulosic wastes to improve ethanol and biogas production: A review. *International Journal of Molecular Sciences* 9:1621–1651.

Tahmooresnejad, L., C. Beaudry, C and A. Schiffauerova. 2015. The role of public funding in nanotechnology scientific production: Where Canada stands in comparison to the United States. *Scientometrics* 102:753–787.

Wilson, D. C. 2007. Development drivers for waste management. *Waste Management & Research* 25:198–207.

Winzer, C. 2012. Conceptualizing energy security. *Energy Policy* 46:36–48.

Yang, B. and C. E. Wyman. 2008. Pretreatment: The key to unlocking low-cost cellulosic ethanol. *Biofuels, Bioproducts and Biorefining* 2:26–40.

Zaldivar, J., J. Nielsen and L. Olsson. 2001. Fuel ethanol production from lignocellulose: A challenge for metabolic engineering and process integration. *Applied Microbiology and Biotechnology* 56:17–34.

60 Second Generation Waste Biomass-based Bioethanol Fuels

Review

Ozcan Konur
(Formerly) Ankara Yildirim Beyazit University

60.1 INTRODUCTION

Crude oil-based gasoline fuels (Ma et al., 2002; Newman and Kenworthy, 1989) have been widely used in the transportation sector since the 1920s. However, there have been great public concerns over the adverse environmental and human impact of these fuels (Hill et al., 2006, 2009). Hence, biomass-based bioethanol fuels (Hill et al., 2006; Konur, 2012, 2015, 2020) have increasingly been used in blending gasoline fuels (Hsieh et al., 2002; Najafi et al., 2009), in fuel cells (Antolini, 2007, 2009), and in biochemical production (Angelici et al., 2013; Morschbacker, 2009) in a biorefinery context (Fernando et al., 2006; Huang et al., 2008).

However, it is necessary to pretreat the biomass (Alvira et al., 2010; Taherzadeh and Karimi, 2008) to enhance the yield of the bioethanol (Hahn-Hagerdal et al., 2006; Sanchez and Cardona, 2008) prior to bioethanol fuel production from the feedstocks through the hydrolysis (Sun and Cheng, 2002; Taherzadeh and Karimi, 2007) and fermentation (Lin and Tanaka, 2006; Olsson and Hahn-Hagerdal, 1996) of the biomass and hydrolysates, respectively.

One of the most studied feedstocks for bioethanol fuels has been waste biomass. In this context, the research in the field of second generation waste biomass-based bioethanol fuels has intensified in the key research fronts of the pretreatment (Eriksson et al., 2002; Kumar et al., 2009) and hydrolysis (Lloyd and Wyman, 2005; Yang and Wyman, 2004) of the waste biomass, fermentation (Laser et al., 2002; Saha et al., 2005) of the waste biomass-based hydrolysates, and production (Kaparaju et al., 2009; Mosier et al., 2005a) and evaluation (Hamelinck et al., 2005; Sheehan et al., 2003) of the second generation waste biomass-based bioethanol fuels.

Research in this field has also intensified for the feedstocks of lignocellulosic biomass at large (Eriksson et al., 2002; Hamelinck et al., 2005), agricultural residues (Kumar et al., 2009; Saha et al., 2005), industrial wastes (Ito et al., 2002; Laser et al., 2002), forestry wastes (Brandt et al., 2010; Duff and Murray, 1996), urban wastes (Guimaraes et al., 2010; Lenihan et al., 2010), and food wastes (Lenihan et al., 2010; Ravindran and Jaiswal, 2016).

However, it is essential to develop efficient incentive structures (North, 1991) for the primary stakeholders to enhance the research in this field (Konur, 2000, 2002a–c, 2006a,b, 2007a,b). Although there has been a number of review papers on waste biomass-based bioethanol fuels (Hendriks and Zeeman, 2009; Mosier et al., 2005a; Sun and Cheng, 2002), there has been no review of the most-cited 25 papers in this field.

Thus, this chapter presents a review of the most-cited 25 articles in the field of waste biomass-based bioethanol fuels. Then, it discusses the key findings of these highly influential papers and comments on the future research priorities in this field.

DOI: 10.1201/9781003226550-80

60.2 MATERIALS AND METHODS

The search for this study was carried out using the Scopus database (Burnham, 2006) in October 2022.

As the first step for the search of the relevant literature, the keywords were selected using the 300 most-cited population papers for each waste biomass. The selected keyword list was then optimized to obtain a representative sample of papers for this research field. These keyword lists were then integrated to obtain the keyword list for this research field (Konur, 2023a–g).

As the second step, a sample dataset was used in this study. The first 25 articles with at least 401 citations each were selected for the review study. Key findings from each paper were taken from the abstracts of these papers and discussed. Additionally, a number of brief conclusions were drawn and a number of relevant recommendations were made to enhance the future research landscape.

60.3 RESULTS

The brief information about the 25 most-cited papers with at least 401 citations each on second generation waste biomass-based bioethanol fuels is given below. The primary research fronts are the hydrolysis of the waste and production and evaluation of waste biomass-based bioethanol fuels, with 13 and 12 highly cited papers (HCPs), respectively.

60.3.1 WASTE BIOMASS HYDROLYSIS

The brief information about the 13 most-cited papers on the hydrolysis of waste biomass with at least 405 citations each is given below (Table 60.1). These papers also cover the research on the pretreatment of the waste.

Eriksson et al. (2002) explored the mechanism of the surfactant effect in the enzymatic hydrolysis of lignocellulosic biomass in a paper with 751 citations. They screened a number of surfactants for their ability to improve the enzymatic hydrolysis of steam-pretreated spruce. They found that non-ionic surfactants were the most effective, and both anionic and non-ionic surfactants reduced enzyme adsorption to the lignocellulose substrate. The approximate reduction of enzyme adsorption was from 90% adsorbed enzyme to 80% with surfactant addition. However, surfactants had only a weak effect on cellulase temperature stability. They explained the improved conversion of lignocellulose with surfactant by the reduction of the unproductive enzyme adsorption to the lignin part of the substrate. This was due to the hydrophobic interaction of surfactant with lignin on the lignocellulose surface, which released unspecifically bound enzymes.

Kumar et al. (2009) characterized corn stover and poplar solids resulting from leading pretreatment technologies in a paper with 723 citations. They performed pretreatments by ammonia fiber expansion (AFEX), ammonia recycled percolation (ARP), controlled pH, dilute acid, flowthrough, lime, and SO_2 technologies. They observed that lime pretreatment removed the most acetyl groups from both corn stover and poplar, while AFEX removed the least. Low pH pretreatments depolymerized cellulose and enhance biomass crystallinity much more than higher pH approaches. Lime-pretreated corn stover solids and flowthrough-pretreated poplar solids had the highest cellulase adsorption capacity, while dilute acid-pretreated corn stover solids and controlled pH-pretreated poplar solids had the least. Furthermore, enzymatically extracted AFEX lignin preparations for both corn stover and poplar had the lowest cellulase adsorption capacity. SO_2-pretreated solids had the highest surface O/C ratio for poplar, but for corn stover, they observed the highest value for dilute acid pretreatment with a Parr reactor. Although dependent on pretreatment and substrate, along with changes in crosslinking and chemical changes, pretreatments might also decrystallize cellulose and change the ratio of crystalline cellulose polymorphs (I_α/I_β).

Lloyd and Wyman (2005) evaluated the combined sugar yields for dilute sulfuric acid pretreatment of corn stover followed by enzymatic hydrolysis of the remaining solids in a paper with 624

TABLE 60.1
The Hydrolysis of Waste Biomass

No.	Papers	Wastes	Prts.	Parameters	Keywords	Lead Authors	Affiliation	Cits.
2	Eriksson et al. (2002)	Lignocellulosic biomass spruce	Surfactants, enzymes, steam	Enzymatic hydrolysis, surfactant effect and mechanism, enzyme adsorption	Lignocellulose, hydrolysis	Tjerneld, Folke 7006446969	Lund Univ. Sweden	751
3	Kumar et al. (2009)	Agricultural residues corn stover	Acid, alkali, ammonia, enzymes, SO_2	Enzymatic hydrolysis, pretreatments, biomass, cellulase adsorption capacity	Corn stover, pretreatments	Wyman, Charles E. 7004396809	Univ. Calif. Riverside USA	723
5	Lloyd and Wyman (2005)	Agricultural residues corn stover	Acids, enzymes	Enzymatic hydrolysis, pretreatment, glucose and xylose yield, enzyme loading	Corn stover, pretreatment, hydrolysis	Wyman, Charles E. 7004396809	Univ. Calif. Riverside USA	624
6	Yang and Wyman (2004)	Agricultural residues corn stover	Acids, enzymes, water	Enzymatic hydrolysis, pretreatment, xylan and lignin removal, batch and flowthrough reactors, enzymatic digestibility	Corn stover, pretreatment, digestibility	Wyman, Charles E. 7004396809	Univ. Calif. Riverside USA	571
8	Kim et al. (2003)	Agricultural residues corn stover	Ammonia, enzymes	Enzymatic hydrolysis, ARP, lignin and hemicellulose removal, enzymatic digestibility, surface area and porosity	Corn stover, pretreatment	Lee, Yoon Y. 8948274900	Auburn Univ. USA	505
9	Zhang et al. (2007)	Lignocellulosic biomass	Solvents, enzymes	Enzymatic hydrolysis, solvents, fractionation, enzymatic digestibility	Lignocellulose, fractionating	Zhang, Yi-Heng P. 34876090400	Tianjin Inst. Ind. Biotechnol. China	497
10	Hsu et al. (2010)	Agricultural residues rice straw	Acids, enzymes	Enzymatic hydrolysis, pretreatment, sugar yield, pore volume	Rice straw, pretreatment, hydrolysis	Guo, Gia-Luen 7402768046	Inst. Nucl. Ener. Res. Taiwan	492

(*Continued*)

TABLE 60.1 (*Continued*)
The Hydrolysis of Waste Biomass

No.	Papers	Wastes	Prts.	Parameters	Keywords	Lead Authors	Affiliation	Cits.
12	Kristensen et al. (2009)	Lignocellulosic biomass	Enzymes	Enzymatic hydrolysis, solids content, conversion yield, cellulase adsorption inhibition	Lignocellulose, hydrolysis	Jorgensen, Henning 7202554496	Univ. Copenhagen Denmark	478
13	Ohgren et al. (2007)	Agricultural residues corn stover	Acids, steam, enzymes	Enzymatic hydrolysis, pretreatments, accessory enzymes, sugar yield, xylanases, pretreatment severity, hemicellulose and lignin removal	Corn stover, pretreated, hydrolysis	Zacchi, Guido 7006727748	Lund Univ. Sweden	473
15	Wyman et al. (2005)	Agricultural residues corn stover	Ammonia, acid, alkali, enzymes	Enzymatic hydrolysis, pretreatments, sugar yield, sugar release	Corn stover, pretreatment	Wyman, Charles E. 7004396809	Univ. Calif. Riverside USA	453
20	Li et al. (2008)	Lignocellulosic biomass corn stalk, rice straw, pine wood, bagasse	Acids	Acid hydrolysis, pretreatments, ILs, sugar yield, biomass	Lignocellulose, hydrolysis	Zhao, Zongbao K. 56972812400	Chinese Acad. Sci. China	425
23	Sun and Cheng (2005)	Agricultural residues rye straw	Acids, enzymes	Enzymatic hydrolysis, pretreatment, acid content, residence time, pretreatment severity	Rye straw, ethanol	Cheng, Jay J. 15046539600	NC State Univ. USA	416
24	Chundawat et al. (2011)	Lignocellulosic biomass	Ammonia	Enzymatic hydrolysis, pretreatment mechanisms, enzyme accessibility, morphology, pore surface area	Lignocellulosic, pretreatment	Chundawat, Shishir P. S. 12803763300	Rutgers, State Univ. N. J. USA	405

Prt., Biomass pretreatments; Cits., Number of citations received for each paper.

citations. They determined the individual xylose and glucose yields as a percentage of the total potential yield of both sugars over a range of sulfuric acid concentrations of 0.22%, 0.49%, and 0.98% w/w at 140°C, 160°C, 180°C, and 200°C. They observed that up to 15% of the total potential sugar in the substrate could be released as glucose during pretreatment, and between 15% and 90+% of the xylose remaining in the solid residue could be recovered in subsequent enzymatic hydrolysis, depending on the enzyme loading. Glucose yields increased from as high as 56% of total maximum potential glucose plus xylose for just enzymatic digestion to 60% when glucose released in pretreatment was included. Xylose yields similarly increased from as high as 34% of total potential sugars for pretreatment alone to between 35% and 37% when credit was taken for xylose released in digestion. However, sugar yields were much lower if no acid was used. Further, conditions that maximized individual sugar yields were often not the same as those that maximized total sugar yields. Overall, up to about 92.5% of the total sugars originally available in the corn stover used could be recovered for coupled dilute acid pretreatment and enzymatic hydrolysis.

Yang and Wyman (2004) evaluated xylan and lignin removal and enzymatic digestibility of cellulose for corn stover pretreated in batch and flowthrough reactors over a range of flow rates between 160°C and 220°C, with water only and also with 0.1 wt% sulfuric acid in a paper with 571 citations. They observed that increasing flow with just water enhanced the xylan dissolution rate, more than doubled total lignin removal, and increased cellulose digestibility. Furthermore, adding dilute sulfuric acid increased the rate of xylan removal for both batch and flowthrough systems. Adding acid also increased the lignin removal rate with the flow, but less lignin was left in the solution when acid was added in batches. Although the enzymatic hydrolysis of pretreated cellulose was related to xylan removal, the digestibility was much better for flowthrough compared with batch systems for the same degree of xylan removal. Similarly, cellulose digestibility for flow-through reactors was related to lignin removal as well. In conclusion, altering lignin also affected the enzymatic digestibility of corn stover.

Kim et al. (2003) evaluated the ARP pretreatment and enzymatic hydrolysis of corn stover in a paper with 505 citations. They pretreated corn stover with aqueous ammonia in a flow-through column reactor. They found that this pretreatment was highly effective in delignifying the biomass, reducing the lignin content by 70%–85%. Most lignin removal occurred within the first 20 min of the process. The ARP process also solubilized 40%–60% of the hemicellulose but left the cellulose intact. The solubilized carbohydrate existed in an oligomeric form. Corn stover treated for 90 min exhibited enzymatic digestibility of 99% with 60 FPU/g of glucan enzyme loading and 92.5% with 10 FPU/g of glucan. The digestibility of ARP-pretreated corn stover was substantially higher than that of α-cellulose. The enzymatic digestibility was related to the removal of lignin and hemicellulose, perhaps due to increased surface area and porosity. The biomass structure was deformed and its fibers were exposed by the pretreatment. Similarly, the crystallinity index increased with pretreatment reflecting the removal of the amorphous portion of the biomass. The crystalline structure of the cellulose in the biomass, however, was not changed by the ARP treatment.

Zhang et al. (2007) fractionated lignocellulosic biomass into cellulose, hemicellulose, lignin, and acetic acid at modest reaction conditions (50°C and atmospheric pressure) in a paper with 496 citations. They used a non-volatile cellulose solvent (concentrated phosphoric acid), a highly volatile organic solvent (acetone), and water. They attributed the highest sugar yields after enzymatic hydrolysis to no sugar degradation during the fractionation and the highest enzymatic cellulose digestibility (~97% in 24 h) during the hydrolysis step at the enzyme loading of 15 filter paper units (FPU) of cellulase and 60 international units (IU) of β-glucosidase per gram of glucan. Further, the isolation of high-value lignocellulose components (lignin, acetic acid, and hemicellulose) would greatly increase the potential revenues of a lignocellulose biorefinery.

Hsu et al. (2010) performed the dilute acid pretreatment of rice straw and explored the effect of the structural properties of the solid residues on the enzymatic hydrolysis in a paper with 492 citations. They obtained a maximal sugar yield of 83% when the rice straw was pretreated with 1% (w/w) sulfuric acid with a reaction time of 1–5 min at 160°C or 180°C, followed by enzymatic hydrolysis. Further, the complete release of sugar increased the pore volume of the pretreated solid residues resulting in an efficiency of 70% for the enzymatic hydrolysis. The extra pore volume was generated by the release of acid-soluble lignin and this resulted in the enzymatic hydrolysis being enhanced by nearly 10%. The increase in the crystallinity index of the pretreated rice straw was limited.

Kristensen et al. (2009) explored the enzymatic hydrolysis of lignocellulosic biomass at high solid concentrations in a paper with 477 citations. They found that the decreasing enzymatic conversion of lignocellulosic biomass at increasing solids concentrations was a generic or intrinsic effect, describing a linear correlation from 5% to 30% initial total solids content (w/w). Neither lignin content nor hemicellulose-derived inhibitors were responsible for the decrease in yields. Product inhibition by glucose and, in particular, cellobiose at the increased concentrations at high solids loading played a role but could not completely account for the decreasing conversion. Adsorption of cellulases decreased at increasing solid concentrations. Hence, there was a strong correlation between the decreasing cellulase adsorption and the enzymatic conversion of biomass, indicating that the inhibition of cellulase adsorption to cellulose was causing the decrease in yield. In conclusion, the inhibition of enzyme adsorption by hydrolysis products was the main cause of the decreasing yields at increasing substrate concentrations in the enzymatic decomposition of cellulosic biomass.

Ohgren et al. (2007) evaluated the effect of hemicellulose and lignin removal on the enzymatic hydrolysis of pretreated corn stover in a paper with 473 citations. They obtained a near-theoretical glucose yield (96%–104%) from acid-catalyzed steam-pretreated corn stover when xylanases were used to supplement cellulases during hydrolysis. Xylanases hydrolyzed residual hemicellulose, thereby improving the access of enzymes to cellulose. Under these conditions, xylose yields reached 70%–74%. When pretreatment severity was reduced by using autocatalysis instead of acid-catalyzed steam explosion pretreatment, xylose yields increased to 80%–86%. The overall glucose yield increased slightly due to delignification, but the overall xylose yield decreased due to hemicellulose loss in the delignification step.

Wyman et al. (2005) evaluated ammonia explosion, aqueous ammonia recycle, controlled pH, dilute acid, flow-through, and lime pretreatments using a single source of corn stover and the same cellulase enzyme to make meaningful comparisons. They found that each pretreatment made it possible to subsequently achieve high yields of glucose from cellulose by cellulase enzymes, and the cellulase formulations used were effective in solubilizing residual xylan left in the solids after each pretreatment. Thus, overall sugar yields from hemicellulose and cellulose in the coupled pretreatment and enzymatic hydrolysis processes were high for all of the pretreatments with corn stover. In addition, high-pH methods offered promise in reducing cellulase use provided hemicellulase activity could be enhanced. However, the substantial differences in sugar release patterns in the pretreatment and enzymatic hydrolysis operations had important implications for the choice of process, enzymes, and fermentative organisms.

Li et al. (2008) used acid in ionic liquid (IL) for the hydrolysis of lignocellulosic biomass in a paper with 425 citations. They showed that this pretreatment was an efficient system for the hydrolysis of lignocellulosic materials with improved total reducing sugar (TRS) yield under mild conditions. TRS yields were up to 66%, 74%, 81%, and 68% for hydrolysis of corn stalk, rice straw, pine wood, and bagasse, respectively, in [C_4mim]Cl in the presence of 7 wt% hydrogen chloride at 100°C under atmospheric pressure within 60 min. Different combinations between ILs, such as [C_6mim]Cl, [C_4mim]Br, [Amim]Cl, [C_4mim]HSO_4, and [Sbmim]HSO_4, and acids, including sulfuric acid, nitric acid, phosphoric acid, as well as maleic acid, afforded similar results, though a longer reaction time was generally required comparing with the combination of [C_4mim]Cl and hydrochloric acid. Further, the modification of lignin occurred during sulfuric acid catalyzed hydrolysis.

Sun and Cheng (2005) evaluated the dilute sulfuric acid pretreatment of rye straw and bermudagrass before enzymatic hydrolysis of cellulose in a paper with 416 citations. They pretreated the biomass at a solid loading rate of 10% at 121°C with different sulfuric acid concentrations (0.6%, 0.9%, 1.2%, and 1.5%, w/w) and residence times (30, 60, and 90 min). They then hydrolyzed the solid residues with cellulases to investigate the enzymatic digestibility. With the increasing acid concentration and residence time, they observed that the amount of arabinose and galactose increased. However, the glucose concentration in the prehydrolysate of rye straw was not significantly influenced by the sulfuric acid concentration and residence time, but it increased in the prehydrolysate of bermudagrass with the increase of pretreatment severity. Further, the xylose concentration increased with the increase of sulfuric acid concentration and residence time. Most of the arabinan, galactan, and xylan in the biomass were hydrolyzed during the acid pretreatment. Cellulose remaining in the pretreated feedstock was highly digestible by cellulases from *Trichoderma reesei*.

Chundawat et al. (2011) characterized corn stover cell walls to elucidate the mechanism of AFEX pretreatment in a paper with 405 citations. They observed that AFEX first dissolved and then extracted, as the ammonia evaporated, redeposited cell wall decomposition products (e.g., amides, arabinoxylan oligomers, lignin-based phenolics) on outer cell wall surfaces. As a result, nanoporous tunnel-like networks were formed within the cell walls. They proposed that this highly porous structure greatly enhanced enzyme accessibility to embedded cellulosic microfibrils. The shape, size (10–1000 nm), and spatial distribution of the pores depended on their location within the cell wall and the pretreatment conditions used. Exposed pore surface area per unit AFEX-pretreated cell wall volume ranged between 0.005 and 0.05 nm^2 per nm^3. AFEX thus resulted in ultrastructural and physicochemical modifications within the cell wall that enhanced enzymatic hydrolysis yield by 4–5 fold over that of untreated cell walls.

60.3.2 Waste Biomass-based Bioethanol Fuels

There are 12 HCPs for the production and evaluation of waste biomass-based bioethanol fuels with at least 401 citations each (Table 60.2). Further, there are nine and three HCPs for the production and evaluation of bioethanol fuels, respectively. As the pretreatment and hydrolysis of the lignocellulosic biomass are the fundamental parts of bioethanol production, these narrated papers often cover these processes too.

60.3.2.1 Waste Biomass-based Bioethanol Production

There are nine HCPs for the production of waste biomass-based bioethanol fuels with at least 418 citations each (Table 60.2). These papers also cover the fermentation of the hydrolysates of the lignocellulosic biomass. As the pretreatment and hydrolysis are the fundamental parts of bioethanol production, these narrated papers often cover these processes too.

Saha et al. (2005) produced ethanol from wheat straw with 49% and 28% cellulose and hemicellulose, respectively, in a paper with 662 citations. They performed dilute acid pretreatment at varied temperature. They found that the maximum yield of monomeric sugars from wheat straw (7.83%, w/v, DS) by dilute H_2SO_4 (0.75%, v/v) pretreatment and enzymatic hydrolysis (45°C, pH 5.0, 72 h) using cellulase, β-glucosidase, xylanase, and esterase was 565 mg/g. Under this condition, there were no measurable quantities of furfural and hydroxymethylfurfural, the fermentation inhibitors. The yield of ethanol (per liter) from acid-pretreated enzyme hydrolyzed wheat straw (78.3 g) hydrolysate by recombinant *Escherichia coli* strain FBR5 was 19 g, with a yield of 0.24 g/g DS. Detoxification of the acid and enzyme-treated wheat straw hydrolysate by overliming reduced the fermentation time from 118 to 39 h in the case of separate hydrolysis and fermentation (SHF) (35°C, pH 6.5), increased the ethanol yield from 13 to 17 g/L, and decreased the fermentation time from 136 to 112 h in the case of simultaneous saccharification and fermentation (SSF) (35°C, pH 6.0).

Kaparaju et al. (2009) produced bioethanol, biohydrogen, and biogas production from wheat straw in a biorefinery concept in a paper with 557 citations. They first fractionated the biomass into a cellulose-rich fiber fraction and a hemicellulose-rich liquid fraction by the hydrothermal pretreatment. They then produced 0.41 g-ethanol/g-glucose by the enzymatic hydrolysis and subsequent fermentation of cellulose, while the dark fermentation of hydrolysate produced 178.0 ml-H_2/g-sugars. They used the effluents from both bioethanol and biohydrogen processes to produce biomethane,

TABLE 60.2
The Production and Evaluation of Waste Biomass-based Bioethanol Fuels

No.	Papers	Wastes	Res. Fronts	Prts.	Yeasts	Parameters	Keywords	Lead Authors	Affiliation	Cits.
1	Hamelinck et al. (2005)	Lignocellulosic biomass	Evaluation	Enzymes	Yeasts	Techno-economics, bioethanol production efficiency and costs, investment costs, hydrolysis, fermentation, biomass costs	Lignocellulosic, biomass, ethanol	Hamelinck, Carlo N. 6603008025	Utrecht Univ. Netherlands	1,214
4	Saha et al. (2005)	Agricultural residues wheat straw	Production	Acids, enzymes	*E. coli*	Ethanol production, pretreatment, enzymatic hydrolysis, fermentation, sugar and ethanol yield, SHF, SSF, detoxification	Wheat straw, pretreatment, saccharification, ethanol, fermentation	Saha, Badal C. 7202946302	USDA Agr. Res. Serv. USA	662
7	Kaparaju et al. (2009)	Agricultural residues wheat straw	Production	Hydrothermal, enzymes	Yeasts	Ethanol production, biohydrogen, biogas, enzymatic hydrolysis, fermentation, ethanol yield, energy balance	Wheat straw, bioethanol	Angelidaki, Irini 6603674728	Technical Univ. Denmark Denmark	557
11	Laser et al. (2002)	Agro-industrial residues sugarcane bagasse	Production	LHW, steam, enzymes	Yeasts	Enzymatic hydrolysis, pretreatments, solid content, SSF conversion, xylan recovery and dissolution, furfural content, hydrolyzate inhibition	Sugar cane bagasse, ethanol, pretreatments	Lynd, Lee R. 35586183800	Dartmouth Coll. USA	485
14	Mosier et al. (2005b)	Agricultural residues corn stover	Production	LHW, enzymes	Yeasts	Ethanol production, pretreatment optimization, enzymatic hydrolysis, fermentation, sugar ethanol yield	Corn stover, pretreatment	Ladisch, Michael R. 7005670397	Purdue Univ. USA	464
16	Sheehan et al. (2003)	Agricultural residues corn stover	Evaluation	Na	Na	Life cycle assessment, E85, corn farming, ethanol production, Soil organic matter and erosion, fossil energy use, GHG emissions, air quality	Corn stover, ethanol	Sheehan, John 7202241794	Colorado State Univ. USA	453
17	Teymouri et al. (2005)	Agricultural residues corn stover	Production	Ammonia, enzymes	Yeast	Enzymatic hydrolysis, pretreatment optimization, sugar and ethanol yield, temperature, moisture content, ammonia loadings treatment time	Corn stover, hydrolysis	Dale, Bruce E. 7201511969	Michigan State Univ. USA	432

(*Continued*)

TABLE 60.2 (*Continued*)
The Production and Evaluation of Waste Biomass-based Bioethanol Fuels

No.	Papers	Wastes	Res. Fronts	Prts.	Yeasts	Parameters	Keywords	Lead Authors	Affiliation	Cits.
18	Larsson et al. (1999)	Lignocellulosic biomass spruce	Production	Acids, enzymes	*S. cerevisiae*	Fermentation, hydrolysate detoxification methods, fermentation inhibitors, sugars	Lignocellulose, hydrolysates	Jonsson, Leif J. 7102349315	Ume Univ. Sweden	430
19	Jorgensen et al. (2007)	Lignocellulosic biomass straw	Production	Enzymes	*S. cerevisiae*	Enzymatic hydrolysis, solids content, glucose content, ethanol yield	Lignocellulose, liquefaction	Jorgensen, Henning 7202554496	Univ. Copenhagen Denmark	429
21	Ito et al. (2005)	Industrial wastes glycerol	Production	Enzymes	*E. aerogenes*	Ethanol and hydrogen production, glycerol and biodiesel waste content, fermentation, ethanol and hydrogen yield	Glycerol, ethanol	Nishio, Naomichi 7005754508	Hiroshima Univ. Japan	421
22	Delgenes et al. (1996)	Lignocellulosic biomass	Production	Na	*S. cerevisiae, Z. mobilis, P. stipitis, C. shehatae*	Hydrolysate fermentation, fermentation inhibitors, yeast strains	Lignocellulose, ethanol, fermentation	Delgenes, Jean P. 7005849678	Univ. Montpellier France	418
25	Kazi et al. (2010)	Agricultural residues corn stover	Evaluation	Acid, ammonia, LHW	Yeasts	Ethanol production, techno-economics, pretreatments, product value, investment cost, pioneer and n^{th} plants	Cprn stover, ethanol	Anex, Robert P. 6701606624	Univ. Wisconsin Madison USA	401

Prt., Biomass pretreatments; Na, non available; Cits., Number of citations received for each paper.

with yields of 0.324 and 0.381 m^3/kg volatile solids, respectively. Further, either the use of wheat straw for biogas production or multi-fuel production was the energetically most efficient process compared to the production of mono-fuel such as bioethanol when fermenting C_6 sugars alone. In conclusion, multiple biofuels production from wheat straw could be more beneficial in terms of energy balance and efficiency.

Laser et al. (2002) compared liquid hot water (LHW) and steam explosion pretreatments of sugarcane bagasse for ethanol production in a paper with 485 citations. Solids concentration ranged from 1% to 8% for LHW pretreatment and was ≥50% for steam pretreatment while reaction temperature and time ranged from 170°C to 230°C and 1–46 min, respectively. They found that the LHW pretreatment achieved ≥80% conversion by SSF, ≥80% xylan recovery, and no hydrolysate inhibition of glucose fermentation yield. However, the combined effectiveness was not as good for steam pretreatment due to low xylan recovery. SSF conversion increased and xylan recovery decreased as xylan dissolution increased for both modes. SSF conversion, xylan dissolution, hydrolysate furfural concentration, and hydrolysate inhibition increased, while xylan recovery and hydrolysate pH decreased, as a function of increasing LHW pretreatment solids concentration (1%–8%).

Mosier et al. (2005b) optimized the pH-controlled LHW pretreatment of corn stover for enzymatic digestibility with respect to processing temperature and time in a paper with 464 citations. The optimized conditions for controlled pH and LHW pretreatment of a 16% slurry of corn stover in water were 190°C for 15 min. At the optimal conditions, 90% of the cellulose was hydrolyzed to glucose by 15 FPU of cellulase per gram of glucan. When the resulting pretreated slurry, in undiluted form, was hydrolyzed by 11 FPU of cellulase per gram of glucan, they obtained a hydrolysate containing 32.5 g/L glucose and 18 g/L xylose. Both the xylose and the glucose in this undiluted hydrolysate were fermented by recombinant yeast 424A(LNH-ST) to ethanol at 88% of the theoretical yield.

Teymouri et al. (2005) optimized the AFEX pretreatment parameters to provide maximum sugar yields by the enzymatic hydrolysis of corn stover in a paper with 432 citations. They varied the AFEX pretreatment conditions (temperature, moisture content, ammonia loadings, and treatment time) for this purpose. Optimal pretreatment conditions for corn stover were temperature, 90°C; ammonia loading, 1.0 kg of ammonia: kg of dry corn stover; moisture content of corn stover, 60% (dry weight basis (dwb)); and residence time (holding at target temperature), 5 min. They obtained around 100% of the theoretical glucose yield and 80% of the theoretical xylose yield during enzymatic hydrolysis of the optimal treated corn stover using 60 FPU of cellulase enzyme/g of glucan. Further, the ethanol yield of optimally AFEX-treated corn stover was increased up to 2.3 times over that of an untreated sample.

Larsson et al. (1999) compared the different methods for the detoxification of lignocellulose hydrolysates of spruce to improve both cell growth and ethanol production in a paper with 430 citations. They used a dilute-acid hydrolysate of spruce with *Saccharomyces cerevisiae* strains. They determined the changes in the concentrations of fermentable sugars and three groups of fermentation inhibitors, aliphatic acids, furan derivatives, and phenolic compound, as well as the fermentability of the detoxified hydrolysate. Further, the applied detoxification methods included treatment with alkali (sodium hydroxide (NaOH) or lime ($Ca(OH)_2$), treatment with sulfite (0.1% [w/v] or 1% [w/v] at pH 5.5 or 10), evaporation of 10% or 90% of the initial volume, anion exchange (at pH 5.5 or 10), enzymatic detoxification with the phenoloxidase laccase, and detoxification with the *T. reesei*. They found that an ion exchange at pH 5.5 or 10, treatment with laccase, treatment with calcium hydroxide, and treatment with *T. reesei* were the most efficient detoxification methods. Evaporation of 10% of the initial volume and treatment with 0.1% sulfite were the least efficient detoxification methods. Treatment with laccase was the only detoxification method that specifically removed only one group of the inhibitors, namely, phenolic compounds. Anion exchange at pH 10 was the most efficient method for removing all three major groups of inhibitory compounds. However, it also resulted in the loss of fermentable sugars.

Jorgensen et al. (2007) produced bioethanol through the enzymatic hydrolysis and fermentation of the lignocellulosic biomass at high-solids concentrations in a paper with 429 citations. They performed the enzymatic liquefaction and saccharification of pretreated wheat straw with up to 40% (w/w) initial dry matter (DM). In <10 h, they observed that the structure of the biomass changed from intact straw particles (length 1–5 cm) into a paste/liquid that could be pumped. There was no significant effect of mixing speed in the range 3.3–11.5 rpm on the glucose conversion after 24 h and ethanol yield after subsequent fermentation for 48 h. Liquefaction and saccharification for 96 h using an enzyme loading of 7 FPU/g·DM and 40% DM resulted in a glucose concentration of 86 g/kg. Experiments conducted at 2%–40% (w/w) initial DM revealed that cellulose and hemicellulose conversion decreased almost linearly with increasing DM. Performing the experiments as SSF also revealed a decrease in ethanol yield at increasing initial DM. *S. cerevisiae* was capable of fermenting hydrolysates up to 40% DM. They obtained the highest ethanol concentration, 48 g/kg, using 35% (w/w) DM.

Ito et al. (2005) produced hydrogen (H_2) and ethanol from glycerol-containing wastes discharged after a manufacturing process for biodiesel fuel (biodiesel wastes) using *Enterobacter aerogenes* HU-101 in a paper with 421 citations. They diluted the biodiesel wastes with a synthetic medium to increase the rate of glycerol utilization and found that the addition of yeast extract and tryptone to the synthetic medium accelerated the production of H_2 and ethanol. Further, the yields of H_2 and ethanol decreased with an increase in the concentrations of biodiesel wastes and commercially available glycerol (pure glycerol). Furthermore, the rates of H_2 and ethanol production from biodiesel wastes were much lower than those at the same concentration of pure glycerol, partially due to high salt content in the wastes. The maximum rate of H_2 production from pure glycerol was 80 mmol/L/h yielding ethanol at 0.8 mol/mol-glycerol, while that from biodiesel wastes was only 30 mmol/L/h. However, using porous ceramics as a support material to fix cells in the reactor, the maximum H_2 production rate from biodiesel wastes reached 63 mmol/L/h obtaining an ethanol yield of 0.85 mol/mol-glycerol.

Delgenes et al. (1996) studied the effects of six fermentation inhibitors on ethanol fermentations of glucose and xylose by four yeast strains in a paper with 418 citations. They used *S. cerevisiae* and *Zymomonas mobilis* for the glucose fermentation and *Pichia stipitis* and *Candida shehatae* for the xylose fermentation in batch cultures. They added the inhibitors in varying concentrations and determined the subsequent inhibitions on growth and ethanol production. They found that vanillin was a strong inhibitor of both growth and ethanol production by xylose fermenting yeasts and *S. cerevisiae* when it was added to the culture media at a concentration of 1 g/L. Further, the fermentative activities of *Z. mobilis* were greatly sensitive to the presence of hydroxybenzaldehyde (0.5 g/L). However, some of the inhibitors, particularly vanillin and furaldehyde, could be assimilated by these strains, which resulted in the partial recovery in both growth and ethanol production processes on prolonged incubation.

60.3.2.2 Waste Biomass-based Bioethanol Evaluation

There are three HCPs for the evaluation of waste biomass-based bioethanol fuels with at least 401 citations each (Table 60.2).

Hamelinck et al. (2005) evaluated ethanol production costs from lignocellulosic biomass in a paper with 1,214 citations. They found that the technology available as of the early 2000s, which was based on dilute acid hydrolysis, had about 35% efficiency (higher heating value, HHV) from biomass to ethanol. The overall efficiency, with bioelectricity co-produced from the lignin, was about 60%. They foresaw that the improvements in pretreatment and advances in biotechnology, especially through process combinations, could bring the ethanol efficiency to 48% and the overall process efficiency to 68%. They estimated investment costs as of the early 2000s at 2.1 k€/kWHHV (at 400 MWHHV input, i.e., a nominal 2,000 tonne dry/day input). However, future technology in a five times larger plant (2 GWHHV) could have investments of 900 k€/kWHHV. They further found that a combined effect of higher hydrolysis and fermentation efficiency, lower specific capital investments, an increase of scale, and cheaper biomass feedstock costs (from 3 to 2 €/GJHHV), could bring the ethanol production costs from 22 €/GJHHV in the next 5 years, to 13 €/GJ over the 10–15 year time scale, and down to 8.7 €/GJ in 20 or more years.

Sheehan et al. (2003) constructed a life-cycle model that described collecting corn stover in the state of Iowa for the production and use of a fuel mixture consisting of 85% ethanol/15% gasoline (E85) in a flexible-fuel light-duty vehicle in a paper with 453 citations. They assumed that all farmers in this state switched from their current cropping and tilling practices to the continuous production of corn and 'no-till' practices. Under these conditions, which maximized the amount of collectible stover, Iowa alone could produce almost 8 billion liters per year of pure stover-derived ethanol (E100) at prices competitive with corn-starch-derived ethanol. Soil organic matter dropped slightly in the early years of stover collection but remained stable over the 90-year time frame studied. Soil erosion was controlled at levels within tolerable soil-loss limits established by the U.S. Department of Agriculture. They found that for each kilometer fueled by the ethanol portion of E85, the vehicle used 95% less petroleum compared to a kilometer driven in the same vehicle on gasoline. Total fossil energy use (coal, oil, and natural gas) and greenhouse gas emissions (fossil CO_2, N_2O, and CH_4) on a life-cycle basis were 102% and 113% lower, respectively. However, air quality impacts were mixed, with emissions of CO, NO_x, and SO_x increasing, whereas hydrocarbon ozone precursors were reduced.

Kazi et al. (2010) performed the technoeconomic comparison of process technologies for biochemical ethanol production from corn stover based on a 5- to 8-year time frame for implementation in a paper with 401 citations. They examined the short-term commercial viability of biochemical ethanol production. They covered four pretreatment technologies (dilute-acid, two-stage dilute-acid, LHW, and AFEX) and three downstream process variations (pervaporation, separate C_5 and C_6 sugar fermentation, and on-site enzyme production). They found that the dilute-acid pretreatment process had the lowest product value (PV) among all process scenarios, which was estimated to be \$1.36/L of gasoline equivalent (LGE) (\$5.13/gal of gasoline equivalent (GGE)). Further, the PV was most sensitive to feedstock, enzyme, and installed equipment costs. A significant fraction of capital costs was related to producing heat and power from the lignin in the biomass. The estimated value of PV for the pioneer plant was substantially larger than for the *n*th plant. The PV for the pioneer plant model with dilute-acid pretreatment was \$2.30/LGE (\$8.72/GGE) for the most probable scenario, and the estimated total capital investment was more than double the *n*th plant cost.

60.4 DISCUSSION

60.4.1 INTRODUCTION

Crude oil-based gasoline fuels have been widely used in the transportation sector since the 1920s. However, there have been great public concerns over the adverse environmental and human impact of these fuels. Hence, biomass-based bioethanol fuels have increasingly been used in blending gasoline and petrodiesel fuels, in fuel cells, and in biochemical production in a biorefinery context.

However, it is necessary to pretreat the biomass to enhance the yield of the bioethanol prior to bioethanol fuel production from the feedstocks through the hydrolysis and fermentation of the biomass and hydrolysates, respectively.

One of the most studied feedstocks for bioethanol fuels has been waste biomass. The research in the field of second generation waste biomass-based bioethanol fuels has intensified in this context in the key research fronts of the pretreatment and hydrolysis of the waste biomass, fermentation of the waste biomass-based hydrolysates, and production and evaluation of second generation waste biomass-based bioethanol fuels. Research in this field has also intensified for the feedstocks of lignocellulosic biomass at large, agricultural residues and industrial, forestry, urban, and food wastes.

However, it is essential to develop efficient incentive structures for the primary stakeholders to enhance research in this field. Although there have been a number of review papers for this field, there has been no review of the 25 most-cited articles in this field.

Thus, this chapter presents a review of the 25 most-cited articles on bioethanol fuel production and evaluation from waste biomass. Then, it discusses the key findings of these highly influential papers and comments on future research priorities in this field.

As a first step for the search of the relevant literature, the keywords were selected using the most-cited first 300 population papers for each waste biomass. The selected keyword list was then optimized to obtain a representative sample of papers for each research field. These keyword lists were then integrated to obtain the keyword list for this research field (Konur, 2023a–g).

As a second step, a sample data set was used for this study. The first 25 articles with at least 401 citations each were selected for the review study. Key findings from each paper were taken from the abstracts of these papers and discussed. Additionally, a number of brief conclusions were drawn and a number of relevant recommendations were made to enhance the future research landscape.

Information about the research fronts for the sample papers in waste biomass-based bioethanol fuels is given in Table 60.3. As this table shows the most prolific research front for this field is the residual starch feedstocks such as corn stover and wheat straw, with 68% of the HCPs, respectively, followed by the lignocellulosic biomass at large, with 36% of these HCPs. The other prolific feedstocks are industrial wastes such as glycerol and residual sugar feedstocks such as sugarcane bagasse, with 12% and 8% of the HCPs, respectively. Further, the most influential research front is the residual starch feedstocks with a 31% surplus, while the lignocellulosic biomass is the least influential feedstock with a 19% deficit.

Information about the thematic research fronts for the sample papers in waste biomass-based bioethanol fuels is given in Table 60.4. As this table shows the most prolific research fronts for this field are the pretreatment and hydrolysis of waste biomass, with 92% and 84% of the HCPs, respectively. The other prolific research fronts are hydrolysate fermentation and bioethanol production, with 40% and 48% of the sample papers, respectively. The other research front is the evaluation of the bioethanol fuels, with 12% of the HCPs. Further, the first four research fronts are influential research fronts with a 16%–33% surplus each. The last research front has the least influence.

60.4.2 Waste Biomass Hydrolysis

The brief information about the 13 most-cited papers on the hydrolysis of waste biomass with at least 405 citations each is given below (Table 60.1). These papers also cover research on the pretreatment of waste biomass. It is notable that, as Table 60.4 shows, 92% and 84% of these HCPs are related to the pretreatments and hydrolysis of the waste biomass, respectively. These findings

TABLE 60.3
The Most Prolific Research Fronts for Waste Biomass-based Bioethanol Fuels

No.	Research Fronts	N Paper (%) Review	N Paper 2 (%) Sample	Surplus (%)	N Paper 3 Population (%)
1	Residual starch feedstocks	68.0	37.1	30.9	29.8
2	Lignocellulosic biomass	36.0	54.9	−18.9	26.5
3	Industrial wastes	12.0	17.5	−5.5	36.2
4	Residual sugar feedstocks	8.0	9.5	−1.5	14.9
5	Urban wastes	0.0	2.2	−2.2	11.9
6	Food wastes	0.0	2.2	−2.2	10.4
7	Forestry wastes	0.0	2.2	−2.2	3.9
	Sample size	25	275		13,728

N paper (%) review: The number of papers in the sample of 25 reviewed papers. N paper (%) sample: The number of papers in the population sample of 275 papers. N paper population: The number of papers in the population sample of 13,728 papers.

TABLE 60.4
The Most Prolific Thematic Research Fronts for Waste Biomass-based Bioethanol Fuels

No.	Research Fronts	N Paper (%) Review	N Paper 2 (%) Sample	Surplus (%)
1	Biomass pretreatments	92.0	75.6	16.4
2	Biomass hydrolysis	84.0	50.9	33.1
3	Hydrolysate fermentation	40.0	22.9	17.1
4	Bioethanol production	48.0	25.5	22.5
5	Bioethanol fuel evaluation	12.0	6.5	5.5

N paper (%) review: The number of papers in the sample of 25 reviewed papers. N paper (%) sample: The number of papers in the population sample of 275 papers.

show that both pretreatments and hydrolysis of the waste biomass are the fundamental processes for bioethanol production from the waste biomass.

These narrated studies highlight the importance of the pretreatment and hydrolysis processes for the production of bioethanol fuels from waste biomass with a high ethanol yield. These pretreatments, primarily enzymatic and chemical pretreatments, fractionate the lignocellulosic biomass and enhance the enzymatic digestibility of the biomass.

Eriksson et al. (2002) explored the mechanism of surfactant effect in enzymatic hydrolysis of lignocellulosic biomass and found that non-ionic surfactants were the most effective, and both anionic and non-ionic surfactants reduced enzyme adsorption to the lignocellulose substrate. Kumar et al. (2009) characterized corn stover and poplar solids resulting from leading pretreatment technologies and observed that lime pretreatment removed the most acetyl groups from both corn stover and poplar, while AFEX removed the least.

Lloyd and Wyman (2005) evaluated the combined sugar yields for dilute sulfuric acid pretreatment of corn stover followed by enzymatic hydrolysis of the remaining solids and observed that up to 15% of the total potential sugar in the substrate could be released as glucose during pretreatment and between 15% and 90+% of the xylose remaining in the solid residue could be recovered in subsequent enzymatic hydrolysis, depending on the enzyme loading. Further, Yang and Wyman (2004) evaluated xylan and lignin removal and enzymatic digestibility of cellulose for corn stover pretreated in batch and flowthrough reactors and observed that increasing flow with just water enhanced the xylan dissolution rate, more than doubled total lignin removal, and increased cellulose digestibility.

Kim et al. (2003) evaluated the ARP pretreatment and enzymatic hydrolysis of corn stover and found that this pretreatment was highly effective in delignifying the biomass. Further, Zhang et al. (2007) fractionated lignocellulosic biomass to cellulose, hemicellulose, lignin, and acetic acid at modest reaction conditions and attributed the highest sugar yields after enzymatic hydrolysis to no sugar degradation during the fractionation and the highest enzymatic cellulose digestibility.

Hsu et al. (2010) performed the dilute acid pretreatment of rice straw, explored the effect of the structural properties of the solid residues on the enzymatic hydrolysis, and obtained a maximal sugar yield of 83%. Further, Kristensen et al. (2009) explored the enzymatic hydrolysis of lignocellulosic biomass at high solid concentrations and found that the decreasing enzymatic conversion of lignocellulosic biomass at increasing solid concentrations was a generic or intrinsic effect.

Ohgren et al. (2007) evaluated the effect of hemicellulose and lignin removal on the enzymatic hydrolysis of pretreated corn stover and obtained a near-theoretical glucose yield from acid-catalyzed steam-pretreated corn stover when xylanases were used to supplement cellulases.

Wyman et al. (2005) evaluated ammonia explosion, aqueous ammonia recycle, controlled pH, dilute acid, flowthrough, and lime pretreatments using a single source of corn stover and the same cellulase enzyme and found that each pretreatment made it possible to subsequently achieve high yields of glucose from cellulose by cellulase enzymes. Further, Li et al. (2008) used acid in IL pretreatment for the hydrolysis of lignocellulosic biomass and showed that this pretreatment was an efficient system for hydrolysis of lignocellulosic materials with improved total reducing sugar yield under mild conditions.

Sun and Cheng (2005) evaluated the dilute sulfuric acid pretreatment of rye straw and bermudagrass before enzymatic hydrolysis of cellulose, and with the increasing acid concentration and residence time, they observed that the amount of arabinose and galactose increased. Chundawat et al. (2011) characterized corn stover cell walls to elucidate the mechanism of AFEX pretreatment and observed that AFEX first dissolved, then extracted and, as the ammonia evaporated, redeposited cell wall decomposition products on outer cell wall surfaces.

60.4.3 Waste Biomass-based Bioethanol Fuels

There are 12 HCPs for the production and evaluation of waste biomass-based bioethanol fuels with at least 401 citations each (Table 60.2). Further, there are nine and three HCPs for the production and evaluation of bioethanol fuels, respectively. As the pretreatment and hydrolysis of the waste biomass, as well as the fermentation of the resulting hydrolysates, are the fundamental parts of bioethanol production, these narrated papers often cover these processes too. It is notable that, as Table 60.4 shows, 48% and 12% of these HCPs are related to the production and evaluation of bioethanol fuels from waste biomass, respectively. However, there is only one HCP on the utilization of these biofuels in diesel or gasoline engines, partially displacing diesel or gasoline fuels.

60.4.3.1 Waste Biomass-based Bioethanol Production

There are nine HCPs for the production of waste biomass-based bioethanol fuels with at least 418 citations each (Table 60.2). These papers also cover the fermentation of the hydrolysates of the lignocellulosic biomass. As pretreatment and hydrolysis are the fundamental parts of bioethanol production, these narrated papers often cover these processes too. It is notable that, as Table 60.4 shows, 48% of these HCPs are related to the production of bioethanol fuels from waste biomass.

These narrated studies highlight the importance of the pretreatment (primarily chemical, enzymatic, or hydrothermal) and hydrolysis (primarily enzymatic or acid) processes, as well as of the fermentation processes (SSF or SHF), on the production of bioethanol fuels from waste biomass with a high ethanol yield. Further, some fermentation studies focus on the detoxification of the lignocellulosic hydrolysates to improve the ethanol yield.

Saha et al. (2005) produced ethanol from wheat straw with 49% and 28% cellulose and hemicellulose, respectively, and found that the yield of ethanol (per liter) was 19 g with a yield of 0.24 g/g DS. Further, Kaparaju et al. (2009) produced bioethanol, biohydrogen, and biogas production from wheat straw in a biorefinery concept and produced 0.41 g-ethanol/g-glucose by the enzymatic hydrolysis and subsequent fermentation of cellulose, while dark fermentation of hydrolysate produced 178.0 ml-H_2/g-sugars.

Laser et al. (2002) compared LHW and steam explosion pretreatments of sugarcane bagasse for ethanol production and found that the LHW pretreatment achieved ≥80% conversion by SSF, ≥80% xylan recovery, and no hydrolysate inhibition of glucose fermentation yield. Further, Mosier et al. (2005b) optimized the pH-controlled LHW pretreatment of corn stover and found that both the xylose and the glucose in this undiluted hydrolysate were fermented by recombinant yeast 424A(LNH-ST) to ethanol at 88% of theoretical yield.

Teymouri et al. (2005) optimized the AFEX pretreatment parameters to provide maximum sugar yields by enzymatic hydrolysis of corn stover and found that the ethanol yield of optimally AFEX-treated corn stover was increased up to 2.3 times over that of an untreated sample. Further, Larsson et al. (1999) compared the different methods for the detoxification of lignocellulose hydrolysates of spruce and found that an ion exchange at pH 5.5 or 10, treatment with laccase, treatment with lime, and treatment with *T. reesei* were the most efficient detoxification methods.

Jorgensen et al. (2007) produced bioethanol through the enzymatic hydrolysis and fermentation of the lignocellulosic biomass at high-solids concentrations and found a decrease in ethanol yield at increasing initial DM. Further, Ito et al. (2005) produced H_2 and ethanol production from glycerol-containing biodiesel wastes and found that the yields of H_2 and ethanol decreased with an increase in the concentrations of biodiesel wastes and pure glycerol. Finally, Delgenes et al. (1996) studied the effects of six fermentation inhibitors on ethanol fermentations of glucose and xylose by four yeast strains and found that vanillin was a strong inhibitor of both growth and ethanol production by xylose-fermenting yeasts and *S. cerevisiae* when it was added to the culture media.

60.4.3.2 Waste Biomass-based Bioethanol Evaluation

There are three HCPs for the evaluation of waste biomass-based bioethanol fuels with at least 401 citations each (Table 60.2). It is notable that, as Table 60.4 shows, 12% of these HCPs are related to the evaluation of the bioethanol fuels from the waste biomass.

These narrated studies often focus on the technoeconomics and environmental impact of bioethanol fuels from lignocellulosic biomass. These technoeconomic studies show that the lignocellulosic ethanol fuels are cost competitive in relation to the crude oil-based gasoline and petrodiesel fuels, in which ethanol fuels partially replace as the ethanol price reached $150 per barrel in 2022 following the invasion of Ukraine by Russia.

Hamelinck et al. (2005) evaluated ethanol production costs from lignocellulosic biomass and found that a combined effect of higher hydrolysis and fermentation efficiency, lower specific capital investments, an increase of scale, and cheaper biomass feedstock costs could bring the ethanol production costs from 22 €/GJHHV in the next 5 years to 13 €/GJ over the 10–15 year time scale, and down to 8.7 €/GJ in 20 or more years.

Sheehan et al. (2003) constructed a life-cycle model that described collecting corn stover in the state of Iowa for the production and use of an E85 blend in a flexible-fuel light-duty vehicle and found that the total fossil energy use and greenhouse gas emissions on a life-cycle basis were 102% and 113% lower, respectively.

Kazi et al. (2010) performed the technoeconomic comparison of process technologies for ethanol production from corn stover based on a 5- to 8-year time frame for implementation and found that the dilute-acid pretreatment process had the lowest PV among all process scenarios, which was estimated to be $1.36/L LGE ($5.13/gal GGE).

60.5 CONCLUSION AND FUTURE RESEARCH

The brief information about the key research fronts covered by the 25 most-cited papers with at least 401 citations each is given under two primary headings: The hydrolysis of the waste biomass and the production and evaluation of bioethanol fuels.

The usual characteristics of these HCPs are that the pretreatments and hydrolysis of the waste biomass and fermentation of the resulting hydrolysates are the primary processes for the bioethanol fuel production from waste biomass to improve the ethanol yield, as waste biomass is one of the most studied feedstocks at large for the bioethanol production, especially for the countries with large farmlands, forests, and crude oil deficiency.

The key findings on these research fronts should be read in the light of the increasing public concerns about climate change, GHG emissions, and global warming as these concerns have been certainly behind the boom in research on waste biomass-based bioethanol fuels as an alternative to crude oil-based gasoline and petrodiesel fuels in the last decades. It is also a sustainable alternative to first generation food crop-based bioethanol fuels, such as corn grain-based bioethanol fuels. The recent supply shocks caused by the COVID-19 pandemics and the Russian invasion of Ukraine also highlight the importance of the production and utilization of bioethanol fuels from lignocecellulosic waste biomass as an alternative to crude oil-based gasoline and petrodiesel fuels.

The most prolific research front for this field is the residual starch feedstocks such as corn stover and wheat straw, followed by the lignocellulosic biomass at large. The other prolific feedstocks are industrial wastes such as glycerol and residual sugar feedstocks such as sugarcane bagasse.

The most prolific thematic research fronts for this field are the pretreatment and hydrolysis of waste biomass. The other prolific research fronts are hydrolysate fermentation and bioethanol production. The other research front is the evaluation of bioethanol fuels.

These studies emphasize the importance of proper incentive structures for the efficient production of waste biomass-based bioethanol fuels in light of North's institutional framework (North, 1991). In this context, the major producers and users of bioethanol fuels such as the USA and Brazil, with vast forests and farmlands, have developed strong incentive structures for the efficient use of waste biomass-based bioethanol fuels. In the light of the recent supply shocks caused primarily by the COVID-19 pandemics and the Russian invasion of Ukraine, it is expected that incentive structures such as public funding would be enhanced to increase the share of bioethanol fuels from lignocellulosic waste biomass in the global fuel portfolio as a strong alternative to crude oil-based gasoline and petrofuels.

In this context, it is expected that the most prolific researchers, institutions, countries, funding bodies, and journals in this field would have a first-mover advantage to benefit from such potential incentives. This is especially true for US and European stakeholders, as the USA and Europe have become global leaders in both the production and utilization of second generation bioethanol fuels from waste biomass.

It is recommended that such review studies are performed for the primary research fronts of waste biomass-based bioethanol fuels.

ACKNOWLEDGMENTS

The contribution of the highly cited researchers in the field of waste biomass-based bioethanol fuels has been gratefully acknowledged.

REFERENCES

Alvira, P., E. Tomas-Pejo, M. Ballesteros and M. J. Negro. 2010. Pretreatment technologies for an efficient bioethanol production process based on enzymatic hydrolysis: A review. *Bioresource Technology* 101:4851–4861.

Angelici, C., B. M. Weckhuysen and P. C. A. Bruijnincx. 2013. Chemocatalytic conversion of ethanol into butadiene and other bulk chemicals. *ChemSusChem* 6:1595–1614.

Antolini, E. 2007. Catalysts for direct ethanol fuel cells. *Journal of Power Sources* 170:1–12.

Antolini, E. 2009. Palladium in fuel cell catalysis. *Energy and Environmental Science* 2:915–931.

Brandt, A., J. P. Hallett, D. J. Leak, R. J. Murphy and T. Welton. 2010. The effect of the ionic liquid anion in the pretreatment of pine wood chips. *Green Chemistry* 12:672–679.

Burnham, J. F. 2006. Scopus database: A review. *Biomedical Digital Libraries* 3:1–8.

Chundawat, S. P. S., B. S. Donohoe and L. da Costa Sousa, et al. 2011. Multi-scale visualization and characterization of lignocellulosic plant cell wall deconstruction during thermochemical pretreatment. *Energy and Environmental Science* 4:973–984.

Delgenes, J. P., R. Moletta and J. M. Navarro. 1996. Effects of lignocellulose degradation products on ethanol fermentations of glucose and xylose by *Saccharomyces cerevisiae*, *Zymomonas mobilis*, *Pichia stipitis*, and *Candida shehatae*. *Enzyme and Microbial Technology* 19:220–225.

Duff, S. J. B. and W. D. Murray. 1996. Bioconversion of forest products industry waste cellulosics to fuel ethanol: A review. *Bioresource Technology* 55:1–33.

Eriksson, T., J. Borjesson and F. Tjerneld. 2002. Mechanism of surfactant effect in enzymatic hydrolysis of lignocellulose. *Enzyme and Microbial Technology* 31:353–364.

Fernando, S., S. Adhikari, C. Chandrapal and M. Murali. 2006. Biorefineries: Current status, challenges, and future direction. *Energy & Fuels* 20:1727–1737.

Guimaraes, P. M. R., J. A. Teixeira and L. Domingues. 2010. Fermentation of lactose to bio-ethanol by yeasts as part of integrated solutions for the valorisation of cheese whey. *Biotechnology Advances* 28:375–384.

Hahn-Hagerdal, B., M. Galbe, M. F. Gorwa-Grauslund, G. Liden and G. Zacchi. 2006. Bio-ethanol: The fuel of tomorrow from the residues of today. *Trends in Biotechnology* 24:549–556.

Hamelinck, C. N., G. van Hooijdonk and A. P. C. Faaij. 2005. Ethanol from lignocellulosic biomass: Techno-economic performance in short-, middle- and long-term. *Biomass and Bioenergy* 28:384–410.

Hendriks, A. T. W. M. and G. Zeeman. 2009. Pretreatments to enhance the digestibility of lignocellulosic biomass. *Bioresource Technology* 100:10–18.

Hill, J., E. Nelson, D. Tilman, S. Polasky and D. Tiffany. 2006. Environmental, economic, and energetic costs and benefits of biodiesel and ethanol biofuels. *Proceedings of the National Academy of Sciences of the United States of America* 103:11206–11210.

Hill, J., S. Polasky and E. Nelson, et al. 2009. Climate change and health costs of air emissions from biofuels and gasoline. *Proceedings of the National Academy of Sciences of the United States of America* 106:2077–2082.

Hsieh, W. D., R. H. Chen, T. L. Wu and T. H. Lin. 2002. Engine performance and pollutant emission of an SI engine using ethanol-gasoline blended fuels. *Atmospheric Environment* 36:403–410.

Hsu, T. C., G. L. Guo, W. H. Chen and W. S. Hwang. 2010. Effect of dilute acid pretreatment of rice straw on structural properties and enzymatic hydrolysis. *Bioresource Technology* 101:4907–4913.

Huang, H. J., S. Ramaswamy, U. W. Tschirner and B. V. Ramarao. 2008. A review of separation technologies in current and future biorefineries. *Separation and Purification Technology* 62:1–21.

Ito, T., Y. Nakashimada, K. Senba, T. Matsui and N. Nishio. 2005. Hydrogen and ethanol production from glycerol-containing wastes discharged after biodiesel manufacturing process. *Journal of Bioscience and Bioengineering* 100:260–265.

Jorgensen, H., J. Vibe-Pedersen, J. Larsen and C. Felby. 2007. Liquefaction of lignocellulose at high-solids concentrations. *Biotechnology and Bioengineering* 96:862–870.

Kaparaju, P., M. Serrano, A. B. Thomsen, P. Kongjan and I. Angelidaki. 2009. Bioethanol, biohydrogen and biogas production from wheat straw in a biorefinery concept. *Bioresource Technology* 100:2562–2568.

Kazi, F. K., J. A. Fortman and R. P. Anex, et al. 2010. Techno-economic comparison of process technologies for biochemical ethanol production from corn stover. *Fuel* 89:S20–S28.

Kim, T. H., J. S. Kim, C. Sunwoo and Y. Y. Lee. 2003. Pretreatment of corn stover by aqueous ammonia. *Bioresource Technology* 90:39–47.

Konur, O. 2000. Creating enforceable civil rights for disabled students in higher education: An institutional theory perspective. *Disability & Society* 15:1041–1063.

Konur, O. 2002a. Access to nursing education by disabled students: Rights and duties of nursing programs. *Nurse Education Today* 22:364–374.

Konur, O. 2002b. Assessment of disabled students in higher education: Current public policy issues. *Assessment and Evaluation in Higher Education* 27:131–152.

Konur, O. 2002c. Access to employment by disabled people in the UK: Is the disability discrimination act working? *International Journal of Discrimination and the Law* 5:247–279.

Konur, O. 2006a. Participation of children with dyslexia in compulsory education: Current public policy issues. *Dyslexia* 12:51–67.

Konur, O. 2006b. Teaching disabled students in higher education. *Teaching in Higher Education* 11:351–363.

Konur, O. 2007a. A judicial outcome analysis of the disability discrimination act: A windfall for the employers? *Disability & Society* 22:187–204.

Konur, O. 2007b. Computer-assisted teaching and assessment of disabled students in higher education: The interface between academic standards and disability rights. *Journal of Computer Assisted Learning* 23:207–219.

Konur, O. 2012. The evaluation of the research on the bioethanol: A scientometric approach. *Energy Education Science and Technology Part A: Energy Science and Research* 28:1051–1064.

Konur, O. 2015. Current state of research on algal bioethanol. In *Marine Bioenergy: Trends and Developments*, Eds. S. K. Kim and C. G. Lee, pp. 217–244. Boca Raton, FL: CRC Press.

Konur, O. 2020. The scientometric analysis of the research on the bioethanol production from green macroalgae. In *Handbook of Algal Science, Technology and Medicine*, Ed. O. Konur, pp. 385–401. London: Academic Press.

Konur, O. 2023a. Second generation bioethanol fuels from residual sugar feedstocks: Scientometric study. In *Feedstock-based Bioethanol Fuels. II. Waste Feedstocks: Agricultural, Food, Industrial, Urban, Forestry, and Lignocellulosic Waste-based Bioethanol Fuels. Handbook of Bioethanol Fuels Volume 4*, Ed. O. Konur, pp. 53–78. Boca Raton, FL: CRC Press.

Konur, O. 2023b. Second generation bioethanol fuels from residual starch feedstocks: *Scientometric study. In Feedstock-based Bioethanol Fuels. II. Waste Feedstocks: Agricultural, Food, Industrial, Urban, Forestry, and Lignocellulosic Waste-based Bioethanol Fuels. Handbook of Bioethanol Fuels Volume 4*, Ed. O. Konur, pp. 79–98. Boca Raton, FL: CRC Press.

Konur, O. 2023c. Second generation food waste-based bioethanol fuels: Scientometric study. In *Feedstock-based Bioethanol Fuels. II. Waste Feedstocks: Agricultural, Food, Industrial, Urban, Forestry, and Lignocellulosic Waste-based Bioethanol Fuels. Handbook of Bioethanol Fuels Volume 4*, Ed. O. Konur, pp. 147–172. Boca Raton, FL: CRC Press.

Konur, O. 2023d. Second generation Industrial waste-based bioethanol fuels: Scientometric study. In *Feedstock-based Bioethanol Fuels. II. Waste Feedstocks: Agricultural, Food, Industrial, Urban, Forestry, and Lignocellulosic Waste-based Bioethanol Fuels. Handbook of Bioethanol Fuels Volume 4*, Ed. O. Konur, pp. 191–216. Boca Raton, FL: CRC Press.

Konur, O. 2023e. Second generation urban waste-based bioethanol fuels: Scientometric study. In *Feedstock-based Bioethanol Fuels. II. Waste Feedstocks: Agricultural, Food, Industrial, Urban, Forestry, and Lignocellulosic Waste-based Bioethanol Fuels. Handbook of Bioethanol Fuels Volume 4*, Ed. O. Konur, pp. 237–261. Boca Raton, FL: CRC Press.

Konur, O. 2023f. Second generation forestry waste-based bioethanol fuels: Scientometric study. In *Feedstock-based Bioethanol Fuels. II. Waste Feedstocks: Agricultural, Food, Industrial, Urban, Forestry, and Lignocellulosic Waste-based Bioethanol Fuels. Handbook of Bioethanol Fuels Volume 4*, Ed. O. Konur, pp. 281–304. Boca Raton, FL: CRC Press.

Konur, O. 2023g. Lignocellulosic biomass-based bioethanol fuels: Scientometric study. In *Feedstock-based Bioethanol Fuels. II. Waste Feedstocks: Agricultural, Food, Industrial, Urban, Forestry, and Lignocellulosic Waste-based Bioethanol Fuels. Handbook of Bioethanol Fuels Volume 4*, Ed. O. Konur, pp. 325–349. Boca Raton, FL: CRC Press.

Kristensen, J. B., C. Felby and H. Jorgensen. 2009. Yield-determining factors in high-solids enzymatic hydrolysis of lignocellulose. *Biotechnology for Biofuels* 2:11.

Kumar, R., G. Mago, V. Balan and C. E. Wyman. 2009. Physical and chemical characterizations of corn stover and poplar solids resulting from leading pretreatment technologies. *Bioresource Technology* 100:3948–3962.

Larsson, S., A. Reimann, N. O. Nilvebrant and L. J. Jonsson. 1999. Comparison of different methods for the detoxification of lignocellulose hydrolyzates of spruce. *Applied Biochemistry and Biotechnology: Part A Enzyme Engineering and Biotechnology* 77–79:91–103.

Laser, M., D. Schulman and S. G. Allen, et al. 2002. A comparison of liquid hot water and steam pretreatments of sugar cane bagasse for bioconversion to ethanol. *Bioresource Technology* 81:33–44.

Lenihan, P., A. Orozco and E. O'Neill, et al. 2010. Dilute acid hydrolysis of lignocellulosic biomass. *Chemical Engineering Journal* 156:395–403.

Li, C., Q. Wang and Z. K. Zhao. 2008. Acid in ionic liquid: An efficient system for hydrolysis of lignocellulose. *Green Chemistry* 10:177–182.

Lin, Y. and S. Tanaka. 2006. Ethanol fermentation from biomass resources: Current state and prospects. *Applied Microbiology and Biotechnology* 69:627–642.

Lloyd, T. A. and C. E. Wyman. 2005. Combined sugar yields for dilute sulfuric acid pretreatment of corn stover followed by enzymatic hydrolysis of the remaining solids. *Bioresource Technology* 96:1967–1977.

Ma, X., L. Sun and C. Song. 2002. A new approach to deep desulfurization of gasoline, diesel fuel and jet fuel by selective adsorption for ultra-clean fuels and for fuel cell applications. *Catalysis Today* 77:107–116.

Morschbacker, A. 2009. Bio-ethanol based ethylene. *Polymer Reviews* 49:79–84.

Mosier, N., C. Wyman and B. Dale, et al. 2005a. Features of promising technologies for pretreatment of lignocellulosic biomass. *Bioresource Technology*, 96:673–686.

Mosier, N., R. Hendrickson, N. Ho, M. Sedlak and M. R. Ladisch. 2005b. Optimization of pH controlled liquid hot water pretreatment of corn stover. *Bioresource Technology* 96:1986–1993.

Najafi, G., B. Ghobadian and T. Tavakoli, et al. 2009. Performance and exhaust emissions of a gasoline engine with ethanol blended gasoline fuels using artificial neural network. *Applied Energy* 86:630–639.

Newman, P. W. G. and J. R. Kenworthy. 1989. Gasoline consumption and cities: A comparison of U.S. cities with a global survey. *Journal of the American Planning Association* 55:24–37.

North, D. C. 1991. Institutions. *Journal of Economic Perspectives* 5:97–112.

Ohgren, K., R. Bura, J. Saddler and G. Zacchi. 2007. Effect of hemicellulose and lignin removal on enzymatic hydrolysis of steam pretreated corn stover. *Bioresource Technology* 98:2503–2510.

Olsson, L. and B. Hahn-Hagerdal. 1996. Fermentation of lignocellulosic hydrolysates for ethanol production. *Enzyme and Microbial Technology* 18:312–331.

Ravindran, R. and A. K. Jaiswal. 2016. A comprehensive review on pre-treatment strategy for lignocellulosic food industry waste: Challenges and opportunities. *Bioresource Technology* 199:92–102.

Saha, B. C., L. B. Iten, M. A. Cotta and Y. V. Wu. 2005. Dilute acid pretreatment, enzymatic saccharification and fermentation of wheat straw to ethanol. *Process Biochemistry* 40:3693–3700.

Sanchez, O. J. and C. A. Cardona. 2008. Trends in biotechnological production of fuel ethanol from different feedstocks. *Bioresource Technology* 99:5270–5295.

Sheehan, J., A. Aden and K. Paustian, et al. 2003. Energy and environmental aspects of using corn stover for fuel ethanol. *Journal of Industrial Ecology* 7:117–146.

Sun, Y. and J. Cheng. 2002. Hydrolysis of lignocellulosic materials for ethanol production: A review. *Bioresource Technology* 83:1–11.

Sun, Y. and J. J. Cheng. 2005. Dilute acid pretreatment of rye straw and bermudagrass for ethanol production. *Bioresource Technology* 96:1599–1606.

Taherzadeh, M. J. and K. Karimi. 2007. Enzyme-based hydrolysis processes for ethanol from lignocellulosic materials: A review. *Bioresources* 2:707–738.

Taherzadeh, M. J. and K. Karimi. 2008. Pretreatment of lignocellulosic wastes to improve ethanol and biogas production: A review. *International Journal of Molecular Sciences* 9:1621–1651.

Teymouri, F., L. Laureano-Perez, H. Alizadeh and B. E. Dale. 2005. Optimization of the ammonia fiber explosion (AFEX) treatment parameters for enzymatic hydrolysis of corn stover. *Bioresource Technology* 96:2014–2018.

Wyman, C. E., B. E. Dale and R. T. Elander, et al. 2005. Comparative sugar recovery data from laboratory scale application of leading pretreatment technologies to corn stover. *Bioresource Technology* 96:2026–2032.

Yang, B. and C. E. Wyman. 2004. Effect of xylan and lignin removal by batch and flowthrough pretreatment on the enzymatic digestibility of corn stover cellulose. *Biotechnology and Bioengineering* 86:88–98.

Zhang, Y. H. P., S. Y. Ding and J. R. Mielenz, et al. 2007. Fractionating recalcitrant lignocellulose at modest reaction conditions. *Biotechnology and Bioengineering* 97:214–223.

Part 17

Second Generation Bioethanol Fuels from Residual Sugar Feedstocks

61 Second Generation Bioethanol Fuels from Residual Sugar Feedstocks
Scientometric Study

Ozcan Konur
(Formerly) Ankara Yildirim Beyazit University

61.1 INTRODUCTION

Crude oil-based gasoline fuels (Ma et al., 2002; Newman and Kenworthy, 1989) have been widely used in transportation sector since the 1920s. However, there have been great public concerns over the adverse environmental and human impact of these fuels (Hill et al., 2006, 2009). Hence, biomass-based bioethanol fuels (Hill et al., 2006; Konur, 2012e, 2015, 2019, 2020a) have increasingly been used in blending gasoline fuels (Hsieh et al., 2002; Najafi et al., 2009), in fuel cells (Antolini, 2007, 2009), and in biochemical production (Angelici et al., 2013; Morschbacker, 2009) in a biorefinery context (Fernando et al., 2006; Huang et al., 2008).

Bioethanol fuels also play a critical role in maintaining energy security (Kruyt et al., 2009; Winzer, 2012) in supply shocks (Kilian, 2008, 2009) related to oil price shocks (Hamilton, 2003, 2009), COVID-19 pandemic (Fauci et al., 2020; Li et al., 2020), or wars (Hamilton, 1983; Jones, 2012) in the aftermath of the Russian invasion of Ukraine (Reeves, 2014).

However, it is necessary to pretreat biomass (Taherzadeh and Karimi, 2008; Yang and Wyman, 2008) to enhance the yield of bioethanol (Hahn-Hagerdal et al., 2006; Sanchez and Cardona, 2008) prior to bioethanol production through hydrolysis (Sun and Cheng, 2002; Taherzadeh and Karimi, 2007) and fermentation (Lin and Tanaka, 2006; Olsson and Hahn-Hagerdal, 1996) of biomass.

The most studied feedstocks for bioethanol fuels have been the residual sugar feedstocks such as sugarcane bagasse and straw, sweet sorghum bagasse, and sugar beet pulp. Research in the field of second generation bioethanol fuels from residual sugar feedstocks has intensified in this context in the key research fronts of the pretreatment of residual sugar feedstocks (Aguilar et al., 2002; Laser et al., 2002; Lavarack et al., 2002), hydrolysis of residual sugar feedstocks (Aguilar et al., 2002; Lavarack et al., 2002; Rezende et al., 2011), fermentation of residual sugar feedstock-based hydrolysates (Chandel et al., 2007; Laser et al., 2002; Martinez et al., 2000), and production (Chandel et al., 2007; da Silva et al., 2010, Rabelo et al., 2011) and evaluation (Dias et al., 2012; MacRelli et al., 2012) of second generation bioethanol fuels from residual sugar feedstocks.

Further, sugarcane bagasse (Aguilar et al., 2002; Laser et al., 2002; Lavarack et al., 2002); sugarcane straw (da Silva et al., 2010; Dias et al., 2009, 2012); sweet sorghum bagasse (Cao et al., 2012; Li et al., 2010); and to a lesser extent sweet sorghum straw (McIntosh and Vancov, 2010), sugar beet pulp (Micard et al., 1996), sugarcane molasses (Nguyen and Gheewela, 2008), and energy cane bagasse (Qiu et al., 2012) have been studied intensively in this context.

However, it is essential to develop efficient incentive structures (North, 1991) for primary stakeholders to enhance research in this field (Konur, 2000, 2002a–c, 2006a,b, 2007a,b).

DOI: 10.1201/9781003226550-82

Scientometric analysis has been used in this context to inform primary stakeholders about the current state of research in a selected research field (Garfield, 1955; Konur, 2011, 2012a–i, 2015, 2018b, 2019, 2020a).

As there have been no published scientometric studies in this field, this book chapter presents a scientometric study of the research in second generation bioethanol fuels from residual sugar feedstocks. It examines scientometric characteristics of both sample and population data presenting scientometric characteristics of these both datasets in order of documents, authors, publication years, institutions, funding bodies, source titles, countries, Scopus subject categories, Scopus keywords, and research fronts.

61.2 MATERIALS AND METHODS

Search for this study was carried out using Scopus database (Burnham, 2006) in July 2022.

As the first step for search of relevant literature, keywords were selected using the first 200 most-cited population papers. The selected keyword list was then optimized to obtain a representative sample of papers for the searched research field. This keyword list was provided in the appendix for future replicative studies.

As the second step, two sets of data were used for this study. First, a population sample of 2,009 papers was used to examine scientometric characteristics of the population data. Secondly, a sample of 201 most-cited papers, corresponding to 10% of the population papers, was used to examine scientometric characteristics of these citation classics.

The scientometric characteristics of both these sample and population datasets were presented in the order of documents, authors, publication years, institutions, funding bodies, source titles, countries, Scopus subject categories, Scopus keywords, and research fronts.

Lastly, the key scientometric findings for both datasets were discussed to highlight research landscape for second generation bioethanol fuels from residual sugar feedstocks. Additionally, a number of brief conclusions were drawn and a number of relevant recommendations were made to enhance the future research landscape.

61.3 RESULTS

61.3.1 The Most Prolific Documents in Second Generation Bioethanol Fuels from Residual Sugar Feedstocks

Information on the types of documents for both datasets is given in Table 61.1. Articles and conference papers, published in journals, dominate both sample (95%) and population (95%) papers as they are equally represented in sample papers in both samples. Further, review papers and short surveys have a 3% surplus as they are over-represented in the sample papers as they constitute 5% and 2% of sample and population papers, respectively.

It is further notable that 97% of population papers were published in journals, while 2% and 1% of them were published in books and book series, respectively. Similarly, 99% and 1% of the sample papers were published in the journals and books, respectively.

61.3.2 The Most Prolific Authors in Second Generation Bioethanol Fuels from Residual Sugar Feedstocks

Information about the most prolific 15 authors with at least 2.5% of sample papers each is given in Table 61.2. The most prolific authors are Carlos Martin and Rubens M. Filho with 4.5% and 4% of sample papers, respectively. Other prolific authors are Zhenhong Yuan, Carlos E. V. Rossell, Marina O. S. Dias, Parameswaran Binod, and Raveendran Sindhu with 3% of sample papers each.

TABLE 61.1
Documents in Second Generation Bioethanol Fuels from Residual Sugar Feedstocks

Documents	Sample Dataset (%)	Population Dataset (%)	Surplus (%)
Article	93.0	93.1	–0.1
Review	4.0	2.1	1.9
Conference paper	2.0	1.9	0.1
Book chapter	0.5	2.3	–1.8
Short Survey	0.5	0.1	0.4
Letter	0.0	0.2	–0.2
Note	0.0	0.1	–0.1
Book	0.0	0.1	–0.1
Editorial	0.0	0.0	0.0
Sample size	201	2,009	

Sample dataset: Number of papers (%) in the set of 201 highly cited papers. Population dataset: Number of papers (%) in the set of 2,009 population papers.

On the other hand, the most influential authors are Carlos Martin and Rubens M. Filho with 3.7% and 3.2% surplus, respectively. Other influential authors are Parameswaran Binod, Raveendran Sindhu, Marina O. S. Dias, and Carlos E. V. Rossell with 2.4%–2.6% surplus each.

The most prolific institution for the sample dataset is the Council for Scientific and Industrial Research (CSIR) of India with four authors, followed by the Chinese Academy of Sciences, State University of Campinas, and University of Sao Paulo with two authors each. In total, eight institutions house these top authors. On the other hand, the most prolific country for the sample dataset is Brazil with seven authors, followed by India and China with four and three authors, respectively. In total, only four countries house these top authors.

There are two primary research fronts for these top authors: Bioethanol fuels based on sugarcane bagasse (SCB) and sugarcane straw (SCS) with 15 and 9 authors, respectively. There are also two highly cited papers (HCPs) for sweet sorghum bagasse (SSB)-based bioethanol fuels.

On the other hand, there is significant gender deficit (Beaudry and Lariviere, 2016) for sample dataset as surprisingly only three of these top researchers are female with a representation rate of 20%.

Additionally, there are other authors with relatively low citation impact and with 0.5%–1.3% of the population papers each: Anuj K. Chandel, Sarita C. Rabelo, George J. M. Rocha, Silvio S. da Silva, Zhanying Zhang, Antonio J. G. da Cruz, Viviane M. Nascimento, Qiong Wang, Felipe A. F. Antunes, Michel Brienzo, Cristiane S. Farinas, Rosana Goldbeck, Leandro V. A. Gurgel, Yunyun Liu, Xuesong Tan, Wenjuan Xiao, Ana C. Costa, Yingxue Gong, Lonnie O. Ingram, Nei Pereira, Walter Carvalho, Verawat Champreda, Adriano V. Ensinas, Erenio Gonzalez, Tassia L. Junqueria, Leyanis Mesa, Carlos A. Rosa, and Kenji Takahashi.

61.3.3 The Most Prolific Research Output by Years in Second Generation Bioethanol Fuels from Residual Sugar Feedstocks

Information about papers published between 1970 and 2022 is given in Figure 61.1. This figure clearly shows that the bulk of research papers in population dataset was published primarily in the 2010s and the early 2020s with 59% and 22% of population dataset, respectively. Similarly, publication rates for the 2000s, 1990s, 1980s, and 1970s were 10%, 4%, 4%, and 0% respectively. Additionally, 0.7% of population papers were published in the pre-1970s.

TABLE 61.2
Most Prolific Authors in Second Generation Bioethanol Fuels from Residual Sugar Feedstocks

No.	Author Name	Author Code	Sample Papers (%)	Population Papers (%)	Surplus	Institution	Country	HI	N	Res. Front
1	Martin, Carlos	56484787200	4.5	0.8	3.7	Inland Norway Univ. Agr. Sci.	Norway	26	66	SCB, SCS
2	Filho, Rubens M.	7003732915	4.0	0.8	3.2	State Univ. Campinas	Brazil	71	699	SCB, SCS
3	Yuan, Zhenhong	15924240400	3.0	1.6	1.4	Chinese Acad. Sci.	China	65	422	SCB, SSB
4	Rossell, Carlos E. V.	12780694800	3.0	0.6	2.4	State Univ. Campinas	Brazil	19	48	SCB
5	Dias, Marina O. S.*	26654773000	3.0	0.5	2.5	Fed. Univ. Sao Paulo	Brazil	20	37	SCB, SCS
6	Binod, Parameswaran	8901704900	3.0	0.4	2.6	CSIR	India	42	245	SCB, SCS
7	Sindhu, Raveendran*	35760609600	3.0	0.4	2.6	CSIR	India	35	208	SCB, SCS
8	Milagres, Adriane M. F.*	6701345269	2.5	1.3	1.2	Univ. Sao Paulo	Brazil	32	110	SCB
9	Yu, Qiang	56924172500	2.5	1.0	1.5	Chinese Acad. Sci.	China	24	107	SCB, SSB
10	Polikarpov, Igor	7006220351	2.5	0.9	1.6	Univ. Sao Paulo	Brazil	48	311	SCB
11	Bonomi, Antonio	7004767629	2.5	0.9	1.6	CTBE	Brazil	33	104	SCB, SCS
12	Sun, Run-Cang	55661525600	2.5	0.5	2.0	Dalian Polytech. Univ.	China	112	1060	SCB
13	Pandey, Ashok	7201771319	2.5	0.4	2.1	CSIR	India	90	857	SCB, SCS
14	Sukumaran, Rajeev K.	9248920200	2.5	0.4	2.1	CSIR	India	43	132	SCB, SCS
15	Jesus, Charles D. F.	23103105400	2.5	0.3	2.2	Natl. Biorenew. Lab.	Brazil	18	31	SCB, SCS

*, Female researchers; Author code: the unique code given by Scopus to authors. Sample papers: number of papers authored in the sample dataset. Population papers: number of papers authored in the population dataset.

SCB, Sugarcane bagasse; SCS, Sugarcane straw; SSB, Sweet sorghum bagasse.

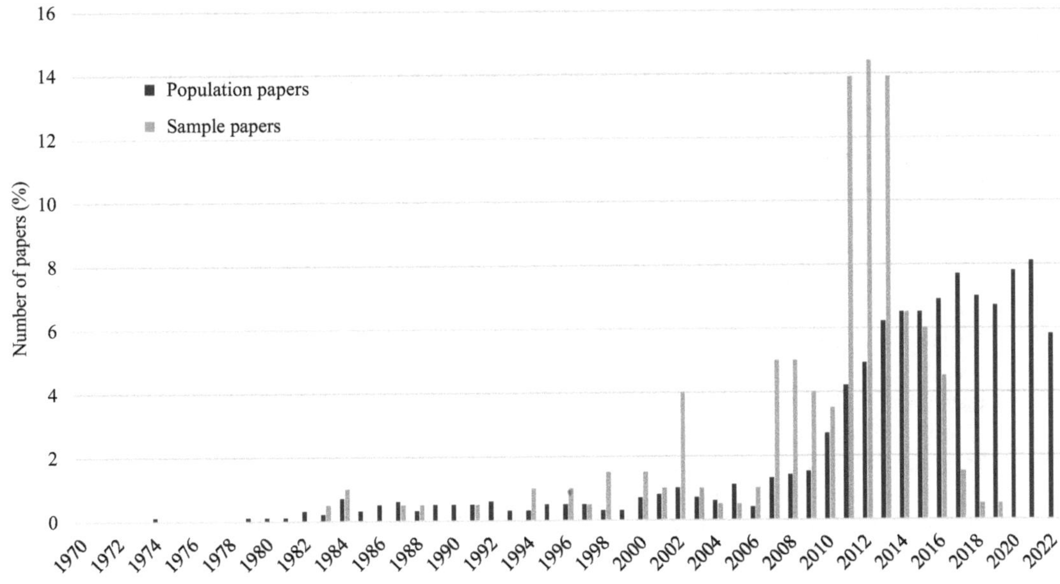

FIGURE 61.1 Research output by years regarding second generation bioethanol fuels from residual sugar feedstocks.

Similarly, the bulk of research papers in sample dataset was published in the 2000s and 2010s with 24% and 65% of the sample dataset, respectively. Similarly, publication rates for the 1990s, 1980s, and 1970s were 5%, 3%, and 0% of the sample papers, respectively.

The most prolific publication years for the population dataset were 2021, 2020, and 2017 with 8.1%, 7.8%, and 7.7% of the dataset, respectively. Further, 78% of population papers were published between 2011 and 2022. Similarly, 77% of sample papers were published between 2007 and 2016, while the most prolific publication years were 2012, 2011, and 2013 with 14.4%, 13.9%, and 13.9% of sample papers each. There was a rising trend for the population papers between 2010 and 2014, and it lost its momentum after 2014. However, there was a sharp rise in 2020 and 2021.

61.3.4 The Most Prolific Institutions in Second Generation Bioethanol Fuels from Residual Sugar Feedstocks

Information about the most prolific 16 institutions publishing papers on second generation bioethanol fuels from residual sugar feedstocks with at least 2.0% of the sample papers each is given in Table 61.3.

The most prolific institution is the University of Sao Paulo with 13.4% of the sample papers, followed by the State University of Campinas and National Biorenewables Laboratory with 8% of the sample papers each. Other prolific institutions are Federal University of Rio de Janeiro, University of Matanzas, South China University of Technology, and Federal University of Sao Carlos with 4%–4.5% of the sample papers each.

Similarly, the top countries for these most prolific institutions are Brazil and China with six and four institutions, respectively. In total, only eight countries house these top institutions.

On the other hand, the institution with the most impact is the National Biorenewables Laboratory and University of Matanzas with 3.9% and 3.8% surplus, respectively. Other influential institutions are University of Sao Paulo and Lund University with 3.1% and 2.8% surplus, respectively.

Additionally, there are other institutions with relatively low citation impact and with 0.6%–3.5% of the population papers each: Sao Paulo State University, National Center for Research in Energy and Materials, Federal University of Vicosa, Guangxi University, Queensland University of Technology,

TABLE 61.3
The Most Prolific Institutions in Second Generation Bioethanol Fuels from Residual Sugar Feedstocks

No.	Institutions	Country	Sample Papers (%)	Population Papers (%)	Surplus (%)
1	Univ. Sao Paulo	Brazil	13.4	10.3	3.1
2	State Univ. Campinas	Brazil	8.0	6.6	1.4
3	Natl. Biorenewables Lab.	Brazil	8.0	4.1	3.9
4	Fed. Univ. Rio de Janeiro	Brazil	4.5	2.4	2.1
5	Univ. Matanzas	Cuba	4.5	0.7	3.8
6	S. China Univ. Technol.	China	4.0	2.6	1.4
7	Fed. Univ. Sao Carlos	Brazil	4.0	1.9	2.1
8	Lund Univ.	Sweden	3.5	0.7	2.8
9	Chinese Acad. Sci.	China	3.0	2.9	0.1
10	Tsinghua Univ.	China	3.0	1.3	1.7
11	Univ. Florida	USA	3.0	0.9	2.1
12	Beijing Forestry Univ.	China	2.5	1.0	1.5
13	AIST	Japan	2.5	0.4	2.1
14	CSIR	India	2.5	0.4	2.1
15	Fed. Univ. Pernambuco	Brazil	2.0	0.7	1.3
16	Auton. Univ. Tamaulipas	Mexico	2.0	0.4	1.6

Federal University of Parana, King Mongkut's University of Technology, Brazilian Agricultural Research Corporation, Stellenbosch University, Federal University of Minas Gerais, South China Agricultural University, Federal University of Santa Catarina, Audubon Sugar Institute, Central University Marta Abreu of Las Villas, Federal University of ABC, Thailand National Center for Genetic Engineering and Biotechnology, Jinan University, Nanjing Forestry University, National Chemical Laboratory India, Indian Institute of Petroleum, CSIC, Mahidol University, Federal University of Rio Grande do Norte, Guangxi University for Nationalities, National Institute of Advanced Industrial Science and Technology, Chulalongkorn University, and National Institute for Interdisciplinary Science and Technology.

61.3.5 The Most Prolific Funding Bodies in Second Generation Bioethanol Fuels from Residual Sugar Feedstocks

Information about the most prolific 13 funding bodies funding at least 2% of the sample papers each is given in Table 61.4. Further, only 54% and 55% of the sample and population papers were funded, respectively.

The most prolific funding bodies are National Council for Scientific and Technological Development and Sao Paulo Research Foundation with 15.4% and 13.4% of the sample papers, respectively, followed by the National Natural Science Foundation of China, CAPES Foundation, Ministry of Science and Technology of China, and Ministry of Science and Technology of India with 5%–7% of the sample papers each.

On the other hand, the most prolific countries for these top funding bodies are Brazil and China with four funding bodies each, followed by India and the EU with three and two funding bodies, respectively. In total, only three countries and the EU house these top funding bodies.

Funding bodies with the most citation impact are Ministry of Science and Technology India and Ministry of Science and Technology of China with 3.2% and 2.9% surplus, respectively, followed by the Technology Information, Forecasting and Assessment Council, National Key Research and Development Program of China, and Sao Paulo Research Foundation with 2.4%–2.7% surplus

TABLE 61.4
The Most Prolific Funding Bodies in Second Generation Bioethanol Fuels from Residual Sugar Feedstocks

No.	Funding Bodies	Country	Sample Paper No. (%)	Population Paper No. (%)	Surplus (%)
1	National Council for Scientific and Technological Development	Brazil	15.4	14.2	1.2
2	Sao Paulo Research Foundation	Brazil	13.4	11.0	2.4
3	National Natural Science Foundation of China	China	7.0	8.4	−1.4
4	CAPES Foundation	Brazil	6.5	10.1	−3.6
5	Ministry of Science and Technology of China	China	5.5	2.6	2.9
6	Ministry of Science and Technology, India	India	5.0	1.8	3.2
7	National Key Research and Development Program of China	China	4.5	1.9	2.6
8	Ministry of Science, Technology and Innovation	Brazil	3.5	4.8	−1.3
9	European Commission	EU	3.0	0.7	2.3
10	Technology Information, Forecasting and Assessment Council	India	3.0	0.3	2.7
11	Seventh Framework Program	EU	2.5	0.8	1.7
12	Council of Scientific and Industrial Research, India	India	2.0	0.9	1.1
13	National High-tech Research and Development Program	China	2.0	0.8	1.2

each. Further, the least influential funding bodies are CAPES Foundation, National Natural Science Foundation of China, and Ministry of Science, Technology and Innovation with 1.3%–3.6% deficit each.

Other funding bodies with relatively low citation impact and with 0.6%–1.6% of the population papers each are Minas Gerais State Agency for Research and Development, Fundamental Research Funds for the Central Universities, Ministry of Education of China, the U.S. Department of Energy, Ministry of Education, Culture, Sports, Science and Technology, Japan Science and Technology Agency, National Council for Science and Technology, Natural Science Foundation of Guangdong Province, Thailand Research Fund, Japan Society for the Promotion of Science, Program for New Century Excellent Talents in University, Science and Technology Planning Project of Guangdong Province, Chinese Academy of Sciences, Ministry of Finance, National Basic Research Program of China (973 Program), U.S. Department of Agriculture, University Grants Commission, Financing of Studies and Projects, National Research Foundation, and Queensland University of Technology.

61.3.6 The Most Prolific Source Titles in Second Generation Bioethanol Fuels from Residual Sugar Feedstocks

Information about the 13 most prolific source titles publishing at least 2% of the sample papers each in second generation bioethanol fuels from residual sugar feedstocks is given in Table 61.5.

The most prolific source title is the Bioresource Technology with 31.3% of the sample papers. Other prolific titles are Applied Energy, Industrial Crops and Products, Biotechnology for Biofuels, and Biomass and Bioenergy with 4.5%–6.5% of the sample papers each.

TABLE 61.5
The Most Prolific Source Titles in Second Generation Bioethanol Fuels from Residual Sugar Feedstocks

No.	Source Titles	Sample Papers (%)	Population Papers (%)	Surplus (%)
1	Bioresource Technology	31.3	12.4	18.9
2	Applied Energy	6.5	1.2	5.3
3	Industrial Crops and Products	6.0	5.2	0.8
4	Biotechnology for Biofuels	5.0	2.1	2.9
5	Biomass and Bioenergy	4.5	3.1	1.4
6	Biotechnology and Bioengineering	3.5	1.0	2.5
7	Enzyme and Microbial Technology	3.5	1.0	2.5
8	Renewable Energy	2.5	2.5	0.0
9	Applied Biochemistry and Biotechnology	2.5	2.2	0.3
10	Applied Biochemistry and Biotechnology Part A Enzyme Engineering and Biotechnology	2.5	0.7	1.8
11	Energy	2.0	1.0	1.0
12	Energy Policy	2.0	0.4	1.6
13	Journal of Food Engineering	2.0	0.2	1.8

On the other hand, source title with the most impact is the Bioresource Technology with 18.9% surplus. Other influential titles are Applied Energy, Biotechnology for Biofuels, Biotechnology and Bioengineering, Enzyme and Microbial Technology with 2.5%–5.3% surplus each. Similarly, source titles with the least impact are Renewable Energy, Applied Biochemistry and Biotechnology, and Industrial Crops and Products with <1% surplus each.

Other source titles with relatively low citation impact with 0.6%–2.2% of the population papers each are Biomass Conversion and Biorefinery, Bioresources, Bioenergy Research, Chemical Engineering Transactions, International Sugar Journal, ACS Sustainable Chemistry and Engineering, Waste and Biomass Valorization, Journal of Cleaner Production, Journal of Chemical Technology and Biotechnology, Fuel, Process Biochemistry, Sugar Tech, Biotechnology Letters, Cellulose, Bioprocess and Biosystems Engineering, Cellulose Chemistry and Technology, Biochemical Engineering Journal, Applied Microbiology and Biotechnology, Energy and Fuels, RSC Advances, Zuckerindustrie (Sugar Industry), Brazilian Journal of Chemical Engineering, and Renewable and Sustainable Energy Reviews.

61.3.7 The Most Prolific Countries in Second Generation Bioethanol Fuels from Residual Sugar Feedstocks

Information about the 14 most prolific countries publishing at least 2.5% of sample papers each in second generation bioethanol fuels from residual sugar feedstocks is given in Table 61.6. Further, four European countries listed in Table 61.6 produce 16% and 8% of the sample and population papers, respectively.

The most prolific country is Brazil with 34.3% of the sample papers, followed by the USA and China with 16.4% and 14.9% of the sample papers, respectively. Other prolific countries are India, Japan, Sweden, and Cuba with 5%–7% of the sample papers each.

On the other hand, the country with the most citation impact is the USA with 7.6% surplus. Other influential countries are Sweden, Brazil, Cuba, Taiwan, and Denmark with 1.8%–3.6% surplus each. Similarly, the country with the least citation impact is India with 4.3% deficit, followed by Thailand and China with 0.6% and 0.4% deficit, respectively.

Additionally, there are other countries with relatively low citation impact and with 0.6%–42.7% of the sample papers each: South Africa, Iran, the UK, France, Egypt, Colombia, Canada, Pakistan,

TABLE 61.6
The Most Prolific Countries in Second Generation Bioethanol Fuels from Residual Sugar Feedstocks

No.	Countries	Sample Papers (%)	Population Papers (%)	Surplus (%)
1	Brazil	34.3	31.2	3.1
2	USA	16.4	8.8	7.6
3	China	14.9	15.3	–0.4
4	India	7.0	11.3	–4.3
5	Japan	5.5	4.1	1.4
6	Sweden	5.5	1.9	3.6
7	Cuba	5.0	2.1	2.9
8	Spain	4.0	2.4	1.6
9	Australia	3.5	3.5	0.0
10	Denmark	3.5	1.7	1.8
11	Thailand	2.5	3.1	–0.6
12	Mexico	2.5	2.2	0.3
13	Germany	2.5	1.5	1.0
14	Taiwan	2.5	0.6	1.9

TABLE 61.7
The Most Prolific Scopus Subject Categories in Second Generation Bioethanol Fuels from Residual Sugar Feedstocks

No.	Scopus Subject Categories	Sample Papers (%)	Population Papers (%)	Surplus (%)
1	Energy	59.7	40.9	18.8
2	Environmental Science	59.2	36.3	22.9
3	Chemical Engineering	58.7	42.9	15.8
4	Biochemistry, Genetics, andMolecular Biology	24.4	22.7	1.7
5	Immunology and Microbiology	21.4	16.4	5.0
6	Agricultural and Biological Sciences	16.9	23.3	–6.4
7	Engineering	11.9	12.7	–0.8
8	Chemistry	8.5	13.5	–5.0

Malaysia, South Korea, Argentina, Indonesia, Italy, Netherlands, Portugal, Finland, Switzerland, Poland, Saudi Arabia, and Turkey.

61.3.8 The Most Prolific Scopus Subject Categories in Second Generation Bioethanol Fuels from Residual Sugar Feedstocks

Information about the eight most prolific Scopus subject categories indexing at least 8.5% of the sample papers each is given in Table 61.7.

The most prolific Scopus subject categories in second generation bioethanol fuels from residual sugar feedstocks are energy, environmental science, and chemical engineering with 60%, 59%, and 59% of the sample papers, respectively. Other prolific subject categories are biochemistry, genetics and molecular biology, immunology and microbiology, and agricultural and biological sciences with 24%, 21%, and 17% of the sample papers, respectively. It is notable that social sciences including economics and business account for 1% and 3% of the sample and population studies, respectively.

On the other hand, Scopus subject categories with the most citation impact are energy, environmental science, and chemical engineering 19%, 23%, and 16% surplus, respectively. Similarly, the least influential subject categories are agricultural and biological sciences and chemistry with 6% and 5% deficit, respectively.

61.3.9 The Most Prolific Keywords in Second Generation Bioethanol Fuels from Residual Sugar Feedstocks

Information about Scopus keywords used with at least 6% or 3.1% of the sample or population papers, respectively, is given in Table 61.8. For this purpose, keywords related to the keyword set given in appendix are selected from a list of the most prolific keyword set provided by Scopus database.

These keywords are grouped under the five headings: biomass, pretreatments, fermentation, hydrolysis and hydrolysates, and products.

The most prolific keywords related to biomass and biomass constituents are bagasse, cellulose, sugar cane, lignin, sugarcane, saccharum, sugarcane bagasse, biomass, sugarcane bagasse, and hemicellulose with 19%–77% of the sample papers each. Further, prolific keywords related to the pretreatments are pretreatment, pre-treatment, ionic liquids, delignification, and sulfuric acid with 16%–29% of the sample papers each.

The most prolific keyword related to the fermentation are fermentation, saccharomyces, and yeast with 13%–35% of the sample papers each. Further, the most prolific keywords related to hydrolysis and hydrolysates are hydrolysis, sugar, enzyme activity, enzymatic hydrolysis, glucose,

TABLE 61.8
The Most Prolific Keywords in Second Generation Bioethanol Fuels from Residual Sugar Feedstocks

No.	Keywords	Sample Papers (%)	Population Papers (%)	Surplus (%)
1.		**Biomass and Biomass Constituents**		
	Bagasse	77.6	60.1	17.5
	Cellulose	61.7	36.1	25.6
	Sugarcane	55.7	27.1	28.6
	Lignin	48.3	28.5	19.8
	Sugar cane	42.8	24.5	18.3
	Saccharum	36.3	19.8	16.5
	Sugarcane bagasse	34.3	24.3	10.0
	Biomass	30.3	22.7	7.6
	Sugar-cane bagasse	29.9	28.9	1.0
	Hemicellulose	19.4	8.6	10.8
	Lignocellulose	17.4	8.4	9.0
	Lignocellulosic biomass	8.5	8.9	−0.4
	Sorghum	8.0	3.7	4.3
	Xylan	7.0	3.6	3.4
	Glucan	6.0	2.8	3.2
	Sweet sorghum	6.0	2.4	3.6
	Molasses	5.5	7.1	−1.6
2.		**Pretreatments**		
	Pretreatment	28.9	14.2	14.7

(Continued)

TABLE 61.8 (*Continued*)
The Most Prolific Keywords in Second Generation Bioethanol Fuels from Residual Sugar Feedstocks

No.	Keywords	Sample Papers (%)	Population Papers (%)	Surplus (%)
	Pre-treatment	26.9	12.0	14.9
	Ionic liquids	16.4	8.6	7.8
	Delignification	15.9	10.2	5.7
	Sulfuric acid	15.9	5.7	10.2
	Water	13.9	4.8	9.1
	Temperature	12.9	6.7	6.2
	Enzymes	11.9	8.9	3.0
	Enzymolysis	11.4	6.9	4.5
	Acetic acid	10.0	4.4	6
	Steam	6.5	2.9	4
	Sodium hydroxide	5.0	6.1	–1
	Carbon dioxide	3.5	3.4	0
3	**Fermentation**			
	Fermentation	34.8	27.0	8
	Saccharomyces	14.9	8.4	7
	Yeast	12.9	10.4	3
	Furfural	8.0	3.3	5
	Fungi		8.6	–9
4.	**Hydrolysis and Hydrolysates**			
	Hydrolysis	59.2	33.7	26
	Sugar	48.3	20.7	28
	Enzyme activity	41.8	21.5	20
	Enzymatic hydrolysis	32.8	25.6	7
	Glucose	24.4	17.6	7
	Saccharification	21.4	18.0	3
	Enzymes	20.4	9.2	11
	Cellulase	19.4	10.0	9
	Xylose	18.4	9.0	9
	Enzymatic digestibility	9.0	3.1	6
	Alkalinity	8.5	5.2	3
	Enzymatic saccharification	15.6	5.2	10
	Fermentable sugars	8.3	3.8	5
	Acid hydrolysis	6.0	2.0	4
5.	**Products**			
	Ethanol	55.2	32.2	23
	Biofuels	45.0	15.0	30
	Bioethanol	36.7	15.9	20.8
	Ethanol production	35.8	8.7	27.1
	Alcohol production	23.9	6.2	17.7
	Bio-ethanol production	16.5	5.7	10.8
	Cellulosic ethanol	16.5	5.2	11.3

and saccharification with 21%–59% of the sample papers each. Finally, the most prolific keywords related to the products are ethanol, biofuels, bioethanol, ethanol production, and alcohol production with 24%–55% of the sample papers each.

On the other hand, the most influential keywords are biofuels, sugar cane, sugar, ethanol production, hydrolysis, cellulose, ethanol, bioethanol, enzyme activity, and lignin with 20%–30% surplus each. Similarly, the most prolific keywords across all of the research fronts are sugar cane, ethanol, sugar, lignin, biofuels, sugarcane, enzyme activity, bioethanol, saccharum, ethanol production, fermentation, sugarcane bagasse, enzymatic hydrolysis, biomass, sugar-cane bagasse, pretreatment, pre-treatment, glucose, alcohol production, saccharification, and enzymes with 20%–78% of the sample papers each.

61.3.10 The Most Prolific Research Fronts in Second Generation Bioethanol Fuels from Residual Sugar Feedstocks

Information about research fronts for the sample papers in second generation bioethanol fuels from residual sugar feedstocks with regard to the residual sugar feedstocks used for the bioethanol production is given in Table 61.9.

As this table shows, the most prolific residual sugar feedstock is the sugarcane residues with 86% of the sample papers. Other residual sugar feedstocks used in these studies are sweet sorghum, sugar beet, and energy cane residues with 8%, 3%, and 1% of the sample papers, respectively. On the other hand, on an individual basis, the most prolific feedstock is sugarcane bagasse with 79% of the sample papers. Other prolific residual feedstocks are sugarcane straw and sweet sorghum bagasse with 12% and 7% of the sample papers, respectively.

Information about the thematic research fronts for the sample papers in second generation bioethanol fuels from residual sugar feedstocks is given in Table 61.10. As this table shows, there are five primary research fronts: biomass pretreatments and hydrolysis, hydrolysate fermentation, and production and evaluation of second generation bioethanol fuels from residual sugar feedstocks with 73%, 66%, 26%, 33%, and 17% of the sample papers, respectively.

TABLE 61.9
The Most Prolific Research Fronts for Second Generation Bioethanol Fuels from Residual Sugar Feedstocks

No.	Research Fronts	N Paper % Sample
1	Sugarcane residues	86.0
	Sugarcane bagasse	78.6
	Sugarcane straw	12.4
	Sugarcane molasses	2.0
	Sugarcane vinasse	1.0
2	Sweet sorghum residues	7.5
	Sweet sorghum bagasse	7.0
	Sweet sorghum straw	0.5
3	Sugar beet residues	3
	Sugar beet pulp	2.0
	Sugar beet molasses	1.0
4	Energy cane residues	1.0
	Energy cane bagasse	1.0

N paper (%) sample: Number of papers in the population sample of 201 papers.

TABLE 61.10
The Most Prolific Thematic Research Fronts for Second Generation Bioethanol Fuels from Residual Sugar Feedstocks

No.	Research Fronts	N Paper % Sample
1	Biomass pretreatments	72.6
1.1	Hydrothermal pretreatments	23.9
	Steam explosion	12.9
	LHW	9.0
	Other pretreatments	2.0
1.2	Chemical pretreatments	61.2
	Acid pretreatment	31.3
	Alkaline pretreatment	22.9
	Solvent pretreatment	5.0
	IL pretreatment	6.0
	Ammonia pretreatment	3.0
	SO_2 pretreatment	1.5
	Ozonolysis	1.5
	Surfactants	3.0
	Other pretreatments	2.5
1.3	Mechanical pretreatments	10.9
	Microwave	6.0
	Milling	1.5
	Ultrasound	3.5
1.4.	Enzymatic pretreatments	56.2
1.5.	Feedstock genetic engineering	2.0
1.6.	Yeast microbial engineering	2.0
2	Biomass hydrolysis	65.7
	Enzymatic hydrolysis	56.2
	Acid hydrolysis	9.5
3	Hydrolysate fermentation	25.9
4	Bioethanol production	33.3
5	Bioethanol fuel evaluation	16.9

N paper (%) sample: Number of papers in the population sample of 201 papers.

For the research front of residual sugar feedstock pretreatments, the most prolific pretreatments are chemical and enzymatic pretreatments of biomass with 61% and 56% of the sample papers. Other prolific pretreatments are hydrothermal and mechanical pretreatments of biomass with 24% and 11% of the sample papers, respectively. Further, other minor pretreatments are genetic engineering of biomass and microbial engineering of yeasts used in fermentation of sugar hydrolysates with 2% of the sample papers each.

On the other hand, on an individual basis, the most prolific research fronts are enzymatic and acid hydrolysis of biomass with 56% and 10% of biomass, while acid, alkaline, steam explosion, and LHW pretreatments of the sugar biomass are other research fronts with 31%, 23%, 13%, and 10% of the sample papers, respectively.

Further, the most prolific feedstock used in all these research fronts is sugarcane bagasse although its content is not shown in this table.

61.4 DISCUSSION

61.4.1 Introduction

Crude oil-based gasoline fuels have been widely used in transportation sector since the 1920s. However, there have been great public concerns over adverse environmental and human impact of these fuels. Hence, biomass-based bioethanol fuels have increasingly been used in blending gasoline fuels, in fuel cells, and in biochemical production in a biorefinery context.

However, it is necessary to pretreat biomass to enhance yield of bioethanol prior to bioethanol production through hydrolysis and fermentation. One of the most studied feedstocks for bioethanol fuels are the residual sugar feedstocks such as sugarcane bagasse and straw, sweet sorghum bagasse, and sugar beet pulp. Research in the field of second generation bioethanol fuels from residual sugar feedstocks has intensified in this context in key research fronts of the pretreatment of residual sugar feedstocks, hydrolysis of residual sugar feedstocks, fermentation of residual sugar feedstock-based hydrolysates, and production and evaluation of second generation bioethanol fuels from residual sugar feedstocks. Further, sugarcane bagasse; sugarcane straw; sweet sorghum bagasse; and to a lesser extent sweet sorghum straw, sugar beet pulp, sugarcane molasses, and energy cane bagasse have been studied intensively in this context.

However, it is essential to develop efficient incentive structures for primary stakeholders to enhance research in this field. This is especially important to maintain energy security in cases of supply shocks such as oil price shocks, war-related shocks as in the case of Russian invasion of Ukraine, or COVID-19 shocks.

The scientometric analysis has been used in this context to inform primary stakeholders about current state of the research in a selected research field. As there has been no scientometric study in this field, this book chapter presents a scientometric study of the research in second generation bioethanol fuels from residual sugar feedstocks. It examines scientometric characteristics of both the sample and population data presenting scientometric characteristics of these both datasets in order of documents, authors, publication years, institutions, funding bodies, source titles, countries, Scopus subject categories, Scopus keywords, and research fronts.

As the first step for search of relevant literature, the keywords were selected using the first 200 most-cited papers. The selected keyword list was then optimized to obtain a representative sample of papers for the searched research field. A copy of this extended keyword list was provided in the Appendix for future replicative studies. Further, a selected list of the keywords was presented in Table 61.8.

As the second step, two sets of data were used for this study. First, a population sample of 2,009 papers was used to examine scientometric characteristics of the population data. Secondly, a sample of 201 most-cited papers, corresponding to 10% of the population data set, was used to examine scientometric characteristics of these citation classics.

The scientometric characteristics of these sample and population datasets were presented in order of documents, authors, publication years, institutions, funding bodies, source titles, countries, Scopus subject categories, Scopus keywords, and research fronts.

Lastly, the key scientometric findings for both datasets were discussed to highlight the research landscape for second generation bioethanol fuels from residual sugar feedstocks. Additionally, a number of brief conclusions were drawn and a number of relevant recommendations were made to enhance future research landscape.

61.4.2 The Most Prolific Documents in Second Generation Bioethanol Fuels from Residual Sugar Feedstocks

Articles (together with conference papers) dominate both sample (95%) and population (95%) papers (Table 61.1). Further, review papers and articles have a surplus (3%) and deficit (0%), respectively. The representation of reviews in sample papers is relatively modest (5%).

Scopus differs from Web of Science database in differentiating and showing articles (93%) and conference papers (2%) published in journals separately. However, it should be noted that these conference papers are also published in journals as articles, compared to those published only in conference proceedings. Hence, the total number of articles and review papers in the sample dataset are 95% and 5%, respectively.

It is observed during the search process that there has been inconsistency in the classification of the documents in Scopus as well as in other databases such as Web of Science. This is especially relevant for the classification of papers as reviews or articles as the papers not involving a literature review may be erroneously classified as a review paper. There is also a case of review papers being classified as articles. For example, the total number of the reviews in the sample data set was manually found as nearly 4% compared to 5% as indexed by Scopus, increasing the number of articles and conference papers to 96% for the sample dataset. A close examination of these papers shows that many evaluative studies such as technoeconomic or life-cycle studies have often been indexed as review papers by Scopus database.

In this context, it would be helpful to provide a classification note for the published papers in books and journals at first instance. It would also be helpful to use document types listed in Table 61.1 for this purpose. Book chapters may also be classified as articles or reviews as an additional classification to differentiate review chapters from experimental chapters as it is done by Web of Science. It would be further helpful to additionally classify conference papers as articles or review papers as well as it is done in Web of Science database.

61.4.3 The Most Prolific Authors in Second Generation Bioethanol Fuels from Residual Sugar Feedstocks

There have been 15 most-prolific authors with at least 2.4% of the sample papers each as given in Table 61.2. These authors have shaped the development of research in this field.

The most prolific authors are Carlos Martin; Rubens M. Filho; and to a lesser extent Zhenhong Yuan, Carlos E. V. Rossell, Marina O. S. Dias, Parameswaran Binod, and Raveendran Sindhu.

It is important to note inconsistencies in indexing of author names in Scopus and other databases. It is especially an issue for names with more than two components such as 'Blake Sam de Hyun Filho'. The probable outcomes are 'Filho, B.S.D.H.', 'de Hyun Filho, B.S.', or 'Hyun Filho, B.S.D.'. The first choice is the gold standard of publishing sector as the last word in the name is taken as the last name. In most of the academic databases, such as PUBMED and EBSCO databases, this version is used predominantly. The second choice is a strong alternative, while the last choice is an undesired outcome as two last words are taken as the last name. It is a good practice to combine the words of the last name by a hyphen: 'Hyun-Filho, B.S.D.'. It is notable that inconsistent indexing of author names may cause substantial inefficiencies in search process for papers as well as allocating credit to authors as there are different author entries for each outcome in the databases.

There are also inconsistencies in shortening Chinese names. For example. 'YangYing Wang' is often shortened as 'Wang, Y.', 'Wang, Y.-Y.', and 'Wang, Y.Y.' as it is done in Web of Science database as well. However, the gold standard in this case is 'Wang, Y' where the last word is taken as the last name and the first word is taken as a single forename. In most of the academic databases such as PUBMED and EBSCO, this first version is used predominantly. However, it makes sense to use the third option to differentiate Chinese names efficiently: 'Wang, Y.Y.'. Therefore, there have been difficulties in locating papers for Chinese authors. In such cases, the use of unique author codes provided for each author by Scopus database has been helpful.

There is also a difficulty in allowing credit for authors especially for authors with common names such as 'Wang, X.' in conducting scientometric studies. These difficulties strongly influence the efficiency of scientometric studies as well as allocating credit to authors as there are same author entries for different authors with the same name. e.g., 'Wang, X.' in the databases.

In this context, coding of authors in Scopus database is a welcome innovation compared to other databases such as Web of Science. In this process, Scopus allocates a unique number to each author in the database (Aman, 2018). However, there might still be substantial inefficiencies in this coding system especially for common names. For example, some of the papers for a certain author maybe allocated to another researcher with a different author code. It is possible that Scopus uses a number of software programs to differentiate author names and the program may not be false-proof (Kim, 2018).

In this context, it does not help that author names are not given in full in some journals and books. This makes it difficult to differentiate authors with common names and makes scientometric studies further difficult in author domain. Therefore, author names should be given in all books and journals at the first instance. There is also a cultural issue where some authors do not use their full names in their papers. Instead, they use initials for their forenames: 'Filho, H.J.', 'Filho', 'Filho, H.', or 'Filho, J.' instead of 'Filho, Hyun Jae'.

There are also inconsistencies in naming of authors with more than two components by the authors themselves in journal papers and book chapters. For example. 'Filho, A.P.C.' might be given as 'Filho, A.' or 'Filho, A.C.' or 'Filho, A.P.' or 'Filho, C.' in journals and books. This also makes scientometric studies difficult in author domain. Hence, contributing authors should use their name consistently in their publications.

Another critical issue regarding author names is inconsistencies in spelling of author names in national spellings (e.g., Şöğütçığıl, Gökçe) rather than in English spellings (e.g., Sogutcigil, Gokce) in Scopus database. Scopus differs from Web of Science database and many other databases in this respect where author names are given only in English spellings. It is observed that national spellings of author names do not help much in conducting scientometric studies as well in allocating credits to authors as sometimes there are different author entries for English and national spellings in Scopus database.

The most prolific institutions for sample dataset are CSIR and to a lesser extent Chinese Academy of Sciences, State University of Campinas, and University of Sao Paulo. Further, the most prolific country for sample dataset is Brazil and to a lesser extent India and China. These findings confirm the dominance of Brazil and to a lesser extent China and India in this field. Primary research fronts are bioethanol fuels based on sugarcane bagasse (SCB) and sugarcane straw (SCS).

It is also notable that there is a significant gender deficit for the sample dataset surprisingly with a representation rate of 20%. This finding is the most thought-provoking with strong public policy implications. Hence, institutions, funding bodies, and policymakers should take efficient measures to reduce the gender deficit in this field as well as other scientific fields with strong gender deficit. In this context, it is worth to note the level of representation of researchers from the minority groups in science on the basis of race, sexuality, age, and disability, besides the gender (Blankenship 1993; Dirth and Branscombe, 2017; Konur, 2000, 2002a–c, 2006a,b, 2007a,b).

61.4.4 The Most Prolific Research Output by Years in Second Generation Bioethanol Fuels from Sugar Feedstocks

The research output observed between 1970 and 2022 is illustrated in Figure 61.1. This figure clearly shows that the bulk of research papers in the population dataset was published primarily in the 2010s and early 2020s. Similarly, the bulk of research papers in the sample dataset was published in the 2000s and 2010s. There was a rising trend for population papers between 2010 and 2014, and it lost its momentum after 2014. However, there was a sharp rise in 2020 and 2021.

These findings suggest that the most prolific sample and population papers were primarily published in the 2010s. These are thought-provoking findings as there has been significant research boom since 2007. In this context, increasing public concerns about climate change (Change, 2007), greenhouse gas emissions (Carlson et al., 2017), and global warming (Kerr, 2007) have been certainly behind the boom in research in this field in the last two decades. Furthermore, supply shocks experienced due to COVID-19 pandemic might also be behind the research boom in this field since 2019.

Based on these findings, the size of the population papers is likely to more than double in the current decade, provided that public concerns about climate change, greenhouse gas emissions, and global warming, as well as the supply shocks are translated efficiently to the research funding in this field.

61.4.5 The Most Prolific Institutions in Second Generation Bioethanol Fuels from Residual Sugar Feedstocks

The 16 most prolific institutions publishing papers on second generation bioethanol fuels from residual sugar feedstocks with at least 2.0% of the sample papers each given in Table 61.3 have shaped the development of research in this field.

The most prolific institutions are the University of Sao Paulo and to a lesser extent State University of Campinas, National Biorenewables Laboratory, Federal University of Rio de Janeiro, University of Matanzas, South China University of Technology, and Federal University of Sao Carlos. Similarly, the top countries for these most prolific institutions are Brazil and China. In total, only eight countries house these top institutions.

On the other hand, the institutions with the most citation impact are National Biorenewables Laboratory, University of Matanzas, and to a lesser extent University of Sao Paulo and Lund University. These findings confirm the dominance of Brazilian and to a lesser extent of Chinese and Mexican institutions for these HCPs.

61.4.6 The Most Prolific Funding Bodies in Second Generation Bioethanol Fuels from Residual Sugar Feedstocks

The 13 most prolific funding bodies funding at least 2% of the sample papers each is given in Table 61.4. It is notable that only 54% and 55% of the sample and population papers were funded, respectively.

The most prolific funding bodies are National Council for Scientific and Technological Development; Sao Paulo Research Foundation; and to a lesser extent the National Natural Science Foundation of China, CAPES Foundation, Ministry of Science and Technology of China, and Ministry of Science and Technology of India. The most prolific countries for these top funding bodies are Brazil, China, and to a lesser extent India and the EU.

These findings on funding of research in this field suggest that the level of the funding, mostly since 2007, is moderately intensive, and it has been largely instrumental in enhancing the research in this field (Ebadi and Schiffauerova, 2016) in the light of North's institutional framework (North, 1991). It is also notable that the funding rate in this field is relatively modest compared to those in other research fronts of bioethanol fuels such as algal bioethanol fuels. Further, it is expected that this funding rate would improve in the light of recent supply shocks. Further, it emerges that Brazil has heavily funded the research on sugarcane-based bioethanol fuels.

61.4.7 The Most Prolific Source Titles in Second Generation Bioethanol Fuels from Residual Sugar Feedstocks

The 13 most prolific source titles publishing at least 12% of the sample papers each in second generation bioethanol fuels from residual sugar feedstocks have shaped the development of research in this field (Table 61.5).

The most prolific source titles are Bioresource Technology and to a lesser extent the Applied Energy, Industrial Crops and Products, Biotechnology for Biofuels, and Biomass and Bioenergy. On the other hand, source titles with the most citation impact are Bioresource Technology and to a lesser extent Applied Energy, Biotechnology for Biofuels, Biotechnology and Bioengineering, Enzyme and Microbial Technology.

It is notable that these top source titles are primarily related to bioresources, energy, biotechnology, and enzymes. This finding suggests that Bioresource Technology and other prolific journals in these fields have significantly shaped the development of the research in this field as they focus primarily on second generation bioethanol fuels from residual sugar feedstocks with a high yield. In this context, the influence of the top journal is quite extraordinary with 31.3% of the sample papers. It is also notable that energy-related journals have also published papers in areas of techno-economics environmental impact, land use change, economics, and labor relations as social science-related journals.

61.4.8 The Most Prolific Countries in Second Generation Bioethanol Fuels from Residual Sugar Feedstocks

The 14 most prolific countries publishing at least 2.5% of the sample papers each have significantly shaped the development of research in this field (Table 61.6).

The most prolific countries are Brazil; the USA; China; and to a lesser extent India, Japan, Sweden, and Cuba. Further, four European countries listed in Table 61.6 produce 16% and 8% of the sample and population papers, respectively.

On the other hand, countries with the most citation impact are the USA and to a lesser extent Sweden, Brazil, Cuba, Taiwan, and Denmark. Similarly, countries with the least impact are India and to a lesser extent Thailand and China.

A close examination of these findings suggests that Brazil; the USA; China; Europe; and to a lesser extent Japan and Cuba are the major producers of research in this field. It is a fact that the USA has been a major player in science (Leydesdorff and Wagner, 2009). The USA has further developed a strong research infrastructure to support its corn and grass-based bioethanol industry (Gillon, 2010).

However, China has been a rising mega star in scientific research in competition with the USA and Europe (Leydesdorff and Zhou, 2005). China is also a major player in this field as a major producer of bioethanol (Fang et al., 2010).

Next, Europe has been a persistent player in the scientific research in competition with both the USA and China (Leydesdorff, 2000). Europe has also been a persistent producer of bioethanol along with the USA and Brazil (Gnansounou, 2010).

Finally, Brazil has been a major producer and user of sugarcane-based bioethanol fuels since the 1970s as a world leader in this respect (Goldemberg et al., 2008; Martinelli and Filoso, 2008).

61.4.9 The Most Prolific Scopus Subject Categories in Second Generation Bioethanol Fuels from Residual Sugar Feedstocks

The eight most prolific Scopus subject categories indexing at least 8.5% of the sample papers each, respectively, given in Table 61.7 have shaped the development of the research in this field.

The most prolific Scopus subject categories in second generation bioethanol fuels from residual sugar feedstocks are energy; environmental science; chemical engineering; and to a lesser extent biochemistry, genetics and molecular biology, immunology and microbiology, and agricultural and biological sciences. It is also notable that social sciences including economics and business have a minimal presence in both sample and population studies.

On the other hand, Scopus subject categories with the most citation impact are energy, environmental science, and chemical engineering. Similarly, the least influential subject categories are agricultural and biological sciences and chemistry.

These findings are thought provoking suggesting that primary subject categories are related to energy, environmental science, chemical engineering, genetics, and microbiology as the core of research in this field concerns with production and utilization of second generation bioethanol fuels from residual sugar feedstocks. Another finding is that social sciences are not well represented in both the sample and population papers as in line with the most fields in bioethanol fuels. The social, environmental, and economics studies account for field of social sciences (Gnansounou et al., 2005; Goldemberg et al., 2008; Martinelli and Filoso, 2008).

61.4.10 The Most Prolific Keywords in Second Generation Bioethanol Fuels from Residual Sugar Feedstocks

A limited number of keywords have shaped the development of research in this field as shown in Table 61.8 and the Appendix. These keywords are grouped under the five headings: biomass, fermentation, hydrolysis and hydrolysates, products, and evaluation.

The most prolific keywords across all of the research fronts are sugar cane, ethanol, sugar, lignin, biofuels, sugarcane, enzyme activity, bioethanol, saccharum, ethanol production, fermentation, sugarcane bagasse, enzymatic hydrolysis, biomass, sugar-cane bagasse, pretreatment, pre-treatment, glucose, alcohol production, saccharification, and enzymes. Similarly, the most influential keywords are biofuels, sugar cane, sugar, ethanol production, hydrolysis, cellulose, ethanol, bioethanol, enzyme activity, and lignin.

These findings suggest that it is necessary to determine the keyword set carefully to locate relevant research in each of these research fronts. Additionally, the size of the samples for each keyword highlights the intensity of research in the relevant research areas. These findings also highlight different spelling of some strategic keywords: sugar cane v. sugarcane v. saccharum, bioethanol v. bio-ethanol, sweet sorghum v. sorghum bicolor, sugar beet v. sugarbeet v. beta vulgaris.

61.4.11 The Most Prolific Research Fronts in Second Generation Bioethanol Fuels from Residual Sugar Feedstocks

Information about research fronts for the sample papers in second generation bioethanol fuels from residual sugar feedstocks with regard to the residual sugar feedstocks used for the bioethanol production is given in Table 61.9. As this table shows, the most prolific residual sugar feedstocks are sugarcane and to a lesser extent sweet sorghum, sugar beet, and energy cane residues. On the other hand, on an individual basis, the most prolific feedstocks are sugarcane bagasse and to a lesser extent sugarcane straw and sweet sorghum bagasse. It is notable that energy cane is a genetically engineered version of sugarcane (Matsuoka et al., 2014).

Information about thematic research fronts for sample papers in second generation bioethanol fuels from residual sugar feedstocks is given in Table 61.10. As this table shows, there are five primary research fronts: biomass pretreatments and hydrolysis and to a lesser extent hydrolysate fermentation, second generation bioethanol fuel production, and evaluation from residual sugar feedstocks.

For research front of residual sugar feedstock pretreatments, the most prolific pretreatments are chemical and enzymatic pretreatments of biomass and to a lesser extent hydrothermal and mechanical pretreatments of biomass, genetic engineering of biomass, and microbial engineering of yeasts used in fermentation of sugar hydrolysates.

On the other hand, on an individual basis, the most prolific research fronts are enzymatic and to a lesser extent acid hydrolysis of biomass, acid, alkaline, steam explosion, and LHW pretreatments of sugar biomass. Further, the most prolific feedstock used in all these research fronts is sugarcane bagasse although its content is not shown in this table.

These findings are thought-provoking in seeking ways to increase residual sugar feedstock-based bioethanol yield at the global scale. It is clear that all of these research fronts have public importance and merit substantial funding and other incentives. Further, it is notable that second generation bioethanol fuels from residual sugar feedstocks have become a core unit of the bioethanol research to make it more competitive with crude oil-based gasoline and diesel fuels, especially for Brazil, the USA, and other countries with vast farmlands.

In comparison to other feedstock-based research fronts, it is notable that pretreatment and hydrolysis of residual sugar feedstocks emerge as a primary research front for this field. This suggests that primary stakeholders have been primarily interested in enzymatic hydrolysis as well as chemical and enzymatic pretreatments of the residual biomass. It is also notable that evaluation of second generation bioethanol fuels from residual sugar feedstocks such as techno-economics, life cycle, economics, social science, land use, labor, and environmental impact-related studies emerges as a case study for the bioethanol fuels together with algal and corn feedstocks in this field.

In this context, Brazil has been the global leader in the production and use of sugarcane-based bioethanol fuels since the 1970s in the aftermath of the global crude oil crisis in the early 1970s. It is also notable research in this field has been more intense compared to the production and evaluation of bioethanol fuels from primary sugar feedstocks such as sugarcane, sweet sorghum, sugar beet, and energy cane with nearly twice sample size.

In the end, these most-cited papers in this field hint that production of second generation bioethanol fuels from residual sugar feedstocks could be optimized using structure, processing, and property relationships of these residual sugar feedstocks such as sugarcane bagasse and straw, sweet sorghum bagasse, and sugar beet pulp in the fronts of feedstock pretreatment and hydrolysis, and hydrolysate fermentation (Formela et al., 2016; Konur, 2018a, 2020b, 2021a–d; Konur and Matthews, 1989).

61.5 CONCLUSION AND FUTURE RESEARCH

The research on second generation bioethanol fuels from residual sugar feedstocks has been mapped through a scientometric study of both sample (201 papers) and population (2,009 papers) datasets.

The critical issue in this study has been to obtain a representative sample of the research as in any other scientometric study. Therefore, the keyword set has been carefully devised and optimized after a number of runs in Scopus database. It is a representative sample of the wider population studies. This keyword set was provided in the Appendix and the relevant keywords are presented in Table 61.8. However, it should be noted that it has been very difficult to compile a representative keyword set since this research field has been connected closely with many other fields. Therefore, it has been necessary to compile a keyword list to exclude papers concerned with the other research fields.

It is notable in this context that research on the production and evaluation of bioethanol fuels from sugar feedstock residues such as sugarcane bagasse and straw is closely related to the research on the bioethanol production from these primary feedstocks themselves, such as sugarcane or sweet sorghum juice. Therefore, it is crucial to collect data on these two interconnected research fronts separately. Hence, studies on production and evaluation of the primary sugar feedstocks for the bioethanol production were presented separately in the third volume.

Another issue has been the selection of a multidisciplinary database to carry out the scientometric study of research in this field. For this purpose, Scopus database has been selected. The journal coverage of this database has been notably wider than that of Web of Science and other multi-subject databases.

The key scientometric properties of the research in this field have been determined and discussed in this book chapter. It is evident that a limited number of documents, authors, institutions, publication years, institutions, funding bodies, source titles, countries, Scopus subject categories, Scopus keywords, and research fronts have shaped the development of the research in this field.

There is ample scope to increase the efficiency of scientometric studies in this field in the author and document domains by developing consistent policies and practices in both domains across all the academic databases. In this respect, it seems that authors, journals, and academic databases have a lot to do. Furthermore, the significant gender deficit as in most scientific fields emerges as a public policy issue. The potential deficits on the basis of age, race, disability, and sexuality need also to be explored in this field as in other scientific fields.

The research in this field has boomed since 2007, possibly promoted by the public concerns about global warming, greenhouse gas emissions, and climate change. Furthermore, the recent COVID-19 pandemic and Russian invasion of Ukraine have resulted in a global supply shock shifting the focus of the stakeholders from crude oil-based fuels to biomass-based fuels such as bioethanol fuels. It is expected that there would be further incentives for key stakeholders to carry out the research for second generation bioethanol fuels from residual sugar feedstocks to increase ethanol yield and to make it more competitive with crude oil-based gasoline and petrodiesel fuels. This might be truer for crude oil- and foreign exchange-deficient countries to maintain the energy and food security at the face of global supply shocks.

The relatively modest funding rate of 54% and 55% for sample and population papers, respectively, suggests that funding in this field significantly enhanced the research in this field primarily since 2007, possibly more than doubling in the current decade. However, it is evident that there is ample room for more funding and other incentives to enhance research in this field further.

The institutions from Brazil and China have mostly shaped the research in this field. Further, Brazil, the USA, China, Europe, and to a lesser extent Japan and Cuba have been the major producers of the research in this field as the major producers and users of bioethanol fuels from different types of biomass materials such as corn, sugarcane, and grass as well as other types of biomass materials. It is evident that these countries have well-developed research infrastructure in bioethanol fuels and their derivatives. It is also notable these major countries mostly have access to large farmlands.

It emerges that ethanol is more popular than bioethanol as a keyword with strong implications for the search strategy. In other words, the search strategy using only bioethanol keyword would not be much helpful. The Scopus keywords are grouped under the five headings: biomass, pretreatments, fermentation, hydrolysis and hydrolysates, and products. Further, there is a need to include Latin terms for biomass in the keyword set.

Sugarcane and to a lesser extent sweet sorghum, sugar beet, and energy cane residues are the prolific feedstocks used in these studies. On the other hand, there are five primary research fronts: biomass pretreatments and hydrolysis and to a lesser extent hydrolysate fermentation, second generation bioethanol fuel production, and evaluation from residual sugar feedstocks.

These findings are thought-provoking in seeking ways to increase bioethanol yield through second generation bioethanol fuels from residual sugar feedstocks at the global scale. It is clear that all of these research fronts have public importance and merit substantial funding and other incentives. Further, it is notable that second generation bioethanol fuels from residual sugar feedstocks, as a second generation biofuel, have become a core unit of the bioethanol research to make it more competitive with the crude oil-based gasoline and diesel fuels, especially for countries with large access to farmlands.

In comparison to other feedstock-based research fronts, it is notable that pretreatment and hydrolysis of the residual sugar feedstocks emerge as a primary research front for this field. This suggests that primary stakeholders have been primarily interested in enzymatic hydrolysis as well as chemical and enzymatic pretreatments of the residual biomass. It is also notable that evaluation of second generation bioethanol fuels from residual sugar feedstocks is a case study for bioethanol fuels together with algal and corn feedstocks in this field. In this context, Brazil has been the global leader in the production and use of the sugarcane-based bioethanol fuels since the 1970s in the aftermath of global crude oil crisis in the early 1970s. It is also notable that research in this field has been more intense compared to production and evaluation of bioethanol fuels from primary sugar feedstocks such as sugarcane, sweet sorghum, sugar beet, and energy cane with nearly twice sample size.

Thus, the scientometric analysis has a great potential to gain valuable insights into the evolution of research in this field as in other scientific fields especially in the aftermath of the significant global supply shocks such as COVID-19 pandemics and the Russian invasion of Ukraine.

It is recommended that further scientometric studies are carried out for the primary research fronts. It is further recommended that reviews of the most-cited papers are carried out for each primary research front to complement these scientometric studies. Next, the scientometric studies of the hot papers in these primary fields are carried out.

ACKNOWLEDGMENTS

The contribution of highly cited researchers in the field of second generation bioethanol fuels from residual sugar feedstocks has been gratefully acknowledged.

APPENDIX: THE KEYWORD SET FOR SECOND GENERATION BIOETHANOL FUELS FROM SUGAR FEEDSTOCKS

((((TITLE (ethanol OR bioethanol OR *saccharification OR *hydrolysis OR digestibili* OR ssf OR shf OR recalcitrance OR hydrolysate* OR hydrolyzate* OR ferment* OR coferment* OR fractionation OR delignification OR depolymerization OR microwave* OR autoclaving OR ultrasound OR sonicat* OR extrusion OR milling OR grinding OR pretreat* OR "pre-treat*" OR bioorganosolve OR "size reduction" OR steam* OR "hot water" OR "hot compressed" OR "sulfuric acid*" OR sulfite* OR sporl OR lime OR "sulfur dioxide" OR so2 OR alkali* OR naoh* OR extruder OR "sodium hydroxide*" OR "ionic liquid*" OR solvent* OR organosolv* OR ammonia OR {wet oxidation} OR "wet explosion" OR flowthrough OR hydrothermolysis OR "supercritical co2" OR "supercritical carbon dioxide" OR surfactant* OR afex* OR "supercritical water" OR tween* OR milled OR hydrothermal OR pelleting OR pelletizing OR {sugar yield} OR precipitation OR "hydrogen peroxide" OR glycerol OR detoxification)) OR ((TITLE (ethanol OR bioethanol) AND TITLE (*diesel OR *hydrogen OR h2 OR *butanol OR biogas OR *methane OR oil OR syngas OR "*synthesis gas" OR anaerobic OR lipids OR biorefinery))) OR (TITLE (*electricity OR *power OR biorefinery OR cogeneration OR gasification OR chp)) OR (TITLE (ethanol OR bioethanol) AND TITLE (brazil* OR paulo))) AND ((TITLE (bagasse)) OR (TITLE (sugarcane OR saccharum OR {sugar beet} OR sugarbeet OR {sugar cane} OR {sweet sorghum} OR {sorghum bicolor} OR {energy cane} OR energycane OR cane OR beet OR {beta vulgaris} OR {sugar juice} OR {sugar mills}) AND TITLE (bagasse OR baggase OR straw OR leaves OR tops OR pulp OR {by-products} OR lignocellulsic OR cellulosic OR cellulose* OR lignocellulose* OR trash OR "second generation" OR residu* OR {co-products} OR molasses OR pomaces OR vinasse OR {distillers' grains} OR ddgs OR solubles OR 2g OR mud OR byproduct OR biomass OR cake OR clones)))) AND NOT (TITLE (*hydrogen OR anaerobic OR lipid* OR pyrolysis OR carbonization OR congo OR adsorption OR lactic OR *butyric OR xylitol OR *butyrate OR blanket* OR levulinic OR *capacitor OR lipase OR citric OR wastewater* OR nanofib* OR succinic OR activated OR pullulan OR "solid state" OR torref* OR *alkanoates OR acer OR stems OR stalks OR lactate OR ferulic OR protein OR anti* OR composites OR vitro OR silage* OR cassava OR *polyester OR agave OR aspergillus OR butanol OR {substrate fermentation} OR apricot OR fumaric OR *bleaching OR liquefaction OR solka OR {Cellulase production}) OR SRCTITLE (composites OR cement* OR water OR nutr* OR animal* OR building OR polymer* OR hydrogen OR dairy OR materials OR macromol* OR zoo* OR grass OR carbohydrate* OR botan* OR cereal* OR chromat* OR soil* OR ceramic* OR carbohydrate* OR fertilizer* OR livestock* OR holzforschung OR cuban OR brewing) OR SUBJAREA (medi OR vete OR phar OR nurs OR heal OR psyc))) AND (LIMIT-TO (SRCTYPE, "j") OR LIMIT-TO (SRCTYPE, "k") OR LIMIT-TO (SRCTYPE, "b")) AND (LIMIT-TO (DOCTYPE, "ar") OR LIMIT-TO (DOCTYPE, "cp") OR LIMIT-TO (DOCTYPE, "ch") OR LIMIT-TO (DOCTYPE, "re") OR LIMIT-TO (DOCTYPE, "le") OR LIMIT-TO (DOCTYPE, "no") OR LIMIT-TO (DOCTYPE, "bk") OR LIMIT-TO (DOCTYPE, "ed") OR LIMIT-TO (DOCTYPE, "sh")) AND (LIMIT-TO (LANGUAGE, "English")).

REFERENCES

Aguilar, R., J. A. Ramirez, G. Garrote and M. Vazquez. 2002. Kinetic study of the acid hydrolysis of sugar cane bagasse. *Journal of Food Engineering* 55:309–318.

Aman, V. 2018. Does the Scopus author ID suffice to track scientific international mobility? A case study based on Leibniz laureates. *Scientometrics* 117:705–720.

Angelici, C., B. M. Weckhuysen and P. C. A. Bruijnincx. 2013. Chemocatalytic conversion of ethanol into butadiene and other bulk chemicals. *ChemSusChem* 6:1595–1614.

Antolini, E. 2007. Catalysts for direct ethanol fuel cells. *Journal of Power Sources* 170:1–12.

Antolini, E. 2009. Palladium in fuel cell catalysis. *Energy and Environmental Science* 2:915–931.

Beaudry, C. and V. Lariviere. 2016. Which gender gap? Factors affecting researchers' scientific impact in science and medicine. *Research Policy* 45:1790–1817.

Blankenship, K. M. 1993. Bringing gender and race in: US employment discrimination policy. *Gender & Society* 7:204–226.

Burnham, J. F. 2006. Scopus database: A review. *Biomedical Digital Libraries* 3:1–8.

Cao, W., C. Sun, R. Liu, R. Yin and X. Wu. 2012. Comparison of the effects of five pretreatment methods on enhancing the enzymatic digestibility and ethanol production from sweet sorghum bagasse. *Bioresource Technology* 111:215–221.

Carlson, K. M., J. S. Gerber and D. Mueller, et al. 2017. Greenhouse gas emissions intensity of global croplands. *Nature Climate Change* 7:63–68.

Chandel, A. K., R. K. Kapoor, A. Singh and R. C. Kuhad. 2007. Detoxification of sugarcane bagasse hydrolysate improves ethanol production by *Candida shehatae* NCIM 3501. Bioresource Technology 98:1947–1950.

Change, C. 2007. Climate change impacts, adaptation and vulnerability. *Science of the Total Environment* 326:95–112.

da Silva, A. S., H. Inoue, T. Endo, S. Yano and E. P. S. Bon. 2010. Milling pretreatment of sugarcane bagasse and straw for enzymatic hydrolysis and ethanol fermentation. *Bioresource Technology* 101:7402–7409.

Dias, M. O. S., A. V. Ensinas and S. A. Nebra, et al. 2009. Production of bioethanol and other bio-based materials from sugarcane bagasse: Integration to conventional bioethanol production process. *Chemical Engineering Research and Design* 87:1206–1216.

Dias, M. O. S., T. L. Junqueira and O. Cavalett, et al. 2012. Integrated versus stand-alone second generation ethanol production from sugarcane bagasse and trash. *Bioresource Technology* 103:152–161.

Dirth, T. P. and N. R. Branscombe. 2017. Disability models affect disability policy support through awareness of structural discrimination. *Journal of Social Issues* 73:413–442.

Ebadi, A. and A. Schiffauerova. 2016. How to boost scientific production? A statistical analysis of research funding and other influencing factors. *Scientometrics* 106:1093–1116.

Fang, X., Y. Shen, J. Zhao, X. Bao and Y. Qu. 2010. Status and prospect of lignocellulosic bioethanol production in China. *Bioresource Technology* 101:4814–4819.

Fauci, A. S., H. C. Lane and R. R. Redfield. 2020. Covid-19-navigating the uncharted. *New England Journal of Medicine* 382:1268–1269.

Fernando, S., S. Adhikari, C. Chandrapal and M. Murali. 2006. Biorefineries: Current status, challenges, and future direction. *Energy & Fuels* 20:1727–1737.

Formela, K., A. Hejna, L. Piszczyk, M. R. Saeb and X. Colom. 2016. Processing and structure-property relationships of natural rubber/wheat bran biocomposites. *Cellulose* 23:3157–3175.

Garfield, E. 1955. Citation indexes for science. *Science* 122:108–111.

Gillon, S. 2010. Fields of dreams: Negotiating an ethanol agenda in the Midwest United States. *Journal of Peasant Studies* 37:723–748.

Gnansounou, E. 2010. Production and use of lignocellulosic bioethanol in Europe: Current situation and perspectives. *Bioresource Technology* 101:4842–4850.

Gnansounou, E., A. Dauriat and C. E. Wyman. 2005. Refining sweet sorghum to ethanol and sugar: Economic trade-offs in the context of North China. *Bioresource Technology* 96:985–1002.

Goldemberg, J., S. T. Coelho and P. Guardabassi. 2008. The sustainability of ethanol production from sugarcane. *Energy Policy* 36:2086–2097.

Hahn-Hagerdal, B., M. Galbe, M. F. Gorwa-Grauslund, G. Liden and G. Zacchi. 2006. Bio-ethanol: The fuel of tomorrow from the residues of today. *Trends in Biotechnology* 24:549–556.

Hamilton, J. D. 1983. Oil and the macroeconomy since World War II. *Journal of Political Economy* 91:228–248.

Hamilton, J. D. 2003. What is an oil shock? *Journal of Econometrics* 113:363–398.

Hamilton, J. D. 2009. Causes and consequences of the oil shock of 2007-08. *Brookings Papers on Economic Activity* 2009:215–261.

Hill, J., E. Nelson, D. Tilman, S. Polasky and D. Tiffany. 2006. Environmental, economic, and energetic costs and benefits of biodiesel and ethanol biofuels. *Proceedings of the National Academy of Sciences of the United States of America* 103:11206–11210.

Hill, J., S. Polasky and E. Nelson, et al. 2009. Climate change and health costs of air emissions from biofuels and gasoline. Proceedings of the National Academy of Sciences of the United States of America 106:2077–2082.

Hsieh, W. D., R. H. Chen, T. L. Wu and T. H. Lin. 2002. Engine performance and pollutant emission of an SI engine using ethanol-gasoline blended fuels. *Atmospheric Environment* 36:403–410.

Huang, H. J., S. Ramaswamy, U. W. Tschirner and B. V. Ramarao. 2008. A review of separation technologies in current and future biorefineries. *Separation and Purification Technology* 62:1–21.

Jones, T. C. 2012. America, oil, and war in the Middle East. *Journal of American History* 99:208–218.

Kerr, R. A. 2007. Global warming is changing the world. *Science* 316:188–190.

Kilian, L. 2008. Exogenous oil supply shocks: How big are they and how much do they matter for the US economy? *Review of Economics and Statistics* 90:216–240.

Kilian, L. 2009. Not all oil price shocks are alike: Disentangling demand and supply shocks in the crude oil market. *American Economic Review* 99:1053–1069.

Kim, J. 2018. Evaluating author name disambiguation for digital libraries: A case of DBLP. *Scientometrics* 116:1867–1886.

Konur, O. 2000. Creating enforceable civil rights for disabled students in higher education: An institutional theory perspective. *Disability & Society* 15:1041–1063.

Konur, O. 2002a. Access to nursing education by disabled students: Rights and duties of nursing programs. *Nurse Education Today* 22:364–374.

Konur, O. 2002b. Assessment of disabled students in higher education: Current public policy issues. *Assessment and Evaluation in Higher Education* 27:131–152.

Konur, O. 2002c. Access to employment by disabled people in the UK: Is the disability discrimination act working? *International Journal of Discrimination and the Law* 5:247–279.

Konur, O. 2006a. Participation of children with dyslexia in compulsory education: Current public policy issues. *Dyslexia* 12:51–67.

Konur, O. 2006b. Teaching disabled students in higher education. *Teaching in Higher Education* 11:351–363.

Konur, O. 2007a. A judicial outcome analysis of the disability discrimination act: A windfall for the employers? *Disability & Society* 22:187–204.

Konur, O. 2007b. Computer-assisted teaching and assessment of disabled students in higher education: The interface between academic standards and disability rights. *Journal of Computer Assisted Learning* 23:207–219.

Konur, O. 2011. The scientometric evaluation of the research on the algae and bio-energy. *Applied Energy* 88:3532–3540.

Konur, O. 2012a. The evaluation of the biogas research: A scientometric approach. *Energy Education Science and Technology Part A: Energy Science and Research* 29:1277–1292.

Konur, O. 2012b. The evaluation of the educational research: A scientometric approach. *Energy Education Science and Technology Part B: Social and Educational Studies* 4:1935–1948.

Konur, O. 2012c. The evaluation of the global energy and fuels research: A scientometric approach. *Energy Education Science and Technology Part A: Energy Science and Research* 30:613–628.

Konur, O. 2012d. The evaluation of the research on the biodiesel: A scientometric approach. *Energy Education Science and Technology Part A: Energy Science and Research* 28:1003–1014.

Konur, O. 2012e. The evaluation of the research on the bioethanol: A scientometric approach. *Energy Education Science and Technology Part A: Energy Science and Research* 28:1051–1064.

Konur, O. 2012f. The evaluation of the research on the biofuels: A scientometric approach. *Energy Education Science and Technology Part A: Energy Science and Research* 28:903–916.

Konur, O. 2012g. The evaluation of the research on the biohydrogen: A scientometric approach. *Energy Education Science and Technology Part A: Energy Science and Research* 29:323–338.

Konur, O. 2012h. The evaluation of the research on the microbial fuel cells: A scientometric approach. *Energy Education Science and Technology Part A: Energy Science and Research* 29:309–322.

Konur, O. 2012i. The scientometric evaluation of the research on the production of bioenergy from biomass. *Biomass and Bioenergy* 47:504–515.

Konur, O. 2015. Current state of research on algal bioethanol. In *Marine Bioenergy: Trends and Developments*, Eds. S. K. Kim and C. G. Lee, pp. 217–244. Boca Raton, FL: CRC Press.

Konur, O., Ed. 2018a. *Bioenergy and Biofuels*. Boca Raton, FL: CRC Press.

Konur, O. 2018b. Bioenergy and biofuels science and technology: Scientometric overview and citation classics. In Bioenergy and Biofuels, Ed. O. Konur, pp. 3–63. Boca Raton, FL: CRC Press.

Konur, O. 2019. Cyanobacterial bioenergy and biofuels science and technology: A scientometric overview. In Cyanobacteria: From Basic Science to Applications, Eds. A. K. Mishra, D. N. Tiwari and A. N. Rai, pp. 419–442. Amsterdam: Elsevier.

Konur, O. 2020a. The scientometric analysis of the research on the bioethanol production from green macroalgae. In *Handbook of Algal Science, Technology and Medicine,* Ed. O. Konur, pp. 385–401. London: Academic Press.

Konur, O., Ed. 2020b. *Handbook of Algal Science, Technology and Medicine.* London: Academic Press.

Konur, O., Ed. 2021a. *Handbook of Biodiesel and Petrodiesel Fuels: Science, Technology, Health, and Environment.* Boca Raton, FL: CRC Press.

Konur, O., Ed. 2021b. Handbook of Biodiesel and Petrodiesel Fuels: Science, Technology, Health, and Environment. Volume 1. Biodiesel Fuels: Science, Technology, Health, and Environment. Boca Raton, FL: CRC Press.

Konur, O., Ed. 2021c. Handbook of Biodiesel and Petrodiesel Fuels: Science, Technology, Health, and Environment. Volume 2. Biodiesel Fuels *Based on the Edible and Nonedible Feedstocks, Wastes, and Algae: Science, Technology, Health, and Environment.* Boca Raton, FL: CRC Press.

Konur, O., Ed. 2021d. Handbook of Biodiesel and Petrodiesel Fuels: Science, Technology, Health, and Environment. Volume 3. Petrodiesel Fuels: Science, Technology, Health, and Environment. Boca Raton, FL: CRC Press.

Konur, O. and F. L. Matthews. 1989. Effect of the properties of the constituents on the fatigue performance of composites: A review. *Composites* 20:317–328.

Kruyt, B., D. P. van Vuuren, H. J. de Vries and H. Groenenberg. 2009. Indicators for energy security. *Energy Policy* 37:2166–2181.

Laser, M., D. Schulman and S. G. Allen, et al. 2002. A comparison of liquid hot water and steam pretreatments of sugar cane bagasse for bioconversion to ethanol. *Bioresource Technology* 81:33–44.

Lavarack, B. P., G. J. Griffin and D. Rodman. 2002. The acid hydrolysis of sugarcane bagasse hemicellulose to produce xylose, arabinose, glucose and other products. *Biomass and Bioenergy* 23:367–380.

Leydesdorff, L. 2000. Is the European Union becoming a single publication system? *Scientometrics* 47:265–280.

Leydesdorff, L. and C. Wagner. 2009. Is the United States losing ground in science? A global perspective on the world science system. *Scientometrics* 78:23–36.

Leydesdorff, L. and P. Zhou. 2005. Are the contributions of China and Korea upsetting the world system of science? *Scientometrics* 63:617–630.

Li, B. Z., V. Balan, Y. J. Yuan and B. E. Dale. 2010. Process optimization to convert forage and sweet sorghum bagasse to ethanol based on ammonia fiber expansion (AFEX) pretreatment. *Bioresource Technology* 101:1285–1292.

Li, H., S. M. Liu, X. H. Yu, S. L. Tang and C. K. Tang. 2020. Coronavirus disease 2019 (COVID-19): Current status and future perspectives. *International Journal of Antimicrobial Agents* 55:105951.

Lin, Y. and S. Tanaka. 2006. Ethanol fermentation from biomass resources: Current state and prospects. *Applied Microbiology and Biotechnology* 69:627–642.

Ma, X., L. Sun and C. Song. 2002. A new approach to deep desulfurization of gasoline, diesel fuel and jet fuel by selective adsorption for ultra-clean fuels and for fuel cell applications. *Catalysis Today* 77:107–116.

MacRelli, S., J. Mogensen and G. Zacchi. 2012. Techno-economic evaluation of 2nd generation bioethanol production from sugar cane bagasse and leaves integrated with the sugar-based ethanol process. *Biotechnology for Biofuels* 5:22.

Martinelli, L. A. and S. Filoso. 2008. Expansion of sugarcane ethanol production in Brazil: Environmental and social challenges. *Ecological Applications* 18:885–898.

Martinez, A., M. E. Rodriguez, S. W. York, J. F. Preston and I. O. Ingram. 2000. Effects of $Ca(OH)_2$ treatments ('overliming') on the composition and toxicity of bagasse hemicellulose hydrolysates. *Biotechnology and Bioengineering* 69:526–536.

Matsuoka, S., A. J. Kennedy, E. G. D. D. Santos, A. L. Tomazela and L. C. S. Rubio. 2014. Energy cane: Its concept, development, characteristics, and prospects. *Advances in Botany* 2014:597275.

McIntosh, S. and T. Vancov. 2010. Enhanced enzyme saccharification of *Sorghum bicolor* straw using dilute alkali pretreatment. *Bioresource Technology* 101:6718–6727.

Micard, V., C. M. G. C. Renard and J. F. Thibault. 1996. Enzymatic saccharification of sugar-beet pulp. *Enzyme and Microbial Technology* 19:162–170.

Morschbacker, A. 2009. Bio-ethanol based ethylene. *Polymer Reviews* 49:79–84.

Najafi, G., B. Ghobadian and T. Tavakoli, et al. 2009. Performance and exhaust emissions of a gasoline engine with ethanol blended gasoline fuels using artificial neural network. *Applied Energy* 86:630–639.

Newman, P. W. G. and J. R. Kenworthy. 1989. Gasoline consumption and cities: A comparison of U.S. cities with a global survey. *Journal of the American Planning Association* 55:24–37.

Nguyen, T. L. T. and S. H. Gheewala. 2008. Life cycle assessment of fuel ethanol from cane molasses in Thailand. *International Journal of Life Cycle Assessment* 13:301–311.

North, D. C. 1991. Institutions. *Journal of Economic Perspectives* 5:97–112.

Olsson, L. and B. Hahn-Hagerdal. 1996. Fermentation of lignocellulosic hydrolysates for ethanol production. *Enzyme and Microbial Technology* 18:312–331.

Qiu, Z., G. M. Aita and M. S. Walker. 2012. Effect of ionic liquid pretreatment on the chemical composition, structure and enzymatic hydrolysis of energy cane bagasse. *Bioresource Technology* 117:251–256.

Rabelo, S. C., H. Carrere, R. M. Filho and A. C. Costa. 2011. Production of bioethanol, methane and heat from sugarcane bagasse in a biorefinery concept. *Bioresource Technology* 102:7887–7895.

Reeves, S. 2014. To Russia with love: How moral arguments for a humanitarian intervention in Syria opened the door for an invasion of the Ukraine. *Michigan State University International Law Review* 23:199.

Rezende, C. A., M. de Lima and P. Maziero, et al. 2011. Chemical and morphological characterization of sugarcane bagasse submitted to a delignification process for enhanced enzymatic digestibility. *Biotechnology for Biofuels* 4:54.

Sanchez, O. J. and C. A. Cardona. 2008. Trends in biotechnological production of fuel ethanol from different feedstocks. *Bioresource Technology* 99:5270–5295.

Sun, Y. and J. Cheng. 2002. Hydrolysis of lignocellulosic materials for ethanol production: A review. *Bioresource Technology* 83:1–11.

Taherzadeh, M. J. and K. Karimi. 2007. Enzyme-based hydrolysis processes for ethanol from lignocellulosic materials: A review. *Bioresources* 2:707–738.

Taherzadeh, M. J. and K. Karimi. 2008. Pretreatment of lignocellulosic wastes to improve ethanol and biogas production: A review. *International Journal of Molecular Sciences* 9:1621–1651.

Winzer, C. 2012. Conceptualizing energy security. *Energy Policy* 46:36–48.

Yang, B. and C. E. Wyman. 2008. Pretreatment: The key to unlocking low-cost cellulosic ethanol. Biofuels, Bi*oproducts and Biorefining* 2:26–40.

62 Second Generation Bioethanol Fuels from Residual Sugar Feedstocks

Review

Ozcan Konur
(Formerly) Ankara Yildirim Beyazit University

62.1 INTRODUCTION

Crude oil-based gasoline fuels (Ma et al., 2002; Newman and Kenworthy, 1989) have been widely used in the transportation sector since the 1920s. However, there have been great public concerns over the adverse environmental and human impact of these fuels (Hill et al., 2006, 2009). Hence, biomass-based bioethanol fuels (Hill et al., 2006; Konur, 2012, 2015, 2019, 2020) have increasingly been used in blending gasoline fuels (Hsieh et al., 2002; Najafi et al., 2009), in fuel cells (Antolini, 2007, 2009), and in biochemical production (Angelici et al., 2013; Morschbacker, 2009) in a biorefinery context (Fernando et al., 2006; Huang et al., 2008).

However, it is necessary to pretreat the biomass (Alvira et al., 2010; Taherzadeh and Karimi, 2008) to enhance the yield of the bioethanol (Hahn-Hagerdal et al., 2006; Sanchez and Cardona, 2008) prior to bioethanol fuel production from residual sugar feedstocks through hydrolysis (Sun and Cheng, 2002; Taherzadeh and Karimi, 2007) and fermentation (Lin and Tanaka, 2006; Olsson and Hahn-Hagerdal, 1996) of the biomass.

One of the most-studied feedstocks for bioethanol fuels has been residual sugar feedstocks such as sugarcane bagasse and straw, sweet sorghum bagasse, and sugar beet pulp. Research in the field of second generation bioethanol fuels from residual sugar feedstocks has intensified in this context in the key research fronts in the pretreatment of the residual sugar feedstocks (Aguilar et al., 2002; Laser et al., 2002; Lavarack et al., 2002), hydrolysis of residual sugar feedstocks (Aguilar et al., 2002; Lavarack et al., 2002; Rezende et al., 2011), fermentation of residual sugar feedstock-based hydrolysates (Chandel et al., 2007; Laser et al., 2002; Martinez et al., 2000), and production (Chandel et al., 2007; da Silva et al., 2010, Rabelo et al., 2011) and evaluation (Dias et al., 2012; MacRelli et al., 2012) of the second generation bioethanol fuels from residual sugar feedstocks. Further, sugarcane bagasse (Aguilar et al., 2002; Laser et al., 2002; Lavarack et al., 2002), sugarcane straw (da Silva et al.2010; Dias et al., 2009, 2012), sweet sorghum bagasse (Cao et al., 2012; Li et al., 2010) and to a lesser extent sweet sorghum straw (McIntosh and Vancov, 2010), sugar beet pulp (Micard et al., 1996), sugarcane molasses (Nguyen and Gheewela, 2008), and energy cane bagasse (Qiu et al., 2012) have been studied intensively in this context.

However, it is essential to develop efficient incentive structures (North, 1991) for the primary stakeholders to enhance the research in this field (Konur, 2000, 2002a–c, 2006a,b, 2007a,b). Although there have been a number of review papers on bioethanol fuel production from residual sugar feedstocks (Bezerra and Ragauskas, 2016; Canilha et al., 2012; Cardona et al., 2010), there has been no review of the 25 most-cited papers in this field.

DOI: 10.1201/9781003226550-83

Thus, this chapter presents a review of the 25 most-cited articles in the field of bioethanol fuel production from residual sugar feedstocks. Then, it discusses the key findings of these highly influential papers and comments on future research priorities in this field.

62.2 MATERIALS AND METHODS

The search for this study was carried out using the Scopus database (Burnham, 2006) in July 2022.

As the first step for the search of the relevant literature, keywords were selected using the 200 most-cited population papers. The selected keyword list was then optimized to obtain a representative sample of papers in this research field. This final keyword set was provided in the appendix of Konur (2023) for future replication studies.

As the second step, a sample dataset was used in this study. Twenty-five articles with at least 157 citations each were selected for the review study. Key findings from each paper were taken from the abstracts of these papers and were discussed. Additionally, a number of brief conclusions were drawn and a number of relevant recommendations were made to enhance the future research landscape.

62.3 RESULTS

Brief information about the 25 most-cited papers with at least 157 citations each on second generation bioethanol fuels from residual sugar feedstocks is given below. The primary research fronts are the hydrolysis of residual sugar feedstocks and the production and evaluation of bioethanol fuels from residual sugar feedstocks with 12 and 12 highly cited papers (HCPs), respectively. There is also one HCP for pretreatments of residual sugar feedstocks. Further, there are six HCPs each for the acid and enzymatic hydrolysis of residual sugar feedstocks. Similarly, there are ten and two HCPs for the production and evaluation of bioethanol fuels from residual sugar feedstocks, respectively.

62.3.1 Biomass Pretreatments and Hydrolysis

Brief information about the 13 most-cited papers on the pretreatment and hydrolysis of residual sugar feedstocks with at least 157 citations each is given below (Table 62.1). There are six HCPs each for the acid and enzymatic hydrolysis of residual sugar feedstocks. Additionally, there is one HCP for the pretreatment of residual sugar feedstocks.

62.3.1.1 Biomass Pretreatments

Chen et al. (2011) studied the impact of dilute sulfuric acid (H_2SO_4) pretreatment on the sugarcane bagasse structure using microwave heating in a paper with 235 citations. They employed three reaction temperatures of 130°C, 160°C, and 190°C with two heating times of 5 and 10 min. They observed that an increase in reaction temperature destroyed the lignocellulosic structure of bagasse in a significant way. Further, when the reaction temperature was as high as 190°C, fragmentation of particles became fairly pronounced so that the specific surface area of the pretreated biomass grew substantially. Meanwhile, almost all hemicellulose was removed from bagasse and the crystalline structure of cellulose disappeared. In contrast, the feature of lignin remained clearly. However, the influence of heating time on the lignocellulosic structure was not significant, indicating that pretreatment with 5 min was sufficiently long.

62.3.1.2 Biomass Hydrolysis

62.3.1.2.1 Acid Hydrolysis

Aguilar et al. (2002) performed a kinetic study of the acid hydrolysis of sugarcane bagasse in a paper with 384 citations. They performed the hydrolysis using H_2SO_4 at several temperatures (100°C, 122°C, and 128°C) and concentrations of the acid (2%, 4%, and 6%). They developed kinetic models

TABLE 62.1
The Pretreatment and Hydrolysis of Residual Sugar Feedstocks

No.	Papers	Biomass/ Hydrolysate	Res. Fronts	Prts.	Parameters	Keywords	Lead Author	Affil.	Cits.
1	Aguilar et al. (2002)	Sugarcane bagasse	Hydrolysis	H_2SO_4	Acid hydrolysis optimization, hydrolysate contents, hydrolysis rate, hemicellulose hydrolysis	Sugar cane bagasse, hydrolysis	Vazquez, Manuel 55188958100	Univ. Santiago Compostela Spain	384
2	Lavarack et al. (2002)	Sugarcane bagasse	Hydrolysis	HCl, H_2SO_4	Acid hydrolysis, xylan hydrolysis, xylose yield, acid type, bagasse particle size	Sugarcane bagasse, hydrolysis	Griffin, Gregory J. 7201836626	RMIT Univ. Australia	368
3	Rezende et al. (2011)	Sugarcane bagasse	Hydrolysis	NaOH, H_2SO_4, enzymes	Enzymatic hydrolysis, cellulose hydrolysis, delignification, bagasse structure	Sugarcane bagasse, digestibility, delignification	Polikarpov, Igor 7006220351	Univ. Sao Paulo Brazil	332
4	Binod et al. (2012)	Sugarcane bagasse	Hydrolysis	Microwave, H_2SO_4, NaOH, enzymes	Pretreatments, enzymatic hydrolysis efficiency, delignification, sugar yield	Sugarcane bagasse, saccharification, fermentable, sugar yield, microwave, pretreatment	Pandey, Ashok 7201771319	CSIR India	279
5	Chen et al. (2011)	Sugarcane bagasse	Pretreatments	Microwave, H_2SO_4	Acid and MW pretreatments, reaction temperature, hemicellulose removal, heating time	Sugarcane bagasse, pretreatment, microwave	Chen, Wei-Hsin 57200873137	Natl. Cheng-Kung Univ. Taiwan	235
6	Gamez et al. (2006)	Sugarcane bagasse	Hydrolysis	H_3PO_4	Acid hydrolysis optimization, hydrolysate content, sugar yield, fermentation inhibitors	Sugar cane bagasse, hydrolysis	Vazquez, Manuel 55188958100	Univ. Santiago Compostela Spain	235
7	Martin et al. (2007)	Sugarcane bagasse	Hydrolysis	Wet oxidation, enzymes	Pretreatment optimization, solubilization efficiency, fermentation inhibitors, sugar yield	Sugarcane bagasse, pretreatment, wet oxidation	Martin, Carlos 56484787200	Inland Norway Univ. Appl. Sci. Norway	219

(Continued)

TABLE 62.1 (*Continued*)
The Pretreatment and Hydrolysis of Residual Sugar Feedstocks

No.	Papers	Biomass/ Hydrolysate	Res. Fronts	Prts.	Parameters	Keywords	Lead Author	Affil.	Cits.
8	McIntosh and Vancov (2010)	Sweet sorghum straw	Hydrolysis	NaOH, β-glucosidase, xylanase	Enzymatic hydrolysis, pretreatment optimization and severity, sugar yield	Sorghum bicolor, saccharification, pretreatment	Vancov, Tony 6508255700	Elizabeth Macarthur Agr. Inst. Australia	212
9	Rodriguez-Chong et al. (2004)	Sugarcane bagasse	Hydrolysis	HNO_3	Acid hydrolysis, sugars, fermentation inhibitors, pretreatment optimization, sugar yield	Sugar cane bagasse, hydrolysis	Vazquez, Manuel 55188958100	Univ. Santiago Compostela Spain	210
10	Qiu et al. (2012)	Energy cane bagasse	Hydrolysis	[EMIM]OAc, Spezyme CP, Novozyme 188	Enzymatic hydrolysis, IL pretreatment optimization, delignification, cellulose and hemicellulose digestibility	Energy cane bagasse, ionic liquid, pretreatment, hydrolysis	Aita, Giovanna M.* 35072415500	Audubon Sugar Inst. USA	204
11	de Moraes Rocha et al. (2011)	Sugarcane bagasse	Hydrolysis	H_2SO_4, CH_3COOH	Acid pretreatment and hydrolysis, hemicellulose removal	Sugarcane bagasse, pretreatment, ethanol	de Moraes Rocha, George J. 36625645900	Natl. Biorenewables Lab. Brazil	186
12	Chen et al. (2012)	Sugarcane bagasse	Hydrolysis	H_2SO_4	Acid hydrolysis optimization, hemicellulose hydrolysis, acid concentration, sugar yield	Sugarcane bagasse, hydrolysis, pretreated, microwave	Chen, Wei-Hsin 57200873137	Natl. Cheng-Kung Univ. Taiwan	170
13	Zhao et al. (2008)	Sugarcane bagasse	Hydrolysis	CH_3CO_3H, enzymes	Enzymatic hydrolysis and digestibility, delignification	Sugarcane bagasse, pretreatment, hydrolysis	Liu, De-Hua 35233867100	Tsinghua Univ. China	157

Prt., Biomass pretreatments; Cits., Number of citations received for each paper; *, Female.

to explain the variation with time of xylose, glucose, acetic acid, and furfural generated in the hydrolysis. They observed that the optimal conditions were 2% H_2SO_4 at 122°C for 24 min, which yielded a solution with 21.6 g xylose/L, 3 g glucose/L, 0.5 g furfural/L, and 3.65 g acetic acid/L. In these conditions, they hydrolyzed above 90% of the hemicelluloses.

Lavarack et al. (2002) performed the acid hydrolysis of sugarcane bagasse hemicellulose to produce xylose, arabinose, glucose, lignin, and furfural in a paper with 368 citations. They used a temperature-controlled digester. The reaction conditions were temperature (80°C–200°C), mass ratio of solid to liquid (1:5 to 1:20), type of bagasse, concentration of acid (0.25%–8% of liquid), type of acid (hydrochloric or sulfuric) and reaction time (10–2,000 min). They found that the most accurate kinetic model of the global reaction for the decomposition of xylan was a simple series hydrolysis of xylan to xylose followed by xylose decomposition. They used similar schemes to model the production of arabinose, glucose, and furfural from hemicellulose. They obtained yields of up to 220 mg xylose/g solid, i.e., about 80% of the theoretical xylose available from the bagasse. Further, the bagasse particle size negligibly affected the rate of hydrolysis while hydrochloric acid was less active for the degradation of xylose compared to H_2SO_4.

Gamez et al. (2006) studied the hydrolysis of sugarcane bagasse using phosphoric acid (H_3PO_4) in a paper with 235 citations. They hydrolyzed it under mild conditions (acid 2%–6%, time 0–300 min and 122°C). They developed kinetic models to describe the course of products of acid hydrolysis and observed that the course of xylose, glucose, arabinose, acetic acid, and furfural were satisfactorily described by the models. The optimal conditions selected were 122°C, 4% H_3PO_4, and 300 min. Under these conditions, they obtained 17.6 g of xylose/L; 2.6 g of arabinose/L; 3.0 g of glucose/L, 1.2 g furfural/L, and 4.0 g acetic acid/L. The efficiency in these conditions was 4.46 g sugars/g inhibitors and the mass fraction of sugars in dissolved solids in the liquid phase was superior to 55%.

Rodriguez-Chong et al. (2004) hydrolyzed sugarcane bagasse using nitric acid (HNO_3) in a paper with 210 citations. They performed this pretreatment at variable concentrations (2%–6%), reaction times (0–300 min), and temperatures (100°C–128°C). They then determined the concentration of sugars released (xylose, glucose, and arabinose) and degradation products (acetic acid and furfural) and obtained the kinetic parameters of mathematical models for predicting them in hydrolysates. Applying the kinetic models obtained, the optimal conditions selected were: 122°C, 6% HNO_3, and 9.3 min. Using these conditions, they obtained 18.6 g xylose/L; 2.04 g arabinose/L; 2.87 g glucose/L; 0.9 g acetic acid/L, and 1.32 g furfural/L. A Comparison of these results with those obtained using sulfuric and hydrochloric acids showed that HNO_3 was the most efficient catalyst for hydrolysis.

de Moraes Rocha et al. (2011) studied the dilute mixed-acid pretreatment of sugarcane bagasse for ethanol production in a paper with 186 citations. They used a mixture of sulfuric and acetic acid (CH_3COOH) and two different solid-to-liquid ratios (1.5:10 and 1:10) in the pretreatment. They observed that both conditions efficiently hydrolyzed the hemicelluloses with removals above 90%. Further, the extractive components were also effectively solubilized, and lignin was only slightly affected. Finally, cellulose degradation was below 15%, which corresponded to the low crystallinity fraction.

Chen et al. (2012) studied the hydrolysis characteristics of sugarcane bagasse pretreated by dilute H_2SO_4 solution in a microwave irradiation environment in a paper with 170 citations. They pretreated the biomass with a dilute H_2SO_4 solution at 180°C for 30 min, with the concentration ranging from 0 to 0.02 M. They observed that a higher acid concentration resulted in the destruction of the bagasse. However, they also observed the buffering capacity possessed by=biomass in the pretreatment. Further, around 40–44 wt% of bagasse was degraded from the pretreatment in which around 80%–98% of hemicellulose was hydrolyzed. In contrast, crystalline cellulose and lignin were hardly affected by pretreatment. The maximum yields of xylose and glucose as well as the minimum furfural selectivity occurred at an acid concentration of 0.005 M.

62.3.1.2.2 Enzymatic Hydrolysis

Rezende et al. (2011) studied the delignification and enzymatic hydrolysis of sugarcane bagasse in a paper with 332 citations. They performed a two-step pretreatment, using diluted H_2SO_4 followed by a delignification process with increasing sodium hydroxide (NaOH) concentrations. They observed that up to 96% and 85% of hemicellulose and lignin fractions, respectively, were removed by this two-step method when NaOH concentrations of 1% (m/v) or higher were used. Further, the efficient lignin removal resulted in an enhanced hydrolysis yield reaching values around 100%. Considering cellulose loss due to the pretreatment (maximum of 30%, depending on the process), the total cellulose conversion increased significantly from 22.0% (value for untreated bagasse) to 72.4%. They also observed the delignification process, with a consequent increase in the cellulose-to-lignin ratio. They further showed that morphological changes contributing to this remarkable improvement occurred as a consequence of lignin removal from the sample. Finally, bagasse dissolution was related to the loss of cohesion between neighboring cell walls, as well as by changes in the inner cell wall structure, such as damaging, hole formation, and loss of mechanical resistance, facilitating liquid and enzyme access to crystalline cellulose.

Binod et al. (2012) performed a short-duration microwave (MW)-assisted pretreatment of sugarcane bagasse to enhance the enzymatic saccharification and fermentable sugar yield in a paper with 279 citations. They compared three types of microwave pretreatments such as microwave-acid, microwave-alkali, and combined microwave-alkali-acid. They found that the MW pretreatment of sugarcane bagasse with 1% NaOH at 600 W for 4 min followed by enzymatic hydrolysis resulted in reducing the sugar yield of 0.665 g/g dry biomass, while combined microwave-alkali-acid treatment with 1% NaOH followed by 1% H_2SO_4, reduced the sugar yield increased to 0.83 g/g dry biomass. Microwave-alkali treatment at 450 W for 5 min resulted in an almost 90% delignification from the bagasse. In conclusion, the combined microwave-alkali-acid treatment for a short duration enhanced the fermentable sugar yield.

Martin et al. (2007) studied the wet oxidation (WO) pretreatment method for enhancing the enzymatic digestibility of sugarcane bagasse in a paper with 219 citations. They observed that WO resulted in an increase in the cellulose content of bagasse as a result of the solubilization of hemicelluloses and lignin. They obtained the highest cellulose content, nearly 70%, in the pretreatment at 195°C, 15 min, and alkaline pH. Pretreatments at 195°C and 15 min solubilized 93%–94% of hemicelluloses and 40%–50% of lignin, while pretreatment at 185°C, 5 min, and alkaline pH solubilized only 30% of hemicelluloses and 20% of lignin. They obtained the highest sugar yield in the liquid fraction, at 16.1 g/100 g at 185°C, 5 min, and acidic pH. The highest formation of carboxylic acids, phenols, and furaldehydes occurred at 195°C, 15 min, and acidic pH. Alkaline pH reduced the formation of furaldehydes, which was irrelevant for most WO conditions. All pretreatment conditions improved the enzymatic convertibility of cellulose and they obtained the highest convertibility, 74.9% in the hydrolysis of the bagasse obtained by pretreatment at 195°C, 15 min, and alkaline pH.

McIntosh and Vancov (2010) studied the enhanced enzymatic hydrolysis of sweet sorghum straw using dilute alkaline pretreatment in a paper with 212 citations. Following pretreatment, they observed that both solids and lignin content was inversely proportional to the severity of the pretreatment while higher temperatures and alkali strength resulted in the maximization of sugar recoveries from the enzymatic hydrolysis of the straw. Total sugar release peaked when sorghum straw was pretreated in 2% NaOH at 121°C for 60 min; representing a 5.6-fold higher yield compared to samples pretreated at 60°C in the absence of alkali. Similarly, 4.3-fold increases in total sugars from samples treated with 2% NaOH at 60°C for 90 min, confirmed the importance of alkali inclusion. The addition of β-glucosidase and xylanase to saccharification mixtures enhanced reaction rates and final sugar yields while reducing cellulase dosage 4-fold. Finally, the hydrolysis efficiency of pretreated solids approached 90% and 95% (w/w) with as little as 2.5 and 5.0 FPU cellulase/g, respectively.

Qiu et al. (2012) studied the effect of ionic liquid (IL) pretreatment on the chemical composition, structure, and enzymatic hydrolysis of energy cane bagasse in a paper with 202 citations. They used 1-ethyl-3-methylimidazolium acetate ([EMIM]OAc). They pretreated the biomass with this IL (5% (w/w)) at 120°C for 30 min followed by hydrolysis with Spezyme CP and Novozyme 188. They observed that the IL-treated biomass resulted in significant lignin removal (32.0%) with slight glucan and xylan losses (8.8% and 14.0%, respectively), and exhibited a much higher enzymatic digestibility (87.0% and 64.3%) than untreated (5.5% and 2.8%) or water-treated (4.0% and 2.1%) energy cane bagasse in terms of both cellulose and hemicellulose digestibilities, respectively. They attributed the enhanced digestibilities of IL-treated biomass to delignification and reduction of cellulose crystallinity.

Zhao et al. (2008) performed the peracetic acid (CH_3CO_3H) pretreatment of sugarcane bagasse for enzymatic hydrolysis in a paper with 157 citations. They analyzed the mechanism of enhancement of enzymatic digestibility caused by this pretreatment. They observed that the delignification of the biomass resulted in an increase in the surface area and a reduction of the irreversible absorption of cellulase, which helped to increase the enzymatic digestibility. Further, the crystallinity of the pretreated samples was increased due to the partial removal of amorphous lignin and hemicelluloses and the probable physical change of cellulose. The effect of acetyl group content on enzymatic digestibility was negligible compared with the degree of delignification and crystallinity. In conclusion, the enhancement of enzymatic digestibility of sugarcane bagasse by this pretreatment was achieved mainly by delignification and an increase in the surface area and exposure of cellulose fibers.

62.3.2 Bioethanol Production and Evaluation

There are 12 HCPs for the production and evaluation of bioethanol fuels from residual sugar feedstocks with at least 169 citations each (Table 62.2). There are 12 and two HCPs for the production and evaluation of bioethanol fuels from residual sugar feedstocks, respectively.

62.3.2.1 Bioethanol Production

Laser et al. (2002) compared the liquid hot water (LHW) and steam explosion (SE) pretreatments of sugarcane bagasse to increase the sugar and ethanol yield in a paper with 475 citations. They performed these pretreatments in a 25 l reactor. Solid concentration ranged from 1% to 8% for LHW pretreatment and was above 50% for SE pretreatment while reaction temperature and time ranged from 170°C to 230°C and 1 to 46 min, respectively. In all cases, they found that LHW pretreatment achieved above 80% conversion by simultaneous saccharification and fermentation (SSF), ≥80% xylan recovery, with no hydrolysate inhibition of glucose fermentation yield. However, the combined effectiveness was not as good for the SE pretreatment due to low xylan recovery. Further, SSF conversion increased and xylan recovery decreased as xylan dissolution increased for both modes. SSF conversion, xylan dissolution, hydrolysate furfural concentration, and hydrolysate inhibition increased, while xylan recovery and hydrolysate pH decreased, as a function of increasing LHW pretreatment solid concentration (1%–8%).

Chandel et al. (2007) detoxified sugarcane bagasse hydrolysate using anion exchange resin, activated charcoal, and laccase to enhance the ethanol yield in a paper with 322 citations. They observed that sugarcane bagasse hydrolysis with 2.5% (v/v) HCl yielded 30.29 g/L total reducing sugars along with various fermentation inhibitors such as furans, phenolics, and acetic acid. They obtained a maximum reduction in furans (63.4%) and total phenolics (75.8%) when they treated the acid hydrolysate with anion exchange resin. Treatment of hydrolysate with activated charcoal caused 38.7% and 57.5% reduction in furans and total phenolics, respectively. Laccase reduced total phenolics (77.5%) without affecting furans and acetic acid content in the hydrolysate. Fermentation of these hydrolysates with *Candida shehatae* NCIM 3501 showed maximum ethanol yield (0.48 g/g)

TABLE 62.2
The Production and Evaluation of Bioethanol Fuels from Residual Sugar Feedstocks

No.	Papers	Biomass/ Hydrolysate	Res. Fronts	Prts.	Yeasts	Parameters	Keywords	Lead Author	Affil.	Cits.
1	Laser et al. (2002)	Sugarcane bagasse	Bioethanol production	LHW, steam	Yeasts	Fermentation efficiency, xylan recovery, fermentation inhibitors, SSF, solid concentration	Sugar cane bagasse, hot water, steam, pretreatments	Lynd, Lee R. 35586183800	Dartmouth Coll. USA	475
2	Chandel et al. (2007)	Sugarcane bagasse	Bioethanol production	HCl	*C. shehatae*	Fermentation inhibitors, hydrolysate detoxification, anion exchange resin, activated charcoal, and laccase, ethanol yield	Sugarcane bagasse, detoxification, hydrolysate, ethanol	Kuhad, Ramesh C. 55663451900	Central Univ. Haryana India	322
3	da Silva et al. (2010)	Sugarcane bagasse and straw	Bioethanol production	Milling, enzymes	*S. cerevisiae*	Milling pretreatment, enzymatic hydrolysis, fermentation, sugar and ethanol yields	Sugarcane bagasse and straw, milling, pretreatment, hydrolysis, ethanol, fermentation.	Bon, Elba P. S.* 7007036976	Fed. Univ. Rio de Janeiro Brazil	268
4	Dias et al. (2012)	Sugarcane bagasse and straw	Bioethanol evaluation	Na	Na	Integrated first- and second generation bioethanol production, technoeconomics	Sugarcane bagasse and trash, ethanol, second generation,	Dias, Marina O. S.* 26654773000	Fed. Univ. Sao Paulo Brazil	260
5	Rabelo et al. (2011)	Sugarcane bagasse	Bioethanol production	Lime, alkaline H_2O_2, enzymes	Na	Bioethanol, biomethane and heat production, pretreatment type, lignin, and bioenergy recovery	Sugarcane bagasse, bioethanol, biorefinery, methane	Rabelo, Sarita C.* 22953880600	Sao Paulo State Univ. Brazil	254
6	Martinez et al. (2000)	Sugarcane bagasse	Bioethanol production	H_2SO_4	*E. coli*	Fermentation inhibitors, hydrolysate detoxification, furfural, HMF	Bagasse, hydrolysates	Ingram, Lonnie O. 7102962097	Univ. Florida USA	245
7	Martin et al. (2002)	Sugarcane bagasse	Bioethanol production	Steam, enzymes	*S. cerevisiae*	Hydrolysate detoxification, fermentation inhibitors, engineered yeasts, ethanol yield, laccase, overliming	Sugarcane bagasse, ethanol, hydrolysates	Jonsson, Leif J. 7102349315	Umea Univ. Sweden	242

(Continued)

TABLE 62.2 (*Continued*)
The Production and Evaluation of Bioethanol Fuels from Residual Sugar Feedstocks

No.	Papers	Biomass/ Hydrolysate	Res. Fronts	Prts.	Yeasts	Parameters	Keywords	Lead Author	Affil.	Cits.
8	Dias et al. (2009)	Sugarcane bagasse and straw	Bioethanol production	Solvent, H_2SO_4	Na	Bioethanol and bioelectricity production, simulation, process integration, fermentation	Sugar cane bagasse, bioethanol	Dias, Marina O. S.* 26654773000	Fed. Univ. Sao Paulo Brazil	217
9	Li et al. (2010)	Sweet sorghum bagasse	Bioethanol production	AFEX, xylanase, cellulase	*S. cerevisiae*	Pretreatment and fermentation optimization, enzymatic hydrolysis efficiency, ethanol yield	Sweet sorghum bagasse, ethanol, ammonia, AFEX, pretreatment	Balan, Venkatesh 15757087100	Univ. Houston USA	192
10	MacRelli et al. (2012)	Sugarcane bagasse and straw	Bioethanol evaluation	Steam, H_3PO_4, enzymes	Yeasts	Technoeconomics of bioethanol production, enzymatic hydrolysis, fermentation, ethanol price,	Sugar cane bagasse and leaves, bioethanol, ethanol process	MacRelli, Stefano 31967691500	Univ. Bologna Italy	189
11	Cao et al. (2012)	Sweet sorghum bagasse	Bioethanol production	H_2O_2, NaOH, autoclaving, enzymes	Yeasts	Pretreatments, enzymatic digestibility, ethanol production, cellulose hydrolysis yield, sugar yield, ethanol concentration	Sweet sorghum bagasse, pretreatment, digestibility, ethanol	Liu, Ronghou 56308244600	Shanghai Jiao Tong Univ. China	169
12	Cheng et al. (2008)	Sugarcane bagasse	Bioethanol production	H_2SO_4	*P. tannophilus*	Acid hydrolysis fermentation, ethanol production, sugar and ethanol yield, hydrolysate detoxification, electrodialysis	Sugarcane bagasse, hydrolysate, ethanol	Zhang, Jian-An 8440790900	East China Univ. Sci. Technol. China	169

Prt., Biomass pretreatments; Na, non available; Cits., Number of citations received for each paper; *, Female.

from ion exchange treated hydrolysate, followed by activated charcoal (0.42 g/g), laccase (0.37 g/g), overliming (0.30 g/g), and neutralized hydrolysate (0.22 g/g).

da Silva et al. (2010) milled sugarcane bagasse and straw for enzymatic hydrolysis and hydrolysate fermentation in a paper with 268 citations. They compared the effectiveness of ball milling (BM) and wet disk milling (WDM). They observed that glucose and xylose hydrolysis yields at optimum conditions for BM-pretreated bagasse and straw were 78.7%, 72.1%, 77.6%, and 56.8%, respectively. Maximum glucose and xylose yields for bagasse and straw using WDM were 49.3%, 36.7%, 68.0%, and 44.9%, respectively. Further, BM improved the enzymatic hydrolysis by decreasing the crystallinity, while the defibrillation effect observed for WDM samples favored enzymatic conversion. They fermented bagasse and straw BM hydrolysates with *Saccharomyces cerevisiae*. They finally observed that ethanol yields using a C6-fermenting strain reached 89.8% and 91.8% for bagasse and straw hydrolysates, respectively, and 82% and 78% when using a C_6/C_5 fermenting strain.

Rabelo et al. (2011) produced bioethanol, biomethane, and heat from sugarcane bagasse in a biorefinery context in a paper with 254 citations. They performed two types of pretreatments: lime ($Ca(OH)_2$) or alkaline hydrogen peroxide (H_2O_2). They pretreated the bagasse, enzymatically hydrolyzed, and used the wastes from pretreatment and hydrolysis to produce biomethane. When the pretreatment was carried out at a bagasse concentration of 4% dry matter (DM), they obtained the highest global biomethane production with peroxide pretreatment: 72.1 L biomethane/kg bagasse. Further, the recovery of lignin from peroxide pretreatment liquor was also the highest, 112.7 g/kg of bagasse. Finally, around 63%–65% of the bioenergy that would be produced by bagasse incineration could be recovered by combining ethanol production with the combustion of lignin and hydrolysis residues, along with the anaerobic digestion of pretreatment liquors, while only 32%–33% of the energy was recovered by bioethanol production alone.

Martinez et al. (2000) optimized the overliming ($Ca(OH)_2$) treatments for the acid-pretreated sugarcane bagasse hydrolysates (primarily pentose sugars) using recombinant *Escherichia coli* LY01 to detoxify them in a paper with 245 citations. They observed a substantial reduction in furfural and hydroxymethylfurfural (HMF) while organic acids (acetic, formic, and levulinic) were not affected. The extent of furan reduction correlated with increasing fermentability. However, furan reduction was not the sole cause for reduced toxicity. After optimal overliming, the bagasse hydrolysate was rapidly and efficiently fermented (above 90% yield) with LY01. They proposed titration of acids, measurement of pH before and after treatment, and furan analyses as relatively simple methods to monitor the reproducibility of hydrolysate preparations and the effectiveness of overliming treatments.

Martin et al. (2002) produced bioethanol from the detoxified enzymatic hydrolysates of sugarcane bagasse using recombinant xylose-utilizing *S. cerevisiae* in a paper with 242 citations. They pretreated bagasse with steam explosion at 205°C and 215°C and hydrolyzed with cellulolytic enzymes. They then detoxified these hydrolysates by enzymatic and chemical detoxification with the phenoloxidase laccase and overliming, respectively. They observed that ~80% of the phenolic compounds were specifically removed by the laccase treatment while overliming partially removed phenolic compounds, but also other fermentation inhibitors such as acetic acid, furfural, and 5-HMF. They then fermented these detoxified hydrolysates with the recombinant xylose-utilizing *S. cerevisiae* laboratory strain TMB 3001. This was a CEN.PK derivative with over-expressed xylulokinase activity and expressing the xylose reductase and xylitol dehydrogenase of *Pichia stipitis*, and the *S. cerevisiae* strain ATCC 96581. They observed that the fermentative performance of the lab strain in the un-detoxified hydrolysate was better than the performance of the industrial strain while they observed an almost two-fold increase in the specific productivity of the strain TMB 3001 in the detoxified hydrolysates compared to the undetoxified hydrolysates. Finally, the ethanol yield in the fermentation of the hydrolysate detoxified by overliming was 0.18 g/g dry bagasse, whereas it reached only 0.13 g/g dry bagasse in the undetoxified hydrolysate.

Dias et al. (2009) produced bioethanol from sugarcane bagasse as an integral part the of conventional bioethanol production process in a paper with 217 citations. They simulated bioethanol production from sugarcane juice and bagasse and considered a typical large-scale production plant: 1000 m^3/day of ethanol production using sugarcane juice. They considered a three-step hydrolysis process (prehydrolysis of hemicellulose, organosolv delignification, and cellulose hydrolysis) of the sugarcane bagasse. They determined the minimum hot utility, obtained with the thermal integration of the plant, in order to find out the maximum availability of bagasse that could be used in the hydrolysis process, taking into consideration the use of 50% of generated sugarcane straw as fuel for bioelectricity and biosteam production. They then analyzed two different cases for the product purification step: conventional and double-effect distillation systems. They found that the double-effect distillation system allowed 90% of generated bagasse to be used as a feedstock in the hydrolysis plant, which accounted for an increase of 26% in bioethanol production, considering exclusively the fermentation of hexoses obtained from the cellulosic fraction.

Li et al. (2010) converted forage and sweet sorghum bagasse to ethanol with the ammonia fiber expansion (AFEX) pretreatment in a paper with 192 citations. They optimized both the pretreatment and fermentation processes. They observed that using xylanase with cellulase during enzymatic hydrolysis increased both glucan and xylan conversion to 90% at 1% glucan loading. They then fermented the high-solid loading hydrolysates using *S. cerevisiae* 424A (LNH-ST) without any external nutrient supplementation or detoxification. They found that this yeast strain utilized xylose at pH 6.0 than at pH 4.8, but glycerol production was higher for the former pH than the latter. Finally, the maximum final bioethanol concentration in the fermentation broth was 30.9 g/L (forage sorghum) and 42.3 g/L (sweet sorghum bagasse).

Cao et al. (2012) compared the effects of five pretreatment methods on enhancing the enzymatic digestibility and ethanol production from sweet sorghum bagasse in a paper with 169 citations. They performed dilute NaOH solution autoclaving pretreatment, high concentration NaOH solution immersing pretreatment, dilute NaOH solution autoclaving and H_2O_2 immersing pretreatment, alkaline peroxide pretreatment, and autoclaving pretreatment. The best result was obtained with the dilute NaOH solution autoclaving and H_2O_2 immersing pretreatment. The highest cellulose hydrolysis yield, total sugar yield, and ethanol concentration were 74.29%, 90.94 g sugar/100 g dry matter, and 6.12 g/L, respectively, which were 5.88, 9.54, and 19.13 times higher than the control. They finally observed significant molecule and surface structure changes in the sweet sorghum bagasse after pretreatments.

Cheng et al. (2008) performed the acid hydrolysis of sugarcane bagasse in an acid recycling process and detoxification of hydrolysate by electrodialysis in a paper with 169 citations. They observed that two cycles of acidic treatments increased the reducing sugar concentration from 28 to 63.5 g/L and H_2SO_4 consumption decreased to 0.056 g/g bagasse. After treatment by electrodialysis, 90% of the acetic acid in the hydrolysate was removed and the recovery ratio of H_2SO_4 was 88%. They then fermented the detoxified hydrolysates, supplemented with nutrient materials, to ethanol using *Pachysolen tannophilus* DW06. A batch culture with the detoxified hydrolysate gave 19 g ethanol/L with a yield of 0.34 g/g sugar and productivity of 0.57 g/L h.

62.3.2.2 Bioethanol Evaluation

Dias et al. (2012) compared the stand-alone second generation bioethanol production from sugarcane bagasse and straw with conventional first generation bioethanol production from sugarcane and with an integrated first- and second generation bioethanol production in a paper with 260 citations. They found that the integrated first- and second generation ethanol production process from sugarcane led to better economic results when compared with the stand-alone plant, especially when advanced hydrolysis technologies and pentose fermentation were included.

MacRelli et al. (2012) performed a technoeconomic evaluation of second generation (2G) bioethanol production from sugarcane bagasse and straw integrated with a sugar-based ethanol process in a paper with 189 citations. They performed simulations to investigate how process integration could affect the minimum ethanol selling price of this 2G process (MESP-2G), as well as to improve the plant's energy efficiency. They pretreated the biomass with steam and phosphoric acid (H_3PO_4). They found that the addition of a steam dryer, doubling of the enzyme dosage in enzymatic hydrolysis, including leaves as raw material in the 2G process, heat integration, and the use of more energy-efficient equipment led to a 37% reduction in MESP-2G compared to the base case. Further, the MESP for 2G ethanol was 0.97 US$/L, while in the future it could be reduced to 0.78 US$/L. In this case, the overall production cost of 1G+2G ethanol would be about 0.40 US$/L with an output of 102 L/ton dry sugar cane including 50% leaves. A 50% decrease in the cost of enzymes, electricity, or leaves would lower the MESP-2G by about 20%, 10%, and 4.5%, respectively. In conclusion, the production of 2G bioethanol from sugarcane bagasse and leaves in Brazil was competitive (without subsidies) compared with 1G starch-based bioethanol production in Europe. Further, 2G bioethanol could be produced at a lower cost if subsidies were used to compensate for the opportunity cost from the sale of excess electricity and if the cost of enzymes continued to fall.

62.4 DISCUSSION

62.4.1 Introduction

Crude oil-based gasoline fuels have been widely used in the transportation sector since the 1920s. However, there have been great public concerns over the adverse environmental and human impact of these fuels. Hence, biomass-based bioethanol fuels have increasingly been used in blending gasoline and petrodiesel fuels, in fuel cells, and in biochemical production in a biorefinery context.

However, it is necessary to pretreat the biomass to enhance the yield of bioethanol prior to bioethanol fuel production from residual sugar feedstocks through the hydrolysis and fermentation of the biomass. One of the most-studied feedstocks for bioethanol fuels has been residual sugar feedstocks such as sugarcane bagasse and straw, sweet sorghum bagasse, and sugar beet pulp. Research in the field of second generation bioethanol fuels from residual sugar feedstocks has intensified in this context in key research fronts for the pretreatment of the residual sugar feedstocks, hydrolysis of the residual sugar feedstocks, fermentation of the residual sugar feedstock-based hydrolysates, and the production and evaluation of second generation bioethanol fuels from residual sugar feedstocks. Further, sugarcane bagasse and straw, sweet sorghum bagasse, and to a lesser extent sweet sorghum straw, sugar beet pulp, sugarcane molasses, and energy cane bagasse have been studied extensively in this context.

However, it is essential to develop efficient incentive structures for the primary stakeholders to enhance the research in this field. Although there have been a number of review papers in this field, there has been no review of the 25 most-cited articles in this field.

Thus, this chapter presents a review of the 25 most-cited articles on bioethanol fuel production and evaluation from residual sugar feedstocks. Then, it discusses the key findings of these highly influential papers and comments on future research priorities in this field.

As the first step for the search of the relevant literature, keywords were selected using the 200 most-cited population papers. The selected keyword list was then optimized to obtain a representative sample of papers from this research field. This keyword list was provided in the appendix of Konur (2023) for future replicative studies.

As the second step, a sample dataset was used in this study. The first 25 articles with at least 157 citations each were selected for the review study. Key findings from each paper were taken from the abstracts of these papers and were discussed. Additionally, a number of brief conclusions were drawn and a number of relevant recommendations were made to enhance the future research landscape.

Information about the research fronts in the sample papers in bioethanol fuel production from residual sugar feedstocks with regard to residual sugar feedstocks used in these processes is given in Table 62.3. As this table shows there are five primary research fronts for this field: sugarcane

TABLE 62.3
The Most Prolific Research Fronts for the Second Generation Bioethanol Fuels from Residual Sugar Feedstocks

No.	Research Fronts	N Paper % Review	N Paper % Sample	Surplus (%)
1	Sugarcane residues	84.0	86.0	–2.0
	Sugarcane bagasse	84.0	78.6	5.4
	Sugarcane straw	16.0	12.4	3.6
	Sugarcane molasses	0.0	2.0	–2.0
	Sugarcane vinasse	0.0	1.0	–1.0
2	Sweet sorghum residues	12.0	7.5	4.5
	Sweet sorghum bagasse	8.0	7.0	1.0
	Sweet sorghum straw	4.0	0.5	3.5
3	Sugar beet residues	0.0	3	–3.0
	Sugar beet pulp	0.0	2.0	–2.0
	Sugar beet molasses	0.0	1.0	–1.0
4	Energy cane residues	4.0	1	3.0
	Energy cane bagasse	4.0	1.0	3.0

N Paper (%) review, The number of papers in the sample of 25 reviewed papers. N paper (%) sample: The number of papers in the population sample of 201 papers.

bagasse and residues, and to a lesser extent sugarcane straw, sweet sorghum bagasse, sweet sorghum straw, and energy cane bagasse residues with 86%, 16%, 8%, 4%, and 4% of the HCPs, respectively.

On the other hand, on an individual basis, sugarcane bagasse and straw are the most prolific residual feedstocks with 84% and 16% of these HCPs, respectively. Further, sugarcane bagasse and straw, sweet sorghum straw, and energy cane bagasse are the most influential residual feedstocks with 4%–5% of the sample papers each.

Information about the thematic research fronts for sample papers in bioethanol fuel production from residual sugar feedstocks is given in Table 62.4. As this table shows there are four primary research fronts for this field: The pretreatment and hydrolysis of residual biomass, fermentation of residual hydrolysates, and bioethanol production with 96%, 64%, 48%, and 36% of these HCPs, respectively. Another minor research front is the evaluation of bioethanol fuels with 8% of these HCPs.

On the other hand, on an individual basis, the most prolific research fronts are chemical pretreatments, acid pretreatment, enzymatic pretreatments, enzymatic hydrolysis, acid hydrolysis, alkaline pretreatment, and hydrothermal pretreatments with 16%–80% of these HCPs.

Further, the most influential research fronts are acid pretreatment, biomass pretreatments, hydrolysate fermentation, chemical pretreatments, and acid hydrolysis with a 15%–29% surplus. Similarly, bioethanol fuel evaluation enzymatic pretreatments, enzymatic hydrolysis, hydrothermal pretreatments, and LHW are the least influential research fronts with a 5%–9% deficit each.

62.4.2 Biomass Pretreatments and Hydrolysis

Brief information about the 13 most-cited papers on the pretreatment and hydrolysis of residual sugar feedstocks with at least 157 citations each is given below (Table 62.1). There are six HCPs each for acid and enzymatic hydrolysis of residual sugar feedstocks. Additionally, there is one HCP for the pretreatment of residual sugar feedstocks.

These HCPs show a sample of the research on the pretreatment and hydrolysis of residual sugar feedstocks. Pretreatment and hydrolysis processes are a fundamental part of the production of

TABLE 62.4
The Most Prolific Thematic Research Fronts for the Residual Sugar Feedstock-based Bioethanol Fuels

No.	Research Fronts	N Paper % Review	N Paper % Sample	Surplus (%)
1	Biomass pretreatments	96.0	72.6	23.4
1.1	Hydrothermal pretreatments	16.0	23.9	–7.9
	Steam explosion	12.0	12.9	–0.9
	LHW	4.0	9.0	–5.0
	Other pretreatments	4.0	2.0	2.0
1.2	Chemical pretreatments	80.0	61.2	18.8
	Acid pretreatment	60.0	31.3	28.7
	Alkaline pretreatment	20.0	22.9	–2.9
	Solvent pretreatment	4.0	5.0	–1.0
	IL pretreatment	4.0	6.0	–2.0
	Ammonia pretreatment	4.0	3.0	1.0
	SO_2 pretreatment	0.0	1.5	–1.5
	Ozonolysis	0.0	1.5	–1.5
	Surfactants	0.0	3.0	–3.0
	Other pretreatments	0.0	2.5	–2.5
1.3	Mechanical pretreatments	12.0	10.9	1.1
	Microwave	8.0	6.0	2.0
	Milling	4.0	1.5	2.5
	Ultrasound	0.0	3.5	–3.5
1.4	Enzymatic pretreatments	48.0	56.2	–8.2
1.5	Feedstock genetic engineering	0.0	2.0	–2.0
1.6	Yeast microbial engineering	4.0	2.0	2.0
2	Biomass hydrolysis	64.0	65.7	–1.7
	Enzymatic hydrolysis	48.0	56.2	–8.2
	Acid hydrolysis	24.0	9.5	14.5
3	Hydrolysate fermentation	48.0	25.9	22.1
4	Bioethanol production	36.0	33.3	2.7
5	Bioethanol fuel evaluation	8.0	16.9	–8.9

N Paper (%) review: The number of papers in the sample of 25 reviewed papers. N paper (%) sample: The number of papers in the population sample of 201 papers.

bioethanol fuels from these feedstocks as they fractionate and hydrolyze these feedstocks, respectively, to produce fermentable sugars with a high sugar yield for bioethanol fuel production.

62.4.2.1 Biomass Pretreatments

Chen et al. (2011) studied the impact of dilute H_2SO_4 pretreatment on the sugarcane bagasse structure using microwave heating and observed that an increase in reaction temperature destroyed the lignocellulosic structure of bagasse in a significant way.

62.4.2.2 Biomass Hydrolysis

62.4.2.2.1 Acid Hydrolysis

Aguilar et al. (2002) performed a kinetic study of the acid hydrolysis of sugarcane bagasse and determined the optimal hydrolysis conditions. Further, Lavarack et al. (2002) performed the acid hydrolysis of sugarcane bagasse hemicellulose and found that the most accurate kinetic model of the global reaction for the decomposition of xylan was a simple series hydrolysis of xylan to xylose followed by xylose decomposition.

Gamez et al. (2006) studied the hydrolysis of sugarcane bagasse using H_3PO_4 and observed that the course of xylose, glucose, arabinose, acetic acid, and furfural were satisfactorily described by the models. Further, Rodriguez-Chong et al. (2004) hydrolyzed sugarcane bagasse using HNO_3 and obtained xylose, arabinose, glucose, acetic acid, and furfural.

de Moraes Rocha et al. (2011) studied the dilute mixed-acid pretreatment of sugarcane bagasse for ethanol production and observed that both conditions efficiently hydrolyzed the hemicelluloses with removals above 90%. Further, Chen et al. (2012) studied the hydrolysis characteristics of sugarcane bagasse pretreated with dilute H_2SO_4 solution in a microwave irradiation environment and observed that a higher acid concentration resulted in the destruction of the bagasse.

62.4.2.2.2 Enzymatic Hydrolysis

Rezende et al. (2011) studied the delignification and enzymatic hydrolysis of sugarcane bagasse and observed that up to 96% and 85% of hemicellulose and lignin fractions, respectively, were removed by this two-step method. Further, Binod et al. (2012) performed a short-duration microwave-assisted pretreatment of sugarcane bagasse and found that the combined microwave-alkali-acid treatment with 1% NaOH followed by 1% H_2SO_4, reduced the sugar yield increased to 0.83 g/g dry biomass.

Martin et al. (2007) studied the wet oxidation pretreatment method for enhancing the enzymatic digestibility of sugarcane bagasse and observed that this pretreatment resulted in an increase in the cellulose content of bagasse as a result of the solubilization of hemicelluloses and lignin. Further, McIntosh and Vancov (2010) studied the enhanced enzymatic hydrolysis of sweet sorghum straw using a dilute alkaline pretreatment and observed that both solids and lignin content were inversely proportional to the severity of the pretreatment.

Qiu et al. (2012) studied the effect of IL pretreatment on the chemical composition, structure, and enzymatic hydrolysis of energy cane bagasse and observed that the IL-treated biomass resulted in significant lignin removal. Further, Zhao et al. (2008) performed the CH_3CO_3H pretreatment of sugarcane bagasse for enzymatic hydrolysis and observed that the delignification of the biomass resulted in an increase in the surface area and reduction of the irreversible absorption of cellulase, which helped to increase the enzymatic digestibility.

62.4.3 Bioethanol Production and Evaluation

There are 12 HCPs for the production and evaluation of bioethanol fuels from residual sugar feedstocks with at least 169 citations each (Table 62.2). There are 12 and two HCPs for the production and evaluation of bioethanol fuels from residual sugar feedstocks, respectively.

These HCPs show a sample of the research on the production and evaluation of bioethanol fuels from residual sugar feedstocks. As the hydrolysates of residual sugar feedstocks include a number of fermentation inhibitors, the production process involves both the detoxification of these hydrolysates to remove these fermentation inhibitors and the fermentation of these hydrolysates by a suitable yeast strain. Thus, the detoxification and fermentation processes are fundamental parts of the bioethanol production process to improve the ethanol yield from residual sugar feedstocks.

Further, as bioethanol fuels have been used in vehicles, blending them with gasoline or diesel fuels, the studies on the evaluation of these bioethanol fuels are a significant research front together with the research front of bioethanol fuel production. In this context, residual sugar feedstocks are used to produce bioethanol and/or bioelectricity and biosteam. These studies are usually concerned with the technoeconomics, environmental, and social impact of the utilization of these bioethanol fuels and bioelectricity. These studies show that the use of residual sugar feedstocks increases the competitiveness of bioethanol fuels compared to crude oil-based gasoline and petrodiesel fuels.

62.4.3.1 Bioethanol Production

Laser et al. (2002) compared the LHW and SE pretreatments of sugarcane bagasse to increase the sugar and ethanol yield and found that LHW pretreatment achieved above 80% conversion by SSF, ≥80% xylan recovery, with no hydrolysate inhibition of glucose fermentation yield. Further, Chandel et al. (2007) detoxified the sugarcane bagasse hydrolysate using anion exchange resin, activated charcoal, and laccase to enhance the ethanol yield and observed that sugarcane bagasse hydrolysis with 2.5% (v/v) HCl yielded 30.29 g/L total reducing sugars along with various fermentation inhibitors.

da Silva et al. (2010) milled sugarcane bagasse and straw and observed that glucose and xylose hydrolysis yields at optimum conditions for BM-pretreated bagasse and straw were 78.7% and 72.1% and 77.6% and 56.8%, respectively. Further, Rabelo et al. (2011) produced bioethanol, biomethane, and heat from sugarcane bagasse in a biorefinery context and obtained the highest global biomethane production with peroxide pretreatment.

Martinez et al. (2000) optimized the overliming pretreatment for the acid-pretreated sugarcane bagasse hydrolysates using recombinant *Escherichia coli* LY01 to detoxify them and observed a substantial reduction in furfural and HMF while organic acids were not affected. Further, Martin et al. (2002) produced bioethanol from detoxified enzymatic hydrolysates of sugarcane bagasse using recombinant xylose-utilizing *S. cerevisiae* and observed that ~80% of the phenolic compounds were specifically removed by the laccase treatment while overliming partially removed phenolic compounds, but also other fermentation inhibitors.

Dias et al. (2009) produced bioethanol from sugarcane bagasse as an integral part of the conventional bioethanol production process and found that the double-effect distillation system allowed 90% of generated bagasse to be used as a feedstock in the hydrolysis plant, which accounted for an increase of 26% in bioethanol production. Further, Li et al. (2010) converted forage and sweet sorghum bagasse to ethanol with the AFEX pretreatment and observed that using xylanase with cellulase increased both glucan and xylan conversion to 90% at 1% glucan loading.

Cao et al. (2012) compared the effects of five pretreatment methods on enhancing the enzymatic digestibility and ethanol production from sweet sorghum bagasse and the best result was obtained with dilute NaOH solution autoclaving and H_2O_2 immersing pretreatment. Further, Cheng et al. (2008) performed the acid hydrolysis of the sugarcane bagasse in an acid recycle process and the detoxification of hydrolysate by electrodialysis and observed that two cycles of acid pretreatments increased the reducing sugar concentration and reduced H_2SO_4 consumption.

62.4.3.2 Bioethanol Evaluation

Dias et al. (2012) compared stand-alone second generation bioethanol production from sugarcane bagasse and straw with conventional first generation bioethanol production from sugarcane and with integrated first- and second generation bioethanol production and found that the integrated first- and second generation ethanol production process from sugarcane led to better economic results when compared with the stand-alone plant. Further, MacRelli et al. (2012) performed a technoeconomic evaluation of second generation (2G) bioethanol production from sugarcane bagasse and straw integrated with the sugar-based ethanol process and found that the addition of a steam dryer, doubling of the enzyme dosage in enzymatic hydrolysis, including leaves as raw material in the 2G process, heat integration, and the use of more energy-efficient equipment led to a 37% reduction in the ethanol price compared to the base case.

62.5 CONCLUSION AND FUTURE RESEARCH

Brief information about the key research fronts covered by the 25 most-cited papers with at least 157 citations each is given under two primary headings: The pretreatment and hydrolysis of residual sugar feedstocks and the production and evaluation of bioethanol fuels.

The usual characteristics of these HCPs are that the pretreatments and hydrolysis of residual sugar feedstocks and the fermentation of the resulting hydrolysates are the primary processes for bioethanol fuel production from residual sugar feedstocks to improve the ethanol yield as residual sugar feedstocks are one of the most-studied feedstocks for bioethanol production, especially for countries with large farmlands and crude oil deficiency such as Brazil.

The key findings on these research fronts should be read in light of the increasing public concerns about climate change, GHG emissions, and global warming as these concerns have certainly been behind the boom in research on bioethanol fuel production from residual sugar feedstocks as an alternative to crude oil-based gasoline and petrodiesel fuels in the past decades. It is also a sustainable alternative to first generation food crop-based bioethanol fuels such as corn grain-based bioethanol fuels. The recent supply shocks caused by the COVID-19 pandemic and the Russian invasion of Ukraine also highlight the importance of the production and utilization of bioethanol fuels from residual sugar feedstocks as an alternative to crude oil-based gasoline and diesel fuels.

There are five primary research fronts for this field: sugarcane bagasse and to a lesser extent sugarcane straw, sweet sorghum bagasse, residues, and sugar beet and energy cane residues. It is notable that energy cane is a genetically engineered version of sugarcane (Matsuoka et al., 2014).

Similarly, there are four primary research fronts for this field: The pretreatment and hydrolysis of the residual biomass, fermentation of the residual hydrolysates, and bioethanol production. Another minor research front is the evaluation of bioethanol fuels. On the other hand, on an individual basis, the most prolific research fronts are chemical pretreatments, acid pretreatment, enzymatic pretreatments, enzymatic hydrolysis, acid hydrolysis, alkaline pretreatment, and hydrothermal pretreatments.

These studies emphasize the importance of proper incentive structures for efficient bioethanol fuel production from residual sugar feedstocks in light of North's institutional framework (North, 1991). In this context, the major producers and users of bioethanol fuels such as the USA, Brazil, and Canada with vast farmlands have developed strong incentive structures for efficient bioethanol fuel production from residual sugar feedstocks and their residues.

In light of the supply shocks caused primarily by the COVID-19 pandemic and the Russian invasion of Ukraine, it is expected that the incentive structures such as public funding would be enhanced to increase the share of bioethanol fuels in the global fuel portfolio as a strong alternative to crude oil-based gasoline and diesel fuels. In this context, it is expected that the most prolific researchers, institutions, countries, funding bodies, and journals in this field would have a first-mover advantage to benefit from such potential incentives. This is especially true for Brazilian stakeholders as Brazil has become the global leader in both the production and utilization of second generation bioethanol fuels from residual sugar feedstocks.

It is recommended that such review studies are performed for the primary research fronts of bioethanol fuel production from residual sugar feedstocks.

ACKNOWLEDGMENTS

The contribution of the highly cited researchers in the field of bioethanol fuel production from residual sugar feedstocks has been gratefully acknowledged.

REFERENCES

Aguilar, R., J. A. Ramirez, G. Garrote and M. Vazquez. 2002. Kinetic study of the acid hydrolysis of sugar cane bagasse. *Journal of Food Engineering* 55:309–318.

Alvira, P., E. Tomas-Pejo, M. Ballesteros and M. J. Negro. 2010. Pretreatment technologies for an efficient bioethanol production process based on enzymatic hydrolysis: A review. *Bioresource Technology* 101:4851–4861.

Angelici, C., B. M. Weckhuysen and P. C. A. Bruijnincx. 2013. Chemocatalytic conversion of ethanol into butadiene and other bulk chemicals. *ChemSusChem* 6:1595–1614.

Antolini, E. 2007. Catalysts for direct ethanol fuel cells. *Journal of Power Sources* 170:1–12.
Antolini, E. 2009. Palladium in fuel cell catalysis. *Energy and Environmental Science* 2:915–931.
Bezerra, T. L. and A. J. Ragauskas. 2016. A review of sugarcane bagasse for second-generation bioethanol and biopower production. *Biofuels, Bioproducts and Biorefining* 10:634–647.
Binod, P., K. Satyanagalakshmi and R. Sindhu, et al. 2012. Short duration microwave assisted pretreatment enhances the enzymatic saccharification and fermentable sugar yield from sugarcane bagasse. *Renewable Energy* 37:109–116.
Burnham, J. F. 2006. Scopus database: A review. *Biomedical Digital Libraries* 3:1–8.
Canilha, L., A. K. Chandel and T. S. dos Santos Milessi, et al. 2012. Bioconversion of sugarcane biomass into ethanol: An overview about composition, pretreatment methods, detoxification of hydrolysates, enzymatic saccharification, and ethanol fermentation. *Journal of Biomedicine and Biotechnology* 2012:989572.
Cao, W., C. Sun, R. Liu, R. Yin and X. Wu. 2012. Comparison of the effects of five pretreatment methods on enhancing the enzymatic digestibility and ethanol production from sweet sorghum bagasse. *Bioresource Technology* 111:215–221.
Cardona, C. A., J. A. Quintero and I. C. Paz. 2010. Production of bioethanol from sugarcane bagasse: Status and perspectives. *Bioresource Technology* 101:4754–4766.
Chandel, A. K., R. K. Kapoor, A. Singh and R. C. Kuhad. 2007. Detoxification of sugarcane bagasse hydrolysate improves ethanol production by *Candida shehatae* NCIM 3501. *Bioresource Technology* 98:1947–1950.
Chen, W. H., S. C. Ye and H. K. Sheen. 2012. Hydrolysis characteristics of sugarcane bagasse pretreated by dilute acid solution in a microwave irradiation environment. *Applied Energy* 93:237–244.
Chen, W. H., Y. J. Tu and H. K. Sheen. 2011. Disruption of sugarcane bagasse lignocellulosic structure by means of dilute sulfuric acid pretreatment with microwave-assisted heating. *Applied Energy* 88:2726–2734.
Cheng, K. K., B. Y. Cai and J. A. Zhang, et al. 2008. Sugarcane bagasse hemicellulose hydrolysate for ethanol production by acid recovery process. *Biochemical Engineering Journal* 38:105–109.
da Silva, A. S., H. Inoue, T. Endo, S. Yano and E. P. S. Bon. 2010. Milling pretreatment of sugarcane bagasse and straw for enzymatic hydrolysis and ethanol fermentation. *Bioresource Technology* 101:7402–7409.
de Moraes Rocha, G. J., C. Martin and I. B. Soares, et al. 2011. Dilute mixed-acid pretreatment of sugarcane bagasse for ethanol production. *Biomass and Bioenergy* 35:663–670.
Dias, M. O. S., A. V. Ensinas and S. A. Nebra, et al. 2009. Production of bioethanol and other bio-based materials from sugarcane bagasse: Integration to conventional bioethanol production process. *Chemical Engineering Research and Design* 87:1206–1216.
Dias, M. O. S., T. L. Junqueira and O. Cavalett, et al. 2012. Integrated versus stand-alone second generation ethanol production from sugarcane bagasse and trash. *Bioresource Technology* 103:152–161.
Fernando, S., S. Adhikari, C. Chandrapal and M. Murali. 2006. Biorefineries: Current status, challenges, and future direction. *Energy & Fuels* 20:1727–1737.
Gamez, S., J. J. Gonzalez-Cabriales, J. A. Ramirez, O. Garrote and M. Vazquez. 2006. Study of the hydrolysis of sugar cane bagasse using phosphoric acid. *Journal of Food Engineering* 74:78–88.
Hahn-Hagerdal, B., M. Galbe, M. F. Gorwa-Grauslund, G. Liden and G. Zacchi. 2006. Bio-ethanol: The fuel of tomorrow from the residues of today. *Trends in Biotechnology* 24:549–556.
Hill, J., E. Nelson, D. Tilman, S. Polasky and D. Tiffany. 2006. Environmental, economic, and energetic costs and benefits of biodiesel and ethanol biofuels. *Proceedings of the National Academy of Sciences of the United States of America* 103:11206–11210.
Hill, J., S. Polasky and E. Nelson, et al. 2009. Climate change and health costs of air emissions from biofuels and gasoline. *Proceedings of the National Academy of Sciences of the United States of America* 106:2077–2082.
Hsieh, W. D., R. H. Chen, T. L. Wu and T. H. Lin. 2002. Engine performance and pollutant emission of an SI engine using ethanol-gasoline blended fuels. *Atmospheric Environment* 36:403–410.
Huang, H. J., S. Ramaswamy, U. W. Tschirner and B. V. Ramarao. 2008. A review of separation technologies in current and future biorefineries. *Separation and Purification Technology* 62:1–21.
Konur, O. 2000. Creating enforceable civil rights for disabled students in higher education: An institutional theory perspective. *Disability & Society* 15:1041–1063.
Konur, O. 2002a. Access to nursing education by disabled students: Rights and duties of nursing programs. *Nurse Education Today* 22:364–374.
Konur, O. 2002b. Assessment of disabled students in higher education: Current public policy issues. *Assessment and Evaluation in Higher Education* 27:131–52.
Konur, O. 2002c. Access to employment by disabled people in the UK: Is the disability discrimination act working? *International Journal of Discrimination and the Law* 5:247–279.
Konur, O. 2006a. Participation of children with dyslexia in compulsory education: Current public policy issues. *Dyslexia* 12:51–67.

Konur, O. 2006b. Teaching disabled students in higher education. *Teaching in Higher Education* 11:351–363.

Konur, O. 2007a. A judicial outcome analysis of the disability discrimination act: A windfall for the employers? *Disability & Society* 22:187–204.

Konur, O. 2007b. Computer-assisted teaching and assessment of disabled students in higher education: The interface between academic standards and disability rights. *Journal of Computer Assisted Learning* 23:207–219.

Konur, O. 2012. The evaluation of the research on the bioethanol: A scientometric approach. *Energy Education Science and Technology Part A: Energy Science and Research* 28:1051–1064.

Konur, O. 2015. Current state of research on algal bioethanol. In *Marine Bioenergy: Trends and Developments*, Eds. S. K. Kim and C. G. Lee, pp. 217–244. Boca Raton, FL: CRC Press.

Konur, O. 2019. Cyanobacterial bioenergy and biofuels science and technology: A scientometric overview. In Cyanobacteria: From Basic Science to Applications, Eds. A. K. Mishra, D. N. Tiwari and A. N. Rai, pp. 419–442. Amsterdam: Elsevier.

Konur, O. 2020. The scientometric analysis of the research on the bioethanol production from green macroalgae. In *Handbook of Algal Science, Technology and Medicine,* Ed. O. Konur, pp. 385–401. London: Academic Press.

Konur, O. 2023. Second generation bioethanol fuels from residual sugar feedstocks: Scientometric study. In *Feedstock-based Bioethanol Fuels. II. Waste Feedstocks: Agricultural, Food, Industrial, Urban, Forestry, and Lignocellulosic Waste-based Bioethanol Fuels. Handbook of Bioethanol Fuels Volume 4*, Ed. O. Konur, pp. 53–78. Boca Raton, FL: CRC Press.

Laser, M., D. Schulman and S. G. Allen, et al. 2002. A comparison of liquid hot water and steam pretreatments of sugar cane bagasse for bioconversion to ethanol. *Bioresource Technology* 81:33–44.

Lavarack, B. P., G. J. Griffin and D. Rodman. 2002. The acid hydrolysis of sugarcane bagasse hemicellulose to produce xylose, arabinose, glucose and other products. *Biomass and Bioenergy* 23:367–380.

Li, B. Z., V. Balan, Y. J. Yuan and B. E. Dale. 2010. Process optimization to convert forage and sweet sorghum bagasse to ethanol based on ammonia fiber expansion (AFEX) pretreatment. *Bioresource Technology* 101:1285–1292.

Lin, Y. and S. Tanaka. 2006. Ethanol fermentation from biomass resources: Current state and prospects. *Applied Microbiology and Biotechnology* 69:627–642.

Ma, X., L. Sun and C. Song. 2002. A new approach to deep desulfurization of gasoline, diesel fuel and jet fuel by selective adsorption for ultra-clean fuels and for fuel cell applications. *Catalysis Today* 77:107–116.

MacRelli, S., J. Mogensen and G. Zacchi. 2012. Techno-economic evaluation of 2nd generation bioethanol production from sugar cane bagasse and leaves integrated with the sugar-based ethanol process. *Biotechnology for Biofuels* 5:22.

Martin, C., H. B. Klinke and A. B. Thomsen. 2007. Wet oxidation as a pretreatment method for enhancing the enzymatic convertibility of sugarcane bagasse. *Enzyme and Microbial Technology* 40:426–432.

Martin, C., M. Galbe, C. F. Wahlbom, B. Hahn-Hagerdal and L. J. Jonsson. 2002. Ethanol production from enzymatic hydrolysates of sugarcane bagasse using recombinant xylose-utilising *Saccharomyces cerevisiae. Enzyme and Microbial Technology* 31:274–282.

Martinez, A., M. E. Rodriguez, S. W. York, J. F. Preston and I. O. Ingram. 2000. Effects of $Ca(OH)_2$ treatments ('overliming') on the composition and toxicity of bagasse hemicellulose hydrolysates. *Biotechnology and Bioengineering* 69:526–536.

Matsuoka, S., A. J. Kennedy, E. G. D. D. Santos, A. L. Tomazela and L. C. S. Rubio. 2014. Energy cane: Its concept, development, characteristics, and prospects. *Advances in Botany* 2014:597275.

McIntosh, S. and T. Vancov. 2010. Enhanced enzyme saccharification of *Sorghum bicolor* straw using dilute alkali pretreatment. *Bioresource Technology* 101:6718–6727.

Micard, V., C. M. G. C. Renard and J. F. Thibault. 1996. Enzymatic saccharification of sugar-beet pulp. *Enzyme and Microbial Technology* 19:162–170.

Morschbacker, A. 2009. Bio-ethanol based ethylene. *Polymer Reviews* 49:79–84.

Najafi, G., B. Ghobadian and T. Tavakoli, et al. 2009. Performance and exhaust emissions of a gasoline engine with ethanol blended gasoline fuels using artificial neural network. *Applied Energy* 86:630–639.

Newman, P. W. G. and J. R. Kenworthy. 1989. Gasoline consumption and cities: A comparison of U.S. cities with a global survey. *Journal of the American Planning Association* 55:24–37.

Nguyen, T. L. T. and S. H. Gheewala. 2008. Life cycle assessment of fuel ethanol from cane molasses in Thailand. *International Journal of Life Cycle Assessment* 13:301–311.

North, D. C. 1991. Institutions. *Journal of Economic Perspectives* 5:97–112.

Olsson, L. and B. Hahn-Hagerdal. 1996. Fermentation of lignocellulosic hydrolysates for ethanol production. *Enzyme and Microbial Technology* 18:312–331.

Qiu, Z., G. M. Aita and M. S. Walker. 2012. Effect of ionic liquid pretreatment on the chemical composition, structure and enzymatic hydrolysis of energy cane bagasse. *Bioresource Technology* 117:251–256.

Rabelo, S. C., H. Carrere, R. M. Filho and A. C. Costa. 2011. Production of bioethanol, methane and heat from sugarcane bagasse in a biorefinery concept. *Bioresource Technology* 102:7887–7895.

Rezende, C. A., M. de Lima and P. Maziero, et al. 2011. Chemical and morphological characterization of sugarcane bagasse submitted to a delignification process for enhanced enzymatic digestibility. *Biotechnology for Biofuels* 4:54.

Rodriguez-Chong, A., J. A. Ramirez, G. Garrote and M. Vazquez. 2004. Hydrolysis of sugar cane bagasse using nitric acid: A kinetic assessment. *Journal of Food Engineering* 61:143–152.

Sanchez, O. J. and C. A. Cardona. 2008. Trends in biotechnological production of fuel ethanol from different feedstocks. *Bioresource Technology* 99:5270–5295.

Sun, Y. and J. Cheng. 2002. Hydrolysis of lignocellulosic materials for ethanol production: A review. *Bioresource Technology* 83:1–11.

Taherzadeh, M. J. and K. Karimi. 2007. Enzyme-based hydrolysis processes for ethanol from lignocellulosic materials: A review. *Bioresources* 2:707–738.

Taherzadeh, M. J. and K. Karimi. 2008. Pretreatment of lignocellulosic wastes to improve ethanol and biogas production: A review. *International Journal of Molecular Sciences* 9:1621–1651.

Zhao, X. B., L. Wang, D. H. Liu. 2008. Peracetic acid pretreatment of sugarcane bagasse for enzymatic hydrolysis: A continued work. *Journal of Chemical Technology and Biotechnology* 83:950–956.

Part 18

Second Generation Bioethanol Fuels from Residual Starch Feedstocks

63 Second Generation Bioethanol Fuels from Residual Starch Feedstocks

Scientometric Study

Ozcan Konur
(Formerly) Ankara Yildirim Beyazit University

63.1 INTRODUCTION

Crude oil-based gasoline fuels (Ma et al., 2002; Newman and Kenworthy, 1989) have been widely used in transportation sector since the 1920s. However, there have been great public concerns over the adverse environmental and human impact of these fuels (Hill et al., 2006, 2009). Hence, biomass-based bioethanol fuels (Hill et al., 2006; Konur, 2012e, 2015, 2019, 2020a) have increasingly been used in blending gasoline fuels (Hsieh et al., 2002; Najafi et al., 2009), in fuel cells (Antolini, 2007, 2009), and in biochemical production (Angelici et al., 2013; Morschbacker, 2009) in a biorefinery context (Fernando et al., 2006; Huang et al., 2008).

Bioethanol fuels also play a critical role in maintaining energy security (Kruyt et al., 2009; Winzer, 2012) in supply shocks (Kilian, 2008, 2009) related to oil price shocks (Hamilton, 2003, 2009), COVID-19 pandemic (Fauci et al., 2020; Li et al., 2020) or wars (Hamilton, 1983; Jones, 2012) in the aftermath of the Russian invasion of Ukraine (Reeves, 2014).

However, it is necessary to pretreat biomass (Taherzadeh and Karimi, 2008; Yang and Wyman, 2008) to enhance yield of bioethanol (Hahn-Hagerdal et al., 2006; Sanchez and Cardona, 2008) prior to bioethanol production through hydrolysis (Sun and Cheng, 2002; Taherzadeh and Karimi, 2007) and fermentation (Lin and Tanaka, 2006; Olsson and Hahn-Hagerdal, 1996) of biomass and hydrolysates, respectively.

One of the most-studied feedstocks for bioethanol fuels have been residual starch feedstocks such as corn stover; wheat straw; and to a lesser extent rice straw, rye straw, and barley straw. Research in the field of second generation bioethanol fuels from residual starch feedstocks has intensified in this context in key research fronts of the pretreatment (Kim et al., 2003; Kumar et al., 2009; Mani et al., 2004) and hydrolysis (Hsu et al., 2010; Lloyd and Wyman, 2005; Yang and Wyman, 2004) of residual starch feedstocks, fermentation (Mosier et al., 2005; Saha et al., 2005; Talebnia et al., 2010) of residual starch feedstock-based hydrolysates, production (Kaparaju et al, 2009; Mosier et al., 2005; Teymouri et al., 2005), and evaluation (Kazi et al., 2010; Sheehan et al., 2003) of second generation bioethanol fuels from residual starch feedstocks. Further, corn stover (Kazi et al., 2010; Mosier et al., 2005; Sheehan et al., 2003), wheat straw (Kabel et al., 2007; Kaparaju et al., 2009; Saha et al., 2005), and to a lesser extent rice straw (Hsu et al., 2010), rye straw (Sun and Cheng, 2005), and barley straw (Mani et al., 2004) have been studied intensively in this context.

However, it is essential to develop efficient incentive structures (North, 1991) for primary stakeholders to enhance research in this field (Konur, 2000, 2002a–c, 2006a,b, 2007a,b). Scientometric analysis has been used in this context to inform primary stakeholders about the current state of research in a selected research field (Garfield, 1955; Konur, 2011, 2012a–i, 2015, 2018b, 2019, 2020a).

DOI: 10.1201/9781003226550-85

As there have been no published scientometric studies in this field, this book chapter presents a scientometric study of research in second generation bioethanol fuels from residual starch feedstocks. It examines scientometric characteristics of both sample and population data presenting scientometric characteristics of these both datasets in the order of documents, authors, publication years, institutions, funding bodies, source titles, countries, Scopus subject categories, Scopus keywords, and research fronts.

63.2 MATERIALS AND METHODS

The search for this study was carried out using Scopus database (Burnham, 2006) in July 2022.

As the first step for search of relevant literature, keywords were selected using the first 200 most-cited population papers. The selected keyword list was then optimized to obtain a representative sample of papers for the searched research field. This keyword list was provided in Appendix for future replicative studies.

As the second step, two sets of data were used for this study. First, a population sample of 4,056 papers was used to examine scientometric characteristics of population data. Secondly, a sample of 203 most-cited papers, corresponding to 5% of the population papers, was used to examine scientometric characteristics of these citation classics.

The scientometric characteristics of these both sample and population datasets were presented in the order of documents, authors, publication years, institutions, funding bodies, source titles, countries, Scopus subject categories, Scopus keywords, and research fronts.

Lastly, key scientometric findings for both datasets were discussed to highlight the research landscape for second generation bioethanol fuels from residual starch feedstocks. Additionally, a number of brief conclusions were drawn and a number of relevant recommendations were made to enhance future research landscape.

63.3 RESULTS

63.3.1 The Most Prolific Documents in Second Generation Bioethanol Fuels from Residual Starch Feedstocks

Information on types of documents for both datasets is given in Table 63.1. Articles and conference papers, published in journals, dominate both the sample (98%) and population (98%) papers as they

TABLE 63.1
Documents in Second Generation Bioethanol Fuels from Residual Starch Feedstocks

Documents	Sample Dataset (%)	Population Dataset (%)	Surplus (%)
Article	90.1	96.1	–6.0
Conference paper	7.9	2.2	5.7
Review	2.0	1.0	1.0
Book chapter	0.0	0.4	–0.4
Letter	0.0	0.1	–0.1
Note	0.0	0.1	–0.1
Short survey	0.0	0.0	0.0
Book	0.0	0.0	0.0
Editorial	0.0	0.0	0.0
Sample size	203	4,056	

Sample dataset: Number of papers (%) in the set of 203 highly cited papers. Population dataset: Number of papers (%) in the set of the 4,056 population papers.

are equallyrepresented in both samples. Further, review papers and short surveys have a 1% surplus as they are over-represented in the sample papers as they constitute 2% and 1% of the sample and population papers, respectively.

It is further notable that 99% of the population papers were published in journals, while 1% of them were published in books and book series. Similarly, 100% of the sample papers were published in journals.

63.3.2 The Most Prolific Authors in Second Generation Bioethanol Fuels from Residual Starch Feedstocks

Information about the 15 most-prolific authors with at least 2.5% of sample papers each is given in Table 63.2. The most prolific authors are Bruce E. Dale and Charles E. Wyman with 6.4% of the sample papers each. Other prolific authors are Guido Zacchi, Badal C. Saha, Michael A. Cotta, and Yoon Y. Lee with 3.9%–4.4% of the sample papers each.

TABLE 63.2
Most-Prolific Authors in Second Generation Bioethanol Fuels from Residual Starch Feedstocks

No.	Author Name	Author Code	Sample Papers (%)	Population Papers (%)	Surplus (%)	Institution	Country	HI	N	Res. Front
1	Dale, Bruce E.	7201511969	6.4	0.8	5.1	Michigan State Univ.	USA	90	430	CS
2	Wyman, Charles E.	7004396809	6.4	0.7	5.7	Univ. Calif. Riverside	USA	80	287	CS
3	Zacchi, Guido	7006727748	4.4	0.6	3.8	Lund Univ.	Sweden	67	204	CS, WS
4	Saha, Badal C.	7202946302	3.9	0.6	3.3	USDA Agr. Res. Serv.	USA	54	160	CS, WS, RH
5	Cotta, Michael A.	7006656876	3.9	0.5	3.4	USDA Agr. Res. Serv.	USA	51	186	CS, WS, RH
6	Lee, Yoon Y.	8948274900	3.9	0.4	3.5	Auburn Univ.	USA	53	115	CS
7	Thomsen, Anne B.*	7102150211	3.4	0.5	2.9	Tech. Univ. Denmark	Denmark	38	62	CS, WS
8	Holtzapple, Mark T.	7004167004	3.4	0.3	3.1	Texas A&M Univ.	USA	47	197	CS, WS
9	Balan, Venkatesh	15757087100	3.0	0.8	2.2	Univ. Houston	USA	55	212	CS
10	Kim, Tae Hyun	57210847338	3.0	0.7	1.6	Hanyang Univ.	S. Korea	23	88	CS
11	Galbe, Mats	7003788758	3.0	0.6	1.9	Lund Univ.	Sweden	50	131	CS, WS
12	Ladisch, Michael R.	7005670397	2.5	0.4	1.8	Purdue Univ.	USA	59	290	CS, CF
13	Himmel, Michael E.	7007125552	2.5	0.2	2.0	Natl. Renew. Ener. Lab.	USA	73	422	CS
14	Mosier, Nathan S.	6602426392	2.5	0.2	2.0	Purdue Univ.	USA	43	117	CS
15	Kumar, Rajeev	57214462045	2.5	0.2	2.1	Univ. Calif. Riverside	USA	45	89	CS

*, Female researchers; Author code: The unique code given by Scopus to the authors. Sample papers: Number of papers authored in the sample dataset. Population papers: number of papers authored in the population dataset.
CS, Corn stover; WS, Wheat straw; RH, Rice hulls; CF, Corn fiber.

On the other hand, the most influential authors are Charles E. Wyman and Bruce E. Dale with 5.7% and 5.1% surplus, respectively. Other influential authors are Guido Zacchi, Yoon Y. Lee, Michael A. Cotta, Badal C. Saha, Mark T. Holtzapple, and Anne B. Thomsen with 2.9%–3.8% surplus each.

The most prolific institutions for sample dataset are Lund University, Purdue University, University of California Riverside, and USDA Agricultural Research Service with two authors each. On the other hand, the most prolific country for sample dataset is the USA with 11 authors, followed by Sweden with two authors. In total, only four countries house these top authors.

There are two primary research fronts for these top authors: Bioethanol fuels based on corn stover (CS) and wheat straw (WS) with 15 and 6 authors, respectively. There are also two HCPs for rice hulls (RH)-based bioethanol fuels.

On the other hand, there is a significant gender deficit (Beaudry and Lariviere, 2016) for the sample dataset as surprisingly only one of these top researchers is female with a representation rate of 7%.

Additionally, there are other authors with relatively low citation impact and with 0.3%–0.8% of the population papers each: Jie Bao, Hongzhang Chen, Keikhosro Karimi, Qiang Yong, Mercedes Ballesteros, Ying-Jin Yuan, Jian Zhang, Lirong Han, Mingjie Jin, Bing-Zhi Li, Anne S. Meyer, Liangcai Pen, Daniel J. Schell, Zhi-Hua Liu, Lei Qin, Verawat Champreda, Bruce S. Dien, Hasan Jameel, Akihiko Kondo, Donghai Wang, Birgitte K. Ahring, Yongcan Jin, Ken Tokuyasu, Jian Xu, Ignacio Ballesteros, Navadol Laosiripojana, Bin Li, Rodney J. Bothast, Claus Felby, Henning Jorgensen, Kasi Muthukumarappan, and Jia-Qing Zhu.

63.3.3 The Most Prolific Research Output by Years in Second Generation Bioethanol Fuels from Residual Starch Feedstocks

Information about papers published between 1970 and 2022 is given in Figure 63.1. This figure clearly shows that the bulk of research papers in population dataset were published primarily in the 2010s and the early 2020s with 61% and 20% of the population dataset, respectively. Similarly, the publication rates for the 2000s, 1990s, 1980s, and 1970s were 12%, 4%, 3%, and 0%, respectively. Additionally, 0.6% of the population papers were published in the pre-1970s.

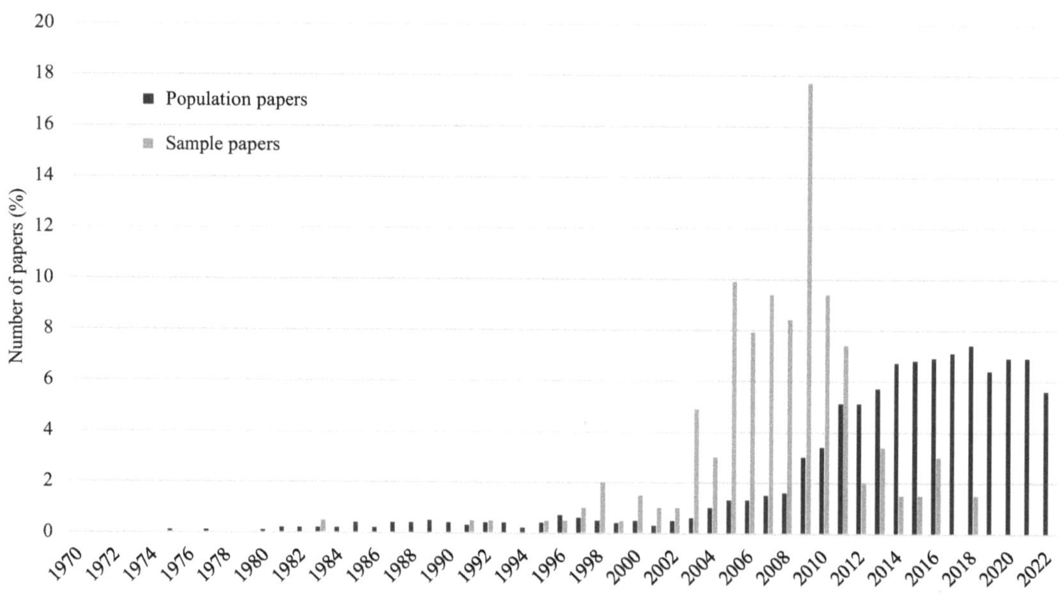

FIGURE 63.1 Research output by years regarding second generation bioethanol fuels from residual starch feedstocks.

Similarly, the bulk of research papers in sample dataset was published in the 2000s and 2010s with 65% and 30% of the sample dataset, respectively. Similarly, publication rates for the 1990s, 1980s, and 1970s were 6%, 1%, and 0% of the sample papers, respectively.

The most prolific publication years for the population dataset were 2018 and 2017 with 7.4% and 7.1% of the dataset, respectively. Further, 83% of the population papers were published between 2009 and 2022. Similarly, 89% of the sample papers were published between 2003 and 2016, while the most prolific publication year was 2009 with 18% of the sample papers. Other prolific publication years were 2005, 2007, and 2010 with 10%, 9%, and 9% of the sample papers, respectively.

On the other hand, number of publications for the population papers rose between 2008 and 2014, and, thereafter, it steadied around 7% of the population papers for each year. Further, there was no sharp rise in research output for population papers in 2020 and 2021 due to supply shocks.

63.3.4 The Most Prolific Institutions in Second Generation Bioethanol Fuels from Residual Starch Feedstocks

Information about the 14 most prolific institutions publishing papers on second generation bioethanol fuels from residual starch feedstocks with at least 2.0% of the sample papers each is given in Table 63.3.

The most prolific institution is National Renewable Energy Laboratory (NREL) with 10.3% of the sample papers, followed by Technical University of Denmark, Michigan State University, and USDA Agricultural Research Service with 8.4%, 7.9%, and 5.4% of the sample papers, respectively. Other prolific institutions are Lund University and Dartmouth College with 4.9% of the sample papers each.

Similarly, the top country for these most prolific institutions is the USA with nine institutions. Another prolific country is China with two institutions. In total, only five countries house these top institutions.

On the other hand, institutions with the most citation impact are NREL, Technical University of Denmark, and Michigan State University with 8.4%, 6.4%, and 6.1% surplus, respectively. Other influential institutions are Dartmouth College, Lund University, and USDA Agricultural Research Service with 4.4%, 3.8%, and 3.6% surplus, respectively.

TABLE 63.3
The Most Prolific Institutions In Second Generation Bioethanol Fuels from Residual Starch Feedstocks

No.	Institutions	Country	Sample Papers (%)	Population Papers (%)	Surplus (%)
1	Natl. Renew. Ener. Lab.	USA	10.3	1.9	8.4
2	Tech. Univ. Denmark	Denmark	8.4	2.0	6.4
3	Michigan State Univ.	USA	7.9	1.8	6.1
4	USDA Agr. Res. Serv.	USA	5.4	1.8	3.6
5	Lund Univ.	Sweden	4.9	1.1	3.8
6	Dartmouth Coll.	USA	4.9	0.5	4.4
7	Purdue Univ.	USA	3.9	0.8	3.1
8	Auburn Univ.	USA	3.9	0.5	3.4
9	Chinese Acad. Sci.	China	3.4	4.2	–0.8
10	Texas A&M Univ.	USA	3.4	0.7	2.7
11	Univ. Calif. Riverside	USA	3.4	0.5	2.9
12	Huazhong Agr. Univ.	China	2.0	0.9	1.1
13	CIEMAT	Spain	2.0	0.8	1.2
14	Univ. British Columbia	USA	2.0	0.6	1.4

Additionally, there are other institutions with relatively low citation impact and with 0.6%–2.0% of the population papers each: Nanjing Forestry University, South China University of Technology, China Agricultural University, Tianjin University, East China University of Science and Technology, Ministry of Agriculture of China, Iowa State University, Wageningen University & Research, Kansas State University, Henan Agricultural University, King Mongkut's University of Technology, Beijing University of Chemical Technology, Isfahan University of Technology, Jiangnan University, Nanjing Tech University, Tianjin University of Science & Technology, Shanghai Jiao Tong University, Copenhagen University, Zhejiang University, Nanjing Agricultural University, Korea University, Shandong University, Beijing Forestry University, National Technical University of Athens, NC State University, Qilu University of Technology, Harbin Institute of Technology, Tsinghua University, University of Minnesota Twin Cities, University of Wisconsin-Madison, and Qingdao Institute of Bioenergy and Bioprocess Technology.

63.3.5 The Most Prolific Funding Bodies in Second Generation Bioethanol Fuels from Residual Starch Feedstocks

Information about the 14 most prolific funding bodies funding at least 1.5% of the sample papers each is given in Table 63.4. Further, only 46% and 52% of the sample and population papers were funded, respectively.

The most prolific funding bodies are U.S. Department of Energy and National Natural Science Foundation of China with 7.9% and 6.9% of the sample papers, respectively. Other prolific funding bodies are Ministry of Science and Technology of China, NREL, Laboratory Directed Research and Development, and National Nuclear Security Administration with 3.4%–4.9% of the sample papers

TABLE 63.4
The Most Prolific Funding Bodies in Second Generation Bioethanol Fuels from Residual Starch Feedstocks

No.	Funding Bodies	Country	Sample Paper No. (%)	Population Paper No. (%)	Surplus (%)
1	U.S. Department of Energy	USA	7.9	2.9	5.0
2	National Natural Science Foundation of China	China	6.9	14.7	–7.8
3	Ministry of Science and Technology of China	China	4.9	4.1	0.8
4	National Renewable Energy Laboratory	USA	3.9	0.7	3.2
5	Laboratory Directed Research and Development	USA	3.4	0.6	2.8
6	National Nuclear Security Administration	USA	3.4	0.6	2.8
7	National Key Research and Development Program of China	China	3.0	3.5	–0.5
8	European Commission	EU	3.0	1.5	1.5
9	National Basic Research Program of China (973 Program)	China	2.0	1.4	0.6
10	National Science Foundation	USA	1.5	1.1	0.4
11	Ministry of Education, Science and Technology	Japan	1.5	0.6	0.9
12	Ford Motor Company	USA	1.5	0.2	1.3
13	National Institute of Standards and Technology	USA	1.5	0.1	1.4
14	University of California, Riverside	USA	1.5	0.1	1.4

each. On the other hand, the most prolific countries for these top funding bodies are the USA and China with eight funding bodies. In total, only three countries and the EU house these top funding bodies.

The funding body with the most citation impact is the U.S. Department of Energy with 5.0% surplus, followed by the NREL, Laboratory Directed Research and Development, and National Nuclear Security Administration with 2.8%–3.2% surplus each. Further, the funding body with the least citation impact is the National Natural Science Foundation of China with 7.8% deficit. It is also notable that this funding body is the largest funder of the population papers with a 14.7% funding rate.

Other funding bodies with relatively low citation impact and with 0.6%–1.8% of the population papers each are Fundamental Research Funds for the Central Universities, Ministry of Education of China, Chinese Academy of Sciences, Priority Academic Program Development of Jiangsu Higher Education Institutions, National High-tech Research and Development Program, Higher Education Personnel Improvement Coordination, Natural Science Foundation of Jiangsu Province, China Postdoctoral Science Foundation, National Council for Scientific and Technological Development, U.S. Department of Agriculture, Ministry of Education, Culture, Sports, Science and Technology, Ministry of Finance, Office of Science, Thailand Research Fund, Japan Society for the Promotion of Science, National Research Foundation of Korea, Natural Sciences and Engineering Research Council of Canada, Seventh Framework Program, China Scholarship Council, European Regional Development Fund, Ministry of Science and Technology India, Natural Science Foundation of Shandong Province, and Swedish Energy Agency.

63.3.6 The Most Prolific Source Titles in Second Generation Bioethanol Fuels from Residual Starch Feedstocks

Information about the 12 most prolific source titles publishing at least 2% of the sample papers each in second generation bioethanol fuels from residual starch feedstocks is given in Table 63.5.

The most prolific source title is the Bioresource Technology with 37.4% of the sample papers. Other prolific titles are Biotechnology and Bioengineering, Biomass and Bioenergy, Applied Biochemistry and Biotechnology Part A, and Applied Biochemistry and Biotechnology with 5.4%–8.9% of the sample papers each.

On the other hand, the source title with the most citation impact is the Bioresource Technology with 21.2% surplus. Other influential titles are Biotechnology and Bioengineering, Biomass and Bioenergy, and Applied Biochemistry and Biotechnology Part A with 4.9%–7.4% surplus each.

TABLE 63.5
The Most Prolific Source Titles in Second Generation Bioethanol Fuels from Residual Starch Feedstocks

No.	Source Titles	Sample Papers (%)	Population Papers (%)	Surplus (%)
1	Bioresource Technology	37.4	16.2	21.2
2	Biotechnology and Bioengineering	8.9	1.5	7.4
3	Biomass and Bioenergy	8.4	3.5	4.9
4	Applied Biochemistry and Biotechnology Part A	5.9	1.0	4.9
5	Applied Biochemistry and Biotechnology	5.4	3.0	2.4
6	Industrial Crops and Products	3.9	4.3	–0.4
7	Process Biochemistry	3.9	1.6	2.3
8	Enzyme and Microbial Technology	3.4	0.7	2.7
9	Biotechnology for Biofuels	3.0	2.9	0.1
10	Biotechnology Progress	3.0	0.7	2.3
11	Industrial and Engineering Chemistry Research	2.0	0.6	1.4
12	Journal of Biotechnology	2.0	0.3	1.7

Other source titles with relatively low citation impact with 0.6%–3.4% of the population papers each are Bioresources, Biomass Conversion and Biorefinery, Renewable Energy, Fuel, Bioenergy Research, ACS Sustainable Chemistry and Engineering, Journal of Chemical Technology and Biotechnology, Waste and Biomass Valorization, Cellulose Chemistry and Technology, RSC Advances, Journal of Cleaner Production, Applied Microbiology and Biotechnology, Cellulose, Energy and Fuels, Bioprocess and Biosystems Engineering, Energies, Biochemical Engineering Journal, Scientific Reports, and Science of the Total Environment.

63.3.7 The Most Prolific Countries in Second Generation Bioethanol Fuels from Residual Starch Feedstocks

Information about the 11 most prolific countries publishing at least 2% of sample papers each in second generation bioethanol fuels from residual starch feedstocks is given in Table 63.6.

The most prolific country is the USA with 43.8% of the sample papers, followed by China with 18.2% of the sample papers. Other prolific countries are Denmark, Sweden, and Spain with 9.9%, 5.9%, and 5.4% of the sample papers, respectively. It is also notable that China is the largest producer of the population papers with a publication rate of 30.6%. Further, five European countries listed in Table 63.6 produce 26% and 13% of the sample and population papers with 13% surplus.

On the other hand, the country with the most citation impact is the USA with 25.8% surplus. Other influential countries are Denmark and Sweden with 7.1% and 3.5% surplus, respectively. Similarly, the country with the least citation impact is China with 12.4% deficit, followed by India and Japan with 5.3% and 1.6% deficit, respectively.

Additionally, there are other countries with relatively low citation impact and with 0.6%–2.9% of the sample papers each: Thailand, Brazil, Iran, Malaysia, France, Germany, Poland, Taiwan, Indonesia, Italy, Nigeria, Turkey, Finland, Egypt, Mexico, Pakistan, Australia, Portugal, South Africa, Austria, Greece, Vietnam, Belgium, and Hungary.

63.3.8 The Most Prolific Scopus Subject Categories in Second Generation Bioethanol Fuels from Residual Starch Feedstocks

Information about the eight most prolific Scopus subject categories indexing at least 3.4% of the sample papers each is given in Table 63.7.

TABLE 63.6
The Most Prolific Countries in Second Generation Bioethanol Fuels from Residual Starch Feedstocks

No.	Countries	Sample Papers (%)	Population Papers (%)	Surplus (%)
1	USA	43.8	18.0	25.8
2	China	18.2	30.6	–12.4
3	Denmark	9.9	2.8	7.1
4	Sweden	5.9	2.4	3.5
5	Spain	5.4	3.6	1.8
6	India	3.9	9.2	–5.3
7	S. Korea	3.9	3.2	0.7
8	Canada	3.4	3.5	–0.1
9	Japan	3.0	4.6	–1.6
10	Netherlands	2.5	1.4	1.1
11	UK	2.0	2.6	–0.6

TABLE 63.7
The Most Prolific Scopus Subject Categories in Second Generation Bioethanol Fuels from Residual Starch Feedstocks

No.	Scopus Subject Categories	Sample Papers (%)	Population Papers (%)	Surplus (%)
1	Chemical Engineering	72.9	46.4	26.5
2	Environmental Science	54.2	42.0	12.2
3	Energy	53.7	40.3	13.4
4	Biochemistry, Genetics and Molecular Biology	37.9	27.7	10.2
5	Immunology and Microbiology	35.5	18.8	16.7
6	Agricultural and Biological Sciences	12.8	18.6	−5.8
7	Chemistry	3.9	14.8	−10.9
8	Engineering	3.4	9.9	−6.5

The most prolific Scopus subject category in second generation bioethanol fuels from residual starch feedstocks is chemical engineering with 73% of the sample papers, followed by environmental science and energy with 54% pf the sample papers each. Other prolific subject categories are biochemistry, genetics and molecular biology and immunology and microbiology with 38% and 36% of the sample papers, respectively. It is notable that social sciences including economics and business account for 0.5% and 2.5% of the sample and population studies, respectively.

On the other hand, the Scopus subject category with the most citation impact is the chemical engineering with 27% surplus, followed by immunology and mjicrobiology with 17% surplus. Other influential subject categories are energy; environmental science; and biochemistry, genetics, and molecular biology with 13%, 12%, and 10% surplus each. Similarly, the least influential subject category is chemistry with 11% deficit, followed by engineering and agricultural and biological sciences with 7% and 6% surplus, respectively.

63.3.9 The Most Prolific Keywords in Second Generation Bioethanol Fuels from Residual Starch Feedstocks

Information about the Scopus keywords used with at least 7.4% or 3.9% of the sample or population papers, respectively, is given in Table 63.8. For this purpose, keywords related to the keyword set given in appendix are selected from a list of the most prolific keyword set provided by Scopus database.

These keywords are grouped under the five headings: biomass, pretreatments, fermentation, hydrolysis and hydrolysates, and products.

The most prolific keywords related to biomass and biomass constituents are cellulose, lignin, *Zea mays* (genus for corn), triticum (genus for wheat), straw, corn stover, hemicellulose, corn, wheat straw, lignocellulose, carbohydrates, wheat, and rice straw with 21%–60% of the sample papers each. Further, the prolific keywords related to the pretreatments are enzymes, cellulases, pretreatments, pre-treatments, and temperature with 21%–38% of the sample papers each.

The most prolific keywords related to the fermentation are fermentation and saccharomyces with 41% and 18% of the sample papers, respectively. Further, the most prolific keywords related to the hydrolysis and hydrolysates are hydrolysis, sugar, enzymatic hydrolysis, enzyme activity, glucose, saccharification, and xylose with 23%–74% of the sample papers each. Finally, the most prolific keywords related to the products are ethanol, bioethanol, and ethanol production with 11%–46% of the sample papers each.

On the other hand, the most influential keywords are hydrolysis, sugar, cellulose, *Zea mays*, triticum, enzymes, corn, cellulases, and lignin with 21%–39% surplus each. Similarly, the most prolific keywords across all of the research fronts are hydrolysis, sugar, cellulose, lignin, *Zea mays*, ethanol,

TABLE 63.8
The Most Prolific Keywords in Second Generation Bioethanol Fuels from Residual Starch Feedstocks

No.	Keywords	Sample Papers (%)	Population Papers (%)	Surplus (%)
1		**Biomass and Biomass Constituents**		
	Cellulose	60.1	32.4	27.7
	Lignin	51.7	30.6	21.1
	Zea mays	46.8	21.2	25.6
	Triticum	40.4	18.0	22.4
	Straw	35.0	24.0	11.0
	Corn stover	29.6	12.8	16.8
	Hemicellulose	29.1	10.9	18.2
	Corn	28.1	6.6	21.5
	Wheat straw	26.6	20.8	5.8
	Lignocellulose	25.1	12.3	12.8
	Carbohydrates	21.7	11.6	10.1
	Wheat	21.7	11.5	10.2
	Rice straw	21.2	18.9	2.3
	Rice	18.2	8.9	9.3
	Xylan	16.3	5.0	11.3
	Maize	15.3	15.0	0.3
	Lignocellulosic biomass	13.3	9.5	3.8
	Oryza	12.3	7.9	4.4
	Glucan	12.3	3.1	9.2
	Rice straw	8.4	6.2	2.2
	Crop residues	7.4		7.4
	Grains	7.4		7.4
	Starch		4.4	–4.4
	Agricultural wastes		3.9	–3.9
2		**Pretreatments**		
	Enzymes	38.4	16.0	22.4
	Cellulase	34.0	12.7	21.3
	Pretreatment	33.0	14.1	18.9
	Pre-treatment	23.6	17.4	6.2
	Temperature	21.2	9.3	11.9
	Sulfuric acid	16.3	6.1	10.2
	Water	14.8	5.9	8.9
	Ammonia	12.8	4.7	8.1
	Steam	9.9	4.1	5.8
	Delignification	8.4	6.7	1.7
	B-glucosidase	8.4	2.6	5.8
	Alkalinity	5.9	4.9	1.0
	Sodium hydroxide		6.0	–6.0
3		**Fermentation**		
	Fermentation	40.9	30.2	10.7

(Continued)

TABLE 63.8 (*Continued*)
The Most Prolific Keywords in Second Generation Bioethanol Fuels from Residual Starch Feedstocks

No.	Keywords	Sample Papers (%)	Population Papers (%)	Surplus (%)
	Saccharomyces	18.2	7.6	10.6
	Furfural	9.4	4.0	5.4
	Yeast	8.9	9.3	–0.4
	SSF	4.9	4.8	0.1
4		**Hydrolysis and Hydrolysates**		
	Hydrolysis	73.9	35.1	38.8
	Sugar	61.1	22.4	38.7
	Enzymatic hydrolysis	43.3	24.1	19.2
	Enzyme activity	37.4	18.5	18.9
	Glucose	34.5	19.4	15.1
	Saccharification	23.2	19.5	3.7
	Xylose	23.2	9.8	13.4
	Enzymolysis	13.3	4.8	8.5
	Fermentable sugars	7.9	3.3	4.6
	Enzymatic saccharification	7.4	5.5	1.9
5		**Products**		
	Ethanol	46.3	31.4	14.9
	Bioethanol	11.3	14.7	–3.4
	Ethanol production	10.8	7.0	3.8
	Cellulosic ethanol	4.9	4.7	0.2
	Bio-ethanol production		6.7	–6.7

enzymatic hydrolysis, fermentation, triticum, enzymes, enzyme activity, straw, glucose, cellulase, pretreatment, corn stover, hemicellulose, corn, wheat straw, lignocellulose, pre-treatment, saccharification, xylose, carbohydrates, and wheat with 22%–74% surplus each.

63.3.10 The Most Prolific Research Fronts in Second Generation Bioethanol Fuels from Residual Starch Feedstocks

Information about research fronts for sample papers in second generation bioethanol fuels from residual starch feedstocks with regard to the residual starch feedstocks used for the bioethanol production is given in Table 63.9.

As this table shows, the most prolific starch feedstock is corn stover with 45% of the sample papers, followed by wheat straw and rice straw with 29% and 16% of the sample papers, respectively. Other residual starch feedstocks used in these studies are barley straw, rye straw, other corn residues, other rice residues, and sorghum residues with 2%–6% of the sample papers each.

Information about the thematic research fronts for the sample papers in second generation bioethanol fuels from residual starch feedstocks is given in Table 63.10. As this table shows, there are five primary research fronts: biomass pretreatments, hydrolysis, hydrolysate fermentation, production, and evaluation of second generation bioethanol fuels from residual starch feedstocks with 95%, 77%, 24%, 27%, and 5% of the sample papers, respectively.

For the research front of residual starch feedstock pretreatments, the most prolific pretreatments are chemical and enzymatic pretreatments of biomass with 87% and 90% of the sample papers,

TABLE 63.9
The Most Prolific Research Fronts for Second Generation Bioethanol Fuels from Residual Starch Feedstocks

No.	Research Fronts	Paper (%)
1	Corn stover	44.8
2	Wheat straw	29.1
3	Rice straw	15.8
4	Other starch residues	17.8
	Other corn residues	5.9
	Other rice residues	5.4
	Barley straw	3.0
	Rye straw	2.0
	Sorghum residues	1.5

Paper (%) sample: Number of papers in the population sample of 203 papers.

respectively. Other pretreatments are hydrothermal and mechanical pretreatments of biomass with 22% and 10% of the sample papers, respectively.

On the other hand, on an individual basis, the most prolific research fronts are enzymatic hydrolysis, acid pretreatments, alkaline pretreatments, ammonia pretreatments, and steam explosion with 13%–75% of the sample papers, respectively.

Further, the most prolific feedstocks used in all these research fronts are corn stover, wheat straw, and rice straw although their content are not shown in this table.

63.4 DISCUSSION

63.4.1 INTRODUCTION

Crude oil-based gasoline fuels have been widely used in transportation sector since the 1920s. However, there have been great public concerns over adverse environmental and human impact of these fuels. Hence, biomass-based bioethanol fuels have increasingly been used in blending gasoline fuels, in fuel cells, and in biochemical production in a biorefinery context.

However, it is necessary to pretreat biomass to enhance the yield of bioethanol prior to bioethanol production through hydrolysis and fermentation. One of the most-studied feedstocks for bioethanol fuels have been residual starch feedstocks such as corn stover, wheat straw, rice straw, rye straw, and barley straw. Research in the field of second generation bioethanol fuels from residual starch feedstocks has intensified in this context in key research fronts of pretreatment and hydrolysis of residual starch feedstocks, fermentation of residual starch feedstock-based hydrolysates, production, and evaluation of second generation bioethanol fuels from residual starch feedstocks. Further, corn stover; wheat straw; and to a lesser extent rice straw, rye straw, and barley straw have been studied intensively in this context.

However, it is essential to develop efficient incentive structures for primary stakeholders to enhance research in this field. This is especially important to maintain energy security in cases of supply shocks such as oil price shocks, war-related shocks as in the case of Russian invasion of Ukraine or COVID-19 shocks.

Scientometric analysis has been used in this context to inform primary stakeholders about the current state of research in a selected research field. As there has been no scientometric study in this field, this book chapter presents a scientometric study of research in second generation bioethanol fuels from residual starch feedstocks. It examines the scientometric characteristics of both sample

TABLE 63.10
The Most Prolific Thematic Research Fronts for Second Generation Bioethanol Fuels from Residual Starch Feedstocks

No.	Research Fronts	Paper (%)
1	Biomass pretreatments	95.1
1.1	Hydrothermal pretreatments	22.2
	Steam explosion	13.3
	LHW	5.4
	Wet oxidation	3.4
	HCW	2.5
1.2	Chemical pretreatments	86.7
	Acid pretreatment	33.5
	Alkaline pretreatment	20.7
	Ammonia pretreatment	14.3
	H_2O_2	5.9
	Solvent pretreatment	5.9
	IL pretreatment	3.9
	SO_2 pretreatment	3.4
	Surfactants	2.5
	Other pretreatments	1.5
1.3	Mechanical pretreatments	9.9
	Milling	4.9
	Microwave	3.0
	Ultrasound	2.0
1.4.	Enzymatic pretreatments	90.1
1.5	Pretreatments in general	2.0
2	Biomass hydrolysis	77.3
	Enzymatic hydrolysis	74.9
	Acid hydrolysis	2.5
3	Hydrolysate fermentation	24.1
4	Bioethanol production	27.1
5	Bioethanol fuel evaluation	4.9

Paper (%) sample: Number of papers in the population sample of 203 papers.

and population data presenting scientometric characteristics of these both datasets in the order of documents, authors, publication years, institutions, funding bodies, source titles, countries, Scopus subject categories, Scopus keywords, and research fronts.

As the first step for the search of the relevant literature, the keywords were selected using the first 200 most-cited papers. The selected keyword list was then optimized to obtain a representative sample of papers for the searched research field. A copy of this extended keyword list was provided in Appendix for future replicative studies. Further, a selected list of the keywords was presented in Table 63.8.

As the second step, two sets of data were used for this study. First, a population sample of 4,056 papers was used to examine scientometric characteristics of the population data. Secondly, a sample of 203 most-cited papers, corresponding to 5% of the population data set, was used to examine scientometric characteristics of these citation classics.

Scientometric characteristics of these sample and population datasets were presented in the order of documents, authors, publication years, institutions, funding bodies, source titles, countries, Scopus subject categories, Scopus keywords, and research fronts.

Lastly, the key scientometric findings for both datasets were discussed to highlight the research landscape for second generation bioethanol fuels from residual starch feedstocks. Additionally, a number of brief conclusions were drawn and a number of relevant recommendations were made to enhance future research landscape.

63.4.2 The Most Prolific Documents in Second Generation Bioethanol Fuels from Residual Starch Feedstocks

Articles (together with conference papers) dominate both sample (98%) and population (98%) papers (Table 63.1). Further, review papers have a surplus (1%). Representation of reviews in the sample papers is relatively modest (2%).

Scopus differs from the Web of Science database in differentiating and showing articles (90%) and conference papers (8%) published in journals separately. However, it should be noted that these conference papers are also published in journals as articles, compared to those published only in the conference proceedings. Hence, the total number of articles and review papers in the sample dataset are 98% and 2%, respectively.

It is observed during the search process that there has been inconsistency in classification of the documents in Scopus as well as in other databases such as Web of Science. This is especially relevant for the classification of papers as reviews or articles as the papers not involving a literature review may be erroneously classified as a review paper. There is also a case of review papers being classified as articles. For example, the total number of the reviews in the sample data set was manually found as nearly 3% compared to 2% as indexed by Scopus, decreasing the number of articles and conference papers to 97% for the sample dataset. A close examination of these papers shows that many evaluative studies such as techno-economic or life-cycle studies have been indexed as the review papers by the Scopus database.

In this context, it would be helpful to provide a classification note for the published papers in books and journals at the first instance. It would also be helpful to use the document types listed in Table 63.1 for this purpose. Book chapters may also be classified as articles or reviews as an additional classification to differentiate review chapters from experimental chapters as it is done by Web of Science. It would be further helpful to additionally classify the conference papers as articles or review papers as well as it is done in Web of Science database.

63.4.3 The Most Prolific Authors in Second Generation Bioethanol Fuels from Residual Starch Feedstocks

There have been 15 most prolific authors with at least 2.5% of the sample papers each as given in Table 63.2. These authors have shaped the development of research in this field.

The most prolific authors are Bruce E. Dale; Charles E. Wyman; and to a lesser extent Guido Zacchi, Badal C. Saha, Michael A. Cotta, and Yoon Y. Lee.

It is important to note the inconsistencies in indexing of the author names in Scopus and other databases. It is especially an issue for names with more than two components such as 'Blake Sam de Hyun Lee'. The probable outcomes are 'Lee, B.S.D.H.', 'de Hyun Lee, B.S.' or 'Hyun Lee, B.S.D.'. The first choice is the gold standard of publishing sector as the last word in the name is taken as the last name. In most of the academic databases such as PUBMED and EBSCO databases, this version is used predominantly. The second choice is a strong alternative, while the last choice is an undesired outcome as two last words are taken as the last name. It is good practice to combine words of the last name with a hyphen: 'Hyun-Lee, B.S.D.'. It is notable that inconsistent indexing of author names may cause substantial inefficiencies in the search process for papers as well as allocating credit to authors as there are different author entries for each outcome in the databases.

There are also inconsistencies in shortening of Chinese names. For example. 'YangYing Lee' is often shortened as 'Lee, Y.', 'Lee, Y.-Y.', and 'Lee, Y.Y.' as it is done in Web of Science database as well. However, the gold standard in this case is 'Lee, Y' where the last word is taken as the last name, and the first word is taken as a single forename. In most of the academic databases such as PUBMED and EBSCO, this first version is used predominantly. However, it makes sense to use the third option to differentiate Chinese names efficiently: 'Lee, Y.Y.'. Therefore, there have been difficulties in locating papers for the Chinese authors. In such cases, use of unique author codes provided for each author by Scopus database has been helpful.

There is also a difficulty in allowing credit for authors especially for authors with common names such as 'Lee, X.' in conducting scientometric studies. These difficulties strongly influence efficiency of scientometric studies as well as allocating credit to authors as there are same author entries for different authors with the same name. e.g., 'Lee, X.' in the databases.

In this context, coding of authors in Scopus database is a welcome innovation compared to other databases such as Web of Science. In this process, Scopus allocates a unique number to each author in the database (Aman, 2018). However, there might still be substantial inefficiencies in this coding system especially for common names. For example, some of the papers for a certain author maybe allocated to another researcher with a different author code. It is possible that Scopus uses a number of software programs to differentiate author names, and the program may not be false-proof (Kim, 2018).

In this context, it does not help that author names are not given in full in some journals and books. This makes difficult to differentiate authors with common names and makes the scientometric studies further difficult in the author domain. Therefore, author names should be given in all books and journals at the first instance. There is also a cultural issue where some authors do not use their full names in their papers. Instead, they use initials for their forenames: 'Lee, H.J.', 'Lee', 'Lee, H.', or 'Lee, J.' instead of 'Lee, Hyun Jae'.

There are also inconsistencies in naming of authors with more than two components by authors themselves in journal papers and book chapters. For example. 'Lee, A.P.C.' might be given as 'Lee, A.' or 'Lee, A.C.' or 'Lee, A.P.' or 'Lee, C.' in journals and books. This also makes scientometric studies difficult in the author domain. Hence, contributing authors should use their name consistently in their publications.

Another critical issue regarding author names is inconsistencies in spelling of author names in national spellings (e.g., Şöğütçığlık, Gökçe) rather than in English spellings (e.g., Sogutciglik, Gokce) in Scopus database. Scopus differs from Web of Science database and many other databases in this respect where author names are given only in English spellings. It is observed that national spellings of author names do not help much in conducting scientometric studies as well as in allocating credits to authors as sometimes there are different author entries for English and National spellings in Scopus database.

The most prolific institutions for the sample dataset are Lund University, Purdue University, University of California Riverside, and USDA Agricultural Research Service. Further, the most prolific country for the sample dataset is the USA and to a lesser extent Sweden. These findings confirm the dominance of the USA and to a lesser extent Sweden in this field. The primary research fronts are bioethanol fuels based on corn stover (CS) and wheat straw (WS).

It is also notable that there is a significant gender deficit for the sample dataset as surprisingly with a representation rate of 7%. This finding is the most thought-provoking with strong public policy implications. Hence, institutions, funding bodies, and policy makers should take efficient measures to reduce the gender deficit in this field as well as other scientific fields with strong gender deficit. In this context, it is worth to note the level of representation of researchers from minority groups in science on the basis of race, sexuality, age, and disability, besides the gender (Blankenship, 1993; Dirth and Branscombe, 2017; Konur, 2000, 2002a–c, 2006a,b, 2007a,b).

63.4.4 The Most Prolific Research Output by Years in Second Generation Bioethanol Fuels from Starch Feedstocks

Research output observed between 1970 and 2022 is illustrated in Figure 63.1. This figure clearly shows that the bulk of research papers in the population dataset were published primarily in the 2010s and early 2020s. Similarly, the bulk of research papers in the sample dataset were published in the 2000s and 2010s.

On the other hand, the number of publications for the population papers rose between 2008 and 2014, and, thereafter, it steadied around 7% of the population papers for each year.

These are the thought-provoking findings as there has been significant research boom since 2009. In this context, the increasing public concerns about climate change (Change, 2007), greenhouse gas emissions (Carlson et al., 2017), and global warming (Kerr, 2007) have been certainly behind the boom in research in this field in the last two decades. Furthermore, the recent supply shocks experienced due to the COVID-19 pandemic might also be behind the research boom in this field since 2019. However, there was no sharp rise in the research output for population papers in 2020 and 2021 due to supply shocks.

Based on these findings, the size of population papers is likely to more than double in the current decade, provided that the public concerns about climate change, greenhouse gas emissions, global warming, and supply shocks are translated efficiently to the research funding in this field.

63.4.5 The Most Prolific Institutions in Second Generation Bioethanol Fuels from Residual Starch Feedstocks

The 14 most prolific institutions publishing papers on second generation bioethanol fuels from residual starch feedstocks with at least 2.0% of the sample papers each given in Table 63.3 have shaped the development of research in this field.

The most prolific institutions are NREL; Technical University of Denmark; Michigan State University; and to a lesser extent USDA Agricultural Research Service, Lund University, and Dartmouth College. Similarly, the top country for these most prolific institutions is the USA and to a lesser extent China. In total, only five countries house these top institutions.

On the other hand, the institutions with the most citation impact are NREL; Technical University of Denmark; Michigan State University; and to a lesser extent Dartmouth College, Lund University, and USDA Agricultural Research Service. These findings confirm the dominance of the USA and to a lesser extent of European institutions for these HCPs.

63.4.6 The Most Prolific Funding Bodies in Second Generation Bioethanol Fuels from Residual Starch Feedstocks

The 14 most prolific funding bodies funding at least 1.5% of the sample papers each is given in Table 63.4. It is notable that only 46% and 52% of the sample and population papers were funded, respectively.

The most prolific funding bodies are U.S. Department of Energy; National Natural Science Foundation of China; and to a lesser extent Ministry of Science and Technology of China, NREL, Laboratory Directed Research and Development, and National Nuclear Security Administration. The most prolific countries for these top funding bodies are the USA and to a lesser extent China. It is also notable that National Natural Science Foundation of China is the largest funder of the population papers with a 14.7% funding rate.

These findings on funding of research in this field suggest that the level of funding, mostly since 2009, is modest, and it has been largely instrumental in enhancing the research in this field (Ebadi and Schiffauerova, 2016) in the light of North's institutional framework (North, 1991). It is also notable that the funding rate in this field is relatively modest compared to those in other research fronts of bioethanol fuels such as algal bioethanol fuels. Further, it is expected that this high funding rate would improve in the light of recent supply shocks. Further, it emerges that the USA has heavily funded research on corn grain- and corn residue-based bioethanol fuels.

63.4.7 The Most Prolific Source Titles in Second Generation Bioethanol Fuels from Residual Starch Feedstocks

The 12 most prolific source titles publishing at least 2% of the sample papers each in second generation bioethanol fuels from residual starch feedstocks have shaped the development of research in this field (Table 63.5).

The most prolific source titles are Bioresource Technology and to a lesser extent the Biotechnology and Bioengineering, Biomass and Bioenergy, Applied Biochemistry and Biotechnology Part A, and Applied Biochemistry and Biotechnology. On the other hand, the source titles with the most citation impact are Bioresource Technology and to a lesser extent Biotechnology and Bioengineering, Biomass and Bioenergy, and Applied Biochemistry and Biotechnology Part A.

It is notable that these top source titles are primarily related to the bioresources, energy, and biotechnology. This finding suggests that Bioresource Technology and the other prolific journals in these fields have significantly shaped the development of research in this field as they focus primarily on second generation bioethanol fuels from residual starch feedstocks with a high yield. In this context, the influence of the top journal is quite extraordinary with 37.4% of the sample papers. It is also notable that energy-related journals have also published papers in areas of techno-economics environmental impact, land use change, economics, and labor relations as social science-related journals.

63.4.8 The Most Prolific Countries in Second Generation Bioethanol Fuels from Residual Starch Feedstocks

The 11 most prolific countries publishing at least 2% of the sample papers each have significantly shaped the development of research in this field (Table 63.6).

The most prolific countries are the USA; China; and to a lesser extent Denmark, Sweden, and Spain. It is also notable that China is the largest producer of the population papers with a publication rate of 30.6%. Further, five European countries listed in Table 63.6 produce 26% and 13% of the sample and population papers with 13% surplus.

On the other hand, countries with the most citation impact are the USA and to a lesser extent Denmark and Sweden. Similarly, countries with the least impact are China and to a lesser extent India and Japan.

A close examination of these findings suggests that the USA, China, and Europe are major producers of research in this field. It is a fact that the USA has been a major player in science (Leydesdorff and Wagner, 2009). The USA has further developed a strong research infrastructure to support its corn- and grass-based bioethanol industry (Gillon, 2010).

However, China has been a rising mega star in scientific research in competition with the USA and Europe (Leydesdorff and Zhou, 2005). China is also a major player in this field as a major producer of bioethanol (Fang et al., 2010).

Next, Europe has been a persistent player in scientific research in competition with both the USA and China (Leydesdorff, 2000). Europe has also been a persistent producer of bioethanol along with the USA and Brazil (Gnansounou, 2010).

63.4.9 The Most Prolific Scopus Subject Categories in Second Generation Bioethanol Fuels from Residual Starch Feedstocks

The eight most prolific Scopus subject categories indexing at least 83.4% of the sample papers each, respectively, given in Table 63.7 have shaped the development of research in this field.

The most prolific Scopus subject categories in second generation bioethanol fuels from residual starch feedstocks are chemical engineering; environmental science; energy; and to a lesser extent biochemistry, genetics and molecular biology, and immunology and microbiology. It is also notable that social sciences including economics and business have a minimal presence in both sample and population studies.

On the other hand, Scopus subject categories with the most citation impact are chemical engineering; immunology and microbiology; and to a lesser extent energy, environmental science, biochemistry, and genetics and molecular biology. Similarly, the least influential subject categories are chemistry, engineering, and agricultural and biological sciences.

These findings are thought provoking suggesting that primary subject categories are related to energy, environmental science, chemical engineering, genetics, and microbiology as the core of research in this field concerns with production and utilization of second generation bioethanol fuels from residual starch feedstocks. Another finding is that social sciences are not well represented in both the sample and population papers as in line with the most fields in bioethanol fuels. The social, environmental, and economics studies account for the field of social sciences.

63.4.10 The Most Prolific Keywords in Second Generation Bioethanol Fuels from Residual Starch Feedstocks

A limited number of keywords have shaped the development of research in this field as shown in Table 63.8 and Appendix. These keywords are grouped under the five headings: biomass, fermentation, hydrolysis and hydrolysates, products, and evaluation.

The most prolific keywords across all of the research fronts are hydrolysis, sugar, cellulose, lignin, *Zea mays*, ethanol, enzymatic hydrolysis, fermentation, triticum, enzymes, enzyme activity, straw, glucose, cellulase, pretreatment, corn stover, hemicellulose, corn, wheat straw, lignocellulose, pre-treatment, saccharification, xylose, carbohydrates, and wheat. Similarly, the most influential keywords are hydrolysis, sugar, cellulose, *Zea mays*, triticum, enzymes, corn, cellulases, and lignin.

These findings suggest that it is necessary to determine the keyword set carefully to locate relevant research in each of these research fronts. Additionally, the size of the samples for each keyword highlights the intensity of research in relevant research areas. These findings also highlight different spelling of some strategic keywords: corn v. maize v. *Zea mays*, ethanol v. bioethanol v. bio-ethanol, wheat v. triticum, and rice v. oryza.

63.4.11 The Most Prolific Research Fronts in Second Generation Bioethanol Fuels from Residual Starch Feedstocks

Information about research fronts for the sample papers in second generation bioethanol fuels from residual starch feedstocks with regard to the residual starch feedstocks used for the bioethanol production is given in Table 63.9. As this table shows, the most prolific residual starch feedstocks are corn stover and to a lesser extent wheat straw and rice straw.

Information about the thematic research fronts for sample papers in second generation bioethanol fuels from residual starch feedstocks is given in Table 63.10. As this table shows, there are five primary research fronts: biomass pretreatments, hydrolysis, hydrolysate fermentation, production, and evaluation of second generation bioethanol fuels from residual starch feedstocks. The first two research fronts dominate the research in this field, compared to the fermentation of the hydrolysates, bioethanol production, and evaluation.

For the research front of residual starch feedstock pretreatments, the most prolific pretreatments are chemical and enzymatic pretreatments of biomass and to a lesser extent hydrothermal and mechanical pretreatments of biomass.

On the other hand, on an individual basis, the most prolific research fronts are enzymatic hydrolysis and to a lesser extent acid, alkaline, ammonia pretreatments, and steam explosion pretreatments of biomass. Further, the most prolific feedstocks used in all these research fronts are corn stover, wheat straw, and rice straw although their contents are not shown in this table.

These findings are thought-provoking in seeking ways to increase residual starch feedstock-based bioethanol yield at the global scale. It is clear that all of these research fronts have public importance and merit substantial funding and other incentives. Further, it is notable that second generation bioethanol fuels from residual starch feedstocks have become a core unit of the bioethanol research to make it more competitive with crude oil-based gasoline and petrodiesel fuels, especially for the USA, Europe, and China.

In comparison to the other feedstock-based research fronts, it is notable that pretreatment and hydrolysis of residual starch feedstocks emerge as a primary research front for this field. This suggests that primary stakeholders have been primarily interested in enzymatic hydrolysis as well as chemical and enzymatic pretreatments of residual biomass. It is also notable that evaluation of second generation bioethanol fuels from residual starch feedstocks such as technoeconomics, life cycle, economics, social, land use, labor, and environment-related studies emerges as a case study for the bioethanol fuels together with algal and sugar feedstocks in this field. In this context, the USA has been the global leader in production and use of corn-based bioethanol fuels since the 1970s in the aftermath of the global crude oil crisis in the early 1970s. It is also notable research in this field has been more intense compared to production and evaluation of the first generation bioethanol fuels from primary starch feedstocks such as corn, wheat, and rice with nearly triple sample size.

In the end, these most-cited papers in this field hint that production of second generation bioethanol fuels from residual starch feedstocks could be optimized using structure, processing, and property relationships of these residual starch feedstocks in fronts of feedstock pretreatment, hydrolysis, and hydrolysate fermentation (Formela et al., 2016; Konur, 2018a, 2020b, 2021a–d; Konur and Matthews, 1989).

63.5 CONCLUSION AND FUTURE RESEARCH

Research on second generation bioethanol fuels from residual starch feedstocks has been mapped through a scientometric study of both sample (203 papers) and population (4,056 papers) datasets.

The critical issue in this study has been to obtain a representative sample of research as in any other scientometric study. Therefore, the keyword set has been carefully devised and optimized after a number of runs in the Scopus database. It is a representative sample of the wider population studies. This keyword set was provided in appendix and the relevant keywords are presented in Table 63.8. However, it should be noted that it has been very difficult to compile a representative keyword set since this research field has been connected closely with many other fields. Therefore, it has been necessary to compile a keyword list to exclude papers concerned with other research fields.

It is notable in this context that research on production and evaluation of bioethanol fuels from starch feedstock residues such as corn stover, wheat straw, and rice straw is closely related to research on first generation bioethanol production from these primary feedstocks themselves, such as corn, wheat, and rice. Therefore, it is crucial to collect data on these two interconnected research fronts separately. Hence, studies on production and evaluation of primary starch feedstocks for bioethanol production were presented separately in third volume.

Another issue has been selection of a multidisciplinary database to carry out the scientometric study of research in this field. For this purpose, Scopus database has been selected. Journal coverage of this database has been notably wider than that of Web of Science and other multi-subject databases.

Key scientometric properties of research in this field have been determined and discussed in this book chapter. It is evident that a limited number of documents, authors, institutions, publication years, institutions, funding bodies, source titles, countries, Scopus subject categories, Scopus keywords, and research fronts have shaped the development of research in this field.

There is ample scope to increase the efficiency of the scientometric studies in this field in author and document domains by developing consistent policies and practices in both domains across all the academic databases. In this respect, it seems that authors, journals, and academic databases have a lot to do. Furthermore, a significant gender deficit as in most scientific fields emerges as a public policy issue. The potential deficits on the basis of age, race, disability, and sexuality need also to be explored in this field as in other scientific fields.

Research in this field has boomed since 2009, possibly promoted by the public concerns on global warming, greenhouse gas emissions, and climate change. Furthermore, the recent COVID-19 pandemic and Russian invasion of Ukraine have resulted in a global supply shocks shifting the focus of the stakeholders from crude oil-based fuels to biomass-based fuels such as bioethanol fuels. It is expected that there would be further incentives for key stakeholders to carry out research for second generation bioethanol fuels from residual starch feedstocks to increase ethanol yield and to make it more competitive with crude oil-based gasoline and petrodiesel fuels. This might be truer for crude oil- and foreign exchange-deficient countries to maintain the energy and food security in the face of the global supply shocks.

The relatively modest funding rate of 46% and 52% for the sample and population papers, respectively, suggests that funding in this field significantly enhanced research in this field primarily since 2009, possibly more than doubling in the current decade. However, it is evident that there is ample room for more funding and other incentives to enhance research in this field further in the light of the recent supply shocks and the stagnation of research output after 2014.

Institutions from the USA and to a lesser extent China have mostly shaped the research in this field. Further, the USA and to a lesser extent China and Europe have been the major producers of research in this field as the major producers and users of bioethanol fuels from different types of biomass such as corn, wheat, and rice as well as other types of biomass. It is evident that these countries have well-developed research infrastructure in bioethanol fuels and their derivatives. It is also notable these major countries mostly have access to the large farmlands.

It emerges that ethanol is more popular than bioethanol as a keyword with strong implications for the search strategy. In other words, the search strategy using only bioethanol keyword would not be much helpful. The Scopus keywords are grouped under the five headings: biomass, pretreatments, fermentation, hydrolysis and hydrolysates, and products. It is also important to use the Latin terms for biomass in the keyword set, besides English terms.

As Table 63.9 shows, corn stover and to a lesser extent wheat and rice straw are most the prolific feedstocks used in these studies. On the other hand, Table 63.10 shows that there are five primary research fronts: Biomass pretreatments and hydrolysis and to a lesser extent hydrolysate fermentation, second generation bioethanol fuel production, and evaluation from residual starch feedstocks.

These findings are thought-provoking in seeking ways to increase bioethanol yield through second generation bioethanol fuels from residual starch feedstocks at the global scale. It is clear that all of these research fronts have public importance and merit substantial funding and other incentives. Further, it is notable that second generation bioethanol fuels from residual starch feedstocks, as a second generation biofuel, have become a core unit of the bioethanol research to make it more competitive with crude oil-based gasoline and petrodiesel fuels, especially for the countries with large access to the farmlands.

In comparison to the other feedstock-based research fronts, it is notable that pretreatment and hydrolysis of residual starch feedstocks emerge as a primary research front for this field. This suggests that primary stakeholders have been primarily interested in enzymatic hydrolysis as well as chemical and enzymatic pretreatments of the residual biomass. It is also notable that evaluation of second generation bioethanol fuels from residual starch feedstocks is a case study for bioethanol fuels together with algal and starch feedstocks in this field. In this context, the USA has been the global leader in production and use of the corn-based bioethanol fuels since the 1970s in the aftermath of the global crude oil crisis in the early 1970s. It is also notable research in this field has been more intense compared to production and evaluation of the first generation bioethanol fuels from the primary starch feedstocks such as corn, wheat, and rice with nearly triple sample size.

Thus, the scientometric analysis has a great potential to gain valuable insights into the evolution of research in this field as in other scientific fields especially in the aftermath of the significant global supply shocks such as COVID-19 pandemic and the Russian invasion of Ukraine.

It is recommended that further scientometric studies are carried out for primary research fronts. It is further recommended that reviews of the most-cited papers are carried out for each primary research front to complement these scientometric studies. Next, scientometric studies of the hot papers in these primary fields are carried out.

ACKNOWLEDGMENTS

Contribution of the highly cited researchers in the field of second generation bioethanol fuels from residual starch feedstocks has been gratefully acknowledged.

APPENDIX: THE KEYWORD SET FOR SECOND GENERATION BIOETHANOL FUELS FROM STARCH FEEDSTOCKS

((((TITLE (ethanol OR bioethanol OR *saccharification OR *hydrolysis OR digestibili* OR ssf OR shf OR recalcitrance OR hydrolysate* OR hydrolyzate* OR ferment* OR coferment* OR fractionation OR delignification OR depolymerization OR microwave* OR autoclaving OR ultrasound OR sonicat* OR extrusion OR grinding OR pretreat* OR "pre-treat*" OR bioorganosolve OR steam* OR "hot water" OR "hot compressed" OR "sulfuric acid*" OR sulfite* OR sporl OR lime OR "sulfur dioxide" OR so2 OR alkali* OR naoh* OR extruder OR "sodium hydroxide*" OR "ionic liquid*" OR solvent* OR organosolv* OR ammonia OR {wet oxidation} OR "wet explosion" OR hydrothermolysis OR "supercritical co2" OR "supercritical carbon dioxide" OR surfactant* OR afex* OR "supercritical water" OR tween* OR hydrothermal OR pelleting OR pelletizing OR {sugar yield} OR precipitation OR "hydrogen peroxide" OR glycerol OR detoxification OR ozonolysis)) AND (TITLE (wheat OR triticum OR corn OR zea OR barley OR hordeum OR rye OR secale OR millet OR sorghum OR penisetum OR potato OR cassava OR manihot OR {sweet potato*} OR ipomea OR maize OR sago OR artichoke OR rice OR triticale) AND TITLE (bagasse OR straw OR leaves OR tops OR pulp OR {by-products} OR *cellulosic OR *cellulose OR trash OR "second generation" OR residu* OR {co-products} OR molasses OR pomaces OR vinasse OR "distillers grains" OR ddgs OR solubles OR 2g OR mud OR byproduct OR biomass OR cake OR clones OR fiber* OR fibre* OR *cobs OR waste OR stover OR peel OR stalks OR stems OR stillage OR cake OR hull OR bran OR coproduct OR {entire corn} OR liquor OR residue OR {germ meal} OR hampas OR slag OR husk))) AND NOT (TITLE ({sweet sorghum} OR ash OR oil OR nano* OR biogas OR anaerobic OR *butanol OR {*hydrogen prod*} OR pyrolysis OR protein* OR {sorghum bicolor} OR *methane OR anti* OR activated OR carboniz* OR adsorp* OR lactic OR xylitol OR amylase OR liquefaction OR *butyrate OR syngas OR levulinate OR dyes OR biochars OR *phenol OR chars OR silica OR seed* OR gasif* OR dispersant OR carboxylic OR *sorption OR "solid state") OR SRCTITLE (food* OR materials OR plant* OR lwt* OR phyto* OR soil OR animal* OR dairy OR livestock OR cereal* OR pyroly* OR poultry OR holz* OR oil* OR polymer* OR grass* OR wood* OR oleo* OR {crop science} OR nutr* OR carbohydrate* OR agric* OR hydrogen OR botan*) OR SUBJAREA (medi OR vete OR phar OR nurs OR heal OR psyc OR neur OR dent)))) OR (TITLE (ethanol OR bioethanol) AND TITLE (*diesel OR *hydrogen OR h2 OR *butanol OR biogas OR *methane OR anaerobic OR biorefinery) AND TITLE (wheat OR triticum OR corn OR zea OR barley OR hordeum OR rye OR secale OR millet OR sorghum OR penisetum OR potato OR cassava OR manihot OR {sweet potato*} OR ipomea OR maize OR sago OR artichoke OR rice OR triticale) AND TITLE (bagasse OR straw OR leaves OR tops OR pulp OR {by-products} OR *cellulosic OR *cellulose OR trash OR "second generation" OR residu* OR {co-products} OR molasses OR pomaces OR vinasse OR "distillers grains" OR ddgs OR solubles OR 2g OR mud OR byproduct OR biomass OR cake OR clones OR fiber* OR fibre* OR *cobs OR waste OR stover OR peel OR stalks OR stems OR stillage OR cake OR hull OR bran OR coproduct OR {entire corn} OR liquor OR residue OR {germ meal} OR hampas OR slag OR husk)) AND (LIMIT-TO (SRCTYPE, "j") OR LIMIT-TO (SRCTYPE, "k") OR LIMIT-TO (SRCTYPE, "b")) AND (LIMIT-TO (DOCTYPE, "ar") OR LIMIT-TO (DOCTYPE, "cp") OR LIMIT-TO (DOCTYPE, "ch") OR LIMIT-TO (DOCTYPE, "re") OR LIMIT-TO (DOCTYPE, "le") OR LIMIT-TO (DOCTYPE, "no") OR LIMIT-TO (DOCTYPE, "bk") OR LIMIT-TO (DOCTYPE, "ed") OR LIMIT-TO (DOCTYPE, "sh")) AND (LIMIT-TO (LANGUAGE, "English)).

REFERENCES

Aman, V. 2018. Does the Scopus author ID suffice to track scientific international mobility? A case study based on Leibniz laureates. *Scientometrics* 117:705–720.

Angelici, C., B. M. Weckhuysen and P. C. A. Bruijnincx. 2013. Chemocatalytic conversion of ethanol into butadiene and other bulk chemicals. *ChemSusChem* 6:1595–1614.

Antolini, E. 2007. Catalysts for direct ethanol fuel cells. *Journal of Power Sources* 170:1–12.

Antolini, E. 2009. Palladium in fuel cell catalysis. *Energy and Environmental Science* 2:915–931.

Beaudry, C. and V. Lariviere. 2016. Which gender gap? Factors affecting researchers' scientific impact in science and medicine. *Research Policy* 45:1790–1817.

Blankenship, K. M. 1993. Bringing gender and race in: US employment discrimination policy. *Gender & Society* 7:204–226.

Burnham, J. F. 2006. Scopus database: A review. *Biomedical Digital Libraries* 3:1–8.

Carlson, K. M., J. S. Gerber and D. Mueller, et al. 2017. Greenhouse gas emissions intensity of global croplands. *Nature Climate Change* 7:63–68.

Change, C. 2007. Climate change impacts, adaptation and vulnerability. *Science of the Total Environment* 326:95–112.

Dirth, T. P. and N. R. Branscombe. 2017. Disability models affect disability policy support through awareness of structural discrimination. *Journal of Social Issues* 73:413–442.

Ebadi, A. and A. Schiffauerova. 2016. How to boost scientific production? A statistical analysis of research funding and other influencing factors. *Scientometrics* 106:1093–1116.

Fang, X., Y. Shen, J. Zhao, X. Bao and Y. Qu. 2010. Status and prospect of lignocellulosic bioethanol production in China. *Bioresource Technology* 101:4814–4819.

Fauci, A. S., H. C. Lane and R. R. Redfield. 2020. Covid-19-navigating the uncharted. *New England Journal of Medicine* 382:1268–1269.

Fernando, S., S. Adhikari, C. Chandrapal and M. Murali. 2006. Biorefineries: Current status, challenges, and future direction. *Energy & Fuels* 20:1727–1737.

Formela, K., A. Hejna, L. Piszczyk, M. R. Saeb and X. Colom. 2016. Processing and structure-property relationships of natural rubber/wheat bran biocomposites. *Cellulose* 23:3157–3175.

Garfield, E. 1955. Citation indexes for science. *Science* 122:108–111.

Gillon, S. 2010. Fields of dreams: Negotiating an ethanol agenda in the Midwest United States. *Journal of Peasant Studies* 37:723–748.

Gnansounou, E. 2010. Production and use of lignocellulosic bioethanol in Europe: Current situation and perspectives. *Bioresource Technology* 101:4842–4850.

Hahn-Hagerdal, B., M. Galbe, M. F. Gorwa-Grauslund, G. Liden and G. Zacchi. 2006. Bio-ethanol: The fuel of tomorrow from the residues of today. *Trends in Biotechnology* 24:549–556.

Hamilton, J. D. 1983. Oil and the macroeconomy since World War II. *Journal of Political Economy* 91:228–248.

Hamilton, J. D. 2003. What is an oil shock? *Journal of Econometrics* 113:363–398.

Hamilton, J. D. 2009. Causes and consequences of the oil shock of 2007-08. *Brookings Papers on Economic Activity* 2009:215–261.

Hill, J., E. Nelson, D. Tilman, S. Polasky and D. Tiffany. 2006. Environmental, economic, and energetic costs and benefits of biodiesel and ethanol biofuels. *Proceedings of the National Academy of Sciences of the United States of America* 103:11206–11210.

Hill, J., S. Polasky and E. Nelson, et al. 2009. Climate change and health costs of air emissions from biofuels and gasoline. *Proceedings of the National Academy of Sciences of the United States of America* 106:2077–2082.

Hsieh, W. D., R. H. Chen, T. L. Wu and T. H. Lin. 2002. Engine performance and pollutant emission of an SI engine using ethanol-gasoline blended fuels. *Atmospheric Environment* 36:403–410.

Hsu, T. C., G. L. Guo, W. H. Chen and W. S. Hwang. 2010. Effect of dilute acid pretreatment of rice straw on structural properties and enzymatic hydrolysis. *Bioresource Technology* 101:4907–4913.

Huang, H. J., S. Ramaswamy, U. W. Tschirner and B. V. Ramarao. 2008. A review of separation technologies in current and future biorefineries. *Separation and Purification Technology* 62:1–21.

Jones, T. C. 2012. America, oil, and war in the Middle East. *Journal of American History* 99:208–218.

Kabel, M. A., G. Bos, J. Zeevalking, A. G. J. Voragen and H. A. Schols. 2007. Effect of pretreatment severity on xylan solubility and enzymatic breakdown of the remaining cellulose from wheat straw. *Bioresource Technology* 98:2034–2042.

Kaparaju, P., M. Serrano, A. B. Thomsen, P. Kongjan and I. Angelidaki. 2009. Bioethanol, biohydrogen and biogas production from wheat straw in a biorefinery concept. *Bioresource Technology* 100:2562–2568.

Kazi, F. K., J. A. Fortman and R. P. Anex, et al. 2010. Techno-economic comparison of process technologies for biochemical ethanol production from corn stover. *Fuel* 89:S20–S28.

Kerr, R. A. 2007. Global warming is changing the world. *Science* 316:188–190.

Kilian, L. 2008. Exogenous oil supply shocks: How big are they and how much do they matter for the US economy? *Review of Economics and Statistics* 90:216–240.

Kilian, L. 2009. Not all oil price shocks are alike: Disentangling demand and supply shocks in the crude oil market. *American Economic Review* 99:1053–1069.

Kim, J. 2018. Evaluating author name disambiguation for digital libraries: A case of DBLP. *Scientometrics* 116:1867–1886.

Kim, T. H., J. S. Kim, C. Sunwoo and Y. Y. Lee. 2003. Pretreatment of corn stover by aqueous ammonia. *Bioresource Technology* 90:39–47.

Konur, O. 2000. Creating enforceable civil rights for disabled students in higher education: An institutional theory perspective. *Disability & Society* 15:1041–1063.

Konur, O. 2002a. Access to nursing education by disabled students: Rights and duties of nursing programs. *Nurse Education Today* 22:364–374.

Konur, O. 2002b. Assessment of disabled students in higher education: Current public policy issues. *Assessment and Evaluation in Higher Education* 27:131–152.

Konur, O. 2002c. Access to employment by disabled people in the UK: Is the disability discrimination act working? *International Journal of Discrimination and the Law* 5:247–279.

Konur, O. 2006a. Participation of children with dyslexia in compulsory education: Current public policy issues. *Dyslexia* 12:51–67.

Konur, O. 2006b. Teaching disabled students in higher education. *Teaching in Higher Education* 11:351–363.

Konur, O. 2007a. A judicial outcome analysis of the disability discrimination act: A windfall for the employers? *Disability & Society* 22:187–204.

Konur, O. 2007b. Computer-assisted teaching and assessment of disabled students in higher education: The interface between academic standards and disability rights. *Journal of Computer Assisted Learning* 23:207–219.

Konur, O. 2011. The scientometric evaluation of the research on the algae and bio-energy. *Applied Energy* 88:3532–3540.

Konur, O. 2012a. The evaluation of the biogas research: A scientometric approach. *Energy Education Science and Technology Part A: Energy Science and Research* 29:1277–1292.

Konur, O. 2012b. The evaluation of the educational research: A scientometric approach. *Energy Education Science and Technology Part B: Social and Educational Studies* 4:1935–1948.

Konur, O. 2012c. The evaluation of the global energy and fuels research: A scientometric approach. *Energy Education Science and Technology Part A: Energy Science and Research* 30:613–628.

Konur, O. 2012d. The evaluation of the research on the biodiesel: A scientometric approach. *Energy Education Science and Technology Part A: Energy Science and Research* 28:1003–1014.

Konur, O. 2012e. The evaluation of the research on the bioethanol: A scientometric approach. *Energy Education Science and Technology Part A: Energy Science and Research* 28:1051–1064.

Konur, O. 2012f. The evaluation of the research on the biofuels: A scientometric approach. *Energy Education Science and Technology Part A: Energy Science and Research* 28:903–916.

Konur, O. 2012g. The evaluation of the research on the biohydrogen: A scientometric approach. *Energy Education Science and Technology Part A: Energy Science and Research* 29:323–338.

Konur, O. 2012h. The evaluation of the research on the microbial fuel cells: A scientometric approach. *Energy Education Science and Technology Part A: Energy Science and Research* 29:309–322.

Konur, O. 2012i. The scientometric evaluation of the research on the production of bioenergy from biomass. *Biomass and Bioenergy* 47:504–515.

Konur, O. 2015. Current state of research on algal bioethanol. In *Marine Bioenergy: Trends and Developments*, Eds. S. K. Kim and C. G. Lee, pp. 217–244. Boca Raton, FL: CRC Press.

Konur, O., Ed. 2018a. *Bioenergy and Biofuels*. Boca Raton, FL: CRC Press.

Konur, O. 2018b. Bioenergy and biofuels science and technology: Scientometric overview and citation classics. In *Bioenergy and Biofuels*, Ed. O. Konur, pp. 3–63. Boca Raton, FL: CRC Press.

Konur, O. 2019. Cyanobacterial bioenergy and biofuels science and technology: A scientometric overview. In *Cyanobacteria: From Basic Science to Applications*, Eds. A. K. Mishra, D. N. Tiwari and A. N. Rai, pp. 419–442. Amsterdam: Elsevier.

Konur, O. 2020a. The scientometric analysis of the research on the bioethanol production from green macroalgae. In *Handbook of Algal Science, Technology and Medicine*, Ed. O. Konur, pp. 385–401. London: Academic Press.

Konur, O., Ed. 2020b. *Handbook of Algal Science, Technology and Medicine*. London: Academic Press.

Konur, O., Ed. 2021a. *Handbook of Biodiesel and Petrodiesel Fuels: Science, Technology, Health, and Environment*. Boca Raton, FL: CRC Press.

Konur, O., Ed. 2021b. *Handbook of Biodiesel and Petrodiesel Fuels: Science, Technology, Health, and Environment. Volume 1. Biodiesel Fuels: Science, Technology, Health, and Environment*. Boca Raton, FL: CRC Press.

Konur, O., Ed. 2021c. *Handbook of Biodiesel and Petrodiesel Fuels: Science, Technology, Health, and Environment. Volume 2. Biodiesel Fuels Based on the Edible and Nonedible Feedstocks, Wastes, and Algae: Science, Technology, Health, and Environment*. Boca Raton, FL: CRC Press.

Konur, O., Ed. 2021d. *Handbook of Biodiesel and Petrodiesel Fuels: Science, Technology, Health, and Environment. Volume 3. Petrodiesel Fuels: Science, Technology, Health, and Environment*. Boca Raton, FL: CRC Press.

Konur, O. and F. L. Matthews. 1989. Effect of the properties of the constituents on the fatigue performance of composites: A review. *Composites* 20:317–328.

Kruyt, B., D. P. van Vuuren, H. J. de Vries and H. Groenenberg. 2009. Indicators for energy security. *Energy Policy* 37:2166–2181.

Kumar, R., G. Mago, G., V. Balan and C. E. Wyman. 2009. Physical and chemical characterizations of corn stover and poplar solids resulting from leading pretreatment technologies. *Bioresource Technology* 100:3948–3962.

Leydesdorff, L. 2000. Is the European Union becoming a single publication system? *Scientometrics* 47:265–280.

Leydesdorff, L. and C. Wagner. 2009. Is the United States losing ground in science? A global perspective on the world science system. *Scientometrics* 78:23–36.

Leydesdorff, L. and P. Zhou. 2005. Are the contributions of China and Korea upsetting the world system of science? *Scientometrics* 63:617–630.

Li, H., S. M. Liu, X. H. Yu, S. L. Tang and C. K. Tang. 2020. Coronavirus disease 2019 (COVID-19): Current status and future perspectives. *International Journal of Antimicrobial Agents* 55:105951.

Lin, Y. and S. Tanaka. 2006. Ethanol fermentation from biomass resources: Current state and prospects. *Applied Microbiology and Biotechnology* 69:627–642.

Lloyd, T. A. and C. E. Wyman. 2005. Combined sugar yields for dilute sulfuric acid pretreatment of corn stover followed by enzymatic hydrolysis of the remaining solids. *Bioresource Technology* 96:1967–1977.

Ma, X., L. Sun and C. Song. 2002. A new approach to deep desulfurization of gasoline, diesel fuel and jet fuel by selective adsorption for ultra-clean fuels and for fuel cell applications. *Catalysis Today* 77:107–116.

Mani, S., L. G. Tabil and S. Sokhansanj. 2004. Grinding performance and physical properties of wheat and barley straws, corn stover and switchgrass. *Biomass and Bioenergy* 27:339–352.

Morschbacker, A. 2009. Bio-ethanol based ethylene. *Polymer Reviews* 49:79–84.

Mosier, N., R. Hendrickson, N. Ho, M. Sedlak and M. R. Ladisch. 2005. Optimization of pH controlled liquid hot water pretreatment of corn stover. *Bioresource Technology* 96:1986–1993.

Najafi, G., B. Ghobadian and T. Tavakoli, et al. 2009. Performance and exhaust emissions of a gasoline engine with ethanol blended gasoline fuels using artificial neural network. *Applied Energy* 86:630–639.

Newman, P. W. G. and J. R. Kenworthy. 1989. Gasoline consumption and cities: A comparison of U.S. cities with a global survey. *Journal of the American Planning Association* 55:24–37.

North, D. C. 1991. Institutions. *Journal of Economic Perspectives* 5:97–112.

Olsson, L. and B. Hahn-Hagerdal. 1996. Fermentation of lignocellulosic hydrolysates for ethanol production. *Enzyme and Microbial Technology* 18:312–331.

Reeves, S. 2014. To Russia with love: How moral arguments for a humanitarian intervention in Syria opened the door for an invasion of the Ukraine. *Michigan State University International Law Review* 23:199.

Saha, B. C., L. B. Iten, M. A. Cotta and Y. V. Wu. 2005. Dilute acid pretreatment, enzymatic saccharification and fermentation of wheat straw to ethanol. *Process Biochemistry* 40:3693–3700.

Sanchez, O. J. and C. A. Cardona. 2008. Trends in biotechnological production of fuel ethanol from different feedstocks. *Bioresource Technology* 99:5270–5295.

Sheehan, J., A. Aden and K. Paustian, et al. 2003. Energy and environmental aspects of using corn stover for fuel ethanol. *Journal of Industrial Ecology* 7:117–146.

Sun, Y. and J. Cheng. 2002. Hydrolysis of lignocellulosic materials for ethanol production: A review. *Bioresource Technology* 83:1–11.

Sun, Y. and J. J. Cheng. 2005. Dilute acid pretreatment of rye straw and bermudagrass for ethanol production. *Bioresource Technology* 96:1599–1606.

Taherzadeh, M. J. and K. Karimi. 2007. Enzyme-based hydrolysis processes for ethanol from lignocellulosic materials: A review. *Bioresources* 2:707–738.

Taherzadeh, M. J. and K. Karimi. 2008. Pretreatment of lignocellulosic wastes to improve ethanol and biogas production: A review. *International Journal of Molecular Sciences* 9:1621–1651.

Talebnia, F., D. Karakashev and I. Angelidaki. 2010. Production of bioethanol from wheat straw: An overview on pretreatment, hydrolysis and fermentation. *Bioresource Technology* 101:4744–4753.

Teymouri, F., L. Laureano-Perez, H. Alizadeh and B. E. Dale. 2005. Optimization of the ammonia fiber explosion (AFEX) treatment parameters for enzymatic hydrolysis of corn stover. *Bioresource Technology* 96:2014–2018.

Winzer, C. 2012. Conceptualizing energy security. *Energy Policy* 46:36–48.

Yang, B. and C. E. Wyman. 2004. Effect of xylan and lignin removal by batch and flowthrough pretreatment on the enzymatic digestibility of corn stover cellulose. *Biotechnology and Bioengineering*, 86:88–98.

Yang, B. and C. E. Wyman. 2008. Pretreatment: The key to unlocking low-cost cellulosic ethanol. *Biofuels, Bioproducts and Biorefining* 2:26–40.

64 Second Generation Bioethanol Fuels from Residual Starch Feedstocks *Review*

Ozcan Konur
(Formerly) Ankara Yildirim Beyazit University

64.1 INTRODUCTION

Crude oil-based gasoline fuels (Ma et al., 2002; Newman and Kenworthy, 1989) have been widely used in the transportation sector since the 1920s. However, there have been great public concerns over the adverse environmental and human impact of these fuels (Hill et al., 2006, 2009). Hence, biomass-based bioethanol fuels (Hill et al., 2006; Konur, 2012, 2015, 2020) have increasingly been used in blending gasoline fuels (Hsieh et al., 2002; Najafi et al., 2009), in fuel cells (Antolini, 2007, 2009), and in biochemical production (Angelici et al., 2013; Morschbacker, 2009) in a biorefinery context (Fernando et al., 2006; Huang et al., 2008).

However, it is necessary to pretreat the biomass (Alvira et al., 2010; Taherzadeh and Karimi, 2008) to enhance the yield of bioethanol (Hahn-Hagerdal et al., 2006; Sanchez and Cardona, 2008) prior to bioethanol fuel production from feedstocks through hydrolysis (Sun and Cheng, 2002; Taherzadeh and Karimi, 2007) and fermentation (Lin and Tanaka, 2006; Olsson and Hahn-Hagerdal, 1996) of the biomass and hydrolysates, respectively.

One of the most-studied feedstocks for bioethanol fuels has been residual starch feedstocks such as corn stover and wheat straw and to a lesser extent rice straw, rye straw, and barley straw. The research in the field of second generation bioethanol fuels from residual starch feedstocks has intensified in this context in key research fronts of the pretreatment (Kim et al., 2003; Kumar et al., 2009; Mani et al., 2004) and hydrolysis (Hsu et al., 2010; Lloyd and Wyman, 2005; Yang and Wyman, 2004) of residual starch feedstocks, the fermentation (Mosier et al., 2005; Saha et al., 2005; Talebnia et al., 2010) of residual starch feedstock-based hydrolysates, and the production (Kaparaju et al., 2009; Mosier et al., 2005; Teymouri et al., 2005) and evaluation (Kazi et al., 2010; Sheehan et al., 2003) of second generation bioethanol fuels from residual starch feedstocks. Further, corn stover (Kazi et al., 2010; Mosier et al., 2005; Sheehan et al., 2003) and wheat straw (Kabel et al., 2007; Kaparaju et al., 2009; Saha et al., 2005) and to a lesser extent rice straw (Hsu et al., 2010), rye straw (Sun and Cheng, 2005), and barley straw (Mani et al., 2004) have been studied intensively in this context.

However, it is essential to develop efficient incentive structures (North, 1991) for the primary stakeholders to enhance research in this field (Konur, 2000, 2002a–c, 2006a,b, 2007a,b). Although there has been a number of review papers on bioethanol fuels from residual starch feedstocks (Binod et al., 2010; Singh et al., 2016; Talebnia et al., 2010), there has been no review of the 25 most-cited papers in this field.

Thus, this chapter presents a review of the 25 most-cited articles in the field of bioethanol fuels from residual starch feedstocks. Then, it discusses the key findings of these highly influential papers and comments on future research priorities in this field.

DOI: 10.1201/9781003226550-86

64.2 MATERIALS AND METHODS

The search for this study was carried out using the Scopus database (Burnham, 2006) in August 2022.

As the first step for the search of the relevant literature, keywords were selected using the 200 most-cited population papers. The selected keyword list was then optimized to obtain a representative sample of papers from this research field. This final keyword set was provided in the appendix of Konur (2023) for future replication studies.

As the second step, a sample data set was used for this study. First, 25 articles with at least 326 citations each were selected for the review study. Key findings from each paper were taken from the abstracts of these papers and were discussed. Additionally, a number of brief conclusions were drawn and a number of relevant recommendations were made to enhance the future research landscape.

64.3 RESULTS

Brief information about the 25 most-cited papers with at least 326 citations each on second generation bioethanol fuels from residual starch feedstocks is given below. The primary research fronts are pretreatments and hydrolysis of residual starch feedstocks and production and evaluation of bioethanol fuels from residual starch feedstocks with 18 and 7 highly cited papers (HCPs), respectively. There are 3 and 15 HCPs for the pretreatments and hydrolysis of residual starch feedstocks, respectively. Similarly, there are five and two HCPs for the production and evaluation of bioethanol fuels from residual starch feedstocks, respectively.

64.3.1 Biomass Pretreatments and Hydrolysis

Brief information about the 18 most-cited papers on the pretreatment and hydrolysis of residual starch feedstocks with at least 326 citations each is given below (Table 64.1). There are 3 and 15 HCPs for pretreatments and hydrolysis of residual starch feedstocks, respectively.

64.3.1.1 Biomass Pretreatments

Kumar et al. (2009) characterized corn stover and poplar solids resulting from leading pretreatment technologies in a paper with 719 citations. They employed ammonia fiber expansion (AFEX), ammonia recycled percolation (ARP), controlled pH, dilute acid, flowthrough, lime, and SO_2 technologies. They observed that lime pretreatment removed the highest acetyl groups from both corn stover and poplar, while AFEX removed the lowest. Low pH pretreatments depolymerized cellulose and enhanced biomass crystallinity much more than higher pH approaches. Lime-pretreated corn stover solids and flowthrough pretreated poplar solids had the highest cellulase adsorption capacity, while dilute acid pretreated corn stover solids and controlled pH pretreated poplar solids had the lowest. Furthermore, enzymatically extracted AFEX lignin preparations for both corn stover and poplar had the lowest cellulase adsorption capacity. SO_2 pretreated solids had the highest surface O/C ratio for poplar, but for corn stover, the highest value was observed for dilute acid pretreatment. Although dependent on pretreatment and substrate, along with changes in cross-linking and chemical changes, these pretreatments might also decrystallize cellulose and change the ratio of crystalline cellulose polymorphs (I_α/I_β).

Mani et al. (2004) studied the grinding performance and physical properties of wheat and barley straws, corn stover, and switchgrass in a paper with 523 citations. They ground these feedstocks at two moisture contents using a hammer mill with three different screen sizes (3.2, 1.6, and 0.8 mm). Among the four materials, they observed that switchgrass had the highest specific energy consumption (27.6 kW h/t), and corn stover had the least specific energy consumption (11.0 kW h/t) at a 3.2 mm screen size. They developed second- or third-order polynomial models relating bulk and

TABLE 64.1
The Pretreatment and Hydrolysis of Residual Starch Feedstocks

No.	Papers	Biomass/ Hydrolysate	Res. Fronts	Prts.	Parameters	Keywords	Lead Author	Affil.	Cits.
1	Kumar et al. (2009)	Corn stover	Pretreatments	Ammonia, acid, alkali, SO_2, cellulase	Pretreatments, enzymatic hydrolysis, characterization	Corn stover, pretreatment	Charles E. Wyman 7004396809	Univ. Calif. Riverside USA	719
2	Lloyd and Wyman (2005)	Corn stover	Hydrolysis	H_2SO_4, cellulase	Pretreatment, enzymatic hydrolysis, sugar yields, glucose, and xylose yields	Corn stover, pretreatments, hydrolysis	Charles E. Wyman 7004396809	Univ. Calif. Riverside USA	624
3	Yang and Wyman (2004)	Corn stover	Hydrolysis	H_2SO_4, enzymes	Enzymatic hydrolysis, pretreatments, xylan, and lignin removal	Corn stover, pretreatment, digestibility	Charles E. Wyman 7004396809	Univ. Calif. Riverside USA	568
4	Mani et al. (2004)	Wheat straw, barley straw, corn stover	Pretreatments	Milling	Milling, feedstocks, specific energy consumption, calorific value, ash content	Wheat, barley straw, corn stover, grinding	Tabil, Lope G 6701349307	Univ. Saskatchewan Canada	523
5	Kim et al. (2003)	Corn stover	Pretreatments	Ammonia, enzyme	Pretreatment, delignification, enzymatic hydrolysis, hemicellulose removal	Corn stover, pretreatment, ammonia	Lee, Yoon Y. 8948274900	Auburn Univ. USA	502
6	Hsu et al. (2010)	Rice straw	Hydrolysis	H_2SO_4, enzymes	Pretreatments, enzymatic hydrolysis, sugar yield, hydrolysis efficiency	Rice straw, pretreatment, hydrolysis	Gia-Luen Guo 7402768046	Inst. Nucl. Ener. Res. Taiwan	485
7	Ohgren et al. (2007)	Corn stover	Hydrolysis	Steam, acid, xylanases, cellulases	Pretreatments, enzymatic hydrolysis, hemicellulose and lignin removal, glucose, and xylose yield, delignification	Corn stover, hydrolysis, steam	Zacchi, Guido 7006727748	Lund Univ. Sweden	469
8	Wyman et al. (2005)	Corn stover	Hydrolysis	Ammonia, acid, alkali, cellulase	Pretreatments, enzymatic hydrolysis, sugar yield	Corn stover, pretreatments	Charles E. Wyman 7004396809	Univ. Calif. Riverside USA	453
9	Sun and Cheng (2005)	Rye straw	Hydrolysis	H_2SO_4, cellulase	Pretreatments, enzymatic hydrolysis, glucose, xylose yield	Rye straw, ethanol, pretreatment	Cheng, Jay J. 15046539600	NC State Univ. USA	409

(Continued)

TABLE 64.1 (*Continued*)
The Pretreatment and Hydrolysis of Residual Starch Feedstocks

No.	Papers	Biomass/ Hydrolysate	Res. Fronts	Prts.	Parameters	Keywords	Lead Author	Affil.	Cits.
10	Kabel et al. (2007)	Wheat straw	Hydrolysis	Pretreatments, enzymes	Pretreatments, enzymatic hydrolysis, pretreatment severity, xylan solubility, cellulose degradability	Wheat straw, pretreatment	Schols, Henk A. 7005663012	Wageningen Univ. Res. Netherlands	395
11	Selig et al. (2007)	Corn stover	Hydrolysis	H_2SO_4, enzymes	Pretreatments, enzymatic hydrolysis, lignin droplet formation	Maize stems, pretreatment, hydrolysis	Selig, Michael J. 13613665100	Cornell Univ. USA	386
12	Kristensen et al. (2008)	Wheat straw	Hydrolysis	Hydrothermal, steam, enzymes	Pretreatments, enzymatic hydrolysis, lignin re-localization, hemicellulose removal	Wheat straw, bioethanol, pretreated	Kristensen, Jan B. 35322181100	Novozymes Denmark	380
13	Laureano-Perez et al. (2005)	Corn stover	Hydrolysis	Ammonia, enzymes	Pretreatments, enzymatic hydrolysis determinants, and hydrolysis initial rate and extent	Corn stover, pretreated, hydrolysis	Dale, Bruce E. 7201511969	Michigan State Univ. USA	379
14	Kim and Holtzapple (2005)	Corn stover	Hydrolysis	Lime, cellulase	Pretreatments, enzymatic hydrolysis determinants, hydrolysis initial rate and extent, glucose and xylose yields	Corn stover, pretreatment, hydrolysis, lime	Holtzapple, Mark T. 7004167004	Texas A&M Univ. USA	370
15	Kim and Holtzapple (2006)	Corn stover	Hydrolysis	Lime, cellulase	Pretreatments, enzymatic hydrolysis determinants, delignification, deacetylation, degree of crystallinity	Corn stover, digestibility	Holtzapple, Mark T. 7004167004	Texas A&M Univ. USA	355
16	Bjerre et al. (1996)	Wheat straw	Hydrolysis	Wet oxidation, alkali, enzymes	Pretreatments, enzymatic hydrolysis, glucose yield	Wheat straw, wet oxidation, alkali, pretreatment	Bjerre, Anne B.* 6701773173	Danish Technol. Inst. Denmark	352
17	Klinke et al. (2002)	Wheat straw	Hydrolysis	Wet oxidation, alkali, enzymes	Pretreatments, enzymatic hydrolysis, glucose yield, fermentation inhibitors, carboxylic acids	Wheat straw, wet oxidation, alkali	Thomsen, Anne B.* 7102150211	Tech. Univ. Denmark Denmark	334
18	Kaar and Holtzapple (2000)	Corn stover	Hydrolysis	Lime, enzymes	Pretreatments, enzymatic hydrolysis, enzyme loading, hydrolysis temperature, conversion rate	Corn stover, hydrolysis, pretreatment, lime	Holtzapple, Mark T. 7004167004	Texas A&M Univ. USA	326

Prt., Biomass pretreatments; Cits., Number of citations received for each paper; *, Female.

particle densities of grinds to geometric mean diameters within the range of 0.18–1.43 mm. Further, switchgrass had the highest calorific value and the lowest ash content for these feedstocks.

Kim et al. (2003) studied the effect of pretreated corn stover with aqueous ammonia in a flowthrough column reactor, ammonia recycled percolation (ARP), in a paper with 502 citations. They found that ARP was highly effective in the delignifying of biomass, reducing the lignin content by 70%–85% where the highest lignin removal occurred within the first 20 min of the process. Further, the ARP process solubilized 40%–60% of the hemicellulose but left the cellulose intact. Corn stover treated for 90 min had enzymatic digestibility of 99% with 60 FPU/g of glucan enzyme loading and 92.5% with 10 FPU/g of glucan. Further, the digestibility of ARP-treated corn stover was substantially higher than that of α-cellulose. Enzymatic digestibility was related to the removal of lignin and hemicellulose, perhaps due to increased surface area and porosity. The biomass structure was deformed and its fibers were exposed by the pretreatment while the crystallinity index increased with pretreatment reflecting the removal of the amorphous portion of the biomass. The crystalline structure of cellulose in the biomass, however, was not changed by the ARP treatment.

64.3.1.2 Biomass Enzymatic Hydrolysis

Lloyd and Wyman (2005) studied the dilute acid pretreatment and enzymatic hydrolysis of corn stover in a paper with 624 citations. They employed sulfuric acid (H_2SO_4) concentrations of 0.22%, 0.49%, and 0.98% w/w at 140°C, 160°C, 180°C, and 200°C. They observed that up to 15% of the total potential sugar in the substrate could be released as glucose during pretreatment and between 15% and 90+% of the xylose remaining in the solid residue could be recovered in subsequent enzymatic hydrolysis, depending on the enzyme loading. Glucose yields increased from as high as 56% of total maximum potential glucose plus xylose for just enzymatic digestion to 60% when glucose released in pretreatment was included. Xylose yields similarly increased from as high as 34% of total potential sugars for pretreatment alone to between 35% and 37% when credit was taken for xylose released in digestion. However, yields were much lower if no acid was used. In conclusion, up to about 92.5% of the total sugars originally available in the corn stover used could be recovered for coupled dilute acid pretreatment and enzymatic hydrolysis. Further, enhanced hemicellulase activity could further improve xylose yields, particularly for low cellulase loadings.

Yang and Wyman (2004) studied the effect of xylan and lignin removal by batch and flowthrough pretreatment on the enzymatic digestibility of corn stover cellulose in a paper with 568 citations. They performed experiments in batch and flowthrough reactors over a range of flow rates between 160°C and 220°C, with water only and also with 0.1 wt% sulfuric acid. They observed that increasing flow with just water enhanced the xylan dissolution rate, more than doubled total lignin removal, and increased cellulose digestibility. Furthermore, adding dilute sulfuric acid increased the rate of xylan removal for both batch and flowthrough systems. Interestingly, adding acid also increased the lignin removal rate with flow, but less lignin was left in the solution when acid was added in batch. Although the enzymatic hydrolysis of pretreated cellulose was related to xylan removal, the digestibility was much better for flowthrough compared with batch systems, for the same degree of xylan removal. Cellulose digestibility for flowthrough reactors was related to lignin removal as well. In conclusion, altering lignin also affected the enzymatic digestibility of corn stover.

Hsu et al. (2010) studied the effect of dilute acid pretreatment of rice straw on structural properties and enzymatic hydrolysis in a paper with 485 citations. They obtained a maximal sugar yield of 83% when the rice straw was pretreated with 1% (w/w) sulfuric acid with a reaction time of 1–5 min at 160°C or 180°C, followed by enzymatic hydrolysis. The complete release of sugar (xylose and glucose) increased the pore volume of the pretreated solid residues and resulted in an efficiency of 70% for enzymatic hydrolysis. Further, extra pore volume was generated by the release of acid-soluble lignin and this resulted in the enzymatic hydrolysis being enhanced by nearly 10%. The increase in the crystallinity index of the pretreated rice straw was limited.

Ohgren et al. (2007) studied the effect of hemicellulose and lignin removal on enzymatic hydrolysis of steam-pretreated corn stover in a paper with 469 citations. They observed that a near-theoretical glucose yield (96%–104%) from acid-catalyzed steam-pretreated corn stover could be obtained if xylanases were used to supplement cellulases during hydrolysis as xylanases hydrolyzed residual hemicellulose, thereby improving the access of enzymes to cellulose. Under these conditions, xylose yields reached 70%–74%. When pretreatment severity was reduced by using autocatalysis instead of acid-catalyzed steam pretreatment, xylose yields were increased to 80%–86%. The overall glucose yield increased slightly due to delignification but the overall xylose yield decreased due to hemicellulose loss during the delignification step.

Wyman et al. (2005) compared sugar recovery data from laboratory-scale applications of leading pretreatment technologies to corn stover in a paper with 453 citations. They employed an ammonia explosion, aqueous ammonia recycle, controlled pH, dilute acid, flowthrough, and lime using a single source of corn stover and the same cellulase enzyme. They observed that each pretreatment achieved high yields of glucose from cellulose with cellulase enzymes, and the cellulase formulations used were effective in solubilizing residual xylan left in the solids after each pretreatment. Thus, overall sugar yields from hemicellulose and cellulose in the coupled pretreatment and enzymatic hydrolysis operations were high for all the pretreatments with corn stover. In addition, high-pH methods had promise in reducing cellulase use provided hemicellulase activity could be enhanced. In conclusion, the substantial differences in sugar release patterns in the pretreatment and enzymatic hydrolysis operations had important implications for the choice of process, enzymes, and fermentative organisms.

Sun and Cheng (2005) studied the dilute acid pretreatment of rye straw and bermudagrass before enzymatic hydrolysis of cellulose for ethanol production in a paper with 405 citations. They pretreated the biomass at a solid loading rate of 10% at 121°C with different sulfuric acid concentrations (0.6%, 0.9%, 1.2%, and 1.5%, w/w) and residence times (30, 60, and 90 min). They then hydrolyzed the solid residues with cellulases. With the increasing acid concentration and residence time, they found that the amount of arabinose and galactose in the filtrates increased. However, the glucose concentration in the prehydrolysate of rye straw was not significantly influenced by the sulfuric acid concentration and residence time, but it increased in the prehydrolysate of bermudagrass with an increase in pretreatment severity. Further, xylose concentration in the filtrates increased with an increase in sulfuric acid concentration and residence time. Most of the arabinan, galactan, and xylan in the biomass were hydrolyzed during the acid pretreatment while the cellulose remaining in the pretreated feedstock was highly digestible with cellulases from Trichoderma reesei.

Kabel et al. (2007) studied the effect of pretreatment severity (R_0) on xylan solubility and the enzymatic breakdown of the remaining cellulose from wheat straw in a paper with 295 citations. They observed that the higher the severity factor the more xylan was released from the wheat straw, but more xylan decomposed and furfural formation occurred. The percentage of residual xylan present after pretreatment was a good indicator concerning cellulose degradability or bioethanol production. Further, cellulose degradation by using commercial enzymes was higher at higher severities corresponding to a lower amount of residual xylan. The presence of (acetylated) xylans with a degree of polymerization (DP) of 9–25 increased slightly from low to medium severity. Finally, the quantification of the DP-distribution of the (acetylated) xylans released was a good tool to predict cellulose degradability.

Selig et al. (2007) studied the effect of dilute acid pretreatment of corn stems on the enzymatic hydrolysis of cellulose in a paper with 386 citations. They observed spherical formations on the surface of the residual biomass after this pretreatment. They hypothesized that these droplet formations were composed of lignins and possible lignin carbohydrate complexes (lignin droplets). They showed that these droplets were produced from corn stover during pretreatment under neutral and acidic pH at and above 130°C and that they could deposit back onto the surface of the residual biomass. The deposition of droplets produced under certain pretreatment conditions (acidic pH; $T > 150$°C) and captured onto pure cellulose had a negative effect (5%–20%) on the enzymatic hydrolysis of this substrate. Further, droplet density (per unit area) was greater and droplet size more variable under conditions where the greatest impact on enzymatic cellulose conversion was observed. In conclusion, this lignin droplet phenomenon had the potential to adversely affect the efficiency of enzymatic conversion in a lignocellulosic biorefinery.

Kristensen et al. (2008) explored the effect of hydrothermal pretreatment on the straw cell-wall matrix and its components in wheat straw pretreated for bioethanol production in a paper with 280 citations. They observed that hydrothermal pretreatment did not degrade the fibrillar structure of cellulose but caused profound lignin relocalization while wax was removed whereas hemicellulose was only partially removed. They found similar changes in wheat straw pretreated with steam explosion. In conclusion, hydrothermal pretreatment increased the digestibility by increasing the accessibility of the cellulose through a re-localization of lignin and a partial removal of hemicellulose, rather than by the disruption of the cell wall.

Laureano-Perez et al. (2005) studied the determinants of the enzymatic hydrolysis of pretreated corn stover in a paper with 379 citations. They developed models of the hydrolysis initial rate at 72 h extent of conversion. They found that the hydrolysis initial rate was most influenced by cellulose crystallinity, while lignin content most influenced the extent of hydrolysis at 72 h. However, they used only crystallinity, lignin, and selected chemical bonds as inputs in the models. They predicted that the incorporation of additional parameters that affect hydrolysis, like pore volume and size and surface area accessibility, would improve the predictive capability of the models.

Kim and Holtzapple (2005) studied the lime pretreatment and enzymatic hydrolysis of corn stover in a paper with 270 citations. They carried out the pretreatment with an excess of lime (0.5 g $Ca(OH)_2$/g raw biomass) in non-oxidative and oxidative conditions at 25°C, 35°C, 45°C, and 55°C. The optimal condition was at 55°C for 4 weeks with aeration. They found that glucan (91.3%) and xylan (51.8%) were converted to glucose and xylose respectively, when the treated corn stover was enzymatically hydrolyzed with 15 FPU/g cellulase. Only 0.073 g $Ca(OH)_2$ was consumed per g of raw corn stover. Of the initial lignin, 87.5% was maximally removed while almost all acetyl groups were removed. The overall yields of glucose and xylose were 93.2% and 79.5% at 15 FPU/g cellulose, respectively, while the pretreatment liquor had no inhibitory effect on ethanol fermentation.

Kim and Holtzapple (2006) studied the determinants of enzymatic digestibility of corn stover in a paper with 355 citations. They pretreated the stover with excess lime (0.5 g $Ca(OH)_2$/g raw biomass) in non-oxidative and oxidative conditions at 25°C, 35°C, 45°C, and 55°C. They observed that the enzymatic digestibility of lime-treated corn stover was affected by the change of structural features (acetylation, lignification, and crystallization) resulting from this pretreatment. Further, extensive delignification required oxidative treatment and additional consumption of lime (up to 0.17 g $Ca(OH)_2$/g biomass). Deacetylation reached a plateau within 1 week and there were no significant differences between non-oxidative and oxidative conditions at 55°C; both conditions removed ~90% of the acetyl groups in 1 week at all temperatures studied. Delignification highly depended on temperature and the presence of oxygen. Lignin and hemicellulose were selectively removed (or solubilized), but cellulose was not affected by lime pretreatment in mild temperatures (25°C–55°C), even though corn stover was in contact with alkali for a long time, 16 weeks. The degree of crystallinity slightly increased from 43% to 60% with delignification because lignin and hemicellulose were removed. However, the increased crystallinity did not negatively affect the 3-d sugar yield of enzymatic hydrolysis. Finally, the oxidative lime pretreatment lowered the acetyl and lignin contents to obtain high digestibility, regardless of crystallinity.

Bjerre et al. (1996) studied the pretreatment of wheat straw using combined wet oxidation and alkaline pretreatments in a paper with 253 citations. They observed that these pretreatments readily oxidized lignin from wheat straw facilitating polysaccharides for enzymatic hydrolysis. By using a specially constructed autoclave system, they optimized the wet oxidation process with respect to both reaction time and temperature. The best conditions (20 g/L straw, 170°C, 5–10 min) gave about 85% w/w yield of converting cellulose to glucose. The process water, containing dissolved hemicellulose and carboxylic acids, was a direct nutrient source for the fungus Aspergillus niger producing exo-β-xylosidase. Finally, furfural and hydroxymethylfurfural (HMF) as fermentation inhibitors were not observed following the wet oxidation pretreatment.

Klinke et al. (2002) characterized the fermentation inhibitors from alkaline wet oxidation of wheat straw in a paper with 334 citations. They observed that this pretreatment was an efficient pretreatment of wheat straw that resulted in solid fractions with high cellulose recovery (96%) and high enzymatic convertibility to glucose (67%). Carbonate and temperature were the most important factors for the fractionation of wheat straw by wet oxidation. Optimal conditions were 10 min at 195°C with the addition of 12 bar oxygen and 6.5g/L Na_2CO_3. At these conditions, the hemicellulose fraction from 100 g straw consisted of soluble hemicellulose (16 g), low molecular weight carboxylic acids (11 g), monomeric phenols (0.48 g), and 2-furoic acid (0.01 g). Formic acid and acetic acid constituted the majority of degradation products (8.5 g). The main phenol monomers were 4-hydroxybenzaldehyde, vanillin, syringaldehyde, acetosyringone (4-hydroxy-3,5-dimethoxy-acetophenone), vanillic acid, and syringic acid, occurring in 0.04–0.12 g per 100 g straw concentrations. High lignin removal from the solid fraction (62%) did not provide a corresponding increase in the phenol monomer content but was correlated to high carboxylic acid concentrations. Degradation products in hemicellulose fractions co-varied with pretreatment conditions according to their chemical structure while aromatic aldehyde formation was correlated to severe conditions with high temperatures and low pH. Apart from CO_2 and water, carboxylic acids were the main degradation products of hemicellulose and lignin.

Kaar and Holtzapple (2000) studied lime pretreatment for the enzymatic hydrolysis of corn stover in a paper with 326 citations. They found that pretreatment with slake lime increased the enzymatic hydrolysis of corn stover nine times as compared to untreated corn stover. Optimal pretreatment conditions were: lime loading 0.075 g $Ca(OH)_2$/(g dry biomass); water loading 5 g H_2O/(g dry biomass); and heating for 4 h at 120°C. The optimal enzyme loading and hydrolysis temperature for the enzymatic hydrolysis of pretreated corn stover was 10 FPU/(g dry biomass) and 40°C, respectively. The enzymatic conversion of corn stover to monosaccharides, when pretreated and saccharified as prescribed for 72 h, was about 60% cellulose, 47% xylan, and 53% total available polysaccharide. Increasing the enzyme loading to 25 FPU/(g dry biomass) and the hydrolysis time to 7 days produced conversions of 88.0%, 87.7%, and 92.1% for glucan, xylan, and arabinan, respectively. In conclusion, pretreatment with lime could lead to corn stover polysaccharide conversions approaching 100% where the success of hydrolysis after lime pretreatment depended on enzyme loading.

64.3.2 Bioethanol Production and Evaluation

There are seven HCPs for the production and evaluation of bioethanol fuels from residual starch feedstocks with at least 385 citations each (Table 64.2). There are five and two HCPs for the production and evaluation of bioethanol fuels from residual starch feedstocks, respectively.

64.3.2.1 Bioethanol Production

Saha et al. (2005) studied the acid pretreatment, enzymatic hydrolysis, and fermentation of wheat straw to ethanol in a paper with 651 citations. They noted that the straw had 48.57% cellulose and 27.70% hemicellulose on a dry solid (DS) basis. They found that the maximum yield of monomeric sugars from wheat straw (7.83%, w/v, DS) by dilute H_2SO_4 (0.75%, v/v) pretreatment and enzymatic hydrolysis (45°C, pH 5.0, 72 h) using cellulase, β-glucosidase, xylanase, and esterase was 565 mg/g. Under this condition, no measurable quantities of furfural and HMF as fermentation inhibitors were produced. The yield of ethanol (per liter) from acid-pretreated enzyme hydrolyzed wheat straw (78.3 g) hydrolysate with recombinant *Escherichia coli* strain FBR5 was 19 g with a yield of 0.24 g/g DS. Detoxification of the acid- and enzyme-treated wheat straw hydrolysate by overliming reduced the fermentation time from 118 to 39 h in the case of separate hydrolysis and fermentation (SHF) (35°C, pH 6.5) whereas it increased the ethanol yield from 13 to 17 g/l and decreased the fermentation time from 136 to 112 h in the case of simultaneous saccharification and fermentation (SSF) (35°C, pH 6.0).

Kaparaju et al. (2009) produced bioethanol, biohydrogen, and biogas from wheat straw in a biorefinery context in a paper with 551 citations. They hydrothermally liberated wheat straw to

TABLE 64.2
The Production and Evaluation of Bioethanol Fuels from Residual Starch Feedstocks

No.	Papers	Biomass/ Hydrolysate	Res. Fronts	Prts.	Yeasts	Parameters	Keywords	Lead Author	Affil.	Cits.
1	Saha et al. (2005)	Wheat straw	Bioethanol production	H_2SO_4, cellulase, β-glucosidase, xylanase, esterase	*E. coli*	Enzymatic hydrolysis, fermentation, pretreatments, detoxification, sugar and, ethanol yield	Wheat straw, ethanol, pretreatment, saccharification, fermentation	Saha, Badal C. 7202946302	USDA Agr. Res. Serv. USA	651
2	Kaparaju et al. (2009)	Wheat straw	Bioethanol production	Hydrothermal, enzymes	Yeasts	Multiple biofuel production, enzymatic hydrolysis, fermentation, pretreatment	Wheat straw, bioethanol, biohydrogen, biogas, biorefinery	Angelidaki, Irini* 6603674728	Tech. Univ. Denmark Denmark	551
3	Mosier et al. (2005)	Corn stover	Bioethanol production	LHW, cellulase	Yeast	Pretreatments, enzymatic hydrolysis, glucose and ethanol yield, fermentation	Corn stover, pretreatment, hot water	Ladisch, Michael R. 7005670397	Purdue Univ. USA	461
4	Sheehan et al. (2003)	Corn stover	Bioethanol evaluation	Na	Na	LCA, energy balance, environmental impact, E85, Iowa, GHG emissions, air quality	Corn stover, ethanol	Sheehan, John 7202241794	Colorado State Univ. USA	449
5	Teymouri et al. (2005)	Corn stover	Bioethanol production	Ammonia, cellulase	Yeasts	Pretreatments, enzymatic hydrolysis, glucose, xylose, and ethanol yield	Corn stover, ammonia, AFEX, hydrolysis	Dale, Bruce E. 7201511969	Michigan State Univ. USA	430
6	Kazi et al. (2010)	Corn stover	Bioethanol evaluation	Acid, LHW, ammonia	Na	Techno-economics, pretreatments, processes, ethanol production cost, total capital investment	Corn stover, ethanol	Anex, Robert P. 6701606624	Univ. Wisconsin Madison USA	396
7	Schell et al. (2003)	Corn stover	Bioethanol production	H_2SO_4, cellulase	Yeasts	Pretreatment, enzymatic hydrolysis, ethanol production, SSF, xylose yield, ethanol conversion	Corn stover, pretreatment, digestibilities	Schell, Daniel J. 35614599300	Natl. Renew. Ener. Lab. USA	385

Prt., Biomass pretreatments; Na, non available; Cits., Number of citations received for each paper; *, Female.

a cellulose-rich fiber fraction and a hemicellulose-rich liquid fraction. They found that the enzymatic hydrolysis and subsequent fermentation of cellulose yielded 0.41 g-ethanol/g-glucose, while dark fermentation of hydrolysate produced 178.0 mL-H_2/g-sugars. They used effluents from both bioethanol and biohydrogen processes to produce biomethane with yields of 0.324 and 0.381 m^3/kg volatile solids, respectively. The use of wheat straw for biogas production or multi-fuel production was the energetically most efficient process compared to the production of mono-fuels such as bioethanol when fermenting C_6 sugars alone. In conclusion, multiple biofuel production from wheat straw could increase the efficiency of material and energy and could be a more economical process for biomass utilization.

Mosier et al. (2005) studied optimized pH-controlled liquid hot water (LHW) pretreatment of corn stover for enzyme digestibility with respect to processing temperature and time in a paper with 461 citations. They found that the optimized conditions for controlled pH, LHW pretreatment of a 16% slurry of corn stover in water were 190°C for 15 min. At the optimal conditions, 90% of the cellulose was hydrolyzed to glucose with 15 FPU of cellulase per gram of glucan. When the resulting pretreated slurry, in undiluted form, was hydrolyzed with 11 FPU of cellulase per gram of glucan, a hydrolysate containing 32.5 g/L glucose and 18 g/L xylose was formed. Both the xylose and the glucose were fermented with recombinant yeast 424A(LNH-ST) to ethanol at 88% of the theoretical yield.

Teymouri et al. (2005) studied optimized AFEX pretreatment for enzymatic hydrolysis of corn stover in a paper with 431 citations. They varied temperature, moisture content, ammonia loadings, and treatment time. Optimal pretreatment conditions for corn stover were 90°C, 1.0 kg of ammonia, 60% moisture content (dry weight basis), and 5 min. residence time. They obtained nearly 100% of the theoretical glucose yield and 80% of the theoretical xylose yield during the enzymatic hydrolysis of optimal treated corn stover using 60 filter paper units (FPU) of cellulase enzyme/g of glucan. The ethanol yield of optimally AFEX-treated corn stover increased up to 2.3 times over that of an untreated sample.

Schell et al. (2003) studied dilute acid pretreatment of corn stover in a pilot-scale reactor in a paper with 385 citations. They pretreated the stover in a continuous 1 t/d reactor at 20% (w/w) solids concentration over a range of conditions encompassing residence times of 3–12 min, temperatures of 165°C–195°C, and H_2SO_4 concentrations of 0.5%–1.4% (w/w). They then tested the pretreated solids in an SSF process to measure the reactivity of their cellulose component to enzymatic digestion with cellulase enzymes. They obtained monomeric xylose yields of 69%–71% and total xylose yields (monomers and oligomers) of 70%–77% with the performance level depending on pretreatment severity. They then obtained cellulose conversion yields in SSF of 80%–87% for some of the most digestible pretreated solids.

64.3.2.2 Bioethanol Evaluation

Sheehan et al. (2003) evaluated the energy and environmental impact of using corn stover for ethanol fuels in a paper with 449 citations. They constructed a life-cycle model that described collecting corn stover in the state of Iowa for the production and use of a fuel mixture consisting of 85% ethanol/15% gasoline (E85) in a flexible-fuel light-duty vehicle. The model incorporated results from individual models of soil carbon dynamics, soil erosion, agronomics of stover collection and transport, and the bioeconomics-version of stover to ethanol. They assumed that all farmers in the state of Iowa switched from their current cropping and tilling practices to continuous production of corn and 'no-till' practices. Under these conditions, which maximized the amount of collectible stover, they found that Iowa alone could produce almost 8 billion liters per year of pure stover-derived ethanol (E100) at prices competitive with today's corn-grain-derived ethanol. Further, for each kilometer fueled by the ethanol portion of E85, the vehicle used 95% less petroleum compared to a kilometer driven in the same vehicle on gasoline. Total fossil energy use (coal, oil, and natural gas) and greenhouse gas emissions (fossil CO_2, N_2O, and CH_4) on a life-cycle basis were 102% and

113% lower, respectively. However, air quality impacts were mixed, with emissions of CO, NO_x, and SO_x increasing, whereas hydrocarbon ozone precursors were reduced.

Kazi et al. (2010) performed a technoeconomic comparison of process technologies for ethanol production from corn stover based on a 5- to 8-year time frame for implementation to examine the short-term commercial viability of ethanol production in a paper with 396 citations. They covered four pretreatment technologies (dilute acid, two-stage dilute acid, LHW, and AFEX); and three downstream process variations (pervaporation, separate C_5 and C_6 sugars fermentation, and on-site enzyme production). They found that the dilute acid pretreatment process had the lowest ethanol production cost with a 10% investment return (PV) with $1.36/L of gasoline equivalent [LGE] ($5.13/gal of gasoline equivalent [GGE]). The PV was most sensitive to feedstock cost, enzyme cost, and installed equipment costs where a significant fraction of capital costs was related to producing heat and power from the lignin in the biomass. Further, the estimated value of PV for the pioneer plant was substantially larger than for the nth plant. The PV for the pioneer plant model with dilute acid pretreatment was $2.30/LGE ($8.72/GGE) for the most probable scenario, and the estimated total capital investment was more than double the nth plant cost.

64.4 DISCUSSION

64.4.1 Introduction

Crude oil-based gasoline fuels have been widely used in the transportation sector since the 1920s. However, there have been great public concerns over the adverse environmental and human impact of these fuels. Hence, biomass-based bioethanol fuels have increasingly been used in blending gasoline and petrodiesel fuels, in fuel cells, and in biochemical production in a biorefinery context.

However, it is necessary to pretreat the biomass to enhance the yield of bioethanol prior to bioethanol fuel production from feedstocks through hydrolysis and fermentation of the biomass. One of the most-studied feedstocks for bioethanol fuels have been residual starch feedstocks such as corn stover and wheat straw and to a lesser extent rice straw, rye straw, and barley straw.

The research in the field of second generation bioethanol fuels from residual starch feedstocks has intensified in this context in the key research fronts of the pretreatment and hydrolysis of residual starch feedstocks, fermentation of residual starch feedstock-based hydrolysates, and the production and evaluation of second generation bioethanol fuels from residual starch feedstocks. Further, the corn stover and wheat straw, and to a lesser extent rice straw, rye straw, and barley straw have been studied intensively in this context.

However, it is essential to develop efficient incentive structures for the primary stakeholders to enhance the research in this field. Although there has been a number of review papers for this field, there has been no review of the 25 most-cited articles in this field.

Thus, this chapter presents a review of the 25 most-cited articles on bioethanol fuel production and evaluation from residual starch feedstocks. Then, it discusses the key findings of these highly influential papers and comments on future research priorities in this field.

As the first step for the search of the relevant literature, keywords were selected using the 200 most-cited population papers. The selected keyword list was then optimized to obtain a representative sample of papers from this research field. This keyword list was provided in the appendix of Konur (2023) for future replicative studies.

As the second step, a sample dataset was used in this study. Twenty-five articles with at least 326 citations each were selected for the review study. Key findings from each paper were taken from the abstracts of these papers and were discussed. Additionally, a number of brief conclusions were drawn and a number of relevant recommendations were made to enhance the future research landscape.

TABLE 64.3
The Most Prolific Research Fronts for the Second Generation Bioethanol Fuels from Residual Starch Feedstocks

No.	Research Fronts	N Paper % Review	N Paper % Sample	Surplus (%)
1	Corn stover	68.0	44.8	23.2
2	Wheat straw	28.0	29.1	−1.1
3	Other starch residues	12.0	33.5	−21.5
	Rice straw	4.0	15.8	−11.8
	Barley straw	4.0	3.0	1.0
	Rye straw	4.0	2.0	2.0
	Other corn residues	0.0	5.9	−5.9
	Other rice residues	0.0	5.4	−5.4
	Sorghum residues	0.0	1.5	−1.5

N Paper (%) review: The number of papers in the sample of 25 reviewed papers. N paper (%) sample: The number of papers in the population sample of 203 papers.

Information about the research fronts for sample papers in bioethanol fuels from residual starch feedstocks with regard to residual starch feedstocks used in these processes is given in Table 64.3. As this table shows there are three primary research fronts for this field: Corn stover and to a lesser extent wheat straw and other starch feedstock residues with 68%, 28%, and 12% of the HCPs, respectively. Further, corn stover is the most influential residual feedstock with a 23% surplus while other starch feedstock residues are the least influential with a 22% deficit whereas rice straw has a 12% deficit.

Information about the thematic research fronts for the sample papers in bioethanol fuels from residual starch feedstocks is given in Table 64.4. As this table shows there are two primary research fronts for this field: The pretreatment and hydrolysis of residual starch biomass with 96% and 88% of the HCPs, respectively. Other research fronts are the fermentation of residual hydrolysates and the production and evaluation of bioethanol fuels with 12%, 20%, and 8% of the HCPs, respectively.

On the other hand, on an individual basis, the most prolific research fronts are enzymatic, chemical, acid, alkaline, hydrothermal, ammonia, and LHW pretreatments with 20%–88% of the HCPs each.

Further, the highest influential research fronts are LHW pretreatment, enzymatic hydrolysis, biomass hydrolysis, acid pretreatment, ammonia pretreatment, and alkaline pretreatment with a 7%–15% surplus each. Similarly, hydrolysate fermentation, chemical pretreatments, bioethanol production, mechanical pretreatments, H_2O_2 pretreatment, and solvent pretreatment are the lowest influential research fronts with a 6%–12% deficit each.

Brief notes about the 25 most-cited papers with at least 326 citations each on second generation bioethanol fuels from residual starch feedstocks are given below. The primary research fronts are pretreatments and hydrolysis of residual starch feedstocks and the production and evaluation of bioethanol fuels from residual starch feedstocks with 18 and 7 highly cited papers (HCPs), respectively. Further, there are 3 and 15 HCPs for the pretreatments and hydrolysis of residual starch feedstocks, respectively. Similarly, there are five and two HCPs for the production and evaluation of bioethanol fuels from residual starch feedstocks, respectively.

These HCPs show a sample of the research on the pretreatment and hydrolysis of residual starch feedstocks. The pretreatment and hydrolysis processes are a fundamental part of the production of bioethanol fuels from these feedstocks as they fractionate and hydrolyze these feedstocks, respectively, to produce fermentable sugars with a high sugar yield for bioethanol fuel production.

TABLE 64.4
The Most Prolific Thematic Research Fronts for the Residual Starch Feedstock-based Bioethanol Fuels

No.	Research Fronts	N Paper % Review	N Paper % Sample	Surplus (%)
1	Biomass pretreatments	96.0	95.1	0.9
1.1	Hydrothermal pretreatments	24.0	22.2	1.8
	Steam explosion	8.0	13.3	–5.3
	LHW	20.0	5.4	14.6
	Wet oxidation	8.0	3.4	4.6
	HCW	0.0	2.5	–2.5
1.2	Chemical pretreatments	76.0	86.7	–10.7
	Acid pretreatment	44.0	33.5	10.5
	Alkaline pretreatment	28.0	20.7	7.3
	Solvent pretreatment	0.0	5.9	–5.9
	IL pretreatment	0.0	3.9	–3.9
	Ammonia pretreatment	24.0	14.3	9.7
	SO_2 pretreatment	4.0	3.4	0.6
	Surfactants	0.0	2.5	–2.5
	H_2O_2 pretreatment	0.0	5.9	–5.9
	Other pretreatments	0.0	1.5	–1.5
1.3	Mechanical pretreatments	4.0	9.9	–5.9
	Microwave	0.0	3.0	–3
	Milling	4.0	4.9	–0.9
	Ultrasound	0.0	2.0	–2.0
1.4.	Enzymatic pretreatments	88.0	90.1	–2.1
1.5	Pretreatments in general	0.0	2.0	–2.0
2	Biomass hydrolysis	88.0	77.3	10.7
	Enzymatic hydrolysis	88.0	74.9	13.1
	Acid hydrolysis	0.0	2.5	–2.5
3	Hydrolysate fermentation	12.0	24.1	–12.1
4	Bioethanol production	20.0	27.1	–7.1
5	Bioethanol fuel evaluation	8.0	4.9	3.1

N Paper (%) review: The number of papers in the sample of 25 reviewed papers. N paper (%) sample: The number of papers in the population sample of 203 papers.

64.4.2 Biomass Pretreatments and Hydrolysis

Brief information about the 18 most-cited papers on the pretreatment and hydrolysis of residual starch feedstocks with at least 326 citations each is given below (Table 64.1). Further, there are 3 and 15 HCPs for the pretreatments and hydrolysis of residual starch feedstocks, respectively

64.4.2.1 Biomass Pretreatments

Kumar et al. (2009) characterized corn stover and poplar solids resulting from leading pretreatment technologies and observed that lime pretreatment removed the highest acetyl groups from both corn stover and poplar, while AFEX pretreatment removed the lowest. Further, Mani et al. (2004) studied the grinding performance and physical properties of wheat and barley straws, corn stover, and switchgrass and observed that switchgrass had the highest specific energy consumption and corn stover had the lowest specific energy consumption. Finally, Kim et al. (2003) pretreated corn stover with ARP pretreatment in a flowthrough column reactor and found that the ARP was highly effective in the delignifying of the biomass, reducing the lignin content by 70%–85%.

64.4.2.2 Biomass Enzymatic Hydrolysis

Lloyd and Wyman (2005) studied dilute acid pretreatment and enzymatic hydrolysis of corn stover and observed that up to 15% of the total potential sugar in the substrate could be released as glucose during pretreatment and between 15% and 90+% of the xylose remaining in the solid residue could be recovered. Further, Yang and Wyman (2004) studied the effect of xylan and lignin removal by batch and flowthrough pretreatment on the enzymatic digestibility of corn stover cellulose and observed that increasing flow with just water enhanced the xylan dissolution rate, more than doubled total lignin removal, and increased cellulose digestibility.

Hsu et al. (2010) studied the effect of dilute acid pretreatment of rice straw on structural properties and enzymatic hydrolysis and obtained a maximal sugar yield of 83%. Further, Ohgren et al. (2007) studied the effect of hemicellulose and lignin removal on the enzymatic hydrolysis of steam-pretreated corn stover and observed that a near-theoretical glucose yield (96%–104%) from acid-catalyzed steam pretreated corn stover could be obtained if xylanases were used to supplement cellulases during hydrolysis as xylanases hydrolyzed residual hemicellulose, thereby improving the access of enzymes to cellulose.

Wyman et al. (2005) compared sugar recovery data from the laboratory-scale application of leading pretreatment technologies of corn stover and observed that each pretreatment achieved high yields of glucose from cellulose with cellulase enzymes. Further, Sun and Cheng (2005) studied dilute acid pretreatment of rye straw and bermudagrass before enzymatic hydrolysis of cellulose for ethanol production, and with the increasing acid concentration and residence time, they found that the amount of arabinose and galactose increased.

Kabel et al. (2007) studied the effect of pretreatment severity on xylan solubility and enzymatic breakdown of the remaining cellulose from wheat straw and observed that the higher the severity factor the more xylan was released from the wheat straw, but more xylan decomposed and furfural formation occurred. Further, Selig et al. (2007) studied the effect of dilute acid pretreatment of corn stems on the enzymatic hydrolysis of cellulose and observed spherical formations on the surface of the residual biomass after this pretreatment.

Kristensen et al. (2008) explored the effect of hydrothermal pretreatment on the straw cell-wall matrix and its components in wheat straw pretreated for bioethanol production and observed that hydrothermal pretreatment did not degrade the fibrillar structure of cellulose but caused profound lignin relocalization. Further, Laureano-Perez et al. (2005) studied the determinants of enzymatic hydrolysis of pretreated corn stover and found that the initial rate of hydrolysis was most influenced by cellulose crystallinity, while lignin content most influenced the extent of hydrolysis.

Kim and Holtzapple (2005) studied lime pretreatment and enzymatic hydrolysis of corn stover and found that glucan (91.3%) and xylan (51.8%) were converted to glucose and xylose respectively when the treated corn stover was enzymatically hydrolyzed with 15 FPU/g cellulase. Further, Kim and Holtzapple (2006) studied the determinants of enzymatic digestibility of corn stover and observed that the enzymatic digestibility of lime-pretreated corn stover was affected by the change of structural features (acetylation, lignification, and crystallization) resulting from this pretreatment.

Bjerre et al. (1996) studied the pretreatment of wheat straw using combined wet oxidation and alkaline pretreatments and observed that these pretreatments readily oxidized lignin from wheat straw facilitating polysaccharides for enzymatic hydrolysis. Further, Klinke et al. (2002) characterized fermentation inhibitors from alkaline wet oxidation of wheat straw and observed that this pretreatment was an efficient pretreatment of wheat straw that resulted in solid fractions with high cellulose recovery (96%) and high enzymatic convertibility to glucose (67%). Finally, Kaar and Holtzapple (2000) studied lime pretreatment for the enzymatic hydrolysis of corn stover and found that pretreatment with slake lime increased the enzymatic hydrolysis of corn stover by nine times compared to untreated corn stover.

64.4.3 Bioethanol Production and Evaluation

There are seven HCPs for the production and evaluation of bioethanol fuels from residual starch feedstocks with at least 385 citations each (Table 64.2). There are five and two HCPs for the production and evaluation of bioethanol fuels from residual starch feedstocks, respectively.

These HCPs show a sample of the research on the production and evaluation of bioethanol fuels from the residual starch feedstocks. As the hydrolysates of residual starch feedstocks include a number of fermentation inhibitors, the production process involves both the detoxification of these hydrolysates to remove these fermentation inhibitors and the fermentation of the hydrolysates by a suitable yeast strain. Thus, the detoxification and fermentation processes are the fundamental parts of the bioethanol production process to improve the ethanol yield from residual starch feedstocks.

Further, as bioethanol fuels have been used in vehicles, blending them with gasoline or petrodiesel fuels, the studies on the evaluation of these bioethanol fuels are a significant research front together with the research front in bioethanol fuel production. In this context, residual starch feedstocks are used to produce bioethanol and/or bioelectricity and biosteam. These studies are usually concerned with the technoeconomics, environmental, and social impact of the utilization of these bioethanol fuels and bioelectricity. These studies show that the use of residual starch feedstocks increases the competitiveness of bioethanol fuels compared to crude oil-based gasoline and petrodiesel fuels.

64.4.3.1 Bioethanol Production

Saha et al. (2005) studied acid pretreatment, enzymatic hydrolysis, and fermentation of wheat straw to ethanol and found that the maximum yield of monomeric sugars from wheat straw with dilute pretreatment and enzymatic hydrolysis using cellulase, β-glucosidase, xylanase, and esterase was 565 mg/g. Further, Kaparaju et al. (2009) produced bioethanol, biohydrogen, and biogas from wheat straw in a biorefinery context and found that the enzymatic hydrolysis and subsequent fermentation of cellulose yielded 0.41 g-ethanol/g-glucose, while dark fermentation of hydrolysate produced 178.0 mL-H_2/g-sugars.

Mosier et al. (2005) optimized pH-controlled LHW pretreatment of corn stover for enzyme digestibility with respect to processing temperature and time and found that the optimized conditions for controlled pH, LHW pretreatment of a 16% slurry of corn stover in water were190°C for 15 min. Further, Teymouri et al. (2005) optimized AFEX pretreatment for the enzymatic hydrolysis of corn stover and obtained nearly 100% of the theoretical glucose yield and 80% of the theoretical xylose yield. Finally, Schell et al. (2003) studied the dilute acid pretreatment of corn stover and obtained monomeric xylose yields of 69%–71% and total xylose yields (monomers and oligomers) of 70%–77% with performance levels depending on pretreatment severity.

64.4.3.2 Bioethanol Evaluation

Sheehan et al. (2003) evaluated the energy and environmental impact of using corn stover for ethanol fuels and found that Iowa alone could produce almost 8 billion liters per year of pure stover-derived ethanol (E100) at prices competitive with today's first generation corn-grain-derived ethanol. Further, Kazi et al. (2010) performed a technoeconomic comparison of process technologies for ethanol production from corn stover based on a 5- to 8-year time frame for implementation to examine the short-term commercial viability of ethanol production and found that the dilute acid pretreatment process had the lowest ethanol production cost with 10% investment return (PV) with \$1.36/L of gasoline equivalent [LGE] (\$5.13/gal of gasoline equivalent [GGE]).

64.5 CONCLUSION AND FUTURE RESEARCH

Brief information about the key research fronts covered by the 25 most-cited papers with at least 326 citations each is given under two primary headings: The pretreatment and hydrolysis of the residual starch feedstocks and the production and evaluation of bioethanol fuels.

The usual characteristics of these HCPs are that the pretreatments and hydrolysis of residual starch feedstocks and fermentation of the resulting hydrolysates are the primary processes for bioethanol fuel production from residual starch feedstocks to improve the ethanol yield as residual starch feedstocks are one of the most studied feedstocks for bioethanol production, especially for the countries with large farmlands and crude oil deficiency.

The key findings on these research fronts should be read in light of the increasing public concerns about climate change, GHG emissions, and global warming as these concerns have certainly been behind the boom in research on bioethanol fuels from residual starch feedstocks as an alternative to crude oil-based gasoline and petrodiesel fuels in the past decades. It is also a sustainable alternative to first generation food crop-based bioethanol fuels such as corn grain-based bioethanol fuels. The recent supply shocks caused by the COVID-19 pandemic and the Russian invasion of Ukraine also highlight the importance of the production and utilization of bioethanol fuels from residual starch feedstocks as an alternative to crude oil-based gasoline and petrodiesel fuels.

As Table 64.3 shows, there are three primary feedstocks for this field: corn stover and to a lesser extent wheat straw and other starch feedstock residues. Similarly, as Table 64.4 shows there are two primary research fronts for this field: The pretreatment and hydrolysis of residual biomass. Other research fronts are the fermentation of residual hydrolysates and the production and evaluation of bioethanol fuels. On the other hand, the most prolific pretreatments are enzymatic, chemical, acid, alkaline, hydrothermal, ammonia, and LHW pretreatments.

These studies emphasize the importance of proper incentive structures for the efficient production and utilization of bioethanol fuels from residual starch feedstocks in light of North's institutional framework (North, 1991). In this context, the major producers and users of bioethanol fuels such as the USA and Canada with vast farmlands have developed strong incentive structures for efficient bioethanol fuels from residual starch feedstocks and their residues.

In light of the recent supply shocks caused primarily by the COVID-19 pandemic and the Russian invasion of Ukraine, it is expected that the incentive structures such as public funding would be enhanced to increase the share of bioethanol fuels in the global fuel portfolio as a strong alternative to crude oil-based gasoline and petrodiesel fuels. In this context, it is expected that the most prolific researchers, institutions, countries, funding bodies, and journals in this field would have a first-mover advantage to benefit from such potential incentives. This is especially true for the US stakeholders as the USA has become the global leader in both the production and utilization of second generation bioethanol fuels from residual starch feedstocks.

It is recommended that such review studies are performed for the primary research fronts of the bioethanol fuels from residual starch feedstocks.

ACKNOWLEDGMENTS

The contribution of the highly cited researchers in the field of bioethanol fuels from residual starch feedstocks has been gratefully acknowledged.

REFERENCES

Alvira, P., E. Tomas-Pejo, M. Ballesteros and M. J. Negro. 2010. Pretreatment technologies for an efficient bioethanol production process based on enzymatic hydrolysis: A review. *Bioresource Technology* 101:4851–4861.

Angelici, C., B. M. Weckhuysen and P. C. A. Bruijnincx. 2013. Chemocatalytic conversion of ethanol into butadiene and other bulk chemicals. *ChemSusChem* 6:1595–1614.

Antolini, E. 2007. Catalysts for direct ethanol fuel cells. *Journal of Power Sources* 170:1–12.

Antolini, E. 2009. Palladium in fuel cell catalysis. *Energy and Environmental Science* 2:915–931.

Binod, P., R. Sindhu and R. R. Singhania, et al. 2010. Bioethanol production from rice straw: An overview. *Bioresource Technology* 101:4767–4774.

Bjerre, A. B., A. B. Olesen, T. Fernqvist, A. Ploger and A. S. Schmidt. 1996. Pretreatment of wheat straw using combined wet oxidation and alkaline hydrolysis resulting in convertible cellulose and hemicellulose. *Biotechnology and Bioengineering* 49:568–577.

Burnham, J. F. 2006. Scopus database: A review. *Biomedical Digital Libraries* 3:1–8.

Fernando, S., S. Adhikari, C. Chandrapal and M. Murali. 2006. Biorefineries: Current status, challenges, and future direction. *Energy & Fuels* 20:1727–1737.

Hahn-Hagerdal, B., M. Galbe, M. F. Gorwa-Grauslund, G. Liden and G. Zacchi. 2006. Bio-ethanol: The fuel of tomorrow from the residues of today. *Trends in Biotechnology* 24:549–556.

Hill, J., E. Nelson, D. Tilman, S. Polasky and D. Tiffany. 2006. Environmental, economic, and energetic costs and benefits of biodiesel and ethanol biofuels. *Proceedings of the National Academy of Sciences of the United States of America* 103:11206–11210.

Hill, J., S. Polasky and E. Nelson, et al. 2009. Climate change and health costs of air emissions from biofuels and gasoline. *Proceedings of the National Academy of Sciences of the United States of America* 106:2077–2082.

Hsieh, W. D., R. H. Chen, T. L. Wu and T. H. Lin. 2002. Engine performance and pollutant emission of an SI engine using ethanol-gasoline blended fuels. *Atmospheric Environment* 36:403–410.

Hsu, T. C., G. L. Guo, W. H. Chen and W. S. Hwang. 2010. Effect of dilute acid pretreatment of rice straw on structural properties and enzymatic hydrolysis. *Bioresource Technology* 101:4907–4913.

Huang, H. J., S. Ramaswamy, U. W. Tschirner and B. V. Ramarao. 2008. A review of separation technologies in current and future biorefineries. *Separation and Purification Technology* 62:1–21.

Kaar, W. E. and M. T. Holtzapple. 2000. Using lime pretreatment to facilitate the enzymic hydrolysis of corn stover. *Biomass and Bioenergy* 18:189–199.

Kabel, M. A., G. Bos, J. Zeevalking, A. G. J. Voragen and H. A. Schols. 2007. Effect of pretreatment severity on xylan solubility and enzymatic breakdown of the remaining cellulose from wheat straw. *Bioresource Technology* 98:2034–2042.

Kaparaju, P., M. Serrano, A. B. Thomsen, P. Kongjan and I. Angelidaki. 2009. Bioethanol, biohydrogen and biogas production from wheat straw in a biorefinery concept. *Bioresource Technology* 100:2562–2568.

Kazi, F. K., J. A. Fortman and R. P. Anex, et al. 2010. Techno-economic comparison of process technologies for biochemical ethanol production from corn stover. *Fuel* 89:S20–S28.

Kim, S. and M. T. Holtzapple. 2005. Lime pretreatment and enzymatic hydrolysis of corn stover. *Bioresource Technology* 96:1994–2006.

Kim, S. and M. T. Holtzapple. 2006. Effect of structural features on enzyme digestibility of corn stover. *Bioresource Technology* 97:583–591.

Kim, T. H., J. S. Kim, C. Sunwoo and Y. Y. Lee. 2003. Pretreatment of corn stover by aqueous ammonia. *Bioresource Technology* 90:39–47.

Klinke, H. B., B. K. Ahring, A. S. Schmidt and A. B. Thomsen. 2002. Characterization of degradation products from alkaline wet oxidation of wheat straw. *Bioresource Technology* 82:15–26.

Konur, O. 2000. Creating enforceable civil rights for disabled students in higher education: An institutional theory perspective. *Disability & Society* 15:1041–1063.

Konur, O. 2002a. Access to nursing education by disabled students: Rights and duties of nursing programs. *Nurse Education Today* 22:364–374.

Konur, O. 2002b. Assessment of disabled students in higher education: Current public policy issues. *Assessment and Evaluation in Higher Education* 27:131–152.

Konur, O. 2002c. Access to employment by disabled people in the UK: Is the Disability Discrimination Act working? *International Journal of Discrimination and the Law* 5:247–279.

Konur, O. 2006a. Participation of children with dyslexia in compulsory education: Current public policy issues. *Dyslexia* 12:51–67.

Konur, O. 2006b. Teaching disabled students in higher education. *Teaching in Higher Education* 11:351–363.

Konur, O. 2007a. A judicial outcome analysis of the Disability Discrimination Act: A windfall for the employers? *Disability & Society* 22:187–204.

Konur, O. 2007b. Computer-assisted teaching and assessment of disabled students in higher education: The interface between academic standards and disability rights. *Journal of Computer Assisted Learning* 23:207–219.

Konur, O. 2012. The evaluation of the research on the bioethanol: A scientometric approach. *Energy Education Science and Technology Part A: Energy Science and Research* 28:1051–1064.

Konur, O. 2015. Current state of research on algal bioethanol. In *Marine Bioenergy: Trends and Developments*, Eds. S. K. Kim and C. G. Lee, pp. 217–244. Boca Raton, FL: CRC Press.

Konur, O. 2020. The scientometric analysis of the research on the bioethanol production from green macroalgae. In *Handbook of Algal Science, Technology and Medicine,* Ed. O. Konur, pp. 385–401. London: Academic Press.

Konur, O. 2023. Second generation bioethanol fuels from residual starch feedstocks: Scientometric study. In *Feedstock-based Bioethanol Fuels. II. Waste Feedstocks: Agricultural, Food, Industrial, Urban, Forestry, and Lignocellulosic Waste-based Bioethanol Fuels. Handbook of Bioethanol Fuels Volume 4,* Ed. O. Konur, pp. 79–98. Boca Raton, FL: CRC Press.

Kristensen, J. B., L. G. Thygesen, C. Felby, H. Jorgensen and T. Elder. 2008. Cell-wall structural changes in wheat straw pretreated for bioethanol production. *Biotechnology for Biofuels* 1:5.

Kumar, R., G. Mago, G., V. Balan and C. E. Wyman. 2009. Physical and chemical characterizations of corn stover and poplar solids resulting from leading pretreatment technologies. *Bioresource Technology* 100:3948–3962.

Laureano-Perez, L., F. Teymouri, H. Alizadeh and B. E. Dale. 2005. Understanding factors that limit enzymatic hydrolysis of biomass: Characterization of pretreated corn stover. *Applied Biochemistry and Biotechnology: Part A, Enzyme Engineering and Biotechnology* 124:1081–1099.

Lin, Y. and S. Tanaka. 2006. Ethanol fermentation from biomass resources: Current state and prospects. *Applied Microbiology and Biotechnology* 69:627–642.

Lloyd, T. A. and C. E. Wyman. 2005. Combined sugar yields for dilute sulfuric acid pretreatment of corn stover followed by enzymatic hydrolysis of the remaining solids. *Bioresource Technology* 96:1967–1977.

Ma, X., L. Sun and C. Song. 2002. A new approach to deep desulfurization of gasoline, diesel fuel and jet fuel by selective adsorption for ultra-clean fuels and for fuel cell applications. *Catalysis Today* 77:107–116.

Mani, S., L. G. Tabil and S. Sokhansanj. 2004. Grinding performance and physical properties of wheat and barley straws, corn stover and switchgrass. *Biomass and Bioenergy* 27:339–352.

Morschbacker, A. 2009. Bio-ethanol based ethylene. *Polymer Reviews* 49:79–84.

Mosier, N., R. Hendrickson, N. Ho, M. Sedlak and M. R. Ladisch. 2005. Optimization of pH controlled liquid hot water pretreatment of corn stover. *Bioresource Technology* 96:1986–1993.

Najafi, G., B. Ghobadian and T. Tavakoli, et al. 2009. Performance and exhaust emissions of a gasoline engine with ethanol blended gasoline fuels using artificial neural network. *Applied Energy* 86:630–639.

Newman, P. W. G. and J. R. Kenworthy. 1989. Gasoline consumption and cities: A comparison of U.S. cities with a global survey. *Journal of the American Planning Association* 55:24–37.

North, D. C. 1991. Institutions. *Journal of Economic Perspectives* 5:97–112.

Ohgren, K., R. Bura, J. Saddler and G. Zacchi. 2007. Effect of hemicellulose and lignin removal on enzymatic hydrolysis of steam pretreated corn stover. *Bioresource Technology* 98:2503–2510.

Olsson, L. and B. Hahn-Hagerdal. 1996. Fermentation of lignocellulosic hydrolysates for ethanol production. *Enzyme and Microbial Technology* 18:312–331.

Saha, B. C., L. B. Iten, M. A. Cotta and Y. V. Wu. 2005. Dilute acid pretreatment, enzymatic saccharification and fermentation of wheat straw to ethanol. *Process Biochemistry* 40:3693–3700.

Sanchez, O. J. and C. A. Cardona. 2008. Trends in biotechnological production of fuel ethanol from different feedstocks. *Bioresource Technology* 99:5270–5295.

Schell, D. J., J. Farmer, M. Newman and J. D. McMillan. 2003. Dilute-sulfuric acid pretreatment of corn stover in pilot-scale reactor: Investigation of yields, kinetics, and enzymatic digestibilities of solids. *Applied Biochemistry and Biotechnology: Part A, Enzyme Engineering and Biotechnology* 108:69–86.

Selig, M. J., S. Viamajala and S. R. Decker, et al. 2007. Deposition of lignin droplets produced during dilute acid pretreatment of maize stems retards enzymatic hydrolysis of cellulose. *Biotechnology Progress* 23:1333–1339.

Sheehan, J., A. Aden and K. Paustian, et al. 2003. Energy and environmental aspects of using corn stover for fuel ethanol. *Journal of Industrial Ecology* 7:117–146.

Singh, R., M. Srivastava and A. Shukla. 2016. Environmental sustainability of bioethanol production from rice straw in India: A review. *Renewable and Sustainable Energy Reviews* 54:202–216.

Sun, Y. and J. Cheng. 2002. Hydrolysis of lignocellulosic materials for ethanol production: A review. *Bioresource Technology* 83:1–11.

Sun, Y. and J. J. Cheng. 2005. Dilute acid pretreatment of rye straw and bermudagrass for ethanol production. *Bioresource Technology* 96:1599–1606.

Taherzadeh, M. J. and K. Karimi. 2007. Enzyme-based hydrolysis processes for ethanol from lignocellulosic materials: A review. *Bioresources* 2:707–738.

Taherzadeh, M. J. and K. Karimi. 2008. Pretreatment of lignocellulosic wastes to improve ethanol and biogas production: A review. *International Journal of Molecular Sciences* 9:1621–1651.

Talebnia, F., D. Karakashev and I. Angelidaki. 2010. Production of bioethanol from wheat straw: An overview on pretreatment, hydrolysis and fermentation. *Bioresource Technology* 101:4744–4753.

Teymouri, F., L. Laureano-Perez, H. Alizadeh and B. E. Dale. 2005. Optimization of the ammonia fiber explosion (AFEX) treatment parameters for enzymatic hydrolysis of corn stover. *Bioresource Technology* 96:2014–2018.

Wyman, C. E., B. E. Dale and R. T. Elander, et al. 2005. Comparative sugar recovery data from laboratory scale application of leading pretreatment technologies to corn stover. *Bioresource Technology* 96:2026–2032.

Yang, B. and C. E. Wyman. 2004. Effect of xylan and lignin removal by batch and flowthrough pretreatment on the enzymatic digestibility of corn stover cellulose. *Biotechnology and Bioengineering* 86:88–98.

Part 19

Second Generation Food Waste-based Bioethanol Fuels

65 Second Generation Food Waste-based Bioethanol Fuels
Scientometric Study

Ozcan Konur
(Formerly) Ankara Yildirim Beyazit University

65.1 INTRODUCTION

Crude oil-based gasoline fuels (Ma et al., 2002; Newman and Kenworthy, 1989) have been widely used in the transportation sector since the 1920s. However, there have been great public concerns over the adverse environmental and human impact of these fuels (Hill et al., 2006, 2009). Hence, biomass-based bioethanol fuels (Hill et al., 2006; Konur, 2012e, 2015, 2019, 2020a) have increasingly been used in blending gasoline fuels (Hsieh et al., 2002; Najafi et al., 2009), in fuel cells (Antolini, 2007, 2009), and in biochemical production (Angelici et al., 2013; Morschbacker, 2009) in a biorefinery context (Fernando et al., 2006; Huang et al., 2008).

Bioethanol fuels also play a critical role in maintaining energy security (Kruyt et al., 2009; Winzer, 2012) in supply shocks (Kilian, 2008, 2009) related to oil price shocks (Hamilton, 2003), COVID-19 pandemics (Fauci et al., 2020; Li et al., 2020), or wars (Hamilton, 1983; Jones, 2012) in the aftermath of the Russian invasion of Ukraine (Reeves, 2014).

However, it is necessary to pretreat the biomass (Taherzadeh and Karimi, 2008; Yang and Wyman, 2008) to enhance the yield of bioethanol (Hahn-Hagerdal et al., 2006; Sanchez and Cardona, 2008) prior to bioethanol production through hydrolysis (Sun and Cheng, 2002; Taherzadeh and Karimi, 2007) and fermentation (Lin and Tanaka, 2006; Olsson and Hahn-Hagerdal, 1996) of biomass and hydrolysates, respectively.

One of the most studied feedstocks for bioethanol fuels has been food wastes such as fruit, vegetable, beverage, and dairy wastes. Research in the field of second generation food waste-based bioethanol fuels has intensified in this context on the key research fronts of pretreatment (Fernandez-Bolanos et al., 2001; Grohmann et al., 1995; Rahman et al., 2007) and hydrolysis (Grohmann et al., 1995; Guerard et al., 2002; Rahman et al., 2007) of the food wastes, fermentation (Kim et al., 2011; Mussatto et al., 2012; Piarpuzan et al., 2011) of the food waste-based hydrolysates, and production (Arapoglou et al., 2010; Mussatto et al., 2012; Piarpuzan et al., 2011) and evaluation (Velasquez-Arredondo et al., 2010) of second generation food waste-based bioethanol fuels. Further, the empty fruit bunches (Kim et al., 2012; Rahman et al., 2007), orange peels (Grohmann et al., 1995; Grohmann and Baldwin, 1992), citrus peels (Choi et al., 2015; Wilkins et al., 2007), potato peels (Arapoglou et al., 2010; Khawla et al., 2014), spent coffee grounds (Kwon et al., 2013; Mussatto et al., 2012), cheese whey (Guimaraes et al., 2010; Silveira et al., 2005), banana peels (Oberoi et al., 2011; Velasquez-Arredondo et al., 2010), apple pomaces (Hang et al., 1982), and olive stones (Fernandez-Bolanos et al., 2001) have been studied intensively in this context.

However, it is essential to develop efficient incentive structures (North, 1991) for the primary stakeholders to enhance the research in this field (Konur, 2000, 2002a–c, 2006a,b, 2007a,b). Scientometric analysis has been used in this context to inform the primary stakeholders about the current state of research in this research field (Garfield, 1955; Konur, 2011, 2012a–i, 2015, 2018b, 2019, 2020a).

DOI: 10.1201/9781003226550-88

As there have been no published scientometric studies in this field, this chapter presents a scientometric study of research on second generation food waste-based bioethanol fuels. It examines the scientometric characteristics of both the sample and population data and presents them in the order of documents, authors, publication years, institutions, funding bodies, source titles, countries, Scopus subject categories, Scopus keywords, and research fronts.

65.2 MATERIALS AND METHODS

The search for this study was carried out using the Scopus database (Burnham, 2006) in July 2022.

As the first step for the search of the relevant literature, keywords were selected using the 200 most-cited population papers. The selected keyword list was then optimized to obtain a representative sample of papers for this research field. This keyword list was provided in the appendix for future replicative studies.

As the second step, two sets of data were used in this study. First, a population sample of 1,418 papers was used to examine the scientometric characteristics of the population data. Second, a sample of 142 most-cited papers, corresponding to 10% of the population papers, was used to examine the scientometric characteristics of these citation classics.

The scientometric characteristics of both these sample and population datasets were presented in the order of documents, authors, publication years, institutions, funding bodies, source titles, countries, Scopus subject categories, Scopus keywords, and research fronts.

Lastly, the key scientometric findings for both datasets were discussed to highlight the research landscape for second generation food waste-based bioethanol fuels. Additionally, a number of brief conclusions were drawn and a number of relevant recommendations were made to enhance the future research landscape.

65.3 RESULTS

65.3.1 The Most Prolific Documents on Second Generation Food Waste-based Bioethanol Fuels

Information on the types of documents for both datasets is given in Table 65.1. Articles and conference papers, published in journals, dominate both the sample (93%) and population (96%) papers, with a 3% deficit. Further, review papers and short surveys have a 5% surplus as they are overrepresented in the sample papers, as they constitute 7% and 2% of the sample and population papers, respectively.

TABLE 65.1
Documents in Second Generation Food Waste-based Bioethanol Fuels

Documents	Sample Dataset (%)	Population Dataset (%)	Surplus (%)
Article	92.3	93.8	−1.5
Review	7.0	2.1	4.9
Conference paper	0.7	2.3	−1.6
Book chapter	0.0	1.3	−1.3
Note	0.0	0.4	−0.4
Letter	0.0	0.1	−0.1
Short Survey	0.0	0.0	0.0
Book	0.0	0.0	0.0
Editorial	0.0	0.0	0.0
Sample size	142	1,418	

Sample dataset: The number of papers (%) in the set of 142 highly cited papers. Population dataset: The number of papers (%) in the set of the 1,418 population papers.

It is further notable that 98% of the population papers were published in journals while 1% of them each were published in books and book series, respectively. Similarly, 100% of the sample papers were published in journals.

65.3.2 The Most Prolific Authors on Second Generation Food Waste-based Bioethanol Fuels

Information about the 15 most prolific authors with at least 2.1% of sample papers each is given in Table 65.2. The most prolific author is Karel Grohmann, with 4.9% of the sample papers, followed by Fikret Kargi and Serpil Ozmihci, with 3.5% of the sample papers each. Other prolific authors are Maria V. P. Rocha, Jose A. Teixeira, Harinder S. Oberoi, and Luciano R. B. Goncalves, with 3.8% of the sample papers each.

On the other hand, the most influential author is Karel Grohmann with 4.2% surplus, followed by Fikret Kargi and Serpil Ozmihci, with 2.9% surplus each. Other influential authors are Maria V. P. Rocha, Maria Jose A. Teixeira, and Harinder S. Oberoi, with a 2.4% surplus each.

The most prolific institutions for the sample dataset are Chonnam National University, Dokuz Eylul University, the Federal University of Ceara, and the USDA Agricultural Research Service, with two authors each. On the other hand, the most prolific country for the sample dataset is the USA with four authors, followed by South Korea with three authors. Other prolific countries are Brazil, India, and Turkey, with two authors each. In total, only seven countries house these top authors.

The most prolific research front for these top authors is bioethanol fuels from fruit wastes such as orange peels and empty fruit bunches, with 13 authors. Other research fronts are bioethanol fuels from dairy wastes such as cheese whey and beverage wastes such as spent coffee grounds, with three and one author, respectively.

On the other hand, there is a significant gender deficit (Beaudry and Lariviere, 2016) for the sample dataset as surprisingly only three of these top researchers are female, with a representation rate of 20%.

Additionally, there are other authors with a relatively low citation impact and with 0.4%–1.1% of the population papers each: Hongzhi Ma, Qunhui Wang, Suraini Abd-Aziz, Md Zahangir Alam, Mohd Ali Hassan, Dimitrios Malamis, Tigrassa H. S. Rodrigues, Claudia Conesa, Ekin Demiray, Gonul Donmez, Abdel E. Ghaly, Keikhosro Karimi, Dimitris Kekos, Kiat M. Lee, Soh K. Loh, Noeli Sellin, Ozair Souza, Mette H. Thomsen, Qunhui Wang, Vincenza Calabro, Eulogio Castro, Asma Choudhary, Manuel Cuevas, Stefano Curcio, Ali Demirci, Widya Fatriasari, Pedro Fito, Ming Gao, Haibo Huang, Gabriele Iorio, Jianguo Jiang, Kyoung-Hyoun Kim, Jalel Labidi, Gerasimos Lyberatos, Cintia Marangoni, Lai Y. Phang, Maria E. Russo, Chayanoot Sangwichien, Jens E. Schmidt, Lucia Segui, Carlos R. Soccol, and Sebastian Sanchez.

65.3.3 The Most Prolific Research Output by Years on Second Generation Food Waste-based Bioethanol Fuels

Information about papers published between 1970 and 2022 is given in Figure 65.1. This figure clearly shows that the bulk of the research papers in the population dataset was published primarily in the 2010s and early 2020s, with 56% and 31% of the population dataset, respectively. Similarly, publication rates for the 2000s, 1990s, 1980s, and 1970s were 7%, 3%, 2%, and 0%, respectively. Additionally, 0.1% of the population papers was published in the pre-1970s.

Similarly, the bulk of the research papers in the sample dataset was published in the 2010s and 2000s, with 67% and 23% of the sample dataset, respectively. Similarly, publication rates for the early 2020s, 1990s, 1980s, and 1970s were 4%, 5%, 1%, and 0% of the sample papers, respectively.

The most prolific publication years for the population dataset were 2021 and 2022, with 10.4% and 10.5% of the dataset, respectively. Further, 88% of the population papers were published between 2009 and 2022. Similarly, 89% of the sample papers were published between 2006 and 2020, while

TABLE 65.2
Most-Prolific Authors on Second Generation Food Waste-based Bioethanol Fuels

No.	Author Name	Author Code	Sample Papers (%)	Population Papers (%)	Surplus (%)	Institution	Country	HI	N	Res. Front
1	Grohmann, Karel	7004503589	4.9	0.7	4.2	Renewable Spirits LLC	USA	49	111	F
2	Kargi, Fikret	57218389979	3.5	0.6	2.9	Dokuz Eylul Univ.	Turkey	54	207	D
3	Ozmihci, Serpil*	6506240734	3.5	0.6	2.9	Dokuz Eylul Univ.	Turkey	20	28	D
4	Rocha, Maria V. P.*	15080896200	2.8	0.4	2.4	Fed. Univ. Ceara	Brazil	25	53	F
5	Teixeira, Jose A.	13402823200	2.8	0.4	2.4	Univ. Minho	Portugal	85	726	D, F
6	Oberoi, Harinder S.	6603479987	2.8	0.4	2.4	Indian Inst. Hort. Res.	India	27	66	F
7	Goncalves, Luciano R. B.	7103169186	2.8	0.9	1.9	Fed. Univ. Ceara	Brazil	46	151	F
8	Taherzadeh, Mohammad J.	6701407496	2.1	0.6	1.5	Univ. Boras	Sweden	66	419	F
9	Bae, Hyeun-Jong	24280549300	2.1	0.4	1.7	Chonnam Natl. Univ.	S. Korea	32	117	B, F
10	Kim, Chul-Ho	57205883973	2.1	0.4	1.7	Korea Res. Inst. Biosci. Biotechnol.	S. Korea	29	113	F
11	Wilkins, Mark R.	56492323200	2.1	0.4	1.7	Univ. Nebraska-Lincoln	USA	28	97	F
12	Vadlani, Praveen V.	24075089500	2.1	0.3	1.8	Sri Sathya Sai Inst. High. Learn.	India	26	80	F
13	Widmer, Wilbur W.	7006097359	2.1	0.3	1.8	USDA Agr. Res. Serv.	USA	20	40	F
14	Baldwin, Elizabeth A.*	7006618084	2.1	0.2	1.9	USDA Agr. Res. Serv.	USA	74	183	F
15	Choi, In Seong	31567492500	2.1	0.2	1.9	Chonnam Natl. Univ.	S. Korea	23	122	F

Author code: the unique code given by Scopus to the authors. Sample papers: the number of papers authored in the sample dataset. Population papers: the number of papers authored in the population dataset. *: Female.

F, Fruit waste; D, Dairy waste; B, Beverage waste.

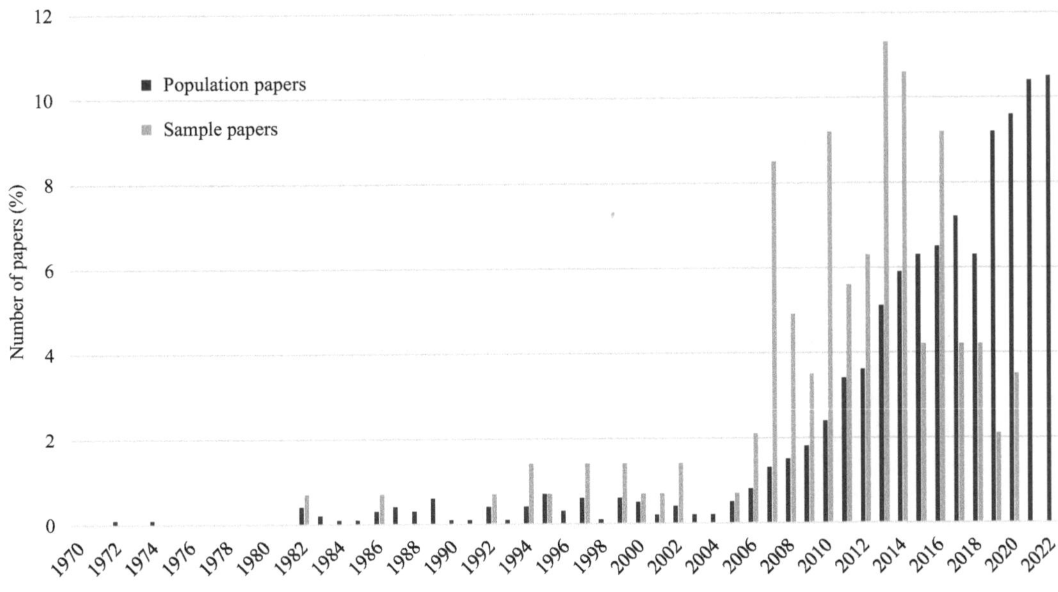

FIGURE 65.1 Research output by years regarding second generation food waste-based bioethanol fuels.

the most prolific publication years were 2013 and 2014, with 11% of the sample papers each. The other prolific publication years were 2016, 2010, and 2007, with 9% of the sample papers each.

There was a rising trend for the population papers starting in 2005, and there was a sharp rise in the research output in 2020 and 2021, possibly due to the supply shocks.

65.3.4 The Most Prolific Institutions on Second Generation Food Waste-based Bioethanol Fuels

Information about the 15 most prolific institutions publishing papers on second generation food waste-based bioethanol fuels with at least 2.1% of the sample papers each is given in Table 65.3.

The most prolific institution is the US Department of Agriculture with 4.9% of the sample papers, followed by Dokuz Eylul University, the University of Sao Paulo, and the Federal University of Ceara, with 3.5% of the sample papers each. Other prolific institutions are the University of Minho and the Korea Research Institute of Bioscience and Biotechnology, with 2.8% of the sample papers each.

Similarly, the top country for these most prolific institutions is the USA with four institutions, followed by Malaysia with three institutions. Other prolific countries are Brazil and South Korea, with two institutions. In total, only eight countries house these top institutions.

On the other hand, institution with the highest citation impact is the US Department of Agriculture with a 4.3% surplus, followed by Dokuz Eylul University and the University of Sao Paulo, with 2.9% surplus each. Other influential institutions are the Federal University of Ceara, the Korea Research Institute of Bioscience and Biotechnology, and the University of Minho, with 2.2%–2.5% surplus each, respectively.

Additionally, there are other institutions with a relatively low citation impact and with 0.6%–1.6% of the population papers each: University of Science and Technology Beijing, University of Kebangsaan Malaysia, University of Technology Petronas, Prince of Songkla University, Chinese Academy of Sciences, Technical University of Denmark, Tsinghua University, University of Malaysia Pahang, University of Jaen, Dalhousie University, International Islamic University Malaysia, National Research Council (CNR), Kasetsart University, Federal University of Rio Grande Norte, Indonesian Institute of Sciences, Kyushu University, University of Malaya, State

TABLE 65.3
The Most Prolific Institutions on Second Generation Food Waste-based Bioethanol Fuels

No.	Institutions	Country	Sample Papers (%)	Population Papers (%)	Surplus (%)
1	US Dept. Agric.	USA	4.9	0.6	4.3
2	Dokuz Eylul Univ.	Turkey	3.5	0.6	2.9
3	Univ. Sao Paulo	Brazil	3.5	0.6	2.9
4	Fed. Univ. Ceara	Brazil	3.5	1.0	2.5
5	Univ. Minho	Portugal	2.8	0.6	2.2
6	Korea Res. Inst. Biosci. Biotechnol.	South Korea	2.8	0.5	2.3
7	Chonnam Natl. Univ.	South Korea	2.1	0.6	1.5
8	Florida Dept. Citrus	USA	2.1	0.4	1.7
9	Oklahoma State Univ.	USA	2.1	0.4	1.7
10	Kansas State Univ.	USA	2.1	0.3	1.8
11	Univ. Putra Malaysia	Malaysia	2.1	2.2	–0.1
12	Natl. Tech. Univ. Athens	Greece	2.1	1.6	0.5
13	Univ. Technol. Malaysia	Malaysia	2.1	1.5	0.6
14	Univ. Boras	Sweden	2.1	0.7	1.4
15	Univ. Sains Malaysia	Malaysia	2.1	0.7	1.4

University of Campinas, Hefei University of Technology, Sichuan University, University of Naples Federico II, Federal University of Santa Catarina, Harbin Institute of Technology, Nanjing Forestry University, Isfahan University of Technology, Chulalongkorn University, University of Indonesia, Ten November Institute of Technology, and Khalifa University of Science and Technology.

65.3.5 The Most Prolific Funding Bodies on Second Generation Food Waste-based Bioethanol Fuels

Information about the 12 most prolific funding bodies funding at least 2.1% of the sample papers each is given in Table 65.4. Further, only 50% and 46% of the sample and population papers were funded, respectively.

The most prolific funding body is the National Council for Scientific and Technological Development with 7.7% of the sample papers, followed by the Higher Education Personnel Improvement Coordination with 6.3% of the sample papers, both from Brazil. Other prolific funding bodies are the Ministry of Science, Technology, and Innovation, the European Commission, the Korea Institute of Energy Technology Evaluation and Planning, the Ministry of Knowledge Economy, and Dokuz Eylul University, with 3.5%–4.9% of the sample papers each. It is notable that the largest funder of the population papers is the National Natural Science Foundation of China with 5% funding rate.

On the other hand, the most prolific countries for these top funding bodies are the South Korea and Brazil, with four and three funding bodies, respectively. Other prolific funding bodies are the European Commission and Turkey, with two funding bodies. In total, only four countries and the EU house these top funding bodies.

The funding body with the highest citation impact is the National Council for Scientific and Technological Development with a 4.6% surplus, followed by the Ministry of Science, Technology, and Innovation with a 3.9% surplus. Other influential funding bodies are the Higher Education Personnel Improvement Coordination, Dokuz Eylul University, the Ministry of Knowledge Economy, and the Korea Institute of Energy Technology Evaluation and Planning, with 2.9%–3.2% surplus each. Further, the funding body with the lowest citation impact is the National Natural Science Foundation of China with 2.2% deficit.

TABLE 65.4
The Most Prolific Funding Bodies on Second Generation Food Waste-based Bioethanol Fuels

No.	Funding Bodies	Country	Sample Paper No. (%)	Population Paper No. (%)	Surplus (%)
1	National Council for Scientific and Technological Development	Brazil	7.7	3.1	4.6
2	Higher Education Personnel Improvement Coordination	Brazil	6.3	3.1	3.2
3	Ministry of Science, Technology, and Innovation	Brazil	4.9	1.0	3.9
4	European Commission	EU	4.2	1.9	2.3
5	Korea Institute of Energy Technology Evaluation and Planning	South Korea	3.5	0.6	2.9
6	Ministry of Knowledge Economy	South Korea	3.5	0.5	3.0
7	Dokuz Eylul University	Turkey	3.5	0.4	3.1
8	National Natural Science Foundation of China	China	2.8	5.0	−2.2
9	Ministry of Education, Science and Technology	South Korea	2.8	0.8	2.0
10	Horizon 2020 Framework Program	EU	2.8	0.7	2.1
11	State Planning Organization	Turkey	2.8	0.4	2.4
12	National Research Foundation of Korea	South Korea	2.1	1.1	1.0

Other funding bodies with a relatively low citation impact and with 0.4%–1.8% of the population papers each are the European Regional Development Fund, Ministry of Higher Education Malaysia, National Key Research and Development Program of China, Ministry of Science and Technology of the People's Republic of China, Fundamental Research Funds for the Central Universities, Foundation for Science and Technology, Ministry of Science and Innovation, National Council of Science and Technology, Research Support Foundation of the State of Sao Paulo, Ministry of Education of China, Japan Society for the Promotion of Science, Prince of Songkla University, Ministry of Finance, University of Technology Malaysia, National Research Council of Thailand, Thailand Research Fund, University of Putra Malaysia, Indian Council of Agricultural Research, Ministry of Education, Culture, Sports, Science and Technology, National Institute of Food and Agriculture, U.S. Department of Agriculture, Department of Science and Technology, Ministry of Science and Technology India, Research Support Foundation of the State of Minas Gerais, Ministry of Science, ICT and Future Planning, Ministry of Trade, Industry and Energy, and National Research Foundation.

65.3.6 The Most Prolific Source Titles on Second Generation Food Waste-based Bioethanol Fuels

Information about the 16 most prolific source titles publishing at least 2.1% of the sample papers each on second generation food waste-based bioethanol fuels is given in Table 65.5.

The most prolific source title is Bioresource Technology with 26.1% of the sample papers, followed by Renewable Energy and Waste Management with 6.3% of the sample papers each. Other prolific titles are Applied Energy, Industrial Crops and Products, Biotechnology for Biofuels, and Enzyme and Microbial Technology, with 2.8%–3.5% of the sample papers each.

On the other hand, the source title with the highest citation impact is Bioresource Technology, with 17.1% surplus, followed by Renewable Energy and Waste Management, with 6.3% surplus each. Other influential titles are Applied Energy, Enzyme and Microbial Technology, and Biotechnology for Biofuels, with 2.8%–3.5% surplus each.

Other source titles with a relatively low citation impact with 0.6%–3.5% of the population papers each are Waste and Biomass Valorization, Biomass Conversion and Biorefinery, Biomass and Bioenergy, Chemical Engineering Transactions, Fermentation, Bioprocess and Biosystems

TABLE 65.5
The Most Prolific Source Titles on Second Generation Food Waste-based Bioethanol Fuels

No.	Source Titles	Sample Papers (%)	Population Papers (%)	Surplus (%)
1	Bioresource Technology	26.1	9.0	17.1
2	Renewable Energy	6.3	1.4	4.9
3	Waste Management	6.3	2.2	4.1
4	Applied Energy	3.5	0.5	3.0
5	Industrial Crops and Products	3.5	2.0	1.5
6	Biotechnology for Biofuels	2.8	0.7	2.1
7	Enzyme and Microbial Technology	2.8	0.4	2.4
8	Applied Biochemistry and Biotechnology	2.1	2.3	–0.2
9	Applied Microbiology and Biotechnology	2.1	0.6	1.5
10	Biochemical Engineering Journal	2.1	0.8	1.3
11	Bioresources	2.1	2.9	–0.8
12	Biotechnology Letters	2.1	0.7	1.4
13	Chemical Engineering Journal	2.1	0.5	1.6
14	Fuel	2.1	1.3	0.8
15	Journal of Agricultural and Food Chemistry	2.1	0.5	1.6
16	Renewable and Sustainable Energy Reviews	2.1	0.3	1.8

Engineering, Journal of Cleaner Production, Energy Sources Part A Recovery Utilization and Environmental Effects, Science of the Total Environment, African Journal of Biotechnology, Journal of Chemical Technology and Biotechnology, Processes, 3 Biotech, Biocatalysis and Agricultural Biotechnology, Bioresource Technology Reports, Preparative Biochemistry and Biotechnology, Applied Sciences Switzerland, Bioenergy Research, Energy, and Energy Conversion and Management.

65.3.7 The Most Prolific Countries on Second Generation Food Waste-based Bioethanol Fuels

Information about the 20 most prolific countries publishing at least 2.1% of sample papers each on second generation food waste-based bioethanol fuels is given in Table 65.6.

The most prolific country is the USA with 14.8% of the sample papers, followed by Brazil, Malaysia, and China with 12.0%, 11.3%, and 10.6% of the sample papers, respectively. Other prolific countries are South Korea and India, with 9.2% and 8.5% of the sample papers, respectively. It is notable that the largest producer of population papers is China with a publication rate of 14.1%, followed by Malaysia and India with 11.5% and 11.3% publication rates, respectively. Further, eight European countries listed in Table 65.6 produce 32% and 21% of the sample and population papers, respectively, with an 11% surplus.

On the other hand, the country with the highest citation impact is the USA with a 7.7% surplus, followed by South Korea and Brazil with 5.3% and 4.3% of the sample papers, respectively. The other influential countries are Sweden, Portugal, Turkey, and Ireland, with 1.6%–2.5% surplus each. Similarly, the country with the lowest citation impact is China with a 3.5% deficit, followed by India with a 2.8% deficit.

Additionally, there are other countries with a relatively low citation impact and with 0.6%–5.9% of the sample papers each: Indonesia, Thailand, Nigeria, Iran, Mexico, the United Kingdom, Pakistan, Egypt, France, Germany, Saudi Arabia, Algeria, Singapore, Poland, the United Arab Emirates, Belgium, and Bangladesh.

TABLE 65.6
The Most Prolific Countries on Second Generation Food Waste-based Bioethanol Fuels

No.	Countries	Sample Papers (%)	Population Papers (%)	Surplus (%)
1	USA	14.8	7.1	7.7
2	Brazil	12.0	7.7	4.3
3	Malaysia	11.3	11.5	–0.2
4	China	10.6	14.1	–3.5
5	South Korea	9.2	4.1	5.1
6	India	8.5	11.3	–2.8
7	Spain	5.6	5.5	0.1
8	Italy	4.9	3.9	1.0
9	Turkey	4.9	3.3	1.6
10	Canada	4.2	2.8	1.4
11	Japan	4.2	4.4	–0.2
12	Sweden	4.2	1.8	2.4
13	Greece	3.5	2.3	1.2
14	Portugal	3.5	1.6	1.9
15	Denmark	2.8	1.7	1.1
16	Australia	2.1	1.0	1.1
17	Colombia	2.1	1.2	0.9
18	Ireland	2.1	0.5	1.6
19	South Africa	2.1	1.2	0.9
20	Taiwan	2.1	1.3	0.8

65.3.8 The Most Prolific Scopus Subject Categories on Second Generation Food Waste-based Bioethanol Fuels

Information about the eight most prolific Scopus subject categories indexing at least 12% of the sample papers each is given in Table 65.7.

The most prolific Scopus subject category on second generation food waste-based bioethanol fuels is Environmental Science, with 56% of the sample papers, followed by Energy and Chemical Engineering, with 55% and 51% of the sample papers, respectively. Other prolific subject categories are Biochemistry, Genetics and Molecular Biology, Immunology and Microbiology, and Agricultural and Biological Sciences, with 15%–25% of the sample papers each. It is notable that Social Sciences, including Economics and Business account for 1% and 4% of the sample and population studies, respectively.

On the other hand, the Scopus subject category with the highest citation impact is energy, with 21% surplus, followed by Environmental Science and Chemical Engineering, with 18% and 14% surplus, respectively. Similarly, the lowest influential subject category is Agricultural and Biological Sciences with a 6% deficit, followed by Chemistry and Engineering with a 1% deficit each.

65.3.9 The Most Prolific Keywords on Second Generation Food Waste-based Bioethanol Fuels

Information about the Scopus keywords used with at least 4.9% or 2.5% of the sample or population papers, respectively, is given in Table 65.8. For this purpose, keywords related to the keyword set given in the appendix are selected from a list of the most prolific keyword sets provided by the Scopus database.

TABLE 65.7
The Most Prolific Scopus Subject Categories on Second Generation Food Waste-based Bioethanol Fuels

No.	Scopus Subject Categories	Sample Papers (%)	Population Papers (%)	Surplus (%)
1	Environmental Science	56.3	38.7	17.6
2	Energy	54.9	33.6	21.3
3	Chemical Engineering	51.4	37.4	14.0
4	Biochemistry, Genetics and Molecular Biology	25.4	25.2	0.2
5	Immunology and Microbiology	19.0	15.2	3.8
6	Agricultural and Biological Sciences	14.8	20.3	–5.5
7	Engineering	12.0	12.8	–0.8
8	Chemistry	10.6	12.0	–1.4

TABLE 65.8
The Most Prolific Keywords on Second Generation Food Waste-based Bioethanol Fuels

No.	Keywords	Sample Papers (%)	Population Papers (%)	Surplus (%)
1	**Food Waste**			
	Fruits	46.5	25.0	21.5
	Cellulose	34.5	18.4	16.1
	Biomass	28.2	16.4	11.8
	Lignin	19.7	13.2	6.5
	Waste	19.7	8.1	11.6
	Carbohydrate	18.3	7.8	10.5
	Empty fruit bunches	14.1	13.3	0.8
	Food waste	14.1	12.3	1.8
	Elaeis (palm)	12.0	4.6	7.4
	Lignocellulose	9.2	5.8	3.4
	Hemicellulose	8.5	4.8	3.7
	Citrus	8.5	2.2	6.3
	Starch	7.7	5.2	2.5
	Lignocellulosic biomass	7.7	5.1	2.6
	Food	7.7	4.4	3.3
	Cheese whey	6.3	3.9	2.4
	Cheese	6.3	3.0	3.3
	Coffee	5.6	2.1	3.5
	Solanum (potato)	4.9	2.2	2.7
2	**Pretreatments**			
	Pre-treatment	19.0	10.6	8.4
	Enzymes	19.0	10.2	8.8
	Sulfuric acid	18.3	5.2	13.1

(Continued)

TABLE 65.8 (*Continued*)
The Most Prolific Keywords on Second Generation Food Waste-based Bioethanol Fuels

No.	Keywords	Sample Papers (%)	Population Papers (%)	Surplus (%)
	Pretreatment	16.2	8.0	8.2
	Temperature	15.5	8.4	7.1
	pH	14.1	10.9	3.2
	Cellulase	14.1	7.1	7.0
	Delignification	6.3	4.5	1.8
	Steam	5.6		5.6
	Alkalinity	4.9	3.3	1.6
	Sodium hydroxide	4.2	4.1	0.1
	Fractionation		2.5	−2.5
3		**Fermentation**		
	Fermentation	68.3	45.5	22.8
	Saccharomyces	32.4	13.0	19.4
	Yeast	30.3	16.9	13.4
	Kluyveromyces	16.9	6.8	10.1
	Ethanol fermentation	11.3	4.7	6.6
	SSF	10.6	5.4	5.2
	Batch fermentation	4.9	2.9	2.0
	Bacteria	3.5	7.8	−4.3
	Fungi		7.2	−7.2
	Fermentation process		2.5	−2.5
4		**Hydrolysis and Hydrolysates**		
	Hydrolysis	54.9	29.0	25.9
	Sugar	49.3	19.0	30.3
	Enzymatic hydrolysis	28.9	18.8	10.1
	Glucose	28.9	14.3	14.6
	Saccharification	28.2	15.4	12.8
	Enzyme activity	26.8	11.8	15.0
	Lactose	7.7	3.5	4.2
	Fermentable sugars	7.0	4.6	2.4
	Xylose	6.3	3.9	2.4
	Reducing sugars	5.6	3.9	1.7
	Enzymatic saccharification	5.6	3.5	2.1
	Enzymolysis	4.9	3.2	1.7
	Acid hydrolysis	4.2	3.2	1.0
5		**Waste Management**		
	Waste management	14.8	5.4	9.4
	Bioreactor	11.3	10.1	1.2
	Waste treatment	11.3	5.9	5.4
	Waste disposal	8.5	5.1	3.4
	Refuse disposal	8.5	4.2	4.3

(*Continued*)

TABLE 65.8 (*Continued*)
The Most Prolific Keywords on Second Generation Food Waste-based Bioethanol Fuels

No.	Keywords	Sample Papers (%)	Population Papers (%)	Surplus (%)
6	**Products**			
	Ethanol	69.0	40.3	28.7
	Biofuels	40.8	18.0	22.8
	Bioethanol	33.1	25.0	8.1
	Bio-ethanol production	19.0	11.6	7.4
	Ethanol concentrations	19.0	5.0	14.0
	Ethanol production	17.6	8.4	9.2
	Biofuel production	9.2	3.5	5.7
	Waste products	7.7	2.5	5.2
	Ethanol yield	7.7	2.4	5.3
	Acetic acid	7.0	4.8	2.2
	Ethanol productivity	4.9		4.9
	Lactic acid		3.5	−3.5

These keywords are grouped under six headings: food waste, pretreatments, fermentation, hydrolysis and hydrolysates, waste management, and products.

The most prolific keywords related to biomass and biomass constituents are fruits, cellulose, and biomass, with 47%, 35%, and 28% of the sample papers, respectively. Other prolific keywords are lignin, waste, carbohydrate, empty fruit bunches, food waste, and elaeis (palm), with 12%–20% of the sample papers each.

Further, the most prolific keywords related to pretreatments are pre-treatments, enzymes, and sulfuric acid, with 19%, 19%, and 18% of the sample papers, respectively. Other prolific keywords are pretreatments, temperature, pH, and cellulases, with 14%–16% of the sample papers each.

The most prolific keywords related to fermentation are fermentation, saccharomyces, and yeast, with 68%, 32%, and 30% of the sample papers, respectively. Other prolific keywords are Kluyveromyces, ethanol fermentation, and simultaneous saccharification and fermentation, with 10%–17% of the sample papers each.

Further, the most prolific keywords related to hydrolysis and hydrolysates are hydrolysis and sugars, with 55% and 49% of the sample papers, respectively. Other prolific keywords are enzymatic hydrolysis, glucose, saccharification, and enzyme activity, with 27%–29% of the sample papers each.

Finally, the most prolific keywords related to the products are ethanol, biofuels, and bioethanol, with 69%, 41%, and 33% of the sample papers, respectively. Other prolific keywords are bio-ethanol production, ethanol concentrations, and ethanol production, with 18%–19% of the sample papers.

On the other hand, the most prolific keywords across all research fronts are ethanol, fermentation, hydrolysis, sugar, fruits, biofuels, cellulose, bioethanol, saccharomyces, yeast, glucose, enzymatic hydrolysis, saccharification, biomass, enzyme activity, waste, and lignin, with 20%–69% of the sample papers each. Similarly, the most influential keywords are sugar, ethanol, hydrolysis, fermentation, biofuels, fruits, saccharomyces, cellulose, enzyme activity, and glucose, with 15%–30% surplus each.

65.3.10 The Most Prolific Research Fronts on Second Generation Food Waste-based Bioethanol Fuels

Information about the research fronts for the sample papers on second generation food waste-based bioethanol fuels with regard to the food waste feedstock used for the bioethanol production is given in Table 65.9.

TABLE 65.9
The Most Prolific Research Fronts for Second Generation Food Waste-based Bioethanol Fuels

No,	Research Fronts	N Paper % Sample
1	Fruit wastes	58.5
	Empty fruit bunches	15.5
	Other fruit wastes	15.5
	Orange peels	9.9
	Apple pomaces	3.5
	Banana peels	3.5
	Cashew apple bagasse	2.8
	Citrus peels	2.8
	Pineapple waste	2.8
	Grapefruit peels	0.7
2	Food wastes in general	24.6
3	Other food wastes	35.9
	Vegetable wastes	12.7
	Dairy wastes	11.3
	Beverage wastes	7.0
	Olive wastes	2.1
	Other wastes	2.1
	Fish wastes	0.7

N Paper (%) review: The number of papers in the sample of 25 reviewed papers. N paper (%) sample: The number of papers in the population sample of 142 papers.

TABLE 65.10
The Most Prolific Thematic Research Fronts for Second Generation Food Waste-based Bioethanol Fuels

No.	Research Fronts	N Paper % Sample
1	Biomass hydrolysis	68.5
2	Biomass pretreatments	65.5
3	Hydrolysate fermentation	62.7
4	Bioethanol fuel production	62.0
5	Bioethanol fuel evaluation	4.2

N Paper (%) review: The number of papers in the sample of 25 reviewed papers. N paper (%) sample: The number of papers in the population sample of 142 papers.

As this table shows, the most prolific food waste feedstock is fruit wastes, with 59% of the sample papers, followed by other food wastes, and food wastes in general, with 36% and 25% of the sample papers, respectively.

Further, the prolific research fronts for the fruit wastes are empty fruit bunches, other fruit wastes, and orange peels, with 16%, 16%, and 10% of the sample papers, respectively. Similarly, the most prolific research fronts for the other food wastes are vegetable wastes, dairy wastes, and beverage wastes, with 13%, 11%, and 7% of the sample papers, respectively.

Information about the thematic research fronts for the sample papers on second generation food waste-based bioethanol fuels is given in Table 65.10. As this table shows, there are four primary research fronts: biomass hydrolysis, biomass pretreatments, hydrolysate fermentation, and bioethanol production, with 69%, 66%, 63%, and 62% of the sample papers, respectively. Another minor research front is the bioethanol fuel evaluation. Further, the most prolific feedstocks used in all these research fronts are fruit wastes although their content are not shown in this table.

65.4 DISCUSSION

65.4.1 Introduction

Crude oil-based gasoline fuels have been widely used in the transportation sector since the 1920s. However, there have been great public concerns over the adverse environmental and human impacts of these fuels. Hence, biomass-based bioethanol fuels have increasingly been used in blending gasoline fuels, in fuel cells, and in biochemical production in a biorefinery context.

However, it is necessary to pretreat the biomass to enhance the yield of bioethanol prior to its bioethanol production through hydrolysis and fermentation. One of the most studied feedstocks for bioethanol fuels has been food wastes such as fruit, vegetable, beverage, and dairy wastes. The research in the field of second generation food waste-based bioethanol fuels has intensified in this context on the key research fronts of the pretreatment and hydrolysis of food wastes, fermentation of the food waste-based hydrolysates, and production and evaluation of second generation food waste-based bioethanol fuels. Further, the empty fruit bunches, orange peels, citrus peels, potato peels, spent coffee grounds, cheese whey, banana peels, apple pomaces, and olive stones have been studied intensively in this context.

However, it is essential to develop efficient incentive structures for the primary stakeholders to enhance research in this field. This is especially important to maintain energy security in the cases of supply shocks such as oil price shocks, war-related shocks as in the case of the Russian invasion of Ukraine, or COVID-19 shocks.

Scientometric analysis has been used in this context to inform the primary stakeholders about the current state of research in this research field. As there has been no scientometric study in this field, this chapter presents a scientometric study of research on second generation food waste-based bioethanol fuels. It examines the scientometric characteristics of both the sample and population data and presents them in the order of documents, authors, publication years, institutions, funding bodies, source titles, countries, Scopus subject categories, Scopus keywords, and research fronts.

As the first step for the search of the relevant literature, keywords were selected using the 200 most-cited papers. The selected keyword list was then optimized to obtain a representative sample of papers in this research field. A copy of this keyword list is provided in the appendix for future replicative studies. Further, a selected list of the keywords is presented in Table 65.8.

As the second step, two sets of data were used in this study. First, a population sample of 1,418 papers was used to examine the scientometric characteristics of the population data. Second, a sample of 142 most-cited papers, corresponding to 10% of the population data set was used to examine the scientometric characteristics of these citation classics.

The scientometric characteristics of these sample and population datasets were presented in the order of documents, authors, publication years, institutions, funding bodies, source titles, countries, Scopus subject categories, Scopus keywords, and research fronts.

Lastly, the key scientometric findings for both datasets were discussed to highlight the research landscape for second generation food waste-based bioethanol fuels. Additionally, a number of brief conclusions were drawn, and a number of relevant recommendations were made to enhance the future research landscape.

65.4.2 The Most Prolific Documents on Second Generation Food Waste-based Bioethanol Fuels

Articles (together with conference papers) dominate both the sample (93%) and population (96%) papers (Table 65.1). Further, review papers have a surplus (5%). The representation of the reviews in the sample papers is relatively modest (7%).

Scopus differs from the Web of Science database in differentiating and showing articles (92%) and conference papers (1%) published in journals separately. However, it should be noted that these conference papers are also published in journals as articles, compared to those published only in conference proceedings. Hence, the total number of articles and review papers in the sample dataset is 93% and 7%, respectively.

It is observed during the search process that there has been inconsistency in the classification of documents in Scopus as well as in other databases such as Web of Science. This is especially relevant for the classification of papers as reviews or articles, as papers not involving a literature review may be erroneously classified as review papers. There is also a case of review papers being classified as articles. For example, the total number of reviews in the sample data set was manually found as nearly 9% compared to 7% as indexed by Scopus, decreasing the number of articles and conference papers to 91% for the sample dataset. The close examination of these papers show that many evaluative studies, such as techno-economic or life-cycle studies, have been indexed as the review papers by the Scopus database.

In this context, it would be helpful to provide a classification note for published papers in books and journals at the first instance. It would also be helpful to use document types listed in Table 65.1 for this purpose. Book chapters may also be classified as articles or reviews as an additional classification to differentiate review chapters from experimental chapters as it is done by the Web of Science. It would be further helpful to additionally classify the conference papers as articles or review papers as well as it is done in the Web of Science database.

65.4.3 The Most Prolific Authors on Second Generation Food Waste-based Bioethanol Fuels

There have been 15 most prolific authors with at least 2.1% of the sample papers each, as given in Table 65.2. These authors have shaped the development of research in this field.

The most prolific authors are Karel Grohmann, Fikret Kargi, Serpil Ozmihci and, to a lesser extent, Maria V. P. Rocha, Jose A. Teixeira, Harinder S. Oberoi, and Luciano R. B. Goncalves.

It is important to note the inconsistencies in the indexing of author names in Scopus and other databases. It is especially an issue for names with more than two components, such as 'Blake Sam de Hyun Rocha'. The probable outcomes are 'Rocha, B.S.D.H.', 'de Hyun Rocha, B.S.' or 'Hyun Rocha, B.S.D.' The first choice is the gold standard of the publishing sector, as the last word in the name is taken as the last name. In most of the academic databases, such as PUBMED and EBSCO databases, this version is used predominantly. The second choice is a strong alternative, while the last choice is an undesired outcome as two last words are taken as the last name. It is good practice to combine the words of the last name with a hyphen: 'Hyun-Rocha, B.S.D.'. It is notable that inconsistent indexing of author names may cause substantial inefficiencies in the search process for papers as well as allocating credit to authors as there are different author entries for each outcome in the databases.

There are also inconsistencies in the shortening of Chinese names. For example, 'YangYing Yang' is often shortened as 'Yang, Y.', 'Yang, Y.-Y.', and 'Yang, Y.Y.', as it is done in the Web of Science database as well. However, the gold standard in this case is 'Lee, Y', where the last word is taken as the last name and the first word is taken as a single forename. In most of the academic databases, such as PUBMED and EBSCO, this first version is used predominantly. However, it makes sense to use the third option to differentiate Chinese names efficiently: 'Yang, Y.Y.'. Therefore, there have been difficulties in locating papers for Chinese authors. In such cases, the use of the unique author codes provided for each author by the Scopus database has been helpful.

There is also a difficulty in allocating credit for authors, especially for those with common names such as 'Yang, X.', in conducting scientometric studies. These difficulties strongly influence the efficiency of scientometric studies as well as allocating credit to authors, as there are the same author entries for different authors with the same name, e.g., 'Yang, X.' in the databases.

In this context, the coding of authors in the Scopus database is a welcome innovation compared to other databases such as Web of Science. In this process, Scopus allocates a unique number to each author in the database (Aman, 2018). However, there might still be substantial inefficiencies in this coding system, especially for common names. For example, some of the papers by a certain author maybe allocated to another researcher with a different author code. It is possible that Scopus uses a number of software programs to differentiate author names, and the program may not be false-proof (Kim, 2018).

In this context, it does not help that author names are not given in full in some journals and books. This makes it difficult to differentiate authors with common names and makes scientometric studies further difficult in the author domain. Therefore, author names should be given in all books and journals at the first instance. There is also a cultural issue where some authors do not use their full names in their papers. Instead, they use initials for their forenames: 'Yang, H.J.', 'Yang, 'Yang, H.', or 'Yang, J.' instead of 'Yang, Hyun Jae'.

There are also inconsistencies in the naming of authors with more than two components by authors themselves in journal papers and book chapters. For example. 'Rocha, A.P.C.' might be given as 'Rocha, A.' or 'Rocha, A.C.' or 'Rocha, A.P.' or 'Rocha, C' in journals and books. This also makes scientometric studies difficult in the author's domain. Hence, contributing authors should use their name consistently in their publications.

Another critical issue regarding author names is the inconsistencies in the spelling of author names in the national spellings (e.g., Şöğütçığsüt, Gökçe) rather than in the English spellings (e.g., Sogutcigsut, Gokce) in the Scopus database. Scopus differs from the Web of Science database and many other databases in this respect, where author names are given only in their English spelling. It is observed that the national spelling of author names do not help much in conducting scientometric studies as well as in allocating credit to authors as sometimes there are different author entries for English and National spellings in the Scopus database.

The most prolific institutions for the sample dataset are Chonnam National University, Dokuz Eylul University, the Federal University of Ceara, and the USDA Agricultural Research Service. Further, the most prolific countries for the sample dataset are the USA, South Korea and, to a lesser extent, Brazil, India, and Turkey. These findings confirm the dominance of the USA and, to a lesser extent, Brazil, India, and Turkey in this field. On the other hand, the primary research fronts are bioethanol fuels based on fruit wastes while the other research fronts are bioethanol fuels from dairy and beverage wastes.

It is also notable that there is a significant gender deficit in the sample dataset surprisingly with a representation rate of 20%. This finding is the most thought-provoking with strong public policy implications. Hence, institutions, funding bodies, and policy makers should take efficient measures to reduce the gender deficit in this field as well as in other scientific fields with a strong gender deficit. In this context, it is worth noting that the level of representation of researchers from minority groups in science on the basis of race, sexuality, age, and disability, besides gender (Blankenship, 1993; Dirth and Branscombe, 2017; Konur, 2000, 2002a–c, 2006a,b, 2007a,b).

65.4.4 The Most Prolific Research Output by Years on Second Generation Food Waste-based Bioethanol Fuels

The research output observed between 1970 and 2022 is illustrated in Figure 65.1. This figure clearly shows that the bulk of the research papers in the population dataset was published primarily in the 2010s and early 2020s. Similarly, the bulk of the research papers in the sample dataset was published in the 2010s and 2000s. These findings suggest that the most prolific sample and population papers were primarily published in the 2010s.

These are thought-provoking findings as there has been a significant research boom since 2009. In this context, the increasing public concerns about climate change (Change, 2007), greenhouse gas emissions (Carlson et al., 2017), and global warming (Kerr, 2007) have been certainly behind the research boom in this field in the last two decades. Furthermore, the recent supply shocks due to the COVID-19 pandemic might also be behind the research boom in this field since 2009.

Based on these findings, the size of the population papers is likely to more than double in the current decade, provided that the public concerns about climate change, greenhouse gas emissions, and global warming, as well as supply shocks are translated efficiently to the research funding in this field.

65.4.5 The Most Prolific Institutions on Second Generation Food Waste-based Bioethanol Fuels

The 15 most prolific institutions publishing papers on second generation food waste-based bioethanol fuels, with at least 2.1% of the sample papers each given in Table 65.3, have shaped the development of research in this field.

The most prolific institutions are the US Department of Agriculture, Dokuz Eylul University, the University of Sao Paulo, the Federal University of Ceara and, to a lesser extent, the University of Minho, and the Korea Research Institute of Bioscience and Biotechnology. Similarly, the top countries for these most prolific institutions are the USA, Malaysia and, to a lesser extent, Brazil and South Korea. In total, only eight countries house these top institutions.

On the other hand, institutions with the highest citation impact are the US Department of Agriculture and, to a lesser extent, Dokuz Eylul University, the University of Sao Paulo, the Federal University of Ceara, the Korea Research Institute of Bioscience and Biotechnology, and the University of Minho.

These findings confirm the dominance of institutions from the USA and, to a lesser extent, from Brazil, Portugal, Malaysia, South Korea, and Turkey.

65.4.6 The Most Prolific Funding Bodies on Second Generation Food Waste-based Bioethanol Fuels

The 12 most prolific funding bodies funding at least 2.1% of the sample papers each are given in Table 65.4. It is notable that only 50% and 46% of the sample and population papers were funded, respectively.

The most prolific funding bodies are the National Council for Scientific and Technological Development, Higher Education Personnel Improvement Coordination, the Ministry of Science, Technology, and Innovation, the European Commission, the Korea Institute of Energy Technology Evaluation and Planning, the Ministry of Knowledge Economy, and Dokuz Eylul University. It is notable that the largest funder of the population papers is the National Natural Science Foundation of China, with 5% funding rate.

The most prolific countries for these top funding bodies are South Korea, Brazil and, to a lesser extent, the European Commission and Turkey. In total, only four countries and the EU house these top funding bodies.

These findings on the funding of the research in this field suggest that the level of funding, mostly since 2009, is modest and it has been largely instrumental in enhancing the research in this field (Ebadi and Schiffauerova, 2016) in light of North's institutional framework (North, 1991). It is also notable that the funding rate in this field is relatively modest compared to those in other research fronts of bioethanol fuels, such as algal bioethanol fuels. Further, it is expected that this this funding rate would improve in light of the recent supply shocks. Further, it emerges that South Korea, Brazil, and Europe have heavily funded the research on food waste-based bioethanol fuels.

65.4.7 The Most Prolific Source Titles on Second Generation Food Waste-based Bioethanol Fuels

The 16 most prolific source titles publishing at least 2.1% of the sample papers each on second generation food waste-based bioethanol fuels have shaped the development of the research in this field (Table 65.5).

The most prolific source titles are Bioresource Technology and, to a lesser extent, Renewable Energy, Waste Management, Applied Energy, Industrial Crops and Products, Biotechnology for Biofuels, and Enzyme and Microbial Technology. On the other hand, the source titles with the most citation impact are Bioresource Technology and, to a lesser extent, Renewable Energy, Waste Management, Applied Energy, Enzyme and Microbial Technology, and Biotechnology for Biofuels.

It is notable that these top source titles are primarily related to bioresources, energy, wastes, biotechnology, and microbial technology. This finding suggests that Bioresource Technology, and other prolific journals in these fields have significantly shaped the development of the research in this field as they focus primarily on second generation food waste-based bioethanol fuels with a high yield. In this context, the influence of top journals is quite extraordinary with 26.1% of the sample papers.

65.4.8 The Most Prolific Countries on Second Generation Food Waste-based Bioethanol Fuels

The 20 most prolific countries publishing at least 2.1% of the sample papers each have significantly shaped the development of research in this field (Table 65.6).

The most prolific countries are the USA, Brazil, Malaysia, China and, to a lesser extent, South Korea and India. It is notable that the largest producer of population papers is China, followed by Malaysia and India. Further, eight European countries listed in Table 65.6 produce 32% and 21% of the sample and population papers, respectively, with an 11% surplus.

On the other hand, countries with the highest citation impact are the USA, South Korea, Brazil and, to a lesser extent, Sweden, Portugal, Turkey, and Ireland. Similarly, countries with the lowest impact are China and India.

A close examination of these findings suggests that the USA, Europe, Brazil, Malaysia, China and, to a lesser extent, South Korea and India are the major producers of research in this field. It is a fact that the USA has been a major player in science (Leydesdorff and Wagner, 2009). The USA has further developed a strong research infrastructure to support its corn and grass-based bioethanol industries (Gillon, 2010).

However, China has been a rising mega star in scientific research in competition with the USA and Europe (Leydesdorff and Zhou, 2005). China is also a major player in this field as a major producer of bioethanol (Fang et al., 2010).

Next, Europe has been a persistent player in scientific research in competition with both the USA and China (Leydesdorff, 2000). Europe has also been a persistent producer of bioethanol along with the USA and Brazil (Gnansounou, 2010).

Malaysia (Goh et al., 2010), Brazil (Soccol et al., 2010), South Korea (Kim et al., 2010), India (Sukumaran et al., 2010), Sweden (Ekman et al., 2013), Turkey (Ozdingis and Kocar, 2018), and Ireland (Murphy and McCarthy, 2005) are the other countries with substantial research activities in bioethanol fuels.

65.4.9 The Most Prolific Scopus Subject Categories on Second Generation Food Waste-based Bioethanol Fuels

The eight most prolific Scopus subject categories indexing at least 12% of the sample papers each, given in Table 65.7, have shaped the development of research in this field.

The most prolific Scopus subject categories on second generation food waste-based bioethanol fuels are Environmental Science, Energy, and Chemical Engineering and, to a lesser extent, Biochemistry, Genetics and Molecular Biology, Immunology and Microbiology, and Agricultural and Biological Science. It is also notable that Social Sciences, including Economics and Business, have a minimal presence in both sample and population studies.

On the other hand, Scopus subject categories with the highest citation impact are Energy and, to a lesser extent, Environmental Science and Chemical Engineering. Similarly, the lowest influential subject categories are Agricultural and Biological Sciences and, to a lesser extent, Chemistry and Engineering.

These findings are thought-provoking, suggesting that the primary subject categories are related to energy, environmental science, chemical engineering, genetics, and microbiology as the core of the research in this field concerns with the production and utilization of second generation food waste-based bioethanol fuels. Another finding is that social sciences are not well represented in both sample and population papers in line with most of the fields in bioethanol fuels. Social, environmental, and economics studies account for the field of social sciences.

65.4.10 The Most Prolific Keywords on Second Generation Food Waste-based Bioethanol Fuels

A limited number of keywords have shaped the development of research in this field, as shown in Table 65.8 and the Appendix. These keywords are grouped under six headings: food waste biomass, fermentation, hydrolysis and hydrolysates, waste management, products, and evaluation.

The most prolific keywords across all research fronts are ethanol, fermentation, hydrolysis, sugar, fruits, biofuels, cellulose, bioethanol, saccharomyces, yeast, glucose, enzymatic hydrolysis, saccharification, biomass, enzyme activity, waste, and lignin. Similarly, the most influential keywords are sugar, ethanol, hydrolysis, fermentation, biofuels, fruits, saccharomyces, cellulose, enzyme activity, and glucose.

These findings suggest that it is necessary to determine the keyword set carefully to locate relevant research in each of these research fronts. Additionally, the size of the samples for each keyword highlights the intensity of research in the relevant research areas for both sample and population datasets. These findings also highlight different spelling of some strategic keywords, such as pretreatment vs. pre-treatment, ethanol vs. bio-ethanol, etc.

65.4.11 The Most Prolific Research Fronts on Second Generation Food Waste-based Bioethanol Fuels

Information about research fronts for sample papers on second generation food waste-based bioethanol fuels with regard to the food waste feedstocks used for the bioethanol production is given in Table 65.9. As this table shows, the most prolific food waste feedstock is fruit wastes, followed by other food wastes, and food wastes in general. Further, the prolific research fronts for fruit wastes are empty fruit bunches, other fruit wastes, and orange peels. Similarly, the prolific research fronts for the other food wastes are vegetable wastes, dairy wastes, and beverage wastes.

Information about the thematic research fronts for the sample papers on second generation food waste-based bioethanol fuels is given in Table 65.10. As this table shows, there are four primary research fronts: biomass hydrolysis, biomass pretreatments, hydrolysate fermentation, and bioethanol production. The other minor research front is bioethanol fuel evaluation. The first four research fronts dominate the research in this field.

These findings are thought-provoking in seeking ways to increase food waste feedstock-based bioethanol yield at the global scale. It is clear that all of these research fronts have public importance and merit substantial funding and other incentives. Further, it is notable that second generation food waste-based bioethanol fuels have become a core unit of bioethanol research to make it more competitive with crude oil-based gasoline and petrodiesel fuels, especially for the USA, Europe, and China.

In comparison to the other feedstock-based research fronts, it is notable that the pretreatment and hydrolysis of wood wastes, hydrolysate fermentation, and bioethanol fermentation emerge as a primary research front for this field. This suggests that the primary stakeholders have been primarily interested in the four key processes of bioethanol production. It is also notable that evaluation of second generation food waste-based bioethanol fuels, such as technoeconomics, life cycle, economics, social, land use, labor, and environment-related studies emerges as a case study for the bioethanol fuels together with algal and sugar feedstocks in this field. In this context, the USA has been the global leader in the production and use of corn-based bioethanol fuels since the 1970s in the aftermath of the global crude oil crisis in the early 1970s.

In the end, these most cited papers in this field hint that the production of second generation food waste-based bioethanol fuels could be optimized using the structure, processing, and property relationships of these food wastes in the fronts of feedstock pretreatment and hydrolysis, and hydrolysate fermentation (Formela et al., 2016; Konur, 2018a, 2020b, 2021a–d; Konur and Matthews, 1989).

65.5 CONCLUSION AND FUTURE RESEARCH

The research on second generation food waste-based bioethanol fuels has been mapped through a scientometric study of both sample (142 papers) and population (1,418 papers) datasets.

The critical issue in this study has been obtaining a representative sample of research as in any other scientometric study. Therefore, the keyword set has been carefully devised and optimized after a number of runs in the Scopus database. It is a representative sample of wider population studies. This keyword set was provided in the Appendix, and the relevant keywords are presented in Table 65.8. However, it should be noted that it has been very difficult to compile a representative keyword set since this research field has been closely connected with many other fields. Therefore, it has been necessary to compile a keyword list to exclude papers concerned with other research fields.

Another issue has been the selection of a multidisciplinary database to carry out the scientometric study of research in this field. For this purpose, the Scopus database has been selected. The journal coverage of this database has been notably wider than that of Web of Science and other multisubject databases.

The key scientometric properties of research in this field have been determined and discussed in this chapter. It is evident that a limited number of documents, authors, institutions, publication years, institutions, funding bodies, source titles, countries, Scopus subject categories, Scopus keywords, and research fronts have shaped the development of research in this field.

There is ample scope to increase the efficiency of scientometric studies in this field in the author and document domains by developing consistent policies and practices in both domains across all academic databases. In this respect, it seems that authors, journals, and academic databases have a lot to do. Furthermore, the significant gender deficit as in most scientific fields, emerges as a public policy issue. Potential deficits on the basis of age, race, disability, and sexuality also need to be explored in this field as in other scientific fields.

Research in this field has boomed since 2007, possibly promoted by public concerns about global warming, greenhouse gas emissions, and climate change. Furthermore, the recent COVID-19 pandemic and the Russian invasion of Ukraine have resulted in a global supply shocks shifting the focus of stakeholders from crude oil-based fuels to biomass-based fuels such as bioethanol fuels. It is expected that there would be further incentives for the key stakeholders to carry out research on second generation food waste-based bioethanol fuels to increase the ethanol yield and to make it more competitive with crude oil-based gasoline and petrodiesel fuels. This might be true for crude oil- and foreign exchange-deficient countries to maintain energy and food security in the face of global supply shocks.

The relatively modest funding rates of 50% and 46% for sample and population papers, respectively, suggests that funding in this field has significantly enhanced research in this field primarily since 2007, possibly more than doubling in the current decade. However, it is evident that there is ample room for more funding and other incentives to enhance research in this field further.

Institutions from the USA, Malaysia and, to a lesser extent, Brazil and South Korea, have mostly shaped the research in this field. Further, the USA, Europe, Brazil, Malaysia, China and, to a lesser extent, South Korea and India have been the major producers of research in this field and users of bioethanol fuels. It is evident that these countries have a well-developed research infrastructure in bioethanol fuels and their derivatives.

It emerges that ethanol is more popular than bioethanol as a keyword with strong implications for the search strategy. In other words, the search strategy using only bioethanol as the keyword would not be much helpful. The Scopus keywords are grouped under six headings: biomass, pretreatments, fermentation, hydrolysis and hydrolysates, waste management, and products. It is also important to include the Latin terms for food waste biomass in the keyword set.

As Table 65.9 shows, the most prolific food waste feedstock is fruit wastes, followed by other food wastes, and food wastes in general. Further, the prolific research fronts for fruit wastes are empty fruit bunches, other fruit wastes, and orange peels. Similarly, the prolific research fronts for the other food wastes are vegetable wastes, dairy wastes, and beverage wastes.

On the other hand, Table 65.10 shows that there are four primary research fronts: biomass hydrolysis, biomass pretreatments, hydrolysate fermentation, and bioethanol production. Another minor research front is bioethanol fuel evaluation. The first four research fronts dominate research in this field.

These findings are thought-provoking in seeking ways to increase food waste feedstock-based bioethanol yield on a global scale. It is clear that all these research fronts have public importance and merit substantial funding and other incentives. Further, it is notable that second generation food waste-based bioethanol fuels have become a core unit of bioethanol research to make it more competitive with crude oil-based gasoline and diesel fuels, especially for the USA, Europe, and China.

Thus, scientometric analysis has a great potential to gain valuable insights into the evolution of research in this field as in other scientific fields especially in the aftermath of significant global supply shocks such as the COVID-19 pandemic and the Russian invasion of Ukraine.

It is recommended that further scientometric studies are carried out on the primary research fronts. It is further recommended that reviews of the most cited papers are carried out for each primary research front to complement these scientometric studies. Next, the scientometric studies of the hot papers in these primary fields need to be carried out.

ACKNOWLEDGMENTS

The contribution of the highly cited researchers in the field of second generation food waste-based bioethanol fuels has been gratefully acknowledged.

APPENDIX: THE KEYWORD SET FOR SECOND GENERATION FOOD WASTE-BASED BIOETHANOL FUELS

(((((TITLE (*food OR cheese OR restaurant OR kitchen OR potato OR grape OR coffee OR cannery OR *fruit OR citrus OR kinnow OR banana OR bread OR tea OR tuna OR vegetable OR mandarin OR olive OR lemon OR *apple OR pear OR mango OR *nut OR orange OR cassava OR almond OR fish OR sago OR poultry OR taro OR salad OR brassica OR date OR dairy OR bean OR pomegranate OR palm OR cocoa OR kiwi* OR egg* OR domestic OR bergamot OR herb* OR jatropha OR pumpkin OR matooke OR papaya OR musambi OR omija OR pongamia OR apricot OR ananas OR seafood OR durian OR kapok OR *melon OR yam OR trilepisium OR rambutan OR milk) AND TITLE (waste OR biowaste OR garbage OR spent OR bunches OR marc OR silverskin OR stone

OR residues OR bagasse OR pomaces OR husk OR whey OR wasted OR "grape seed" OR dregs OR *shells OR scrap* OR "grape stalks" OR pith OR permeate)) OR TITLE (peel* OR "spent mushroom")) AND TITLE (ethanol OR bioethanol OR etoh OR c2h5oh)) AND NOT (TITLE (wine OR extract* OR stem OR precipitation OR anti* OR "agricultural waste" OR "agricultural banana" OR ester* OR "organic waste" OR "pre fermentation" OR pruning OR "agro-waste" OR caproate OR "leaf waste" OR stillage OR solvent OR combustion OR preferment* OR adsorption OR transester* OR dehydration OR cooking OR rats OR hydroyxmethylfurfural OR sensor OR steel OR trunk OR lycoris OR biosorbent OR dehydrogenation OR equilibria OR zno OR "crop residues" OR triterpene OR fat OR shallot OR "non food" OR aerogels OR preservation OR packaging OR "agricultural resid*" OR cadmium) OR SRCTITLE (toxic* OR data OR thermo* OR separat*)))) OR (((((TITLE (*food OR cheese OR restaurant OR kitchen OR potato OR grape OR coffee OR cannery OR *fruit OR citrus OR kinnow OR banana OR bread OR tea OR tuna OR vegetable OR mandarin OR olive OR lemon OR *apple OR pear OR mango OR *nut OR orange OR cassava OR almond OR fish OR sago OR poultry OR taro OR salad OR brassica OR date OR dairy OR bean OR pomegranate OR palm OR cocoa OR kiwi* OR egg* OR domestic OR bergamot OR herb* OR jatropha) AND TITLE (waste OR biowaste OR garbage OR spent OR bunches OR marc OR silverskin OR stone OR residues OR bagasse OR pomaces OR husk OR whey OR wasted OR "grape seed" OR dregs OR *shells OR scrap* OR "grape stalks" OR pith OR permeate)) OR TITLE (peel* OR "spent mushroom")) AND TITLE (*saccharification OR *hydrolysis OR digestibili* OR ssf OR hydrolysate* OR hydrolyzate* OR ferment* OR coferment* OR fractionation OR delignification OR depolymerization OR pretreat* OR "pre-treat*" OR "consolidated processing")) AND NOT (TITLE (*hydrogen OR anaerobic OR methan* OR biomethan* OR groundwater OR "solid state" OR h2 OR ch4 OR "fatty acid*" OR *butanol OR succin* OR citric OR chitin OR furfural OR "solid substrate" OR lactic OR *hythane OR faecal OR tannase OR xylitol OR wastewater OR nanowh* OR "waste water" OR biosorption OR pyrolysis OR biogas OR *butyrate OR "fuel cells" OR *alkanoates OR anti* OR activated OR *glucanase OR fumaric OR syngas OR *lipid OR beverage OR "enzyme* production" OR amino OR gasification OR *cyanidins OR "bio-oil" OR liquefaction OR odor OR *amylase OR lycopene OR protein OR cadmium OR *saccharide OR carboxylic OR collagen OR "agricultural waste" OR *phenol OR *phenolic OR "agro waste" OR carbamate OR h2 OR vfa OR properties OR pruning OR surfactant OR {agricultural banana} OR ester* OR {tree wood} OR microfib* OR decanter OR distil* OR volumetr* OR *digestion OR nanofib* OR "chemical prod*" OR "acid prod*" OR xanthan OR baker* OR leaf OR foul* OR *diesel OR carbonization OR *manure OR phb OR naring* OR surfactin OR demineral* OR sludge OR adsorp* OR tower OR levul* OR pig OR feed* OR mombin OR combust* OR basid* OR biocatal* OR bromelain OR "solid media" OR "ethanol type" OR "*cane bagasse") OR SRCTITLE (nutr* OR food* OR poultry OR dairy OR carbohydrate OR hydrogen OR animal OR lwt OR livestock OR materials OR macromol* OR cereal OR aqua* OR ruminant OR brew* OR plant* OR fluids OR water OR meat OR chromat* OR peptide* OR water OR archaeo* OR oil* OR carcinogen* OR zoo* OR hort* OR agron* OR fish* OR analy* OR phar* OR lebens* OR enol* OR color OR composites OR polymer OR pyrol* OR wood*) OR SUBJAREA (medi OR phar OR vete OR nurs OR neur OR dent)))) AND (LIMIT-TO (SRCTYPE, "j") OR LIMIT-TO (SRCTYPE, "k") OR LIMIT-TO (SRCTYPE, "b")) AND (LIMIT-TO (LANGUAGE, "English")) AND (LIMIT-TO (DOCTYPE, "ar") OR LIMIT-TO (DOCTYPE, "cp") OR LIMIT-TO (DOCTYPE, "re") OR LIMIT-TO (DOCTYPE, "ch") OR LIMIT-TO (DOCTYPE, "le") OR LIMIT-TO (DOCTYPE, "sh") OR LIMIT-TO (DOCTYPE, "no")).

REFERENCES

Aman, V. 2018. Does the Scopus author ID suffice to track scientific international mobility? A case study based on Leibniz laureates. *Scientometrics* 117:705–720.

Angelici, C., B. M. Weckhuysen and P. C. A. Bruijnincx. 2013. Chemocatalytic conversion of ethanol into butadiene and other bulk chemicals. *ChemSusChem* 6:1595–1614.

Antolini, E. 2007. Catalysts for direct ethanol fuel cells. *Journal of Power Sources* 170:1–12.

Antolini, E. 2009. Palladium in fuel cell catalysis. *Energy and Environmental Science* 2:915–931.

Arapoglou, D., T. Varzakas, A. Vlyssides and C. Israilides. 2010. Ethanol production from potato peel waste (PPW). *Waste Management* 30:1898–1902.

Beaudry, C. and V. Lariviere. 2016. Which gender gap? Factors affecting researchers' scientific impact in science and medicine. *Research Policy* 45:1790–1817.

Blankenship, K. M. 1993. Bringing gender and race in: US employment discrimination policy. *Gender & Society* 7:204–226.

Burnham, J. F. 2006. Scopus database: A review. *Biomedical Digital Libraries* 3:1–8.

Carlson, K. M., J. S. Gerber and D. Mueller, et al. 2017. Greenhouse gas emissions intensity of global croplands. *Nature Climate Change* 7:63–68.

Change, C. 2007. Climate change impacts, adaptation and vulnerability. *Science of the Total Environment* 326:95–112.

Choi, I. S., Y. G. Lee, S. K. Khanal, B. J. Park and H. J. Bae. 2015. A low-energy, cost-effective approach to fruit and citrus peel waste processing for bioethanol production. *Applied Energy* 140:65–74.

Dirth, T. P. and N. R. Branscombe. 2017. Disability models affect disability policy support through awareness of structural discrimination. *Journal of Social Issues* 73:413–442.

Ebadi, A. and A. Schiffauerova. 2016. How to boost scientific production? A statistical analysis of research funding and other influencing factors. *Scientometrics* 106:1093–1116.

Ekman, A., O. Wallberg, E. Joelsson and P. Borjesson. 2013. Possibilities for sustainable biorefineries based on agricultural residues: A case study of potential straw-based ethanol production in Sweden. *Applied Energy* 102:299–308.

Fang, X., Y. Shen, J. Zhao, X. Bao and Y. Qu. 2010. Status and prospect of lignocellulosic bioethanol production in China. *Bioresource Technology* 101:4814–4819.

Fauci, A. S., H. C. Lane and R. R. Redfield. 2020. Covid-19-navigating the uncharted. *New England Journal of Medicine* 382:1268–1269.

Fernandez-Bolanos, J., B. Felizon and A. Heredia, et al. 2001. Steam-explosion of olive stones: Hemicellulose solubilization and enhancement of enzymatic hydrolysis of cellulose. *Bioresource Technology* 79:53–61.

Fernando, S., S. Adhikari, C. Chandrapal and M. Murali. 2006. Biorefineries: Current status, challenges, and future direction. *Energy & Fuels* 20:1727–1737.

Formela, K., A. Hejna, L. Piszczyk, M. R. Saeb and X. Colom. 2016. Processing and structure-property relationships of natural rubber/wheat bran biocomposites. *Cellulose* 23:3157–3175.

Garfield, E. 1955. Citation indexes for science. *Science* 122:108–111.

Gillon, S. 2010. Fields of dreams: Negotiating an ethanol agenda in the Midwest United States. *Journal of Peasant Studies* 37:723–748.

Gnansounou, E. 2010. Production and use of lignocellulosic bioethanol in Europe: Current situation and perspectives. *Bioresource Technology* 101:4842–4850.

Goh, C. S., K. T. Tan, K. T. Lee and S. Bhatia. 2010. Bio-ethanol from lignocellulose: Status, perspectives and challenges in Malaysia. *Bioresource Technology* 101:4834–4841.

Grohmann, K. and E. A. Baldwin. 1992. Hydrolysis of orange peel with pectinase and cellulase enzymes. *Biotechnology Letters* 14:1169–1174.

Grohmann, K., R. G. Cameron and B. S. Buslig. 1995. Fractionation and pretreatment of orange peel by dilute acid hydrolysis. *Bioresource Technology* 54:129–141.

Guerard, F., L. Guimas and A. Binet. 2002. Production of tuna waste hydrolysates by a commercial neutral protease preparation. *Journal of Molecular Catalysis B: Enzymatic* 19:489–498.

Guimaraes, P. M. R., J. A. Teixeira and L. Domingues. 2010. Fermentation of lactose to bio-ethanol by yeasts as part of integrated solutions for the valorisation of cheese whey. *Biotechnology Advances* 28:375–384.

Hahn-Hagerdal, B., M. Galbe, M. F. Gorwa-Grauslund, G. Liden and G. Zacchi. 2006. Bio-ethanol: The fuel of tomorrow from the residues of today. *Trends in Biotechnology* 24:549–556.

Hamilton, J. D. 1983. Oil and the macroeconomy since World War II. *Journal of Political Economy* 91:228–248.

Hamilton, J. D. 2003. What is an oil shock? *Journal of Econometrics* 113:363–398.

Hang, Y. D., C. Y. Lee and E. E. Woodams. 1982. A solid state fermentation system for production of ethanol from apple pomace. *Journal of Food Science* 47:1851–1852.

Hill, J., E. Nelson, D. Tilman, S. Polasky and D. Tiffany. 2006. Environmental, economic, and energetic costs and benefits of biodiesel and ethanol biofuels. *Proceedings of the National Academy of Sciences of the United States of America* 103:11206–11210.

Hill, J., S. Polasky and E. Nelson, et al. 2009. Climate change and health costs of air emissions from biofuels and gasoline. *Proceedings of the National Academy of Sciences of the United States of America* 106:2077–2082.

Hsieh, W. D., R. H. Chen, T. L. Wu and T. H. Lin. 2002. Engine performance and pollutant emission of an SI engine using ethanol-gasoline blended fuels. *Atmospheric Environment* 36:403–410.

Huang, H. J., S. Ramaswamy, U. W. Tschirner and B. V. Ramarao. 2008. A review of separation technologies in current and future biorefineries. *Separation and Purification Technology* 62:1–21.

Jones, T. C. 2012. America, oil, and war in the Middle East. *Journal of American History* 99:208–218.

Kerr, R. A. 2007. Global warming is changing the world. *Science* 316:188–190.

Khawla, B. J., M. Sameh and G. Imen, et al. 2014. Potato peel as feedstock for bioethanol production: A comparison of acidic and enzymatic hydrolysis. *Industrial Crops and Products* 52:144–149.

Kilian, L. 2008. Exogenous oil supply shocks: How big are they and how much do they matter for the US economy? *Review of Economics and Statistics* 90:216–240.

Kilian, L. 2009. Not all oil price shocks are alike: Disentangling demand and supply shocks in the crude oil market. *American Economic Review*, 99:1053–1069.

Kim, J. 2018. Evaluating author name disambiguation for digital libraries: A case of DBLP. *Scientometrics* 116:1867–1886.

Kim, J. H., J. C. Lee and D. Pak. 2011. Feasibility of producing ethanol from food waste. *Waste Management* 31:2121–2125.

Kim, J. S., S. C. Park and J. W. Kim, et al. 2010. Production of bioethanol from lignocellulose: Status and perspectives in Korea. *Bioresource Technology* 101:4801–4805.

Kim, S., J. M. Park, J. W. Seo and C. H. Kim. 2012. Sequential acid-/alkali-pretreatment of empty palm fruit bunch fiber. *Bioresource Technology* 109:229–233.

Konur, O. 2000. Creating enforceable civil rights for disabled students in higher education: An institutional theory perspective. *Disability & Society* 15:1041–1063.

Konur, O. 2002a. Access to nursing education by disabled students: Rights and duties of nursing programs. *Nurse Education Today* 22:364–374.

Konur, O. 2002b. Assessment of disabled students in higher education: Current public policy issues. *Assessment and Evaluation in Higher Education* 27:131–152.

Konur, O. 2002c. Access to employment by disabled people in the UK: Is the Disability Discrimination Act working? *International Journal of Discrimination and the Law* 5:247–279.

Konur, O. 2006a. Participation of children with dyslexia in compulsory education: Current public policy issues. *Dyslexia* 12:51–67.

Konur, O. 2006b. Teaching disabled students in higher education. *Teaching in Higher Education* 11:351–363.

Konur, O. 2007a. A judicial outcome analysis of the Disability Discrimination Act: A windfall for the employers? *Disability & Society* 22:187–204.

Konur, O. 2007b. Computer-assisted teaching and assessment of disabled students in higher education: The interface between academic standards and disability rights. *Journal of Computer Assisted Learning* 23:207–219.

Konur, O. 2011. The scientometric evaluation of the research on the algae and bio-energy. *Applied Energy* 88:3532–3540.

Konur, O. 2012a. The evaluation of the biogas research: A scientometric approach. *Energy Education Science and Technology Part A: Energy Science and Research* 29:1277–1292.

Konur, O. 2012b. The evaluation of the educational research: A scientometric approach. *Energy Education Science and Technology Part B: Social and Educational Studies* 4:1935–1948.

Konur, O. 2012c. The evaluation of the global energy and fuels research: A scientometric approach. *Energy Education Science and Technology Part A: Energy Science and Research* 30:613–628.

Konur, O. 2012d. The evaluation of the research on the biodiesel: A scientometric approach. *Energy Education Science and Technology Part A: Energy Science and Research* 28:1003–1014.

Konur, O. 2012e. The evaluation of the research on the bioethanol: A scientometric approach. *Energy Education Science and Technology Part A: Energy Science and Research* 28:1051–1064.

Konur, O. 2012f. The evaluation of the research on the biofuels: A scientometric approach. *Energy Education Science and Technology Part A: Energy Science and Research* 28:903–916.

Konur, O. 2012g. The evaluation of the research on the biohydrogen: A scientometric approach. *Energy Education Science and Technology Part A: Energy Science and Research* 29:323–338.

Konur, O. 2012h. The evaluation of the research on the microbial fuel cells: A scientometric approach. *Energy Education Science and Technology Part A: Energy Science and Research* 29:309–322.

Konur, O. 2012i. The scientometric evaluation of the research on the production of bioenergy from biomass. *Biomass and Bioenergy* 47:504–515.

Konur, O. 2015. Current state of research on algal bioethanol. In *Marine Bioenergy: Trends and Developments*, Eds. S. K. Kim and C. G. Lee, pp. 217–244. Boca Raton, FL: CRC Press.

Konur, O., Ed. 2018a. *Bioenergy and Biofuels*. Boca Raton, FL: CRC Press.

Konur, O. 2018b. Bioenergy and biofuels science and technology: Scientometric overview and citation classics. In *Bioenergy and Biofuels*, Ed. O. Konur, pp. 3–63. Boca Raton, FL: CRC Press.

Konur, O. 2019. Cyanobacterial bioenergy and biofuels science and technology: A scientometric overview. In *Cyanobacteria: From Basic Science to Applications*, Eds. A. K. Mishra, D. N. Tiwari and A. N. Rai, pp. 419–442. Amsterdam: Elsevier.

Konur, O. 2020a. The scientometric analysis of the research on the bioethanol production from green macroalgae. In *Handbook of Algal Science, Technology and Medicine,* Ed. O. Konur, pp. 385–401. London: Academic Press.

Konur, O., Ed. 2020b. *Handbook of Algal Science, Technology and Medicine.* London: Academic Press.

Konur, O., Ed. 2021a. *Handbook of Biodiesel and Petrodiesel Fuels: Science, Technology, Health, and Environment.* Boca Raton, FL: CRC Press.

Konur, O., Ed. 2021b. *Handbook of Biodiesel and Petrodiesel Fuels: Science, Technology, Health, and Environment. Volume 1. Biodiesel Fuels: Science, Technology, Health, and Environment.* Boca Raton, FL: CRC Press.

Konur, O., Ed. 2021c. *Handbook of Biodiesel and Petrodiesel Fuels: Science, Technology, Health, and Environment. Volume 2. Biodiesel Fuels Based on the Edible and Nonedible Feedstocks, Wastes, and Algae: Science, Technology, Health, and Environment.* Boca Raton, FL: CRC Press.

Konur, O., Ed. 2021d. *Handbook of Biodiesel and Petrodiesel Fuels: Science, Technology, Health, and Environment. Volume 3. Petrodiesel Fuels: Science, Technology, Health, and Environment.* Boca Raton, FL: CRC Press.

Konur, O. and F. L. Matthews. 1989. Effect of the properties of the constituents on the fatigue performance of composites: A review. *Composites* 20:317–328.

Kruyt, B., D. P. van Vuuren, H. J. de Vries and H. Groenenberg. 2009. Indicators for energy security. *Energy Policy* 37:2166–2181.

Kwon, E. E., H. Yi and Y. J. Jeon. 2013. Sequential co-production of biodiesel and bioethanol with spent coffee grounds. *Bioresource Technology* 136:475–480.

Leydesdorff, L. 2000. Is the European Union becoming a single publication system? *Scientometrics* 47:265–280.

Leydesdorff, L. and C. Wagner. 2009. Is the United States losing ground in science? A global perspective on the world science system. *Scientometrics* 78:23–36.

Leydesdorff, L. and P. Zhou. 2005. Are the contributions of China and Korea upsetting the world system of science? *Scientometrics* 63:617–630.

Li, H., S. M. Liu, X. H. Yu, S. L. Tang and C. K. Tang. 2020. Coronavirus disease 2019 (COVID-19): Current status and future perspectives. *International Journal of Antimicrobial Agents* 55:105951.

Lin, Y. and S. Tanaka. 2006. Ethanol fermentation from biomass resources: Current state and prospects. *Applied Microbiology and Biotechnology* 69:627–642.

Ma, X., L. Sun and C. Song. 2002. A new approach to deep desulfurization of gasoline, diesel fuel and jet fuel by selective adsorption for ultra-clean fuels and for fuel cell applications. *Catalysis Today* 77:107–116.

Morschbacker, A. 2009. Bio-ethanol based ethylene. *Polymer Reviews* 49:79–84.

Murphy, J. D. and K. McCarthy. 2005. Ethanol production from energy crops and wastes for use as a transport fuel in Ireland. *Applied Energy* 82:148–166.

Mussatto, S. I., E. M. S. Machado, L. M. Carneiro and J. A. Teixeira. 2012. Sugars metabolism and ethanol production by different yeast strains from coffee industry wastes hydrolysates. *Applied Energy* 92:763–768.

Najafi, G., B. Ghobadian and T. Tavakoli, et al. 2009. Performance and exhaust emissions of a gasoline engine with ethanol blended gasoline fuels using artificial neural network. *Applied Energy* 86:630–639.

Newman, P. W. G. and J. R. Kenworthy. 1989. Gasoline consumption and cities: A comparison of U.S. cities with a global survey. *Journal of the American Planning Association* 55:24–37.

North, D. C. 1991. Institutions. *Journal of Economic Perspectives* 5:97–112.

Oberoi, H. S., P. V. Vadlani, L. Saida, S. Bansal and J. D. Hughes. 2011. Ethanol production from banana peels using statistically optimized simultaneous saccharification and fermentation process. *Waste Management* 31:1576–1584.

Olsson, L. and B. Hahn-Hagerdal. 1996. Fermentation of lignocellulosic hydrolysates for ethanol production. *Enzyme and Microbial Technology* 18:312–331.

Ozdingis, A. G. B. and G. Kocar, G. 2018. Current and future aspects of bioethanol production and utilization in Turkey. *Renewable and Sustainable Energy Reviews* 81:2196–2203.

Piarpuzan, D., J. A. Quintero and C. A. Cardona. 2011. Empty fruit bunches from oil palm as a potential raw material for fuel ethanol production. *Biomass and Bioenergy* 35:1130–1137.

Rahman, S. H. A., J. P. Choudhury, A. L. Ahmad and A. H. Kamaruddin. 2007. Optimization studies on acid hydrolysis of oil palm empty fruit bunch fiber for production of xylose. *Bioresource Technology* 98:554–559.

Reeves, S. 2014. To Russia with love: How moral arguments for a humanitarian intervention in Syria opened the door for an invasion of the Ukraine. *Michigan State University International Law Review* 23:199.

Sanchez, O. J. and C. A. Cardona. 2008. Trends in biotechnological production of fuel ethanol from different feedstocks. *Bioresource Technology* 99:5270–5295.

Silveira, W. B., F. J. V. Passos, H. C. Mantovani and F. M. L. Passos. 2005. Ethanol production from cheese whey permeate by *Kluyveromyces marxianus* UFV-3: A flux analysis of oxido-reductive metabolism as a function of lactose concentration and oxygen levels. *Enzyme and Microbial Technology* 36:930–936.

Soccol, C. R., L. P. de Souza Vandenberghe and A. B. P. Medeiros, et al. 2010. Bioethanol from lignocelluloses: Status and perspectives in Brazil. *Bioresource Technology* 101:4820–4825.

Sukumaran, R. K., V. J. Surender and R. Sindhu, et al. 2010. Lignocellulosic ethanol in India: Prospects, challenges and feedstock availability. *Bioresource Technology* 101:4826–4833.

Sun, Y. and J. Cheng. 2002. Hydrolysis of lignocellulosic materials for ethanol production: A review. *Bioresource Technology* 83:1–11.

Taherzadeh, M. J. and K. Karimi. 2007. Enzyme-based hydrolysis processes for ethanol from lignocellulosic materials: A review. *Bioresources* 2:707–738.

Taherzadeh, M. J. and K. Karimi. 2008. Pretreatment of lignocellulosic wastes to improve ethanol and biogas production: A review. *International Journal of Molecular Sciences* 9:1621–1651.

Velasquez-Arredondo, H. I., A. A. Ruiz-Colorado and S. de Oliveira. 2010. Ethanol production process from banana fruit and its lignocellulosic residues: Energy analysis. *Energy* 35:3081–3087.

Wilkins, M. R., W. W. Widmer and K. Grohmann. 2007. Simultaneous saccharification and fermentation of citrus peel waste by *Saccharomyces cerevisiae* to produce ethanol. *Process Biochemistry* 42:1614–1619.

Winzer, C. 2012. Conceptualizing energy security. *Energy Policy* 46:36–48.

Yang, B. and C. E. Wyman. 2008. Pretreatment: The key to unlocking low-cost cellulosic ethanol. *Biofuels, Bioproducts and Biorefining* 2:26–40.

66 Second Generation Food Waste-based Bioethanol Fuels

Review

Ozcan Konur
(Formerly) Ankara Yildirim Beyazit University

66.1 INTRODUCTION

The crude oil-based gasoline fuels (Ma et al., 2002; Newman and Kenworthy, 1989) have been widely used in the transportation sector since the 1920s. However, there have been great public concerns over the adverse environmental and human impact of these fuels (Hill et al., 2006, 2009). Hence, biomass-based bioethanol fuels (Hill et al., 2006; Konur, 2012, 2015, 2019, 2020) have increasingly been used in blending gasoline fuels (Hsieh et al., 2002; Najafi et al., 2009), in the fuel cells (Antolini, 2007, 2009), and in the biochemical production (Angelici et al., 2013; Morschbacker, 2009) in a biorefinery context (Fernando et al., 2006; Huang et al., 2008).

However, it is necessary to pretreat the biomass (Alvira et al., 2010; Taherzadeh and Karimi, 2008) to enhance the yield of the bioethanol (Hahn-Hagerdal et al., 2006; Sanchez and Cardona, 2008) prior to the bioethanol fuel production from the feedstocks through the hydrolysis (Sun and Cheng, 2002; Taherzadeh and Karimi, 2007) and fermentation (Lin and Tanaka, 2006; Olsson and Hahn-Hagerdal, 1996) of the biomass and hydrolysates, respectively.

One of the most-studied feedstocks for the bioethanol fuels has been the food wastes such as fruit, vegetable, beverage, and dairy wastes. The research in the field of the second generation food waste-based bioethanol fuels has intensified in this context in the key research fronts of the pretreatment (Fernandez-Bolanos et al., 2001; Grohmann et al., 1995; Rahman et al., 2007) and hydrolysis (Grohmann et al., 1995; Guerard et al., 2002; Rahman et al., 2007) of the food wastes, fermentation (Kim et al., 2011; Mussatto et al., 2012; Piarpuzan et al., 2011) of the food waste-based hydrolysates, and production (Arapoglou et al., 2010; Mussatto, et al., 2012; Piarpuzan et al., 2011) and evaluation (Velasquez-Arredondo et al., 2010) of the second generation food waste-based bioethanol fuels.

Further, the empty fruit bunches (Kim et al., 2012; Rahman et al., 2007), orange peels (Grohmann et al., 1995; Grohmann and Baldwin, 1992), citrus peels (Choi et al., 2015; Wilkins et al., 2007b), potato peels (Arapoglou et al., 2010; Khawla et al., 2014), spent coffee grounds (Kwon et al., 2013; Mussatto et al., 2012), cheese whey (Guimaraes et al., 2010; Silveira et al., 2005), banana peels (Oberoi et al., 2011; Velasquez-Arredondo et al., 2010), apple pomaces (Hang et al., 1982), and olive stones (Fernandez-Bolanos et al., 2001) have been studied intensively in this context.

However, it is essential to develop efficient incentive structures (North, 1991) for the primary stakeholders to enhance the research in this field (Konur, 2000, 2002a–c, 2006a,b, 2007a,b). Although there have been a number of review papers on the food waste-based bioethanol fuels (Guimaraes et al., 2010; Hafid et al., 2017; Ravindran and Jaiswal, 2016), there has been no review of the most-cited 25 papers in this field.

Thus, this book chapter presents a review of the most-cited 25 articles in the field of the food waste-based bioethanol fuels. Then, it discusses the key findings of these highly influential papers and comments on the future research priorities in this field.

DOI: 10.1201/9781003226550-89

66.2 MATERIALS AND METHODS

The search for this study was carried out using Scopus database (Burnham, 2006) in September 2022.

As a first step for the search of the relevant literature, the keywords were selected using the most-cited first 200 population papers. The selected keyword list was then optimized to obtain a representative sample of papers for the searched research field. This final keyword set was provided in the appendix of Konur (2023) for future replication studies.

As a second step, a sample dataset was used for this study. The first 25 articles with at least 97 citations each were selected for the review study. Key findings from each paper were taken from the abstracts of these papers and were discussed. Additionally, a number of brief conclusions were drawn, and a number of relevant recommendations were made to enhance the future research landscape.

66.3 RESULTS

The brief information about 25 most-cited papers with at least 97 citations each on the second generation food waste-based bioethanol fuels is given below. The primary research fronts are the hydrolysis of the food wastes and production of the food waste-based bioethanol fuels with 11 and 14 highly cited papers (HCPs), respectively.

66.3.1 Food Waste Hydrolysis

The brief information about 11 most-cited papers on the hydrolysis of food wastes with at least 97 citations each is given below (Table 66.1).

Guerard et al. (2002) hydrolyzed tuna waste by umamizyme in a paper with 221 citations. They performed the hydrolysis in a 1-l batch reactor at pH 7 and 45°C with enzyme/protein substrate ratio ranging from 0.1% to 1.5% (w/w) protein. They obtained a degree of hydrolysis up to 22.5% with an enzyme/substrate ratio of 1.5%, after 4h of hydrolysis. There was a linear correlation between the hydrolysis degree and the N recovery. This enzyme performed as effectively as Alcalase® 2, 4L for the tuna waste solubilization. However, the umamizyme stability was lower than this commercial enzyme.

Wilkins et al. (2007a) hydrolyzed grapefruit peel waste with cellulase and pectinase enzymes in a paper with 174 citations. They tested enzyme loadings of 0, 1, 2, 5, and 10mg protein/g peel dry matter at 45°C. They supplemented them with 2.1 mg β-glucosidase protein/g peel dry matter. They observed that five mg pectinase/g peel dry matter and 2mg cellulase/g peel dry matter were the lowest loadings to yield the most glucose. Further, optimum pH was 4.8, while cellulose, pectin, and hemicellulose in this waste could be hydrolyzed by pectinase and cellulase enzymes to monomer sugars.

Rahman et al. (2007) optimized acid hydrolysis of oil palm empty fruit bunch fiber for the production of xylose in a paper with 157 citations. They determined the effect of H_2SO_4 concentration, reaction temperature, and reaction time for the production of xylose by carrying out batch reactions under various reaction temperature, reaction time, and acid concentrations. They found that the optimum reaction temperature, reaction time, and acid concentration were 119°C, 60min, and 2%, respectively, resulting in xylose yield and selectivity of 91.27% and 17.97 g/g, respectively.

Fernandez-Bolanos et al. (2001) hydrolyzed olive stones pretreated with steam explosion and acids in a paper with 129 citations. They processed olive stones by steam explosion under different experimental conditions of temperature and time, 200°C–236°C for 2–4min, with or without previous acid impregnation with 0.1% H_2SO_4 (w/w). They obtained the maximum yield of the pentosan in the water solution as 63% pentose in the starting material for seed husk treated at 200°C for 2min prior to acid-impregnation, or at 215°C for 2min without acid, compared to 39% of the potential yield for whole stones preimpregnated with acid under more severe conditions. This showed that the autohydrolysis of hemicellulose in seed husks when compared to whole stones was enhanced, and the depolymerization of hemicelluloses was a function of the severity of the pretreatment. In conclusion, steam explosion improved the accessibility of the cellulose and increased the enzymatic hydrolysis yield after steam explosion with respect to material without steam explosion.

TABLE 66.1
The Hydrolysis of Food Wastes

No.	Papers	Biomass/Hydrolysate	Prts.	Parameters	Keywords	Lead Author	Affil.	Cits.
1	Guerard et al. (2002)	Fish waste Tuna waste	Umamizyme	Enzymatic hydrolysis, enzyme/ substrate ratio, hydrolysis rate	Waste, hydrolysates, tuna	Guerard, Fabienne* 6603404306	Univ. Brest France	221
2	Wilkins et al. (2007a)	Fruit waste Grapefruit peels	Cellulase, pectinase, glucosidase	Enzymatic hydrolysis, enzyme/ substrate ratio, hydrolysis rate, enzyme type	Hydrolysis, grapefruit, peel, waste	Wilkins, Mark R. 56492323200	Univ. Nebraska-Lincoln USA	174
3	Rahman et al. (2007)	Fruit waste Empty palm fruit bunch	H_2SO_4	Acid hydrolysis, acid concentration, reaction temperature and time, optimization	Hydrolysis, fruit bunch	Choudhury, Jyoti P. 57198146735	Univ. Sains Malaysia Malaysia	157
4	Fernandez-Bolanos et al. (2001)	Olive waste Olive stones	Steam explosion, H_2SO_4, cellulase	Enzymatic hydrolysis, hydrolysis rate, pretreatment	Olive stones, hydrolysis, cellulose	Heredia, Antonia* 35560421500	CSIC Spain	129
5	Grohmann et al. (1995)	Fruit waste Orange peels	H_2SO_4, cellulolytic and pectinolytic enzymes.	Enzymatic hydrolysis, pretreatment, hydrolysis rate	Fractionation, pretreatment, orange peel, hydrolysis	Grohmann, Karel 7004503589	Renewable Spirits LLC USA	126
6	Correia et al. (2013)	Fruit waste Cashew apple bagasse	Alkaline H_2O_2, cellulase	Enzymatic hydrolysis, pretreatment, concentration of H_2O_2, biomass loading, pretreatment duration	Pretreatment, cashew apple bagasse, ethanol	Rocha, Maria V. P.* 15080896200	Fed. Univ. Ceara Brazil	109
7	Kim et al. (2012)	Fruit waste Empty palm fruit bunch	H_2SO_4, NaOH	Enzymatic hydrolysis, pretreatments, delignification, enzymatic digestibility	Pretreatment, empty palm fruit bunch	Kim, Seonghun 8505178100	Korea Res. Inst. Biosci. Biotechnol. S. Korea	107
8	Yunus et al. (2010)	Fruit waste Empty palm fruit bunch	Ultrasonics, H_2SO_4	Acid hydrolysis, pretreatment, xylose yield	Pre-treatment, hydrolysis, palm empty fruit bunch	Yunus, Robiah* 6603243672	Univ. Putra Malaysia Malaysia	104
9	Hamzah et al. (2011)	Fruit waste Empty palm fruit bunch	Cellulase, glucosidase	Enzymatic hydrolysis, pH, temperature and substrate loading	Hydrolysis empty palm fruit bunches	Idris, Ani* 23090578300	Univ. Technol. Malaysia Malaysia	100
10	Del Campo et al. (2006)	Vegetable waste	H_2SO_4	Acid hydrolysis, pretreatment, sugar recovery, waste type	Hydrolysis, pretreatment, food wastes, bioethanol	Del Campo, Ines* 15131397400	Natl. Renew. Ener. Ctr. Spain	99
11	Grohmann and Baldwin (1992)	Fruit waste Orange peels	Pectinase, cellulase	Enzymatic hydrolysis, enzyme type	Hydrolysis, orange peels	Grohmann, Karel 7004503589	Renewable Spirits LLC USA	97

Prt., biomass pretreatments; Cits., number of citations received for each paper; *, female.

Grohmann et al. (1995) solubilized and depolymerized carbohydrates by the pretreatment of orange peels with dilute (0.06% and 0.5%) sulfuric acid at 100°C, 120°C, and 140°C in a paper with 126 citations. They observed that the acid treatments solubilized a large portion of total carbohydrates in orange peel. However, only soluble sugars and sugars derived from hydrolysis of hemicelluloses were efficiently released by the pretreatment with hot dilute sulfuric acid. Further, cellulose and segments of pectin containing galacturonic acid units were very resistant to acid-catalyzed hydrolysis. Finally, the treatment with dilute sulfuric acid had a positive effect on the rate of subsequent enzymatic hydrolysis of orange peel by a mixture of cellulolytic and pectinolytic enzymes.

Correia et al. (2013) performed alkaline hydrogen peroxide (H_2O_2) pretreatment of cashew apple bagasse in a paper with 109 citations. They evaluated the effects of the concentration of H_2O_2 at pH 11.5, the biomass loading, and the pretreatment duration performed at 35°C and 250 rpm after the subsequent enzymatic hydrolysis of the pretreated biomass using a commercial cellulase enzyme. The waste contained 20.56% cellulose, 10.17% hemicellulose, and 35.26% lignin. They observed that the pretreatment resulted in a reduced lignin content in the residual solids. Increasing the H_2O_2 concentration (0%–4.3% v/v) resulted in a higher rate of enzymatic hydrolysis, while lower biomass loadings gave higher glucose yields. In addition, no measurable furfural and hydroxymethylfurfural (HMF) were produced in the liquid fraction during the pretreatment. In conclusion, this pretreatment was effective for the pretreatment of this waste.

Kim et al. (2012) performed the sequential acid/alkali pretreatment of empty palm fruit bunch fiber in a paper with 107 citations. They used dilute H_2SO_4 in the first step, which removed 90% of the hemicellulose and 32% of the lignin, but left most of the cellulose under the optimum pretreatment condition. They then applied sodium hydroxide (NaOH) in the second step, which extracted lignin effectively with a 70% delignification yield, partially disrupting the ordered fibrils of the waste and thus enhancing the enzymatic digestibility of the cellulose. They found that the sequentially pretreated biomass consisted of 82% cellulose, <1% hemicellulose, and 30% lignin content afterward. Further, the pretreated biomasses morphologically had rough, porous, and irregularly ordered surfaces for enhancing enzymatic digestibility. In conclusion, the sequentially acid/alkali pretreatments were effective enhancing the enzymatic digestibility of the waste.

Yunus et al. (2010) studied the effect of ultrasonic pretreatment on the acid hydrolysis of empty palm fruit bunch at low temperature and pressure in a paper with 104 citations. They employed 2% H_2SO_4, 1:25 solid liquid ratio, and 100°C operating temperature. They obtained a maximum xylose yield of 58% when the waste was ultrasonicated at 90% amplitude for 45 min. In the absence of ultrasonic pretreatment, only 22% of xylose was obtained. However, they observed no substantial increase of xylose formation for acid hydrolysis at higher temperatures of 120°C and 140°C on the ultrasonicated waste. Further, there were interesting morphological changes within the waste for different acid hydrolysis conditions.

Hamzah et al. (2011) studied the enzymatic hydrolysis of treated empty palm fruit bunches in a paper with 100 citations. They used a combination of cellulase and β 1–4 glucosidase. They found that a combination of both enzymes with the ratio of 5:1 hydrolyzed more cellulose from treated waste and gave highest soluble glucose concentration up to 4 g/L. Further, as pH and temperature were increased, the glucose produced also increased until pH 4.8 and 50°C, while beyond these values, the reverse occurred. They observed that glucose produced in the reaction increased with the increment in the substrate loading, and they obtained maximum glucose concentration (2.7 g/L) when 8% (w/v) treated waste was used as a substrate.

Del Campo et al. (2006) performed the diluted acid hydrolysis pretreatment of fresh and processed vegetable wastes in a paper with 99 citations. They found that the maximum ratios of single sugar recovery in the liquid fraction from dilute acid hydrolysis assays were 40.29 and 50.20% (w/w) for tomato and red pepper residues, respectively. However, more intensive pretreatments were necessary to maximize sugar recovery in the case of pulse food (legumes such as beans, lentils, and chickpeas) and artichoke residues because of their high starch and inulin content. Finally, for cardoon residues, they obtained a maximum single sugar recovery of 78.18% (w/w) in the liquid fraction. In conclusion, because the sugars in these wastes were widely available and easily obtainable, they could be considered as potential feedstocks for bioethanol production.

Grohmann and Baldwin (1992) performed the enzymatic hydrolysis of orange peels with pectinase and cellulase enzymes in a paper with 97 citations. They observed high levels of conversion to monomeric sugars after treatment with pectinase enzyme, but cellulase enzyme achieved only limited solubilization. In conclusion, the combination of cellulase and pectinase enzymes was a most efficient system for enzymatic hydrolysis of polysaccharides in orange peel.

66.3.2 Food Waste-based Bioethanol Production

There are 14 HCPs for the production the food waste-based bioethanol fuels with at least 105 citations each (Table 66.2).

Arapoglou et al. (2010) produced ethanol from potato peel waste in a paper with 199 citations. They hydrolyzed this waste with various enzymes and/or acid, and fermented by *Saccharomyces cerevisiae* var. bayanus to determine fermentability and ethanol production. They found that the enzymatic hydrolysis with a combination of three enzymes released 18.5 g/L reducing sugar and produced 7.6 g/L of ethanol after fermentation. In conclusion, this waste had a high potential for ethanol production.

Mussatto et al. (2012) produced ethanol from spent coffee grounds (SCG) in a paper with 148 citations. They used *Saccharomyces cerevisiae, Pichia stipites,* and *Kluyveromyces fragilis* to ferment hydrolysates of coffee silverskin (CS) and spent coffee grounds (SCG) produced by acid hydrolysis of CS and SCG. They observed that *S. cerevisiae* provided the best ethanol production from SCG hydrolysate (11.7 g/L, 50.2% efficiency). On the other hand, they obtained insignificant (≤1.0 g/L) ethanol production from CS hydrolysate, for all the evaluated yeast strains, probably due to the low sugars concentration present in this medium (~22 g/L). In conclusion, SCG was the best feedstock for ethanol production using *S. cerevisiae.*

Wilkins et al. (2007b) produced ethanol from citrus peel waste in a paper with 142 citations. They performed the simultaneous saccharification and fermentation (SSF) of this waste using *S. cerevisiae* at 37°C. They first performed steam explosion pretreatment to remove more than 90% of the initial d-limonene present in this waste. They observed that ethanol concentrations after 24 h were reduced in fermentations with initial d-limonene concentrations greater than or equal to 0.33% (v/v) and final (24 h) d-limonene concentrations greater than or equal to 0.14% (v/v). Further, ethanol production was reduced when enzyme loadings were (IU or FPU/g peel dry solids) less than 25, pectinase; 0.02, cellulase; and 13, β-glucosidase. Finally, ethanol production was greatest when the initial pH of the peel waste was adjusted to 6.0.

Kwon et al. (2013) coproduced biodiesel and bioethanol from spent coffee grounds in a paper with 139 citations. They found that the direct conversion of bioethanol from this waste was not a desirable option because of the relatively slow enzymatic saccharification behavior in the presence of triglycerides and the free fatty acids (FFAs) present in the waste. Similarly, the direct transformation of this waste into ethanol without first extracting lipids was not a feasible alternative. They converted the crude lipids extracted from this waste into fatty acid methyl ester (FAME) and fatty acid ethyl ester (FAEE) via the non-catalytic biodiesel transesterification reaction. The yields of bioethanol and biodiesel were 0.46 g/g and 97.5%, based on consumed sugar and lipids extracted from this waste, respectively. In conclusion, this waste was a strong candidate for the coproduction of bioethanol and biodiesel.

Kim et al. (2011) produced ethanol from food waste in a paper with 128 citations. They hydrolyzed the waste with carbohydrase with glucose yield of 0.63 g glucose/g total solid and fermented the hydrolysate with *S. cerevisiae* in the batch mode. For separate hydrolysis and fermentation (SHF), they observed that ethanol concentration reached at the level corresponding to an ethanol yield of 0.43 g ethanol/g total solids. For SSF, the ethanol yield was 0.31 g ethanol/g total solids. During the continuous operation of SHF, the volumetric ethanol production rate was 1.18 g/L h with

TABLE 66.2
The Production of Food Waste-based Bioethanol Fuels

No.	Papers	Biomass/Hydrolysate	Prts.	Yeasts	Parameters	Keywords	Lead Author	Affil.	Cits.
1	Arapoglou et al. (2010)	Vegetable waste Potato peels	Enzymes	*S. cerevisiae*	Enzymatic hydrolysis, fermentation, sugar and ethanol yield	Ethanol, potato, peel, waste	Varzakas, Theodoros 6603098855	Univ. Peloponnese Greece	199
2	Mussatto et al. (2012)	Beverage waste Spent coffee grounds	Acid	*S. cerevisiae, P. stipitis, K. fragilis*	Fermentation, yeast type, waste type, ethanol productivity and yield	Ethanol, coffee, wastes, hydrolysates	Mussatto, Solange I* 6602643634	Tech. Univ. Denmark Denmark	148
3	Wilkins et al. (2007b)	Fruit waste Citrus peels	Cellulase, pectinase, glucosidase, steam explosion	*S. cerevisiae*	SSF, d-limonene concentration, enzyme loading, and pH, ethanol productivity	Saccharification, fermentation, citrus, peel, ethanol	Wilkins, Mark R. 56492323200	Univ. Nebraska-Lincoln USA	142
4	Kwon et al. (2013)	Beverage waste Spent coffee grounds	Enzymes	Yeasts	Biodiesel and bioethanol co-production, bioethanol and biodiesel yields	Bioethanol, spent, coffee, grounds	Yeon, Young Jae 7201888480	Pukyong Natl. Univ. S. Korea	139
5	Kim et al. (2011)	Food waste	Carbohydrase	*S. cerevisiae*	Enzymatic hydrolysis, fermentation, sugar and ethanol yield, SSF, SHF	Ethanol, food, waste	Pak, Daewon 7005142765	Seoul Natl. Univ. Sci. Technol. S. Korea	128
6	Piarpuzan et al. (2011)	Fruit waste Empty palm fruit bunch	Alkali, autoclaving, enzymes	*S. cerevisiae*	Enzymatic hydrolysis, fermentation, ethanol yield, energy balance	Empty fruit bunches, ethanol	Cardona, Carlos A. 57214443163	Natl. Univ. Colombia Colombia	125
7	Choi et al. (2015)	Fruit waste Citrus peels, banana peel, apple pomace, and pear waste	Enzymes	Yeasts	Enzymatic hydrolysis, fermentation, sugar and ethanol yield, inhibitor removal	Fruit, citrus peel, waste, bioethanol	Bae, Hyeun-Jong 24280549300	Chonnam Natl. Univ. S. Korea	123

(Continued)

TABLE 66.2 (*Continued*)
The Production of Food Waste-based Bioethanol Fuels

No.	Papers	Biomass/Hydrolysate	Prts.	Yeasts	Parameters	Keywords	Lead Author	Affil.	Cits.
8	Le Man et al. (2010)	Food waste	Enzymes	Yeasts	Ethanol production optimization, temperature, pH and reducing sugar concentration, ethanol yield	Ethanol, food waste	Park, Hung-Suck 7601570293	Univ. Ulsan S. Korea	123
9	Oberoi et al. (2011)	Fruit waste Orange peels	Cellulase, pectinase, hydrothermal	Yeasts	Enzymatic hydrolysis, fermentation, SSF, ethanol productivity, yield, optimization, temperature, time, enzyme concentration	Ethanol, banana peels, saccharification, fermentation	Oberoi, Harinder S. 6603479987	Indian Inst. Hort. Sci. India	114
10	Silveira et al. (2005)	Dairy waste Cheese whey	Na	*K. marxianus*	Fermentation, ethanol productivity and yield, lactose concentration and oxygen level	Ethanol, cheese whey	Passos, Flavia M. L.* 7004639509	Fed. Univ. Vicosa Brazil	113
11	Wilkins et al. (2007c)	Fruit waste Orange peel	Na	*K. marxianus, S. cerevisiae*	Fermentation, ethanol productivity and yield, inhibitory peel oil concentration, yeast type	Ethanol, orange peel	Wilkins, Mark R. 56492323200	Univ. Nebraska-Lincoln USA	109
12	Nigam (2000)	Fruit waste Pineapple cannery waste	Na	*S. cerevisiae*	Fermentation, ethanol productivity, yeast immobilization	Ethanol, pineapple cannery waste	Nigam, J.N. 7006307015	Harcourt Butler Res. Inst. India	107
13	Widmer et al. (2010)	Fruit waste Orange processing waste	Steam explosion, acid, alkali, enzymes	*S. cerevisiae*	Enzymatic hydrolysis, fermentation, inhibitory compound, times, pH, and temperatures, ethanol yield	Pretreatment, orange, waste, ethanol, saccharification, fermentation	Widmer, Wilbur 7006097359	USDA Agr. Res. Serv. USA	106
14	Kim et al. (2008)	Food waste	Enzymes	Yeasts	Enzymatic hydrolysis, fermentation, optimization, reducing sugar and ethanol yield	Saccharification, ethanol, fermentation, food waste	Kim, Si Wouk 56689100900	Chosun Univ. S. Korea	105

Prt., biomass pretreatments. Na, nonavailable; Cits., number of citations received for each paper; *, female.

an ethanol yield of 0.3 g ethanol/g total solids. For SSF process, the volumetric ethanol production rate was 0.8 g/L h with an ethanol yield of 0.2 g ethanol/g total solids.

Piarpuzan et al. (2011) produced ethanol from the empty palm fruit bunches using alkaline pretreatment, enzymatic hydrolysis, and fermentation with *S. cerevisiae* in a paper with 125 citations. They found that coupling alkaline pretreatment with a later autoclaving improved the sugars yield in enzymatic hydrolysis. They obtained better results for enzymatic hydrolysis when sodium acetate buffer was used. Ethanol yield obtained from both experiments and simulation were very similar: 66.50 and 65.84 dm^3 of ethanol per each t of empty fruit bunches, respectively. They obtained these low ethanol yields since the native *S. cerevisiae* strain did not assimilate all reducing sugars, suggesting that those sugars were pentoses. Simulated alkaline and autoclaving pretreatment contributed only with 2% of the total energy consumption (198.4 GJ/m^3 ethanol), while product recovery represented 57% of the total energy.

Choi et al. (2015) produced ethanol from fruit waste in a paper with 123 citations. They developed a novel approach for converting citrus peel waste (i.e., orange, mandarin, grapefruit, lemon, or lime) or waste in combination with other fruit waste (i.e., banana peel, apple pomace, and pear waste) to produce bioethanol. They produced two in-house enzymes from Avicel and waste and tested them with fruit waste at 12%–15% (w/v) solid loading. They observed that the rates of enzymatic conversion of fruit waste to fermentable sugars were nearly 90% for all feedstocks after 48 h. They also designed a d-limonene removal column (LRC) that successfully removed this inhibitor from the fruit waste. When the LRC was coupled with an immobilized cell reactor (ICR), yeast fermentation resulted in ethanol concentrations (14.4–29.5 g/L) and yields (90.2%–93.1%) that were 12-fold greater than products from ICR fermentation alone.

Le Man et al. (2010) produced ethanol from food waste leachate in a paper with 123 citations. The reducing sugar concentration of this leachate was 75 g/L. They obtained the maximum ethanol concentration of 24.2 g/L at the optimum condition of temperature (38°C), pH (5.45), and reducing sugar concentration (75 g/L). This experimental value agreed very well with the predicted one (23.7 g/L), indicating the suitability of the model employed and the success of response surface methodology in optimizing the conditions of ethanol production from this leachate. In conclusion, an ethanol yield of 0.32 g ethanol/g reducing sugar demonstrated the potential of this leachate for the production of ethanol.

Oberoi et al. (2011) produced ethanol from banana peels using statistically optimized SSF process after hydrothermal pretreatment in a paper with 114 citations. They employed cellulase, pectinase, temperature, and time of nine cellulase filter paper unit/gram cellulose (FPU/g-cellulose), 72 international units/gram pectin (IU/g-pectin), 37°C and 15 h, respectively. They observed that the experiment using optimized parameters in batch fermenter not only resulted in higher ethanol concentration than the one predicted by the model equation but also saved fermentation time. Thus, both hydrothermal pretreatment and SSF could be successfully carried out in a single vessel, and the use of optimized process parameters helped achieve significant ethanol productivity: ethanol concentration and productivity of 28.2 g/L and 2.3 g/L/h, respectively.

Silveira et al. (2005) produced ethanol from cheese whey permeate in a paper with 113 citations. They investigated the effect of lactose concentration and oxygen level on the growth and metabolism of *K. marxianus* UFV-3 in this permeate with lactose at initial concentration ranging from 1 to 240 g/L. They observed that the increase in lactose concentration increased ethanol yield and volumetric productivity and reduced cell yield. When lactose concentration was equal or above 50 g/L and the oxygen levels were low, the ethanol yield was close to its theoretical value. They obtained the maximum ethanol concentrations of 76 and 80 g/L in hipoxia and anoxia, respectively. The lactose consumption rate in anoxia was greater than in aerobiosis and hipoxia. However, under anoxia, the lactose consumption rate of *K. marxianus* followed a saturation kinetics, which was not observed in hypoxia and aerobiosis. All oxygen levels investigated showed a tendency for saturation of the ethanol production rate above 65 g/L lactose. Ethanol production rate was also higher on anoxia.

Wilkins et al. (2007c) produced ethanol production from orange peel waste in the presence of inhibitory orange peel oil in a paper with 109 citations. They used *S. cerevisiae* and *K. marxianus* to ferment sugar solutions at 37°C. They added orange peel oil in various amounts and determined the minimum peel oil concentration that inhibited ethanol production after 24, 48, and 72 h. They found that the minimum inhibitory peel oil concentrations for ethanol production were 0.05% at 24 h, 0.10% at 48 h, and 0.15% at 72 h for both yeasts. *S. cerevisiae* produced more ethanol than *K. marxianus* at each time point.

Nigam (2000) produced ethanol from pineapple cannery waste using immobilized yeast cells in a paper with 107 citations. They immobilized the cells of *S. cerevisiae* ATCC 24553, in k-carrageenan, and packed in a tapered glass column reactor at temperature 30°C and pH 4.5. They obtained the maximum productivity of 42.8 g ethanol/L h at a dilution rate of 1.5/h. Further, the volumetric ethanol productivity of the immobilized cells was ca. 11.5 times higher than the free cells. They operated the immobilized cell reactor over a period of 87 days at a dilution rate of 1.0/h, without any loss in the immobilized cell activity. The maximum specific ethanol productivity and specific sugar uptake rate of the immobilized cells were 1.2 g ethanol/g dry wt. cell h and 2.6 g sugar/g dry wt. cell h, respectively, at a dilution rate of 1.5/h.

Widmer et al. (2010) produced ethanol from orange processing waste in a paper with 106 citations. They pretreated this waste under different times, pH, and temperatures. They found that pretreatments at 160°C for longer than 4 min with steam purging were needed to remove limonene to below 0.1%. While hemicelluloses were solubilized well following all pretreatments at 160°C, just 70% of the pectin was solubilized in natural waste compared to over 80% after pretreatments using acid modified waste (pH 2.8). Pretreatments at 160°C on base-modified waste (initial pH 6.8) quickly destroyed pectin, had significantly lower dissolved solids, and were excessively viscous. Total sugars fermentable by *S. cerevisiae* were not changed after pretreatment at 160°C for up to 8 min in the waste between pH 2.2 and 8.2. Ethanol yields based on sugar content after enzymatic hydrolysis after 48 h SSF ranged from 76% to 94%.

Kim et al. (2008) optimized enzymatic saccharification and fermentation of food waste in a paper with 105 citations. Optimum conditions were saccharification pH of 5.20, enzyme reaction temperature of 46.3°C, enzyme concentration of 0.16% (v/v), fermentation pH of 6.85, fermentation temperature of 35.3°C, and fermentation time of 14 h. The model predicted that maximum concentration of reducing sugar and ethanol under these optimum conditions was 117.0 g reducing sugar/L and 57.6 g ethanol/L, respectively. Experimental results were in close agreement with model prediction with 120.1 g reducing sugar/L and 57.5 g ethanol/L, respectively.

66.4 DISCUSSION

66.4.1 INTRODUCTION

The crude oil-based gasoline fuels have been widely used in the transportation sector since the 1920s. However, there have been great public concerns over the adverse environmental and human impact of these fuels. Hence, biomass-based bioethanol fuels have increasingly been used in blending gasoline and petrodiesel fuels, in the fuel cells, and in the biochemical production in a biorefinery context.

However, it is necessary to pretreat the biomass to enhance the yield of the bioethanol prior to the bioethanol fuel production from the feedstocks through the hydrolysis and fermentation of the biomass.

One of the most-studied feedstocks for the bioethanol fuels has been the food wastes such as fruit, vegetable, beverage, and dairy wastes. The research in the field of the second generation food waste-based bioethanol fuels has intensified in this context in the key research fronts of the pretreatment and hydrolysis of the food wastes, fermentation of the food waste-based hydrolysates, and production and evaluation of the second generation food waste-based bioethanol fuels. Further, the empty fruit bunches, orange peels, citrus peels, potato peels, spent coffee grounds, cheese whey, banana peels, apple pomaces, and olive stones have been studied intensively in this context.

However, it is essential to develop efficient incentive structures for the primary stakeholders to enhance the research in this field. Although there have been a number of review papers for this field, there has been no review of the most-cited 25 articles in this field.

Thus, this book chapter presents a review of the most-cited 25 articles on the bioethanol fuel production and evaluation from the food wastes. Then, it discusses the key findings of these highly influential papers and comments on the future research priorities in this field.

As a first step for the search of the relevant literature, the keywords were selected using the most-cited first 200 population papers. The selected keyword list was then optimized to obtain a representative sample of papers for the searched research field. This keyword list was provided in the appendix of Konur (2023) for future replicative studies.

As a second step, a sample dataset was used for this study. The first 25 articles with at least 97 citations each were selected for the review study. Key findings from each paper were taken from the abstracts of these papers and were discussed. Additionally, a number of brief conclusions were drawn, and a number of relevant recommendations were made to enhance the future research landscape.

Information about the research fronts for the sample papers in the food waste-based bioethanol fuels with regard to the feedstocks used in these processes is given in Table 66.3. As this table shows, the most-prolific research front for this field is the fruit wastes with 60% of the HCPs. Further, the other prolific research fronts are food wastes in general, beverage wastes, and vegetable wastes with 12%, 8%, and 8% of the sample papers, respectively. Further, for the first research front, the most-prolific research subfronts are the empty fruit bunches and orange peels with 20% of the HCPs each.

TABLE 66.3
The Most-Prolific Research Fronts for the Second Generation Food Waste-based Bioethanol Fuels

No.	Research Fronts	N Paper % Review	N Paper % Sample	Surplus (%)
1	Fruit wastes	60.0	58.5	1.5
	Empty fruit bunches	20.0	15.5	4.5
	Orange peels	20.0	9.9	10.1
	Citrus peels	8.0	2.8	5.2
	Cashew apple bagasse	4.0	2.8	1.2
	Pineapple waste	4.0	2.8	1.2
	Grapefruit peels	4.0	0.7	3.3
	Other fruit wastes	0.0	15.5	−15.5
	Apple pomaces	0.0	3.5	−3.5
	Banana peels	0.0	3.5	−3.5
2	Food wastes in general	12.0	24.6	−12.6
3	Other food wastes	28.0	35.9	−7.9
	Beverage wastes	8.0	7.0	1.0
	Vegetable wastes	8.0	12.7	−4.7
	Olive wastes	4.0	2.1	1.9
	Fish wastes	4.0	0.7	3.3
	Dairy wastes	4.0	11.3	−7.3
	Other wastes	0.0	2.1	−2.1

N paper (%) review: the number of papers in the sample of 25 reviewed papers. N paper (%) sample: the number of papers in the population sample of 142 papers.

TABLE 66.4
The Most-Prolific Thematic Research Fronts for the Food Waste-based Bioethanol Fuels

No.	Research Fronts	N Paper % Review	N Paper % Sample	Surplus (%)
1	Biomass pretreatments	88.0	65.5	22.5
2	Biomass hydrolysis	76.0	68.5	7.5
3	Hydrolysate fermentation	48.0	62.7	−14.7
4	Bioethanol production	56.0	62.0	−6.0
5	Bioethanol fuel evaluation	0.0	4.2	−4.2

N paper (%) review: the number of papers in the sample of 25 reviewed papers. N paper (%) sample: the number of papers in the population sample of 142 papers.

On the other hand, the most influential feedstock is the orange peels with 10.1% surplus, followed by the citrus peels and empty fruit bunches with 5.2% and 4.5% surplus, respectively. Similarly, the least influential research front is the other fruit wastes with 15.5% deficit, followed by the food wastes in general with 12.6% deficit. The other least influential feedstocks are dairy wastes, vegetable wastes, apple pomaces, and banana peels with 3.5%–7.3% deficit each.

Information about the thematic research fronts for the sample papers in the food waste-based bioethanol fuels is given in Table 66.4. As this table shows, there are four primary research fronts for this field: the pretreatment and hydrolysis of the food wastes fermentation of the food-waste hydrolysates, and the production of bioethanol with 88%, 76%, 48%, and 56% of the HCPs, respectively.

Further, the most influential research fronts are pretreatment and hydrolysis of the food wastes with 22.5% and 7.5% surplus, respectively. Similarly, hydrolysate fermentation is the least influential research front with 14.7% deficit. The other lest influential research fronts are the bioethanol production and bioethanol evaluation with 6% and 4.2% deficits, respectively.

66.4.2 Food Waste Hydrolysis

The brief information about 11 most-cited papers on the hydrolysis of food wastes with at least 97 citations each is given below (Table 66.1).

Guerard et al. (2002) hydrolyzed tuna waste by umamizyme and obtained a degree of hydrolysis up to 22.5% with an enzyme/substrate ratio of 1.5%, after 4 h of hydrolysis. Further, Wilkins et al. (2007a) hydrolyzed grapefruit peel waste with cellulase and pectinase enzymes and observed that five mg pectinase/g peel dry matter and 2 mg cellulase/g peel dry matter were the lowest loadings to yield the most glucose.

Rahman et al. (2007) optimized acid hydrolysis of oil palm empty fruit bunch fiber for production of xylose and found that the optimum reaction temperature, reaction time, and acid concentration were 119°C, 60 min, and 2%, respectively, resulting in xylose yield and selectivity of 91.27% and 17.97 g/g, respectively. Further, Fernandez-Bolanos et al. (2001) hydrolyzed olive stones pretreated with steam explosion and acids and found that steam explosion pretreatment improved the accessibility of the cellulose and increased the enzymatic hydrolysis yield after steam explosion with respect to material without steam explosion.

Grohmann et al. (1995) solubilized and depolymerized carbohydrates by treatment of orange peel with dilute (0.06% and 0.5%) sulfuric acid at 100°C, 120°C, and 140°C and observed that the acid pretreatments solubilized a large portion of total carbohydrates in orange peel. Further, Correia et al. (2013) performed alkaline H_2O_2 pretreatment of cashew apple bagasse and observed that the pretreatment resulted in a reduced lignin content in the residual solids.

Kim et al. (2012) performed the sequential acid/alkali pretreatment of empty palm fruit bunch fiber and found that the sequentially acid/alkali pretreatments were effective enhancing the enzymatic digestibility of the waste. Further, Yunus et al. (2010) studied the effect of ultrasonic pretreatment on acid hydrolysis of empty palm fruit bunch at low temperature and pressure and obtained a maximum xylose yield of 58% when the waste was ultrasonicated at 90% amplitude for 45 min.

Hamzah et al. (2011) studied the enzymatic hydrolysis of treated empty palm fruit bunches and found that a combination of both enzymes with the ratio of 5:1 hydrolyzed more cellulose from treated waste and gave the highest soluble glucose concentration up to 4 g/L. Further, Del Campo et al. (2006) performed the diluted acid hydrolysis pretreatment of fresh and processed vegetable wastes and found that the maximum ratios of single sugar recovery in the liquid fraction from dilute acid hydrolysis assays were 40.29% and 50.20% (w/w) for tomato and red pepper residues, respectively. Finally, Grohmann and Baldwin (1992) performed the enzymatic hydrolysis of orange peels with pectinase and cellulase enzymes and observed high levels of conversion to monomeric sugars after treatment with pectinase enzyme, but cellulase enzyme achieved only limited solubilization.

66.4.3 Food Waste-based Bioethanol Production

There are 14 HCPs for the production the food waste-based bioethanol fuels with at least 105 citations each (Table 66.2).

Arapoglou et al. (2010) produced ethanol from potato peel waste and found that the enzymatic hydrolysis with a combination of three enzymes released 18.5 g/L reducing sugar and produced 7.6 g/L of ethanol after fermentation. Further, Mussatto et al. (2012) produced ethanol from spent coffee ground and observed that *S. cerevisiae* provided the best ethanol production from this hydrolysate (11.7 g/L, 50.2% efficiency).

Wilkins et al. (2007b) produced ethanol from citrus peel waste and observed that ethanol concentrations after 24 h were reduced in fermentations with initial d-limonene concentrations greater than or equal to 0.33% (v/v) and final (24 h) d-limonene concentrations greater than or equal to 0.14% (v/v). Further, Kwon et al. (2013) coproduced biodiesel and bioethanol from spent coffee grounds and found that this waste was a strong candidate for the coproduction of bioethanol and biodiesel.

Kim et al. (2011) produced ethanol from food waste and for SHF they observed that ethanol concentration reached at the level corresponding to an ethanol yield of 0.43 g ethanol/g total solids. Further, Piarpuzan et al. (2011) produced ethanol from the empty palm fruit bunches using alkaline pretreatment, enzymatic hydrolysis, and fermentation with *S. cerevisiae* and found that coupling alkaline pretreatment with a later autoclaving improved the sugars yield in enzymatic hydrolysis.

Choi et al. (2015) produced ethanol from fruit waste and observed that the rates of enzymatic conversion of fruit waste to fermentable sugars were nearly 90% for all feedstocks after 48 h. Further, Le Man et al. (2010) produced ethanol from food waste leachate and obtained the maximum ethanol concentration of 24.2 g/L at the optimum condition of temperature (38°C), pH (5.45), and reducing sugar concentration (75 g/L).

Oberoi et al. (2011) produced ethanol from banana peels using statistically optimized SSF process after hydrothermal pretreatment and observed that the experiment using optimized parameters in batch fermenter not only resulted in higher ethanol concentration than the one predicted by the model equation but also saved fermentation time. Further, Silveira et al. (2005) produced ethanol from cheese whey permeate and observed that the increase in lactose concentration increased ethanol yield and volumetric productivity and reduced cell yield.

Wilkins et al. (2007c) produced ethanol production from orange peel waste and found that the minimum inhibitory orange peel oil concentrations for ethanol production were 0.05% at 24 h, 0.10% at 48 h, and 0.15% at 72 h for both yeasts. Further, Nigam (2000) produced ethanol from pineapple cannery waste using immobilized yeast cells and obtained the maximum productivity of 42.8 g ethanol/L h at a dilution rate of 1.5/h.

Widmer et al. (2010) produced ethanol from orange processing waste and found that ethanol yields based on sugar content after enzymatic hydrolysis after 48 h SSF ranged from 76% to 94%. Further, Kim et al. (2008) optimized enzymatic saccharification and fermentation of food waste and found that the optimum conditions were saccharification pH of 5.20, enzyme reaction temperature of 46.3°C, enzyme concentration of 0.16% (v/v), fermentation pH of 6.85, fermentation temperature of 35.3°C, and fermentation time of 14 h.

66.5 CONCLUSION AND FUTURE RESEARCH

The brief information about the key research fronts covered by the 25 most-cited papers with at least 97 citations each is given under two primary headings: the hydrolysis of the food wastes and production of the bioethanol fuels.

The usual characteristics of these HCPs are that the pretreatments and hydrolysis of the food wastes and fermentation of the resulting hydrolysates are the primary processes for the bioethanol fuel production from food wastes to improve the ethanol yield as the food wastes are one of the most studied feedstocks for the bioethanol production especially for the countries with the large farmlands and crude oil deficiency.

The key findings on these research fronts should be read in the light of the increasing public concerns about climate change, GHG emissions, and global warming as these concerns have been certainly behind the boom in the research on the food waste-based bioethanol fuels as an alternative to crude oil-based gasoline and diesel fuels in the last decades. It is also a sustainable alternative to first generation food crop-based bioethanol fuels such as corn grain-based bioethanol fuels. The recent supply shocks caused by the COVID-19 pandemics and the Russian invasion of Ukraine also highlight the importance of the production and utilization of the bioethanol fuels from food wastes as an alternative to the crude oil-based gasoline and petrodiesel fuels.

As Table 66.3 shows, the most-prolific research front for this field is the fruit wastes with 60% of the HCPs. Further, the other prolific research fronts are food wastes in general, beverage wastes, and vegetable wastes. Additionally, for the first research front, the most-prolific research subfronts are the empty fruit bunches and orange peels.

Similarly, as Table 66.4 shows, there are four primary research front for this field: the pretreatment and hydrolysis of the food wastes, fermentation of the food-waste hydrolysates, and the production of bioethanol.

These studies emphasize the importance of proper incentive structures for the efficient production of food waste-based bioethanol fuels in the light of North's institutional framework (North, 1991). In this context, the major producers and users of bioethanol fuels such as the USA and Canada with vast farmlands have developed strong incentive structures for the efficient food waste-based bioethanol fuels. In the light of the recent supply shocks caused primarily by the COVID-19 pandemics and Russian invasion of Ukraine, it is expected that the incentive structures such as public funding would be enhanced to increase the share of bioethanol fuels in the global fuel portfolio as a strong alternative to crude oil-based gasoline and diesel fuels.

In this context, it is expected that the most-prolific researchers, institutions, countries, funding bodies, and journals in this field would have a first-mover advantage to benefit from such potential incentives. This is especially true for the US stakeholders as the USA has become the global leader in both the production and utilization of second generation bioethanol fuels from the food wastes.

It is recommended that such review studies are performed for the primary research fronts of the food waste-based bioethanol fuels.

ACKNOWLEDGMENTS

The contribution of the highly cited researchers in the field of the food waste-based bioethanol fuels has been gratefully acknowledged.

REFERENCES

Alvira, P., E. Tomas-Pejo, M. Ballesteros and M. J. Negro. 2010. Pretreatment technologies for an efficient bioethanol production process based on enzymatic hydrolysis: A review. *Bioresource Technology* 101:4851–4861.

Angelici, C., B. M. Weckhuysen and P. C. A. Bruijnincx. 2013. Chemocatalytic conversion of ethanol into butadiene and other bulk chemicals. *ChemSusChem* 6:1595–1614.

Antolini, E. 2007. Catalysts for direct ethanol fuel cells. *Journal of Power Sources* 170:1–12.

Antolini, E. 2009. Palladium in fuel cell catalysis. *Energy and Environmental Science* 2:915–931.

Arapoglou, D., T. Varzakas, A. Vlyssides and C. Israilides. 2010. Ethanol production from potato peel waste (PPW). *Waste Management* 30:1898–1902.

Burnham, J. F. 2006. Scopus database: A review. *Biomedical Digital Libraries* 3:1–8.

Choi, I. S., Y. G. Lee, S. K. Khanal, B. J. Park and H. J. Bae. 2015. A low-energy, cost-effective approach to fruit and citrus peel waste processing for bioethanol production. *Applied Energy* 140:65–74.

Correia, J. A. D. C., J. E. M. Junior, L. R. B. Goncalves and M. V. P. Rocha. 2013. Alkaline hydrogen peroxide pretreatment of cashew apple bagasse for ethanol production: Study of parameters. *Bioresource Technology* 139:249–256.

Del Campo, I., I. Alegria, M. Zazpe, M. Echeverria and I. Echeverria. 2006. Diluted acid hydrolysis pretreatment of agri-food wastes for bioethanol production. *Industrial Crops and Products* 24:214–221.

Fernandez-Bolanos, J., B. Felizon and A. Heredia, et al. 2001. Steam-explosion of olive stones: Hemicellulose solubilization and enhancement of enzymatic hydrolysis of cellulose. *Bioresource Technology* 79:53–61.

Fernando, S., S. Adhikari, C. Chandrapal and M. Murali. 2006. Biorefineries: Current status, challenges, and future direction. *Energy & Fuels* 20:1727–1737.

Grohmann, K. and E. A. Baldwin. 1992. Hydrolysis of orange peel with pectinase and cellulase enzymes. *Biotechnology Letters* 14:1169–1174.

Grohmann, K., R. G. Cameron and B. S. Buslig. 1995. Fractionation and pretreatment of orange peel by dilute acid hydrolysis. *Bioresource Technology* 54:129–141.

Guerard, F., L. Guimas and A. Binet. 2002. Production of tuna waste hydrolysates by a commercial neutral protease preparation. *Journal of Molecular Catalysis B: Enzymatic* 19:489–498.

Guimaraes, P. M. R., J. A. Teixeira and L. Domingues. 2010. Fermentation of lactose to bio-ethanol by yeasts as part of integrated solutions for the valorisation of cheese whey. *Biotechnology Advances* 28:375–384.

Hafid, H. S., N. A. A. Rahman, U. K. M. Shah, A. S. Baharuddin and A. B. Ariff. 2017. Feasibility of using kitchen waste as future substrate for bioethanol production: A review. *Renewable and Sustainable Energy Reviews* 74:671–686.

Hahn-Hagerdal, B., M. Galbe, M. F. Gorwa-Grauslund, G. Liden and G. Zacchi. 2006. Bio-ethanol: The fuel of tomorrow from the residues of today. *Trends in Biotechnology* 24:549–556.

Hamzah, F., A. Idris and T. K. Shuan. 2011. Preliminary study on enzymatic hydrolysis of treated oil palm (*Elaeis*) empty fruit bunches fibre (EFB) by using combination of cellulase and β 1-4 glucosidase. *Biomass and Bioenergy* 35:1055–1059.

Hang, Y. D., C. Y. Lee and E. E. Woodams. 1982. A solid state fermentation system for production of ethanol from apple pomace. *Journal of Food Science* 47:1851–1852.

Hill, J., E. Nelson, D. Tilman, S. Polasky and D. Tiffany. 2006. Environmental, economic, and energetic costs and benefits of biodiesel and ethanol biofuels. *Proceedings of the National Academy of Sciences of the United States of America* 103:11206–11210.

Hill, J., S. Polasky and E. Nelson, et al. 2009. Climate change and health costs of air emissions from biofuels and gasoline. *Proceedings of the National Academy of Sciences of the United States of America* 106:2077–2082.

Hsieh, W. D., R. H. Chen, T. L. Wu and T. H. Lin. 2002. Engine performance and pollutant emission of an SI engine using ethanol-gasoline blended fuels. *Atmospheric Environment* 36:403–410.

Huang, H. J., S. Ramaswamy, U. W. Tschirner and B. V. Ramarao. 2008. A review of separation technologies in current and future biorefineries. *Separation and Purification Technology* 62:1–21.

Khawla, B. J., M. Sameh and G. Imen, et al. 2014. Potato peel as feedstock for bioethanol production: A comparison of acidic and enzymatic hydrolysis. *Industrial Crops and Products* 52:144–149.

Kim, J. H., J. C. Lee and D. Pak. 2011. Feasibility of producing ethanol from food waste. *Waste Management* 31:2121–2125.

Kim, J. K., B. R. Oh, H. J. Shin, C. Y. Eom and S. W. Kim. 2008. Statistical optimization of enzymatic saccharification and ethanol fermentation using food waste. *Process Biochemistry* 43:1308–1312.

Kim, S., J. M. Park, J. W. Seo and C. H. Kim. 2012. Sequential acid-/alkali-pretreatment of empty palm fruit bunch fiber. *Bioresource Technology* 109:229–233.

Konur, O. 2000. Creating enforceable civil rights for disabled students in higher education: An institutional theory perspective. *Disability & Society* 15:1041–1063.

Konur, O. 2002a. Access to nursing education by disabled students: Rights and duties of nursing programs. *Nurse Education Today* 22:364–374.

Konur, O. 2002b. Assessment of disabled students in higher education: Current public policy issues. *Assessment and Evaluation in Higher Education* 27:131–152.

Konur, O. 2002c. Access to employment by disabled people in the UK: Is the Disability Discrimination Act working? *International Journal of Discrimination and the Law* 5:247–279.

Konur, O. 2006a. Participation of children with dyslexia in compulsory education: Current public policy issues. *Dyslexia* 12:51–67.

Konur, O. 2006b. Teaching disabled students in higher education. *Teaching in Higher Education* 11:351–363.

Konur, O. 2007a. A judicial outcome analysis of the Disability Discrimination Act: A windfall for the employers? *Disability & Society* 22:187–204.

Konur, O. 2007b. Computer-assisted teaching and assessment of disabled students in higher education: The interface between academic standards and disability rights. *Journal of Computer Assisted Learning* 23:207–219.

Konur, O. 2012. The evaluation of the research on the bioethanol: A scientometric approach. *Energy Education Science and Technology Part A: Energy Science and Research* 28:1051–1064.

Konur, O. 2015. Current state of research on algal bioethanol. In *Marine Bioenergy: Trends and Developments*, Eds. S. K. Kim and C. G. Lee, pp. 217–244. Boca Raton, FL: CRC Press.

Konur, O. 2019. Cyanobacterial bioenergy and biofuels science and technology: A scientometric overview. In *Cyanobacteria: From Basic Science to Applications*, Eds. A. K. Mishra, D. N. Tiwari and A. N. Rai, pp. 419–442. Amsterdam: Elsevier.

Konur, O. 2020. The scientometric analysis of the research on the bioethanol production from green macroalgae. In *Handbook of Algal Science, Technology and Medicine,* Ed. O. Konur, pp. 385–401. London: Academic Press.

Konur, O. 2023. Second generation food waste-based bioethanol fuels: Scientometric study. In *Feedstock-based Bioethanol Fuels. II. Waste Feedstocks: Agricultural, Food, Industrial, Urban, Forestry, and Lignocellulosic Waste-based Bioethanol Fuels. Handbook of Bioethanol Fuels Volume 4*, Ed. O. Konur, pp. 147–172. Boca Raton, FL: CRC Press.

Kwon, E. E., H. Yi and Y. J. Jeon. 2013. Sequential co-production of biodiesel and bioethanol with spent coffee grounds. *Bioresource Technology* 136:475–480.

Le Man, H., S. K. Behera and H. S. Park. 2010. Optimization of operational parameters for ethanol production from Korean food waste leachate. *International Journal of Environmental Science and Technology* 7:157–164.

Lin, Y. and S. Tanaka. 2006. Ethanol fermentation from biomass resources: Current state and prospects. *Applied Microbiology and Biotechnology* 69:627–642.

Ma, X., L. Sun and C. Song. 2002. A new approach to deep desulfurization of gasoline, diesel fuel and jet fuel by selective adsorption for ultra-clean fuels and for fuel cell applications. *Catalysis Today* 77:107–116.

Morschbacker, A. 2009. Bio-ethanol based ethylene. *Polymer Reviews* 49:79–84.

Mussatto, S. I., E. M. S. Machado, L. M. Carneiro and J. A. Teixeira. 2012. Sugars metabolism and ethanol production by different yeast strains from coffee industry wastes hydrolysates. *Applied Energy* 92:763–768.

Najafi, G., B. Ghobadian and T. Tavakoli, et al. 2009. Performance and exhaust emissions of a gasoline engine with ethanol blended gasoline fuels using artificial neural network. *Applied Energy* 86:630–639.

Newman, P. W. G. and J. R. Kenworthy. 1989. Gasoline consumption and cities: A comparison of U.S. cities with a global survey. *Journal of the American Planning Association* 55:24–37.

Nigam, J. N. 2000. Continuous ethanol production from pineapple cannery waste using immobilized yeast cells. *Journal of Biotechnology* 80:189–193.

North, D. C. 1991. Institutions. *Journal of Economic Perspectives* 5:97–112.

Oberoi, H. S., P. V. Vadlani, L. Saida, S. Bansal and J. D. Hughes. 2011. Ethanol production from banana peels using statistically optimized simultaneous saccharification and fermentation process. *Waste Management* 31:1576–1584.

Olsson, L. and B. Hahn-Hagerdal. 1996. Fermentation of lignocellulosic hydrolysates for ethanol production. *Enzyme and Microbial Technology* 18:312–331.

Piarpuzan, D., J. A. Quintero and C. A. Cardona. 2011. Empty fruit bunches from oil palm as a potential raw material for fuel ethanol production. *Biomass and Bioenergy* 35:1130–1137.

Rahman, S. H. A., J. P. Choudhury, A. L. Ahmad and A. H. Kamaruddin. 2007. Optimization studies on acid hydrolysis of oil palm empty fruit bunch fiber for production of xylose. *Bioresource Technology* 98:554–559.

Ravindran, R. and A. K. Jaiswal. 2016. A comprehensive review on pre-treatment strategy for lignocellulosic food industry waste: Challenges and opportunities. *Bioresource Technology* 199:92–102.

Sanchez, O. J. and C. A. Cardona. 2008. Trends in biotechnological production of fuel ethanol from different feedstocks. *Bioresource Technology* 99:5270–5295.

Silveira, W. B., F. J. V. Passos, H. C. Mantovani and F. M. L. Passos. 2005. Ethanol production from cheese whey permeate by *Kluyveromyces marxianus* UFV-3: A flux analysis of oxido-reductive metabolism as a function of lactose concentration and oxygen levels. *Enzyme and Microbial Technology* 36:930–936.

Sun, Y. and J. Cheng. 2002. Hydrolysis of lignocellulosic materials for ethanol production: A review. *Bioresource Technology* 83:1–11.

Taherzadeh, M. J. and K. Karimi. 2007. Enzyme-based hydrolysis processes for ethanol from lignocellulosic materials: A review. *Bioresources* 2:707–738.

Taherzadeh, M. J. and K. Karimi. 2008. Pretreatment of lignocellulosic wastes to improve ethanol and biogas production: A review. *International Journal of Molecular Sciences* 9:1621–1651.

Velasquez-Arredondo, H. I., A. A. Ruiz-Colorado and S. de Oliveira. 2010. Ethanol production process from banana fruit and its lignocellulosic residues: Energy analysis. *Energy* 35:3081–3087.

Widmer, W., W. Zhou and K. Grohmann. 2010. Pretreatment effects on orange processing waste for making ethanol by simultaneous saccharification and fermentation. *Bioresource Technology* 101:5242–5249.

Wilkins, M. R., L. Suryawati, N. O. Maness and D. Chrz. 2007c. Ethanol production by *Saccharomyces cerevisiae* and *Kluyveromyces marxianus* in the presence of orange-peel oil. *World Journal of Microbiology and Biotechnology* 23:1161–1168.

Wilkins, M. R., W. W. Widmer and K. Grohmann. 2007b. Simultaneous saccharification and fermentation of citrus peel waste by *Saccharomyces cerevisiae* to produce ethanol. *Process Biochemistry* 42:1614–1619.

Wilkins, M. R., W. W. Widmer, K. Grohmann and R. G. Cameron. 2007a. Hydrolysis of grapefruit peel waste with cellulase and pectinase enzymes. *Bioresource Technology* 98:1596–1601.

Yunus, R., S. F. Salleh, N. Abdullah and D. R. A. Biak. 2010. Effect of ultrasonic pre-treatment on low temperature acid hydrolysis of oil palm empty fruit bunch. *Bioresource Technology* 101:9792–9796.

Part 20

Second Generation Industrial Waste-based Bioethanol Fuels

67 Second Generation Industrial Waste-based Bioethanol Fuels

Scientometric Study

Ozcan Konur
(Formerly) Ankara Yildirim Beyazit University

67.1 INTRODUCTION

The crude oil-based gasoline fuels (Ma et al., 2002; Newman and Kenworthy, 1989) have been widely used in the transportation sector since the 1920s. However, there have been great public concerns over the adverse environmental and human impact of these fuels (Hill et al., 2006, 2009). Hence, biomass-based bioethanol fuels (Hill et al., 2006; Konur, 2012e, 2015, 2019, 2020a) have increasingly been used in blending gasoline fuels (Hsieh et al., 2002; Najafi et al., 2009), in the fuel cells (Antolini, 2007, 2009), and in the biochemical production (Angelici et al., 2013; Morschbacker, 2009) in a biorefinery context (Fernando et al., 2006; Huang et al., 2008).

Bioethanol fuels also play a critical role in maintaining the energy security (Kruyt et al., 2009; Winzer, 2012) in the supply shocks (Kilian, 2008, 2009) related to oil price shocks (Hamilton, 1983, 2003), COVID-19 pandemics (Fauci et al., 2020; Li et al., 2020), or wars (Hamilton, 1983; Jones, 2012) in the aftermath of the Russian invasion of Ukraine (Reeves, 2014).

However, it is necessary to pretreat the biomass (Taherzadeh and Karimi, 2008; Yang and Wyman, 2008) to enhance the yield of the bioethanol (Hahn-Hagerdal et al., 2006; Sanchez and Cardona, 2008) prior to the bioethanol production through the hydrolysis (Sun and Cheng, 2002; Taherzadeh and Karimi, 2007) and fermentation (Lin and Tanaka, 2006; Olsson and Hahn-Hagerdal, 1996) of the biomass and hydrolysates, respectively.

One of the most-studied feedstocks for the bioethanol fuels has been the industrial wastes such as milling industrial wastes, food processing industrial wastes, wood industrial wastes, and other industrial wastes such as biodiesel industrial wastes. The research in the field of the second generation industrial waste-based bioethanol fuels has intensified in this context in the key research fronts of the pretreatment (Laser et al., 2002; Ravindran and Jaiswal, 2016) and hydrolysis (Grohmann et al., 1995; Lavarack et al., 2002) of the industrial wastes, fermentation (Kadar et al., 2004; Martín et al., 2002) of the industrial waste-based hydrolysates, and production (Duff and Murray, 1996; Koutinas et al., 2014) and evaluation (Dias et al., 2012; MacRelli et al., 2012) of the second generation industrial waste-based bioethanol fuels. Further, milling industrial wastes (Cardona et al., 2010; Laser et al., 2002), food processing industrial wastes (Grohmann et al., 1995; Ravindran and Jaiswal, 2016), wood and paper industrial wastes (Brandt et al., 2010; Fan et al., 2003), and other industrial wastes such as biodiesel industrial wastes (Ito et al., 2005; Yazdani and Gonzalez, 2008) have been studied intensively in this context.

However, it is essential to develop efficient incentive structures (North, 1991) for the primary stakeholders to enhance the research in this field (Konur, 2000, 2002a–c, 2006a,b, 2007a,b). The scientometric analysis has been used in this context to inform the primary stakeholders about the current state of the research in a selected research field (Garfield, 1955; Konur, 2011, 2012a–i, 2015, 2018b, 2019, 2020a).

DOI: 10.1201/9781003226550-91

As there have been no published scientometric studies in this field, this book chapter presents a scientometric study of the research in the second generation industrial waste-based bioethanol fuels. It examines the scientometric characteristics of both the sample and population data presenting scientometric characteristics of these both datasets in the order of documents, authors, publication years, institutions, funding bodies, source titles, countries, Scopus subject categories, Scopus keywords, and research fronts.

67.2 MATERIALS AND METHODS

The search for this study was carried out using Scopus database (Burnham, 2006) in October 2022.

As a first step for the search of the relevant literature, the keywords were selected using the first most-cited 300 population papers. The selected keyword list was then optimized to obtain a representative sample of papers for the searched research field. This keyword list was provided in the appendix for future replicative studies.

As a second step, two sets of data were used for this study. First, a population sample of 5,000 papers was used to examine the scientometric characteristics of the population data. Second, a sample of 250 most-cited papers, corresponding to 5% of the population papers, was used to examine the scientometric characteristics of these citation classics.

The scientometric characteristics of these both sample and population datasets were presented in the order of documents, authors, publication years, institutions, funding bodies, source titles, countries, Scopus subject categories, Scopus keywords, and research fronts.

Lastly, the key scientometric findings for both datasets were discussed to highlight the research landscape for the second generation industrial waste-based bioethanol fuels. Additionally, a number of brief conclusions were drawn, and a number of relevant recommendations were made to enhance the future research landscape.

67.3 RESULTS

67.3.1 The Most-Prolific Documents in the Second Generation Industrial Waste-based Bioethanol Fuels

The information on the types of documents for both datasets is given in Table 67.1. The articles and conference papers, published in journals, dominate both the sample (96%) and population (97%) papers with 1% deficit. Further, review papers and short surveys have a 1% surplus as they are overrepresented in the sample papers as they constitute 3% and 2% of the sample and population papers, respectively.

It is further notable that 98% of the population papers were published in journals, while 2% of them were published in books and book series. Similarly, 100% of the sample papers were published in the journals.

67.3.2 The Most-Prolific Authors in the Second Generation Industrial Waste-based Bioethanol Fuels

The information about the most-prolific 21 authors with at least 1.6% of sample papers each is given in Table 67.2. The most-prolific authors are Carlos Martin and Karel Grohmann with 3.2% of the sample papers, followed by Rubens M. Filho, John N. Saddler, Michael R. Ladisch, and Guido Zacchi with 2.4% of the sample papers each. The other prolific authors are Lonnie O. Ingram, George J. M. Rocha, Mats Galbe, and Michael A. Cotta with 2.0% of the sample papers each.

On the other hand, the most influential author is Karel Grohmann with 3.0% surplus, followed by Carlos Martin with 2.7% surplus. The other influential authors are Michael R. Ladisch, Guido Zacchi, John N. Saddler, and Rubens M. Filho with 2.0%–2.2% surplus each.

TABLE 67.1
Documents in the Second Generation Industrial Waste-based Bioethanol Fuels

Documents	Sample Dataset (%)	Population Dataset (%)	Surplus (%)
Article	92.8	94.5	−1.7
Conference paper	3.2	2.6	0.6
Review	2.8	1.4	1.4
Book chapter	0.4	1.0	−0.6
Letter	0.0	0.2	−0.2
Note	0.4	0.2	0.2
Short Survey	0.4	0.1	0.3
Book	0.0	0.0	0.0
Editorial	0.0	0.0	0.0
Sample size	250	5,000	

Sample dataset: the number of papers (%) in the set of 250 highly cited papers. Population dataset: the number of papers (%) in the set of the 5,000 population papers.

The most-prolific institutions for the sample dataset are the Lund University, State University of Campinas, University of Sao Paulo, and USDA Agricultural Research Service with two authors each. On the other hand, the most-prolific countries for the sample dataset are Brazil and the USA with seven authors each, followed by Sweden and India with three and two authors, respectively. In total, only six countries house these top authors.

The most-prolific research front for these top authors is the bioethanol fuels from milling industrial wastes with 19 authors. The other research fronts are the bioethanol fuels from the ethanol industry, wood industry, fruit industry, and brewing industry with four, three, three, and one authors, respectively.

On the other hand, there is a significant gender deficit (Beaudry and Lariviere, 2016) for the sample dataset as surprisingly only three of these top researchers are female with a representation rate of 14%.

Additionally, there are other authors with the relatively low citation impact and with 0.3%–0.7% of the population papers each: Zhenhong Yuan, Anuj K. Chandel, Carlos R. Soccol, Qiang Yu, Wen Wang, Wei Qi, Xinshu Zhuang, Maria C. Area, Mohammad J. Taherzadeh, Zhanying Zhang, Keikhesro Karimi, Ian M. O'Hara, Jun Xie, Jingliang Xu, Silvio S. da Silva, Marco A. Z. Ayub, Dehua Liu, Xuebing Zhao, Eulogio Castro, Carlos A. Rosa, Qiong Wang, Hongdan Zhang, Yu Zhang, Suraini Abd-Aziz, Fernando E. Felissia, Andre Ferraz, Leandro V. A. Gurgel, Cheng J. F. Gorgens, Marcio A. Mazutti, Hector A. Ruiz, Walter Borzani, William O. S. Doherty, Valeria M. Guimaraes, Nei Pereira, Marcos L. Corazza, Jianxin Jiang, Navadol Laosiripojana, Yunyun Liu, and Solange I. Mussatto.

67.3.3 The Most-Prolific Research Output by Years in the Second Generation Industrial Waste-based Bioethanol Fuels

Information about papers published between 1970 and 2022 is given in Figure 67.1. This figure clearly shows that the bulk of the research papers in the population dataset were published primarily in the 2010s and the early 2020s with 55% and 23% of the population dataset, respectively. Similarly, the publication rates for the 2000s, 1990s, 1980s, and 1970s were 10%, 5%, 4%, and 1% respectively. Additionally, 1% of the population papers were published in the pre-1970s.

Similarly, the bulk of the research papers in the sample dataset were published in the 2010s and 2000s with 47% and 41% of the sample dataset, respectively. Similarly, the publication rates for the early 2020s, 1990s, 1980s, and 1970s were 0.4%, 8%, 4%, and 0% of the sample papers, respectively.

TABLE 67.2
The Most-Prolific Authors in the Second Generation Industrial Waste-based Bioethanol Fuels

No.	Author Name	Author Code	Sample Papers (%)	Population Papers (%)	Surplus (%)	Institution	Country	HI	N	Res. Front
1	Martin, Carlos	56484787200	3.2	0.5	2.7	Inland Norway Univ. Appl. Sci.	Norway	26	66	M
2	Grohmann, Karel	7004503589	3.2	0.2	3.0	Renewable Spirits LLC	USA	49	111	F
3	Filho, Rubens M.	7003732915	2.4	0.4	2.0	State Univ. Campinas	Brazil	73	704	M
4	Saddler, John N.	57202481615 7005297559	2.4	0.3	2.1	Univ. British Columbia	Canada	99	408	M, W
5	Ladisch, Michael R.	7005670397	2.4	0.2	2.2	Purdue Univ.	USA	59	290	W, M, E
6	Zacchi, Guido	7006727748	2.4	0.2	2.2	Lund Univ.	Sweden	68	204	M
7	Ingram, Lonnie O.	7102962097	2.0	0.4	1.6	Univ. Florida	USA	73	281	M
8	Rocha, George J. M.	14012497200	2.0	0.3	1.7	Univ. Sao Paulo	Brazil	32	75	M
9	Galbe, Mats	7003788758	2.0	0.2	1.8	Lund Univ.	Sweden	51	131	M
10	Cotta, Michael A.	7006656876	2.0	0.2	1.8	USDA Agr. Res. Serv.	USA	53	186	E, M
11	Rabelo, Sarita C.*	22953880600	1.6	0.5	1.1	Paulista State Univ.	Brazil	24	70	M
12	Ramos, Luiz P.	7202180586	1.6	0.5	1.1	Fed. Univ. Parana	Brazil	45	181	M, W
13	Milagres, Adriane M. F.*	6701345269	1.6	0.4	1.2	Univ. Sao Paulo	Brazil	32	109	B, M
14	Polikarpov, Igor	7006220351	1.6	0.3	1.3	Univ. Sao Paulo	Brazil	49	312	M
15	Costa, Aline C.*	57203868309	1.6	0.3	1.3	State Univ. Campinas	Brazil	34	114	M
16	Dale, Bruce E.	7201511969	1.6	0.3	1.3	Michigan State Univ.	USA	92	430	M, E
17	Balan, Venkatesh	15757087100	1.6	0.2	1.4	Univ. Houston	USA	56	212	E, M
18	Vadlani, Praveen V.	24075089500	1.6	0.2	1.4	Sri Sathya Sai Inst. High. Learn	India	26	80	F, M
19	Jonsson, Leif J.	7102349315	1.6	0.2	1.4	Umea Univ.	Sweden	41	148	M
20	Oberoi, Harinder S.	6603479987	1.6	0.2	1.4	Indian Inst. Hort. Res.	India	27	66	F
21	Saha, Badal C.	7202946302	1.6	0.1	1.5	USDA Agr. Res. Serv.	USA	55	160	M

Author code: the unique code given by Scopus to the authors. Sample papers: the number of papers authored in the sample dataset. Population papers: the number of papers authored in the population dataset.

*, Female; M, milling wastes; W, wood wastes; E, ethanol wastes; F, fruit wastes; and B, brewing wastes.

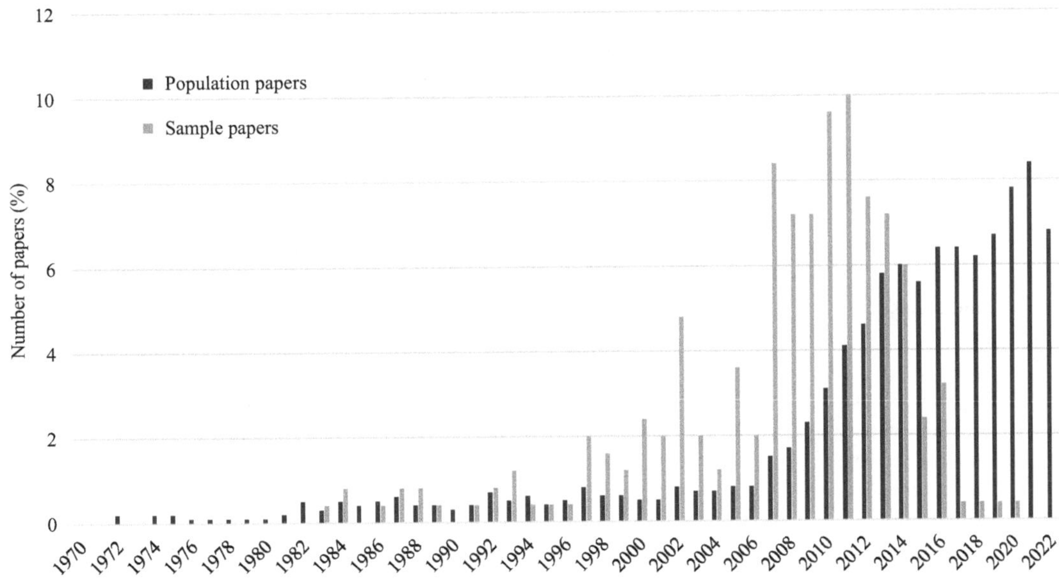

FIGURE 67.1 The research output by years regarding the second generation industrial waste-based bioethanol fuels.

The most-prolific publication years for the population dataset were 2021 and 2020 with 8.4% and 7.8% of the dataset, respectively, while 78% of the population papers were published between 2010 and 2022. Similarly, 74% of the sample papers were published between 2005 and 2016, while the most-prolific publication years were 2010, 2007, and 2016 with 9.6%, 8.4%, and 7.6% of the sample papers, respectively.

There was a rising trend for the population papers between 2006 and 2014, and it lost its momentum in part after 2014. However, there was a sharp rise in the research output in 2020 and 2021 possibly due to the supply shocks.

67.3.4 The Most-Prolific Institutions in the Second Generation Industrial Waste-based Bioethanol Fuels

Information about the most-prolific 15 institutions publishing papers on the second generation industrial waste-based bioethanol fuels with at least 2% of the sample papers each is given in Table 67.3.

The most-prolific institution is the University of Sao Paulo with 7.6% of the sample papers, followed by the National Biorenewables Laboratory, USDA Agricultural Research Service, and Lund University with 3.6% of the sample papers each. The other prolific institutions are the University of Matanzas, State University of Campinas, and Purdue University with 3.2%, 2.8%, and 2.8% of the sample papers, respectively. Similarly, the top countries for these most-prolific institutions are the USA and Brazil with five institutions each. In total, only seven countries house these top institutions.

On the other hand, the institution with the most citation impact is the University of Sao Paulo with 3.7% surplus, followed by Lund University, USDA Agricultural Research Service, and University of Matanzas with 2.7%–2.9% surplus each. The other influential institutions are the National Biorenewables Laboratory and Purdue University with 2.4% and 2.2% surplus, respectively.

Additionally, there are other institutions with the relatively low citation impact and with 0.5%–1.5% of the population papers each: State University of Paulista, South China University of Technology, Nanjing Forestry University, State Key Laboratory of Pulp and Paper Engineering, Federal University of Vicosa, Federal University of Sao Carlos, Technical University of Denmark, Guangxi University,

TABLE 67.3
The Most-Prolific Institutions in the Second Generation Industrial Waste-based Bioethanol Fuels

No.	Institutions	Country	Sample Papers (%)	Population Papers (%)	Surplus (%)
1	Univ. Sao Paulo	Brazil	7.6	4.1	3.5
2	Natl. Biorenew. Lab.	Brazil	3.6	1.2	2.4
3	USDA Agr. Res. Serv.	USA	3.6	0.8	2.8
4	Lund Univ.	Sweden	3.6	0.7	2.9
5	Univ. Matanzas	Cuba	3.2	0.5	2.7
6	State Univ. Campinas	Brazil	2.8	1.9	0.9
7	Purdue Univ.	USA	2.8	0.6	2.2
8	Fed. Univ. Rio de Janeiro	Brazil	2.4	0.8	1.6
9	Chinese Acad. Sci.	China	2.0	1.9	0.1
10	Fed. Univ. Parana	Brazil	2.0	1.2	0.8
11	Univ. British Columbia	Canada	2.0	0.6	1.4
12	Univ. Florida	USA	2.0	0.5	1.5
13	Natl. Renew. Ener. Lab.	USA	2.0	0.4	1.6
14	Natl. Inst. Adv. Ind. Sci. Technol.	Japan	2.0	0.3	1.7
15	Univ. Hawaii	USA	2.0	0.3	1.7

University of Putra Malaysia, Tsinghua University, National Scientific Research Center (CNRS), Stellenbosch University, King Mongkut's University of Technology, Beijing Forestry University, Brazilian Agricultural Research Corporation (EMBRAPA), University of Minho, Chulalongkorn University, National Center for Research in Energy and Materials, Queensland University of Technology, University of Technology Malaysia, Kasetsart University, Federal University of Rio Grande do Sul, Jiangnan University, National Research Centre, National Council for Scientific and Technical Research, Korea University, University of Vigo, Federal University of Minas Gerais, South China Agricultural University, University of Kebangsaan Malaysia, Federal University of Santa Catarina, Tianjin University of Science & Technology, and Federal University of Santa Maria.

67.3.5 The Most-Prolific Funding Bodies in the Second Generation Industrial Waste-based Bioethanol Fuels

Information about the most-prolific 11 funding bodies funding at least 2% of the sample papers each is given in Table 67.4. Further, only 48% and 47% of the sample and population papers were funded, respectively.

The most-prolific funding body is the National Council for Scientific and Technological Development of Brazil with 12% of the sample papers, followed by the Research Support Foundation of the State of Sao Paulo and Higher Education Personnel Improvement Coordination, both of Brazil with 6.8% and 6.0% of the sample papers, respectively. The other prolific funding bodies are the Ministry of Science, Technology and Innovation, National Natural Science Foundation of China, and the U.S. Department of Energy with 3.2%–4.4% of the sample papers each.

On the other hand, the most-prolific country for these top funding bodies is Brazil with four funding bodies, followed by China and the USA with three and two funding bodies, respectively. In total, only four countries and the EU house these top funding bodies.

The funding body with the most citation impact is the National Council for Scientific and Technological Development of Brazil with 5.3% surplus, followed by the Research Support Foundation of the State of Sao Paulo and the U.S. Department of Energy with 3.0% and 2.3% surplus, respectively. Further, the funding body with the least citation impact is the National Natural

TABLE 67.4
The Most-Prolific Funding Bodies in the Second Generation Industrial Waste-based Bioethanol Fuels

No.	Funding Bodies	Country	Sample Paper No. (%)	Population Paper No. (%)	Surplus (%)
1	National Council for Scientific and Technological Development	Brazil	12.0	6.7	5.3
2	Research Support Foundation of the State of Sao Paulo	Brazil	6.8	3.8	3.0
3	Higher Education Personnel Improvement Coordination	Brazil	6.0	5.3	0.7
4	Ministry of Science, Technology and Innovation	Brazil	4.4	2.8	1.6
5	National Natural Science Foundation of China	China	3.2	6.6	–3.4
6	U.S. Department of Energy	USA	3.2	0.9	2.3
7	Ministry of Science and Technology, India	India	2.8	1.3	1.5
8	Ministry of Science and Technology of China	China	2.4	1.7	0.7
9	National Key Research and Development Program of China	China	2.4	1.5	0.9
10	European Commission	EU	2.4	1.2	1.2
11	National Key Research and Development Program of China	USA	2.0	0.6	1.4

Science Foundation of China with 3.4% deficit. This funding body is the largest second funder of the population papers after Brazil.

The other funding bodies with the relatively low citation impact and with 0.5%–1.4% of the population papers each are the European Regional Development Fund, National Council of Science and Technology, Fundamental Research Funds for the Central Universities, Research Support Foundation of the State of Minas Gerais, Ministry of Education of China, Japan Society for the Promotion of Science, Ministry of Education, Culture, Sports, Science and Technology, Natural Sciences and Engineering Research Council of Canada, Thailand Research Fund, Ministry of Higher Education Malaysia, Foundation for Science and Technology, National Research Foundation of Korea, Seventh Framework Program, Ministry of Education, Japan Science and Technology Agency, Ministry of Finance, National Basic Research Program of China (973 Program), Natural Science Foundation of Jiangsu Province, Priority Academic Program Development of Jiangsu Higher Education Institutions, Council of Scientific and Industrial Research India, Research Support Foundation of the State of Rio Grande do Sul, University Grants Commission, and Ministry of Education, Science and Technology.

67.3.6 The Most-Prolific Source Titles in the Second Generation Industrial Waste-based Bioethanol Fuels

Information about the most-prolific 16 source titles publishing at least 1.2% of the sample papers each in the second generation industrial waste-based bioethanol fuels is given in Table 67.5.

The most-prolific source title is the Bioresource Technology with 32% of the sample papers. The other prolific source titles are Biomass and Bioenergy, Enzyme and Microbial Technology, Biotechnology and Bioengineering, Applied Biochemistry and Biotechnology Part A Enzyme Engineering and Biotechnology, and Industrial Crops and Products with 4.0%–5.6% of the sample papers each.

On the other hand, the source title with the most citation impact is the Bioresource Technology with 22.6% surplus. The other influential source titles are Enzyme and Microbial Technology, Biotechnology and Bioengineering, Applied Biochemistry and Biotechnology Part A Enzyme Engineering and Biotechnology, and Biomass and Bioenergy with 3.3%–4.2% surplus each.

The other source titles with the relatively low citation impact with 0.5%–2.3% of the population papers each are Bioresources, Biomass Conversion and Biorefinery, Renewable Energy, Waste

TABLE 67.5
The Most-Prolific Source Titles in the Second Generation Industrial Waste-based Bioethanol Fuels

No.	Source Titles	Sample Papers (%)	Population Papers (%)	Surplus (%)
1	Bioresource Technology	32.0	9.4	22.6
2	Biomass and Bioenergy	5.6	2.3	3.3
3	Enzyme and Microbial Technology	5.2	1.0	4.2
4	Biotechnology and Bioengineering	4.8	0.9	3.9
5	Applied Biochemistry and Biotechnology Part A Enzyme Engineering and Biotechnology	4.4	0.8	3.6
6	Industrial Crops and Products	4.0	3.2	0.8
7	Applied Energy	3.6	0.7	2.9
8	Biotechnology for Biofuels	3.2	1.3	1.9
9	Applied Biochemistry and Biotechnology	2.8	2.3	0.5
10	Process Biochemistry	2.8	1.0	1.8
11	Journal of Chemical Technology and Biotechnology	1.6	1.1	0.5
12	Applied Microbiology and Biotechnology	1.2	0.9	0.3
13	Industrial and Engineering Chemistry Research	1.2	0.5	0.7
14	Journal of Biotechnology	1.2	0.3	0.9
15	Brazilian Journal of Chemical Engineering	1.2	0.3	0.9
16	Journal of Food Engineering	1.2	0.2	1.0

and Biomass Valorization, Fuel, Biotechnology Letters, Chemical Engineering Transactions, ACS Sustainable Chemistry and Engineering, Bioenergy Research, Bioprocess and Biosystems Engineering, Journal of Cleaner Production, Fermentation, Cellulose, RSC Advances, Journal of Industrial Microbiology and Biotechnology, Biochemical Engineering Journal, International Journal of Biological Macromolecules, 3 Biotech, Waste Management, Biocatalysis and Agricultural Biotechnology, Biotechnology Progress, Cellulose Chemistry and Technology, Chemical Engineering Journal, Energies, and Energy Sources Part A Recovery Utilization and Environmental Effects.

67.3.7 The Most-Prolific Countries in the Second Generation Industrial Waste-based Bioethanol Fuels

Information about the most-prolific 15 countries publishing at least 2.4% of sample papers each in the second generation industrial waste-based bioethanol fuels is given in Table 67.6.

The most-prolific country is the USA with 24% of the sample papers, followed by Brazil with 22% of the sample papers. The other prolific countries are China, India, Spain, Sweden, and Japan with 5.6%–8.8% of the sample papers, respectively. Further, four European countries listed in Table 67.6 produce 17% and 8% of the sample and population papers, respectively.

On the other hand, the country with the most citation impact is the USA with 13.6% surplus, followed by Brazil with 7.3% surplus. The other influential countries are Sweden, Cuba, Spain, and Greece with 2.0%–3.9% surplus each. Similarly, the country with the least citation impact is China with 5.9% deficit, followed by India and Malaysia with 2.6% and 1.8% deficits, respectively.

Additionally, there are other countries with relatively low citation impact and with 0.5%–3.4% of the sample papers each: Thailand, Indonesia, the UK, Italy, Turkey, Iran, France, Taiwan, Pakistan, Denmark, Egypt, Finland, Colombia, Argentina, Poland, Belgium, Russia, Saudi Arabia, Netherlands, and Vietnam.

TABLE 67.6
The Most-Prolific Countries in the Second Generation Industrial Waste-based Bioethanol Fuels

No.	Countries	Sample Papers (%)	Population Papers (%)	Surplus (%)
1	USA	24.0	10.4	13.6
2	Brazil	22.0	14.7	7.3
3	China	8.8	14.7	–5.9
4	India	7.6	10.2	–2.6
5	Spain	6.0	3.3	2.7
6	Sweden	6.0	2.1	3.9
7	Japan	5.6	5.1	0.5
8	Cuba	4.0	1.1	2.9
9	S. Korea	3.6	3.5	0.1
10	Canada	3.6	2.7	0.9
11	Australia	3.2	2.0	1.2
12	Greece	3.2	1.2	2.0
13	Mexico	2.8	2.6	0.2
14	Malaysia	2.4	4.2	–1.8
15	Portugal	2.4	1.7	0.7

67.3.8 The Most-Prolific Scopus Subject Categories in the Second Generation Industrial Waste-based Bioethanol Fuels

Information about the most-prolific eight Scopus subject categories indexing at least 7.6% of the sample papers each is given in Table 67.7.

The most-prolific Scopus subject category in the second generation industrial waste-based bioethanol fuels is the Chemical Engineering with 66% of the sample papers, followed by Environmental Science and Energy with 53% and 52% of the sample papers, respectively. The other prolific subject categories are Biochemistry, Genetics and Molecular Biology, and Immunology and Microbiology with 35% and 30% of the sample papers, respectively. It is notable that Social Sciences including Economics and Business account for 1% and 2% of the sample and population studies, respectively.

On the other hand, the Scopus subject category with the most citation impact is the Chemical Engineering with 26% surplus, followed by Environmental Science and Energy with 22% and 21%

TABLE 67.7
The Most-Prolific Scopus Subject Categories in the Second Generation Industrial Waste-based Bioethanol Fuels

No.	Scopus Subject Categories	Sample Papers (%)	Population Papers (%)	Surplus (%)
1	Chemical Engineering	66.4	40.9	25.5
2	Environmental Science	53.2	31.6	21.6
3	Energy	51.6	30.4	21.2
4	Biochemistry, Genetics, and Molecular Biology	35.2	28.5	6.7
5	Immunology and Microbiology	30.4	18.4	12.0
6	Agricultural and Biological Sciences	13.2	21.8	–8.6
7	Chemistry	10.0	17.3	–7.3
8	Engineering	7.6	11.5	–3.9

surplus, respectively. Similarly, the least influential subject categories are Agricultural and Biological Sciences with 9% deficit, followed by Chemistry and Engineering with 7% and 4% deficits, respectively.

67.3.9 The Most-Prolific Keywords in the Second Generation Industrial Waste-based Bioethanol Fuels

Information about the Scopus keywords used with at least 6.4% or 3.3% of the sample or population papers, respectively, is given in Table 67.8. For this purpose, keywords related to the keyword set given in the appendix are selected from a list of the most-prolific keyword set provided by Scopus database.

TABLE 67.8
The Most-Prolific Keywords in the Second Generation Industrial Waste-based Bioethanol Fuels

No.	Keywords	Sample Papers (%)	Population Papers (%)	Surplus (%)
1	**Industrial Waste**			
	Cellulose	58.8	25.6	33.2
	Bagasse	41.6	21.7	19.9
	Lignin	40.8	19.8	21.0
	Biomass	28.0	16.1	11.9
	Sugar cane	26.4	8.1	18.3
	Sugarcane	22.8	8.1	14.7
	Sugarcane bagasse	20.4	9.4	11.0
	Hemicellulose	20.4	6.5	13.9
	Carbohydrate	18.4	8.3	10.1
	Lignocellulose	17.2	6.9	10.3
	Sugar cane bagasse	16.0	10.3	5.7
	Fruits	10.4	7.9	2.5
	Industrial waste	7.6	3.0	4.6
	Lignocellulosic biomass	6.4	6.1	0.3
	Xylan	6.4	2.7	3.7
	Glycerol	5.6	7.7	−2.1
	Molasses		4.7	−4.7
2	**Pretreatments**			
	Pretreatment	28.0	12.1	15.9
	Sulfuric acid	25.6	5.2	20.4
	Pre-treatment	24.8	10.4	14.4
	Enzymes	23.2	14.0	9.2
	Temperature	15.6	6.8	8.8
	Delignification	11.2	7.0	4.2
	Water	10.8	3.6	7.2
	pH	8.8	6.3	2.5
	Ionic liquids	8.4	8.1	0.3
	Steam	8.0	1.9	6.1
	Alkalinity	7.6	3.4	4.2
	Steam explosion	6.4		6.4
	Sodium hydroxide	4.8	4.6	0.2

(Continued)

TABLE 67.8 (*Continued*)
The Most-Prolific Keywords in the Second Generation Industrial Waste-based Bioethanol Fuels

No.	Keywords	Sample Papers (%)	Population Papers (%)	Surplus (%)
3		Fermentation		
	Fermentation	50.0	32.8	17.2
	Yeast	20.4	12.8	7.6
	Saccharomyces	18.4	10.3	8.1
	Bioreactors	11.6	7.5	4.1
	Acetic acids	8.8	4.0	4.8
	Furfural	8.0	2.6	5.4
4		Hydrolysis and Hydrolysates		
	Hydrolysis	62.4	31.7	30.7
	Sugar	51.2	19.2	32.0
	Enzyme activity	40.8	16.4	24.4
	Enzymatic hydrolysis	39.2	20.4	18.8
	Glucose	30.0	16.0	14.0
	Cellulases	23.2	8.9	14.3
	Xylose	20.8	6.5	14.3
	Saccharification	20.4	14.9	5.5
	Enzymolysis	10.8	4.5	6.3
	Acid hydrolysis	7.2	2.6	4.6
	Arabinose	6.8	2.1	4.7
	Enzymatic digestibility	6.4	1.9	4.5
	Enzymatic saccharification	5.6	3.8	1.8
5		Products		
	Ethanol	61.6	35.4	26.2
	Biofuels	20.4	13.5	6.9
	Bioethanol	18.4	16.1	2.3
	Ethanol production	14.8	7.8	7.0
	Bioethanol production	7.6	6.4	1.2

These keywords are grouped under the five headings: industrial waste biomass, pretreatments, fermentation, hydrolysis and hydrolysates, and products.

The most-prolific keywords related to the biomass and biomass constituents are cellulose, bagasse, and lignin with 59%, 42%, and 41% of the sample papers, respectively. The other prolific keywords are biomass, sugar cane, sugarcane, sugarcane bagasse, and hemicellulose with 20%–28% of the sample papers each.

Further, the prolific keywords related to the pretreatments are the pretreatments, sulfuric acids, pretreatments, and enzymes with 23%–28% of the sample papers each. The other prolific keywords are the temperature, delignification, and water with 11%–16% of the sample papers each.

The most-prolific keyword related to the fermentation is fermentation with 50% of the sample papers. The other prolific keywords are yeasts and saccharomyces with 20% and 18% of the sample papers, respectively.

Further, the most-prolific keywords related to the hydrolysis and hydrolysates are hydrolysis, sugars, enzyme activity, and enzymatic hydrolysis with 39%–62% of the sample papers each. The other prolific keywords are glucose, cellulases, xylose, and saccharification with 20%–30% of the sample papers each.

Finally, the most-prolific keyword related to the products is ethanol with 62% of the sample papers. The other prolific keywords are biofuels and bioethanol with 20% and 18% of the sample papers, respectively.

On the other hand, the most-prolific keywords across all of the research fronts are hydrolysis, ethanol, celluloses, sugars, fermentation, bagasse, lignins, enzyme activity, and enzymatic hydrolysis with 39%–62% surplus each. Similarly, the most influential keywords are cellulose, sugars, hydrolysis, ethanol, enzyme activity, lignins, sulfuric acids, bagasse, and enzymatic hydrolysis with 19%–33% surplus each.

67.3.10 The Most-Prolific Research Fronts in the Second Generation Industrial Waste-based Bioethanol Fuels

Information about the research fronts for the sample papers in the second generation industrial waste-based bioethanol fuels with regard to the industrial waste feedstocks used for the bioethanol production is given in Table 67.9.

As this table shows, the most-prolific industrial waste feedstock is the milling industrial wastes with 63% of the sample papers, followed by food processing industrial wastes with 22% of the sample papers. The other research fronts are the wood and paper, biodiesel, ethanol, brewing, plastic, textile, and leather industrial wastes with 0.4%–2.8% of the sample papers. Further, sugarcane bagasse and other milling wastes form 42% and 21% of the milling industrial wastes, respectively. The primary wood and paper industrial wastes are wood chips, sawdust, and paper sludge. Similarly, the primary food industrial wastes are the fruit, beverage, dairy, and vegetable wastes. The most-prolific milling industrial waste has been sugarcane bagasse.

Information about the thematic research fronts for the sample papers in the second generation industrial waste-based bioethanol fuels is given in Table 67.10. As this table shows, the most prolific research fronts are the pretreatments and hydrolysis of the industrial waste feedstocks with 67% and 66% of the sample papers, respectively. The other research fronts are the bioethanol fuel production and hydrolysate fermentation with 43% and 38% of the sample papers, respectively. The minor research front is bioethanol fuel evaluation and utilization with 4% of the sample papers.

TABLE 67.9
The Most-Prolific Research Fronts for the Second Generation Industrial Waste-based Bioethanol Fuels

No.	Research Fronts	N Paper % Sample
1	Milling industrial wastes	63.2
2	Food processing industrial wastes	21.6
3	Wood and paper industrial wastes	8.0
4	Other industrial wastes	7.6
	Biodiesel industrial wastes	2.8
	Ethanol industrial wastes	1.6
	Brewing industrial wastes	1.2
	Plastic industrial wastes	0.8
	Textile industrial wastes	0.8
	Leather industrial wastes	0.4

N paper (%) sample: the number of papers in the population sample of 250 papers.

TABLE 67.10
The Most-Prolific Thematic Research Fronts for the Second Generation Industrial Waste-based Bioethanol Fuels

No.	Research Fronts	N Paper % Sample
1	Biomass hydrolysis	66.8
2	Biomass pretreatments	66.4
3	Bioethanol production	43.2
4	Hydrolysate fermentation	37.6
5	Bioethanol fuel evaluation and utilization	4.4

N paper (%) sample: the number of papers in the population sample of 250 papers.

67.4 DISCUSSION

67.4.1 Introduction

The crude oil-based gasoline fuels have been widely used in the transportation sector since the 1920s. However, there have been great public concerns over the adverse environmental and human impact of these fuels. Hence, biomass-based bioethanol fuels have increasingly been used in blending gasoline fuels, in the fuel cells, and in the biochemical production in a biorefinery context.

However, it is necessary to pretreat the biomass to enhance the yield of the bioethanol prior to the bioethanol production through the hydrolysis and fermentation. One of the most-studied feedstocks for the bioethanol fuels has been the industrial wastes such as milling, food processing, wood and paper, and other industrial wastes such as biodiesel industrial wastes. The research in the field of the second generation industrial waste-based bioethanol fuels has intensified in this context in the key research fronts of the pretreatment and hydrolysis of the industrial wastes, fermentation of the industrial waste-based hydrolysates, and production and evaluation of the second generation industrial waste-based bioethanol fuels. Further, milling, food processing, wood and paper, and other industrial wastes such as biodiesel industrial wastes have been studied intensively in this context.

However, it is essential to develop efficient incentive structures for the primary stakeholders to enhance the research in this field. This is especially important to maintain energy security in the cases of supply shocks such as oil price shocks, war-related shocks as in the case of Russian invasion of Ukraine or COVID-19 shocks.

The scientometric analysis has been used in this context to inform the primary stakeholders about the current state of the research in a selected research field. As there has been no scientometric study in this field, this book chapter presents a scientometric study of the research in the second generation industrial waste-based bioethanol fuels. It examines the scientometric characteristics of both the sample and population data presenting scientometric characteristics of these both datasets in the order of documents, authors, publication years, institutions, funding bodies, source titles, countries, Scopus subject categories, Scopus keywords, and research fronts.

As a first step for the search of the relevant literature, the keywords were selected using the first most-cited 300 papers. The selected keyword list was then optimized to obtain a representative sample of papers for the searched research field. A copy of this keyword list was provided in the appendix for future replicative studies. Further, a selected list of the keywords was presented in Table 67.8.

As a second step, two sets of data were used for this study. First, a population sample of 5,000 papers was used to examine the scientometric characteristics of the population data. Second, a sample of 250 most-cited papers, corresponding to 5% of the population dataset, was used to examine the scientometric characteristics of these citation classics.

The scientometric characteristics of these sample and population datasets were presented in the order of documents, authors, publication years, institutions, funding bodies, source titles, countries, Scopus subject categories, Scopus keywords, and research fronts.

Lastly, the key scientometric findings for both datasets were discussed to highlight the research landscape for second generation industrial waste-based bioethanol fuels. Additionally, a number of brief conclusions were drawn, and a number of relevant recommendations were made to enhance the future research landscape.

67.4.2 The Most-Prolific Documents in the Second Generation Industrial Waste-based Bioethanol Fuels

Articles (together with conference papers) dominate both the sample (96%) and population (97%) papers (Table 67.1). Further, review papers have a surplus (1%). The representation of the reviews in the sample papers is relatively modest (3%).

Scopus differs from the Web of Science database in differentiating and showing articles (93%) and conference papers (3%) published in the journals separately. However, it should be noted that these conference papers are also published in journals as articles, compared to those published only in the conference proceedings. Hence, the total number of articles and review papers in the sample dataset are 96% and 3%, respectively.

It is observed during the search process that there has been inconsistency in the classification of the documents in Scopus as well as in other databases such as Web of Science. This is especially relevant for the classification of papers as reviews or articles as the papers not involving a literature review may be erroneously classified as a review paper. There is also a case of review papers being classified as articles. For example, the total number of the reviews in the sample data set was manually found as nearly 4% compared to 3% as indexed by Scopus, decreasing the number of articles and conference papers to 95% for the sample dataset.

In this context, it would be helpful to provide a classification note for the published papers in the books and journals at the first instance. It would also be helpful to use the document types listed in Table 67.1 for this purpose. Book chapters may also be classified as articles or reviews as an additional classification to differentiate review chapters from the experimental chapters as it is done by the Web of Science. It would be further helpful to additionally classify the conference papers as articles or review papers as well as it is done in the Web of Science database.

67.4.3 The Most-Prolific Authors in the Second Generation Industrial Waste-based Bioethanol Fuels

There have been most-prolific 21 authors with at least 1.6% of the sample papers each as given in Table 67.2. These authors have shaped the development of the research in this field.

The most-prolific authors are Carlos Martin, Karel Grohmann, and to a lesser extent Rubens M. Filho, John N. Saddler, Michael R. Ladisch, Guido Zacchi, Lonnie O. Ingram, George J. M. Rocha, Mats Galbe, and Michael A. Cotta.

It is important to note the inconsistencies in indexing of the author names in Scopus and other databases. It is especially an issue for the names with more than two components such as 'Blake Sam de Hyun Grohmann'. The probable outcomes are 'Grohmann, B.S.D.H.', 'de Hyun Grohmann, B.S.', or 'Hyun Grohmann, B.S.D.'. The first choice is the gold standard of the publishing sector as the last word in the name is taken as the last name. In most of the academic databases such as PUBMED and EBSCO databases, this version is used predominantly. The second choice is a strong alternative, while the last choice is an undesired outcome as two last words are taken as the last name. It is a good practice to combine the words of the last name by a hyphen: 'Hyun- Grohmann, B.S.D.'. It is notable that inconsistent indexing of the author names may cause substantial inefficiencies in the search process for the papers as well as allocating credit to the authors as there are different author entries for each outcome in the databases.

There are also inconsistencies in the shortening Chinese names. For example, 'YangYing Zhuang' is often shortened as 'Zhuang, Y.', 'Zhuang, Y.-Y.', and 'Zhuang, Y.Y.', as it is done in the Web of Science database as well. However, the gold standard in this case is 'Zhuang, Y' where the last word is taken as the last name and the first word is taken as a single forename. In most of the academic databases such as PUBMED and EBSCO, this first version is used predominantly. However, it makes sense to use the third option to differentiate Chinese names efficiently: 'Zhuang, Y.Y.'. Therefore, there have been difficulties in locating papers for the Chinese authors. In such cases, the use of the unique author codes provided for each author by the Scopus database has been helpful.

There is also a difficulty in allowing credit for the authors especially for the authors with common names such as 'Zhuang, X.' in conducting scientometric studies. These difficulties strongly influence the efficiency of the scientometric studies as well as allocating credit to the authors as there are the same author entries for different authors with the same name, e.g., 'Zhuang, X.' in the databases.

In this context, the coding of authors in Scopus database is a welcome innovation compared to the other databases such as Web of Science. In this process, Scopus allocates a unique number to each author in the database (Aman, 2018). However, there might still be substantial inefficiencies in this coding system especially for common names. For example, some of the papers for a certain author maybe allocated to another researcher with a different author code. It is possible that Scopus uses a number of software programs to differentiate the author names, and the program may not be false-proof (Kim, 2018).

In this context, it does not help that author names are not given in full in some journals and books. This makes difficult to differentiate authors with common names and makes the scientometric studies further difficult in the author domain. Therefore, the author names should be given in all books and journals at the first instance. There is also a cultural issue where some authors do not use their full names in their papers. Instead they use initials for their forenames: 'Ladisch, H.J.', 'Ladisch, H.', or 'Ladisch, J.' instead of 'Ladisch, Hyun Jae'.

There are also inconsistencies in naming of the authors with more than two components by the authors themselves in journal papers and book chapters. For example, 'Ladisch, A.P.C.' might be given as 'Ladisch, A.' or 'Ladisch, A.C.' or 'Ladisch, A.P.' or 'Ladisch, C' in the journals and books. This also makes the scientometric studies difficult in the author domain. Hence, contributing authors should use their name consistently in their publications.

The other critical issue regarding the author names is the inconsistencies in the spelling of the author names in the national spellings (e.g., Şöğütçağla, Çiğdem) rather than in the English spellings (e.g., Sogutcagla, Cigdem) in Scopus database. Scopus differs from the Web of Science database and many other databases in this respect where the author names are given only in the English spellings. It is observed that national spellings of the author names do not help much in conducting scientometric studies as well in allocating credits to the authors as sometimes there are different author entries for the English and National spellings in the Scopus database.

The most-prolific institutions for the sample dataset are Lund University, State University of Campinas, University of Sao Paulo, and USDA Agricultural Research Service. Further, the most prolific countries for the sample dataset are Brazil, the USA, and to a lesser extent Sweden and India. These findings confirm the dominance of Brazil, the USA, and to a lesser extent the Sweden and India in this field.

On the other hand, the primary research front is the bioethanol fuels from milling industrial wastes, while the other research fronts are the bioethanol fuels from the ethanol, wood and paper, fruit, and brewing industry.

It is also notable that there is a significant gender deficit for the sample dataset as surprisingly with a representation rate of 14%. This finding is the most thought-provoking with strong public policy implications. Hence, institutions, funding bodies, and policymakers should take efficient measures to reduce the gender deficit in this field as well as other scientific fields with strong gender deficit. In this context, it is worth to note the level of representation of the researchers from the minority groups in science on the basis of race, sexuality, age, and disability, besides the gender (Blankenship, 1993; Dirth and Branscombe, 2017; Konur, 2000, 2002a–c, 2006a,b, 2007a,b).

67.4.4 The Most-Prolific Research Output by Years in the Second Generation Industrial Waste-based Bioethanol Fuels

The research output observed between 1970 and 2022 is illustrated in Figure 67.1. This figure clearly shows that the bulk of the research papers in the population dataset were published primarily in the 2010s and early 2020s. Similarly, the bulk of the research papers in the sample dataset were published in the 2010s and 2000s.

There was a rising trend for the population papers between 2006 and 2014, but it lost its momentum after 2014. However, there was a sharp rise in the research output between 2020 and 2021 possibly due to the supply shocks.

These findings suggest that the most-prolific sample and population papers were primarily published in the 2010s. Further, a significant portion of the sample and population papers were published in the early 2020s and 2000s, respectively.

These are the thought-provoking findings as there has been a significant research boom since 2007. In this context, the increasing public concerns about climate change (Change, 2007), greenhouse gas emissions (Carlson et al., 2017), and global warming (Kerr, 2007) have been certainly behind the boom in the research in this field in the last two decades. Furthermore, the recent supply shocks experienced due to the COVID-19 pandemics and the Ukrainian war might also be behind the research boom in this field since 2019.

Based on these findings, the size of the population papers is likely to more than double in the current decade, provided that the public concerns about climate change, greenhouse gas emissions, and global warming, as well as the supply shocks are translated efficiently to the research funding in this field.

67.4.5 The Most-Prolific Institutions in the Second Generation Industrial Waste-based Bioethanol Fuels

The most-prolific 15 institutions publishing papers on the second generation industrial waste-based bioethanol fuels with at least 2% of the sample papers each given in Table 67.3 have shaped the development of the research in this field.

The most-prolific institutions are the University of Sao Paulo and to a lesser extent National Biorenewables Laboratory, USDA Agricultural Research Service, Lund University, University of Matanzas, State University of Campinas, and Purdue University. Similarly, the top countries for these most-prolific institutions are the USA and Brazil. In total, only seven countries house these top institutions.

On the other hand, the institutions with the most citation impact are the University of Sao Paulo and to a lesser extent Lund University, USDA Agricultural Research Service, University of Matanzas, National Biorenewables Laboratory, and Purdue University.

These findings confirm the dominance of the institutions from the USA, Brazil, and to a lesser extent from Sweden in this field.

67.4.6 The Most-Prolific Funding Bodies in the Second Generation Industrial Waste-based Bioethanol Fuels

The most-prolific 11 funding bodies funding at least 2% of the sample papers each is given in Table 67.4. It is notable that only 48% and 47% of the sample and population papers were funded, respectively.

The most-prolific funding bodies are the National Council for Scientific and Technological Development, and to a lesser extent Research Support Foundation of the State of Sao Paulo, Higher Education Personnel Improvement Coordination, Ministry of Science, Technology and Innovation, National Natural Science Foundation of China, and the U.S. Department of Energy. Further, the most-prolific countries for these top funding bodies are Brazil and to a lesser extent China and the USA. In total, only four countries and the EU house these top funding bodies.

The funding bodies with the most citation impact are the National Council for Scientific and Technological Development and to a lesser extent the Research Support Foundation of the State of Sao Paulo and the U.S. Department of Energy.

These findings on the funding of the research in this field suggest that the level of the funding, mostly since 2007, has been largely instrumental in enhancing the research in this field (Ebadi and Schiffauerova, 2016) in the light of North's institutional framework (North, 1991). It is also notable that the funding rate in this field is relatively modest compared to those in the other research fronts of the bioethanol fuels such as algal bioethanol fuels. Further, it is expected that this high funding rate would improve in the light of the recent supply shocks. Further, it emerges that China, the USA, Brazil, and Europe have heavily funded the research on the industrial waste-based bioethanol fuels.

67.4.7 The Most-Prolific Source Titles in the Second Generation Industrial Waste-based Bioethanol Fuels

The most-prolific 16 source titles publishing at least 1.2% of the sample papers each in the second generation industrial waste-based bioethanol fuels have shaped the development of the research in this field (Table 67.5).

The most-prolific source titles are the Bioresource Technology and to a lesser extent Biomass and Bioenergy, Enzyme and Microbial Technology, Biotechnology and Bioengineering, Applied Biochemistry and Biotechnology Part A Enzyme Engineering and Biotechnology, and Industrial Crops and Products. On the other hand, the source titles with the most citation impact are the Bioresource Technology and to a lesser extent Enzyme and Microbial Technology, Biotechnology and Bioengineering, Applied Biochemistry and Biotechnology Part A Enzyme Engineering and Biotechnology, and Biomass and Bioenergy.

It is notable that these top source titles are primarily related to the bioresources, biotechnology, and energy. This finding suggests that Bioresource Technology and the other prolific journals in these fields have significantly shaped the development of the research in this field as they focus primarily on the second generation industrial waste-based bioethanol fuels with a high yield. In this context, the influence of the top journal is quite extraordinary.

67.4.8 The Most-Prolific Countries in the Second Generation Industrial Waste-based Bioethanol Fuels

The most-prolific 15 countries publishing at least 2.4% of the sample papers each have significantly shaped the development of the research in this field (Table 67.6).

The most-prolific countries are the USA, Brazil, and to a lesser extent China, India, Spain, Sweden, and Japan. Further, four European countries listed in Table 67.6 produce 17% and 8% of the sample and population papers, respectively.

On the other hand, the countries with the most citation impact are the USA, Brazil, and to a lesser extent Sweden, Cuba, Spain, and Greece. Similarly, the countries with the least impact are China and to a lesser extent India and Malaysia.

The close examination of these findings suggests that the USA, Brazil, Europe, and to a lesser extent China, India, Japan, and Cuba are the major producers of the research in this field. It is a fact that the USA has been a major player in science (Leydesdorff and Wagner, 2009). The USA has further developed a strong research infrastructure to support its corn- and grass-based bioethanol industry (Gillon, 2010).

However, China has been a rising mega star in scientific research in competition with the USA and Europe (Leydesdorff and Zhou, 2005). China is also a major player in this field as a major producer of bioethanol (Fang et al., 2010).

Next, Europe has been a persistent player in the scientific research in competition with both the USA and China (Leydesdorff, 2000). Europe has also been a persistent producer of bioethanol along with the USA and Brazil (Gnansounou, 2010).

Further, Brazil (Gnansounou, 2010; Soccol et al., 2010), Japan (Fukuzawa and Ida, 2016), India (Basu and Kumar, 2000), Malaysia (Kumar and Jan, 2014), and Cuba (Arencibia-Jorge and de Moya-Anegon, 2010) are the other countries with substantial research activities in bioethanol fuels.

67.4.9 The Most-Prolific Scopus Subject Categories in the Second Generation Industrial Waste-based Bioethanol Fuels

The most-prolific eight Scopus subject categories indexing at least 7.6% of the sample papers each, given in Table 67.7, have shaped the development of the research in this field.

The most-prolific Scopus subject categories in the second generation industrial waste-based bioethanol fuels are Chemical Engineering, Environmental Science, Energy, and to a lesser extent Biochemistry, Genetics and Molecular Biology, and Immunology and Microbiology. It is also notable that Social Sciences including Economics and Business have a minimal presence in both sample and population studies.

On the other hand, the Scopus subject categories with the most citation impact are Chemical Engineering and to a lesser extent Energy and Environmental Science. Similarly, the least influential subject categories are Agricultural and Biological Sciences, Chemistry, and Engineering.

These findings are thought-provoking, suggesting that the primary subject categories are related to energy, chemical engineering, and environmental sciences as the core of the research in this field concerns with the production and utilization of the second generation industrial waste-based bioethanol fuels. The other finding is that social sciences are not well represented in both the sample and population papers as in line with the most fields in bioethanol fuels. The social, environmental, and economics studies account for the field of social sciences.

67.4.10 The Most-Prolific Keywords in the Second Generation Industrial Waste-based Bioethanol Fuels

A limited number of keywords have shaped the development of the research in this field as shown in Table 67.8 and the appendix. These keywords are grouped under five headings: industrial waste biomass, pretreatments, fermentation, hydrolysis and hydrolysates, and products.

The most-prolific keywords across all the research fronts are hydrolysis, ethanol, celluloses, sugars, fermentation, bagasse, lignins, enzyme activity, and enzymatic hydrolysis. Similarly, the most influential keywords are cellulose, sugars, hydrolysis, ethanol, enzyme activity, lignins, sulfuric acids, bagasse, and enzymatic hydrolysis.

These findings suggest that it is necessary to determine the keyword set carefully to locate the relevant research in each of these research fronts. Additionally, the size of the samples for each keyword highlights the intensity of the research in the relevant research areas for both sample and population datasets. These findings also highlight different spellings of some strategic keywords such as pretreatment v. pre-treatment and ethanol v. bio-ethanol, etc.

67.4.11 The Most-Prolific Research Fronts in the Second Generation Industrial Waste-based Bioethanol Fuels

Information about the research fronts for the sample papers in the second generation industrial waste-based bioethanol fuels with regard to the industrial waste feedstocks used for the bioethanol production is given in Table 67.9. As this table shows, the most-prolific industrial waste feedstock is the milling industrial wastes. The other prolific feedstocks are the food processing, wood and paper, and other industrial wastes.

The other industrial wastes include biodiesel, ethanol, brewing, plastic, textile, and leather industrial wastes. Further, the primary milling industrial wastes are sugarcane bagasse and other milling wastes, while primary wood and paper industrial wastes are wood chips, sawdust, and paper sludge. Similarly, the primary food industrial wastes are the fruit, beverage, dairy, and vegetable wastes.

Information about the thematic research fronts for the sample papers in the second generation industrial waste-based bioethanol fuels is given in Table 67.10. As this table shows, the most prolific research fronts are the pretreatments and hydrolysis of the industrial waste feedstock, while the other prolific research fronts are the bioethanol fuel production and hydrolysate fermentation. The minor research front is bioethanol fuel evaluation and utilization.

These findings are thought-provoking in seeking ways to increase industrial waste feedstock-based bioethanol yield at the global scale. It is clear that all of these research fronts have public importance and merit substantial funding and other incentives. Further, it is notable that second generation industrial waste-based bioethanol fuels have become a core unit of the bioethanol research to make it more competitive with the crude oil-based gasoline and diesel fuels, especially for the USA, Europe, Brazil, and China, the major producers and consumers of the bioethanol fuels.

In comparison to the other feedstock-based research fronts, it is notable that the pretreatment and hydrolysis of the industrial wastes emerge as primary research fronts for this field. This suggests that the primary stakeholders have been primarily interested in these key processes of the bioethanol production. It is also notable that evaluation of the second generation industrial waste-based bioethanol fuels such as technoeconomics, life cycle, economics, social, land use, labor, and environment-related studies emerges as a case study for the bioethanol fuels. In this context, the USA has been the global leader in the production and use of the corn-based bioethanol fuels since the 1970s in the aftermath of the global crude oil crisis in the early 1970s.

It is further notable that the research on the industrial waste-based bioethanol fuels complements the research on the first generation bioethanol fuels, extracting further value from the first generation feedstocks.

In the end, these most-cited papers in this field hint that the production of second generation industrial waste-based bioethanol fuels could be optimized using the structure, processing, and property relationships of these industrial wastes in the fronts of the feedstock pretreatment and hydrolysis and hydrolysate fermentation (Formela et al., 2016; Konur, 2018a, 2020b, 2021a–d; Konur and Matthews, 1989).

67.5 CONCLUSION AND FUTURE RESEARCH

The research on the second generation industrial waste-based bioethanol fuels has been mapped through a scientometric study of both sample (250 papers) and population (5,000 papers) datasets.

The critical issue in this study has been to obtain a representative sample of the research as in any other scientometric study. Therefore, the keyword set has been carefully devised and optimized after a number of runs in the Scopus database. It is a representative sample of the wider population studies. This keyword set was provided in the appendix, and the relevant keywords are presented in Table 67.8. However, it should be noted that it has been very difficult to compile a representative keyword set since this research field has been connected closely with many other fields. Therefore, it has been necessary to compile a keyword list to exclude papers concerned with the other research fields.

The other issue has been the selection of a multidisciplinary database to carry out the scientometric study of the research in this field. For this purpose, Scopus database has been selected. The journal coverage of this database has been notably wider than that of Web of Science and other multisubject databases.

The key scientometric properties of the research in this field have been determined and discussed in this book chapter. It is evident that a limited number of documents, authors, institutions, publication years, institutions, funding bodies, source titles, countries, Scopus subject categories, Scopus keywords, and research fronts have shaped the development of the research in this field.

There is ample scope to increase the efficiency of the scientometric studies in this field in the author and document domains by developing consistent policies and practices in both domains across all the academic databases. In this respect, it seems that authors, journals, and academic databases have a lot to do. Furthermore, the significant gender deficit as in most scientific fields emerges as a public policy issue. The potential deficits on the basis of age, race, disability, and sexuality need also to be explored in this field as in other scientific fields.

The research in this field has boomed since 2007, possibly promoted by the public concerns on global warming, greenhouse gas emissions, and climate change. Furthermore, the recent COVID-19 pandemics and Russian invasion of Ukraine have resulted in a global supply shocks shifting the focus of the stakeholders from the crude oil-based fuels to biomass-based fuels such as bioethanol fuels. It is expected that there would be further incentives for the key stakeholders to carry out the research for the second generation industrial waste-based bioethanol fuels to increase the ethanol yield and to make it more competitive with the crude oil-based gasoline and petrodiesel fuels. This might be truer for the crude oil- and foreign exchange-deficient countries to maintain the energy and food security at the face of the global supply shocks.

The relatively modest funding rate of 48% and 47% for the sample and population papers, respectively, suggests that funding in this field significantly enhanced the research in this field primarily since 2007, possibly more than doubling in the current decade. However, it is evident that there is ample room for more funding and other incentives to enhance the research in this field further.

The institutions from the USA and to a lesser extent Spain and China have mostly shaped the research in this field. Further, the USA, Europe, and Brazil have been the major producers of the research in this field as the major producers and users of bioethanol fuels. It is evident that these countries have well-developed research infrastructure in bioethanol fuels and their derivatives.

It emerges that ethanol is more popular than bioethanol as a keyword with strong implications for the search strategy. In other words, the search strategy using only bioethanol keyword would not be much helpful. The Scopus keywords are grouped under the five headings: biomass, pretreatments, fermentation, hydrolysis and hydrolysates, and products. It is also important to include the Latin terms for the industrial waste biomass in the keyword set.

As Table 67.9 shows, the most-prolific industrial waste feedstocks are milling industrial wastes and to a lesser extent the food processing, wood and paper, and other industry wastes. The other industry wastes include biodiesel, ethanol, brewing, plastic, textile, and leather industrial wastes. Further, the primary milling industrial wastes are sugarcane bagasse and other milling wastes, while the primary wood and paper industrial wastes are wood chips, sawdust, and paper sludge. Similarly, the primary food industrial wastes are the fruit, beverage, dairy, and vegetable industrial wastes.

On the other hand, Table 67.10 shows that the most-prolific thematic research fronts are the pretreatments and hydrolysis of the industrial waste feedstocks and to a lesser extent bioethanol fuel production, hydrolysate fermentation, and bioethanol fuel evaluation and utilization. The first four research fronts dominate the research in this field.

These findings are thought-provoking in seeking ways to increase industrial waste feedstock-based bioethanol yield at the global scale. It is clear that all of these research fronts have public importance and merit substantial funding and other incentives. Further, it is notable that second generation industrial waste-based bioethanol fuels have become a core unit of the bioethanol research to make it more competitive with the crude oil-based gasoline and petrodiesel fuels, especially for the USA, Europe, and China. It is further notable that the research on the industrial waste-based bioethanol complements the research on the first generation bioethanol fuels, extracting further value from the first generation feedstocks.

Thus, the scientometric analysis has a great potential to gain valuable insights into the evolution of the research in this field as in other scientific fields especially in the aftermath of the significant global supply shocks such as COVID-19 pandemics and the Russian invasion of Ukraine.

It is recommended that further scientometric studies are carried out for the primary research fronts. It is further recommended that reviews of the most-cited papers are carried out for each primary research front to complement these scientometric studies. Next, the scientometric studies of the hot papers in these primary fields are carried out.

ACKNOWLEDGMENTS

The contribution of the highly cited researchers in the field of the second generation industrial waste-based bioethanol fuels has been gratefully acknowledged.

APPENDIX: THE KEYWORD SET FOR SECOND GENERATION INDUSTRIAL WASTE-BASED BIOETHANOL FUELS

((((((TITLE (industr* OR *paper OR textile* OR *boards OR boxes OR tire OR tyre OR processing OR plastic* OR pet OR coffee OR terephthalate OR leather OR wool OR tannery OR polycarbonate OR polyurethane OR distillery OR brewery OR rice OR wheat OR *nut OR cheese OR potato OR grape OR cannery OR citrus OR kinnow OR banana OR bread OR tea OR tuna OR vegetable OR mandarin OR lemon OR *apple OR pear OR mango OR orange OR cassava OR almond OR fish OR sago OR poultry OR taro OR salad OR date OR dairy OR *bean OR pomegranate OR cocoa OR kiwi* OR egg* OR bergamot OR herb* OR *wood OR woody OR eucalyptus OR pine OR pinus OR aspen* OR populus OR cedar OR spruce OR beech OR oak OR willow* OR salix OR cypress OR prosopis OR birch OR bamboo* OR maple* OR fir OR sorghum OR rye OR barley OR corn OR cassava OR maize OR *cane OR *beet OR agave) AND TITLE (waste OR residue OR whey OR *bran OR spent OR hulls OR husks OR chips)) OR TITLE (peel* OR "saw dust" OR sawdust OR bagasse OR "spent grain" OR "distiller's grains" OR ddgs OR "corn fib*" OR molasses OR "paper sludge" OR glycerol OR glycerin* OR pomace* OR marc OR "fruit bunch*" OR "olive stone*" OR {ethanol residue*})) AND TITLE (ethanol OR bioethanol OR *saccharification OR *hydrolysis OR digestibili* OR ssf OR hydrolysate* OR hydrolyzate* OR ferment* OR fractionation OR delignification OR pretreat* OR "pre-treat*" OR "ionic liquid*" OR "consolidated processing")) AND NOT (TITLE (dyes OR {glycerol *production} OR "activated carbon" OR *hydrogen OR reforming OR *diesel OR "solid state" OR anaerobic OR docosahexaenoic OR carboxylic OR (glycerol AND ferment*) OR h2 OR pyrolysis OR *lipid OR *butanol OR coumaric OR lactic OR *polyester OR xylitol OR "cellulose production" OR {h 2} OR p3hb OR *amylase OR rat OR levul* OR "waste water" OR *capacitor OR pectinase OR lipase OR *butyrate OR *butyric OR citric OR "bio-oil" OR "product recovery" OR biogas OR ether* OR *sorption OR "cellulase prod*" OR hydrogenolysis OR "solid substrate" OR aerogel* OR leach* OR ester* OR *oligosaccharide OR succin* OR *alkanoate OR discharge OR nano* OR furfural OR "am(iii)" OR propanediol OR keratin OR wines OR human OR erythritol OR fumaric OR {rice straw} OR methan* OR biomethan* OR protein* OR {wheat straw} OR {wood resid*} OR lignac* OR "corn stover" OR arsenic OR {bamboo residues} OR {corn residues} OR {forest residues} OR "cell oil*" OR meal OR {agricultur* waste} OR wastewater OR {*cellulosic residues} OR {grandis residues} OR "fatty acid*" OR silica OR dehydrogenation OR struvite OR binders OR *alkanoates OR *amylose OR pullulan OR anti* OR amino OR {*cane residues} OR stem OR endoglucanase OR bromelain OR nutr* OR beverage OR {enzyme prod*} OR prunings OR "dry grind" OR stillage OR recalcitrant OR vinasse OR entrainer OR house OR *ether OR levan OR jar OR soil) OR SRCTITLE (cement* OR food* OR materials OR lwt OR animal* OR water OR fluids OR brew* OR chemo* OR dairy OR hydrogen OR chromat* OR genom* OR poult* OR carbohyd* OR cereal* OR meat OR aqua* OR powder) OR SUBJAREA (medi OR phar OR nurs OR eart OR vete OR heal OR neur))))) OR ((TITLE (ethanol OR bioethanol OR hydrolysis) AND TITLE (bagasse OR molasses OR "distiller's grains" OR whey OR *bran OR "paper sludge" OR hulls OR husks OR glycerol OR pomace OR "corn fib*" OR chips OR marc OR peel* OR "saw dust" OR sawdust OR "spent grain" OR glycerin* OR "fruit bunch*" OR "olive stone*" OR "spent coffee" OR {ethanol resid*})) AND NOT TITLE ("fuel cells" OR reforming OR biosorption OR furfural OR precipitation OR h2 OR methanol OR "corn stover" OR (glycerol AND ferment*) OR "waste gases" OR "h 2" OR solvent* OR corncob OR rats OR anti* OR soil* OR munic* OR nanowh* OR dehydrogenation OR supercritical OR ether* OR wine* OR sorption OR oxid* OR adhesive OR silk OR dehydrate OR receptor* OR fat OR liquefaction OR *cyanins OR levan OR kitchen OR harvest OR assay* OR syngas OR entrainer OR xylinum OR formamide OR protein* OR subcritical OR ferulic OR ester OR lactic OR levulinic OR oleochem* OR xylitol OR amylases OR {glycerol pretreatment} OR nanofib* OR silica OR carotene OR cereal* OR nanocry* OR {agric* waste} OR {*cellulosic resid*} OR "wheat straw")) AND (LIMIT-TO (SRCTYPE, "j") OR LIMIT-TO (SRCTYPE, "k") OR LIMIT-TO (SRCTYPE, "b")) AND (LIMIT-TO (DOCTYPE, "ar") OR LIMIT-TO (DOCTYPE, "cp") OR LIMIT-TO (DOCTYPE, "re") OR LIMIT-TO (DOCTYPE, "ch") OR LIMIT-TO (DOCTYPE, "no") OR LIMIT-TO (DOCTYPE, "le") OR LIMIT-TO (DOCTYPE, "sh")) AND (LIMIT-TO (LANGUAGE, "English")).

REFERENCES

Aman, V. 2018. Does the Scopus author ID suffice to track scientific international mobility? A case study based on Leibniz laureates. *Scientometrics* 117:705–720.

Angelici, C., B. M. Weckhuysen and P. C. A. Bruijnincx. 2013. Chemocatalytic conversion of ethanol into butadiene and other bulk chemicals. *ChemSusChem* 6:1595–1614.

Antolini, E. 2007. Catalysts for direct ethanol fuel cells. *Journal of Power Sources* 170:1–12.

Antolini, E. 2009. Palladium in fuel cell catalysis. *Energy and Environmental Science* 2:915–931.

Arencibia-Jorge, R. and F. de Moya-Anegon. 2010. Challenges in the study of Cuban scientific output. *Scientometrics* 83:723–737.

Basu, A. and B. V. Kumar. 2000. International collaboration in Indian scientific papers. *Scientometrics* 48:381–402.

Beaudry, C. and V. Lariviere. 2016. Which gender gap? Factors affecting researchers' scientific impact in science and medicine. *Research Policy* 45:1790–1817.

Blankenship, K. M. 1993. Bringing gender and race in: US employment discrimination policy. *Gender & Society* 7:204–226.

Brandt, A., J. P. Hallett, D. J. Leak, R. J. Murphy and T. Welton. 2010. The effect of the ionic liquid anion in the pretreatment of pine wood chips. *Green Chemistry* 12:672–679.

Burnham, J. F. 2006. Scopus database: A review. *Biomedical Digital Libraries* 3:1–8.

Cardona, C. A., J. A. Quintero and I. C. Paz. 2010. Production of bioethanol from sugarcane bagasse: Status and perspectives. *Bioresource Technology* 101:4754–4766.

Carlson, K. M., J. S. Gerber and D. Mueller, et al. 2017. Greenhouse gas emissions intensity of global croplands. *Nature Climate Change* 7:63–68.

Change, C. 2007. Climate change impacts, adaptation and vulnerability. *Science of the Total Environment* 326:95–112.

Dias, M. O. S., T. L. Junqueira and O. Cavalett, et al. 2012. Integrated versus stand-alone second generation ethanol production from sugarcane bagasse and trash. *Bioresource Technology* 103:152–161.

Dirth, T. P. and N. R. Branscombe. 2017. Disability models affect disability policy support through awareness of structural discrimination. *Journal of Social Issues* 73:413–442.

Duff, S. J. B. and W. D. Murray. 1996. Bioconversion of forest products industry waste cellulosics to fuel ethanol: A review. *Bioresource Technology* 55:1–33.

Ebadi, A. and A. Schiffauerova. 2016. How to boost scientific production? A statistical analysis of research funding and other influencing factors. *Scientometrics* 106:1093–1116.

Fan, Z., C. South and K. Lyford, et al. 2003. Conversion of paper sludge to ethanol in a semicontinuous solids-fed reactor. *Bioprocess and Biosystems Engineering* 26:93–101.

Fang, X., Y. Shen, J. Zhao, X. Bao and Y. Qu. 2010. Status and prospect of lignocellulosic bioethanol production in China. *Bioresource Technology* 101:4814–4819.

Fauci, A. S., H. C. Lane and R. R. Redfield. 2020. Covid-19-navigating the uncharted. *New England Journal of Medicine* 382:1268–1269.

Fernando, S., S. Adhikari, C. Chandrapal and M. Murali. 2006. Biorefineries: Current status, challenges, and future direction. *Energy & Fuels* 20:1727–1737.

Formela, K., A. Hejna, L. Piszczyk, M. R. Saeb and X. Colom. 2016. Processing and structure-property relationships of natural rubber/wheat bran biocomposites. *Cellulose* 23:3157–3175.

Fukuzawa, N. and T. Ida. 2016. Science linkages between scientific articles and patents for leading scientists in the life and medical sciences field: The case of Japan. *Scientometrics* 106:629–644.

Garfield, E. 1955. Citation indexes for science. *Science* 122:108–111.

Gillon, S. 2010. Fields of dreams: Negotiating an ethanol agenda in the Midwest United States. *Journal of Peasant Studies* 37:723–748.

Gnansounou, E. 2010. Production and use of lignocellulosic bioethanol in Europe: Current situation and perspectives. *Bioresource Technology* 101:4842–4850.

Grohmann, K., R. G. Cameron and B. S. Buslig. 1995. Fractionation and pretreatment of orange peel by dilute acid hydrolysis. *Bioresource Technology* 54:129–141.

Hahn-Hagerdal, B., M. Galbe, M. F. Gorwa-Grauslund, G. Liden and G. Zacchi. 2006. Bio-ethanol: The fuel of tomorrow from the residues of today. *Trends in Biotechnology* 24:549–556.

Hamilton, J. D. 1983. Oil and the macroeconomy since World War II. *Journal of Political Economy* 91:228–248.

Hamilton, J. D. 2003. What is an oil shock? *Journal of Econometrics* 113:363–398.

Hill, J., E. Nelson, D. Tilman, S. Polasky and D. Tiffany. 2006. Environmental, economic, and energetic costs and benefits of biodiesel and ethanol biofuels. *Proceedings of the National Academy of Sciences of the United States of America* 103:11206–11210.

Hill, J., S. Polasky and E. Nelson, et al. 2009. Climate change and health costs of air emissions from biofuels and gasoline. *Proceedings of the National Academy of Sciences of the United States of America* 106:2077–2082.

Hsieh, W. D., R. H. Chen, T. L. Wu and T. H. Lin. 2002. Engine performance and pollutant emission of an SI engine using ethanol-gasoline blended fuels. *Atmospheric Environment* 36:403–410.

Huang, H. J., S. Ramaswamy, U. W. Tschirner and B. V. Ramarao. 2008. A review of separation technologies in current and future biorefineries. *Separation and Purification Technology* 62:1–21.

Ito, T., Y. Nakashimada, K. Senba, T. Matsui and N. Nishio. 2005. Hydrogen and ethanol production from glycerol-containing wastes discharged after biodiesel manufacturing process. *Journal of Bioscience and Bioengineering* 100:260–265.

Jones, T. C. 2012. America, oil, and war in the Middle East. *Journal of American History* 99:208–218.

Kadar, Z., Z. Szengyel and K. Reczey. 2004. Simultaneous saccharification and fermentation (SSF) of industrial wastes for the production of ethanol. *Industrial Crops and Products* 20:103–110.

Kerr, R. A. 2007. Global warming is changing the world. *Science* 316:188–190.

Kilian, L. 2008. Exogenous oil supply shocks: How big are they and how much do they matter for the US economy? *Review of Economics and Statistics* 90:216–240.

Kilian, L. 2009. Not all oil price shocks are alike: Disentangling demand and supply shocks in the crude oil market. *American Economic Review* 99:1053–1069.

Kim, J. 2018. Evaluating author name disambiguation for digital libraries: A case of DBLP. *Scientometrics* 116:1867–1886.

Konur, O. 2000. Creating enforceable civil rights for disabled students in higher education: An institutional theory perspective. *Disability & Society* 15:1041–1063.

Konur, O. 2002a. Access to nursing education by disabled students: Rights and duties of nursing programs. *Nurse Education Today* 22:364–374.

Konur, O. 2002b. Assessment of disabled students in higher education: Current public policy issues. *Assessment and Evaluation in Higher Education* 27:131–152.

Konur, O. 2002c. Access to employment by disabled people in the UK: Is the Disability Discrimination Act working? *International Journal of Discrimination and the Law* 5:247–279.

Konur, O. 2006a. Participation of children with dyslexia in compulsory education: Current public policy issues. *Dyslexia* 12:51–67.

Konur, O. 2006b. Teaching disabled students in higher education. *Teaching in Higher Education* 11:351–363.

Konur, O. 2007a. A judicial outcome analysis of the Disability Discrimination Act: A windfall for the employers? *Disability & Society* 22:187–204.

Konur, O. 2007b. Computer-assisted teaching and assessment of disabled students in higher education: The interface between academic standards and disability rights. *Journal of Computer Assisted Learning* 23:207–219.

Konur, O. 2011. The scientometric evaluation of the research on the algae and bio-energy. *Applied Energy* 88:3532–3540.

Konur, O. 2012a. The evaluation of the biogas research: A scientometric approach. *Energy Education Science and Technology Part A: Energy Science and Research* 29:1277–1292.

Konur, O. 2012b. The evaluation of the educational research: A scientometric approach. *Energy Education Science and Technology Part B: Social and Educational Studies* 4:1935–1948.

Konur, O. 2012c. The evaluation of the global energy and fuels research: A scientometric approach. *Energy Education Science and Technology Part A: Energy Science and Research* 30:613–628.

Konur, O. 2012d. The evaluation of the research on the biodiesel: A scientometric approach. *Energy Education Science and Technology Part A: Energy Science and Research* 28:1003–1014.

Konur, O. 2012e. The evaluation of the research on the bioethanol: A scientometric approach. *Energy Education Science and Technology Part A: Energy Science and Research* 28:1051–1064.

Konur, O. 2012f. The evaluation of the research on the biofuels: A scientometric approach. *Energy Education Science and Technology Part A: Energy Science and Research* 28:903–916.

Konur, O. 2012g. The evaluation of the research on the biohydrogen: A scientometric approach. *Energy Education Science and Technology Part A: Energy Science and Research* 29:323–338.

Konur, O. 2012h. The evaluation of the research on the microbial fuel cells: A scientometric approach. *Energy Education Science and Technology Part A: Energy Science and Research* 29:309–322.

Konur, O. 2012i. The scientometric evaluation of the research on the production of bioenergy from biomass. *Biomass and Bioenergy* 47:504–515.

Konur, O. 2015. Current state of research on algal bioethanol. In *Marine Bioenergy: Trends and Developments*, Eds. S. K. Kim and C. G. Lee, pp. 217–244. Boca Raton, FL: CRC Press.

Konur, O., Ed. 2018a. *Bioenergy and Biofuels*. Boca Raton, FL: CRC Press.

Konur, O. 2018b. Bioenergy and biofuels science and technology: Scientometric overview and citation classics. In *Bioenergy and Biofuels*, Ed. O. Konur, pp. 3–63. Boca Raton: CRC Press.

Konur, O. 2019. Cyanobacterial bioenergy and biofuels science and technology: A scientometric overview. In *Cyanobacteria: From Basic Science to Applications*, Eds. A. K. Mishra, D. N. Tiwari and A. N. Rai, pp. 419–442. Amsterdam: Elsevier.

Konur, O. 2020a. The scientometric analysis of the research on the bioethanol production from green macroalgae. In *Handbook of Algal Science, Technology and Medicine,* Ed. O. Konur, pp. 385–401. London: Academic Press.

Konur, O., Ed. 2020b. *Handbook of Algal Science, Technology and Medicine.* London: Academic Press.

Konur, O., Ed. 2021a. *Handbook of Biodiesel and Petrodiesel Fuels: Science, Technology, Health, and Environment*. Boca Raton, FL: CRC Press.

Konur, O., Ed. 2021b. *Handbook of Biodiesel and Petrodiesel Fuels: Science, Technology, Health, and Environment. Volume 1. Biodiesel Fuels: Science, Technology, Health, and Environment.* Boca Raton, FL: CRC Press.

Konur, O., Ed. 2021c. *Handbook of Biodiesel and Petrodiesel Fuels: Science, Technology, Health, and Environment. Volume 2. Biodiesel Fuels based on the Edible and Nonedible Feedstocks, Wastes, and Algae: Science, Technology, Health, and Environment.* Boca Raton, FL: CRC Press.

Konur, O., Ed. 2021d. *Handbook of Biodiesel and Petrodiesel Fuels: Science, Technology, Health, and Environment. Volume 3. Petrodiesel Fuels: Science, Technology, Health, and Environment.* Boca Raton, FL: CRC Press.

Konur, O. and F. L. Matthews. 1989. Effect of the properties of the constituents on the fatigue performance of composites: A review. *Composites* 20:317–328.

Koutinas, A. A., A. Vlysidis and D. Pleissner, et al. 2014. Valorization of industrial waste and by-product streams via fermentation for the production of chemicals and biopolymers. *Chemical Society Reviews* 43:2587–2627.

Kruyt, B., D. P. van Vuuren, H. J. de Vries and H. Groenenberg. 2009. Indicators for energy security. *Energy Policy* 37:2166–2181.

Kumar, S. and J. M. Jan. 2014. Research collaboration networks of two OIC nations: Comparative study between Turkey and Malaysia in the field of 'Energy Fuels', 2009-2011. *Scientometrics* 98:387–414.

Laser, M., D. Schulman and S G. Allen, et al. 2002. A comparison of liquid hot water and steam pretreatments of sugar cane bagasse for bioconversion to ethanol. *Bioresource Technology* 81:33–44.

Lavarack, B. P., G. J. Griffin and D. Rodman. 2002. The acid hydrolysis of sugarcane bagasse hemicellulose to produce xylose, arabinose, glucose and other products. *Biomass and Bioenergy* 23:367–380.

Leydesdorff, L. 2000. Is the European Union becoming a single publication system? *Scientometrics* 47:265–280.

Leydesdorff, L. and C. Wagner. 2009. Is the United States losing ground in science? A global perspective on the world science system. *Scientometrics* 78:23–36.

Leydesdorff, L. and P. Zhou. 2005. Are the contributions of China and Korea upsetting the world system of science? *Scientometrics* 63:617–630.

Li, H., S. M. Liu, X. H. Yu, S. L. Tang and C. K. Tang. 2020. Coronavirus disease 2019 (COVID-19): Current status and future perspectives. *International Journal of Antimicrobial Agents* 55:105951.

Lin, Y. and S. Tanaka. 2006. Ethanol fermentation from biomass resources: Current state and prospects. *Applied Microbiology and Biotechnology* 69:627–642.

Ma, X., L. Sun and C. Song. 2002. A new approach to deep desulfurization of gasoline, diesel fuel and jet fuel by selective adsorption for ultra-clean fuels and for fuel cell applications. *Catalysis Today* 77:107–116.

MacRelli, S., J. Mogensen and G. Zacchi. 2012. Techno-economic evaluation of 2nd generation bioethanol production from sugar cane bagasse and leaves integrated with the sugar-based ethanol process. *Biotechnology for Biofuels* 5:22.

Martín, C., M. Galbe, C. F. Wahlbom, B. Hahn-Hagerdal and L. J. Jonsson. 2002. Ethanol production from enzymatic hydrolysates of sugarcane bagasse using recombinant xylose-utilising *Saccharomyces cerevisiae*. *Enzyme and Microbial Technology* 31:274–282.

Morschbacker, A. 2009. Bio-ethanol based ethylene. *Polymer Reviews* 49:79–84.

Najafi, G., B. Ghobadian and T. Tavakoli, et al. 2009. Performance and exhaust emissions of a gasoline engine with ethanol blended gasoline fuels using artificial neural network. *Applied Energy* 86:630–639.

Newman, P. W. G. and J. R. Kenworthy. 1989. Gasoline consumption and cities: A comparison of U.S. cities with a global survey. *Journal of the American Planning Association* 55:24–37.

North, D. C. 1991. Institutions. *Journal of Economic Perspectives* 5:97–112.

Olsson, L. and B. Hahn-Hagerdal. 1996. Fermentation of lignocellulosic hydrolysates for ethanol production. *Enzyme and Microbial Technology* 18:312–331.

Ravindran, R. and A. K. Jaiswal. 2016. A comprehensive review on pre-treatment strategy for lignocellulosic food industry waste: Challenges and opportunities. *Bioresource Technology* 199:92–102.

Reeves, S. 2014. To Russia with love: How moral arguments for a humanitarian intervention in Syria opened the door for an invasion of the Ukraine. *Michigan State University International Law Review* 23:199.

Sanchez, O. J. and C. A. Cardona. 2008. Trends in biotechnological production of fuel ethanol from different feedstocks. *Bioresource Technology* 99:5270–5295.

Soccol, C. R., L. P. de Souza Vandenberghe and A. B. P. Medeiros, et al. 2010. Bioethanol from lignocelluloses: Status and perspectives in Brazil. *Bioresource Technology* 101:4820–4825.

Sun, Y. and J. Cheng. 2002. Hydrolysis of lignocellulosic materials for ethanol production: A review. *Bioresource Technology* 83:1–11.

Taherzadeh, M. J. and K. Karimi. 2007. Enzyme-based hydrolysis processes for ethanol from lignocellulosic materials: A review. *Bioresources* 2:707–738.

Taherzadeh, M. J. and K. Karimi. 2008. Pretreatment of lignocellulosic wastes to improve ethanol and biogas production: A review. *International Journal of Molecular Sciences* 9:1621–1651.

Winzer, C. 2012. Conceptualizing energy security. *Energy Policy* 46:36–48.

Yang, B. and C. E. Wyman. 2008. Pretreatment: The key to unlocking low-cost cellulosic ethanol. *Biofuels, Bioproducts and Biorefining* 2:26–40.

Yazdani, S. S. and R. Gonzalez. 2008. Engineering *Escherichia coli* for the efficient conversion of glycerol to ethanol and co-products. *Metabolic Engineering* 10:340–351.

68 Second Generation Industrial Waste-based Bioethanol Fuels

Review

Ozcan Konur
(Formerly) Ankara Yildirim Beyazit University

68.1 INTRODUCTION

Crude oil-based gasoline fuels (Ma et al., 2002; Newman and Kenworthy, 1989) have been widely used in the transportation sector since the 1920s. However, there have been great public concerns over the adverse environmental and human impact of these fuels (Hill et al., 2006, 2009). Hence, biomass-based bioethanol fuels (Hill et al., 2006; Konur, 2012, 2015, 2020) have increasingly been used in blending gasoline fuels (Hsieh et al., 2002; Najafi et al., 2009), in fuel cells (Antolini, 2007, 2009), and in biochemical production (Angelici et al., 2013; Morschbacker, 2009) in a biorefinery context (Fernando et al., 2006; Huang et al., 2008).

However, it is necessary to pretreat the biomass (Alvira et al., 2010; Taherzadeh and Karimi, 2008) to enhance the yield of the bioethanol (Hahn-Hagerdal et al., 2006; Sanchez and Cardona, 2008) before the bioethanol fuel production from the feedstocks through the hydrolysis (Sun and Cheng, 2002; Taherzadeh and Karimi, 2007) and fermentation (Lin and Tanaka, 2006; Olsson and Hahn-Hagerdal, 1996) of the biomass and hydrolysates, respectively.

One of the most-studied feedstocks for bioethanol fuels has been industrial wastes such as milling industrial wastes, food processing industrial wastes, wood industrial wastes, and other industrial wastes such as biodiesel industrial wastes. The research in the field of the second generation industrial waste-based bioethanol fuels has intensified in this context in the key research fronts of the pretreatment (Binod et al., 2012; Laser et al., 2002) and hydrolysis (Aguilar et al., 2002; Lavarack et al., 2002) of industrial wastes, fermentation (Kadar et al., 2004; Martin et al., 2002) of industrial waste-based hydrolysates, and production (Chandel et al., 2007; Ito et al., 2005) and evaluation (Dias et al., 2012; Rabelo et al., 2011) of the second generation industrial waste-based bioethanol fuels.

Furthermore, milling industrial wastes (Aguilar et al., 2002; Laser et al., 2002), food processing industrial wastes (Grohmann et al., 1995; Mussatto et al., 2012), wood and paper industrial wastes (Brandt et al., 2010; Fan et al., 2003), and other industrial wastes such as biodiesel industrial wastes (Ito et al., 2005; Yazdani and Gonzalez, 2008) have been studied intensively in this context.

However, it is essential to develop efficient incentive structures (North, 1991) for the primary stakeholders to enhance the research in this field (Konur, 2000, 2002a–c, 2006a,b, 2007a,b). Although there have been several review papers on industrial waste-based bioethanol fuels (Cardona et al., 2010; Duff and Murray, 1996; Koutinas et al., 2014), there has been no review of the 25 most-cited papers in this field.

Thus, this book chapter presents a review of the 25 most-cited articles in the field of industrial waste-based bioethanol fuels. Then, it discusses the key findings of these highly influential papers and comments on future research priorities in this field.

DOI: 10.1201/9781003226550-92

68.2 MATERIALS AND METHODS

The search for this study was carried out using the Scopus database (Burnham, 2006) in September 2022.

As a first step for the search of the relevant literature, the keywords were selected using the first 300 most-cited population papers. The selected keyword list was then optimized to obtain a representative sample of papers for the searched research field. This final keyword set was provided in the appendix of Konur (2023) for future replication studies.

As a second step, a sample dataset was used for this study. The first 25 articles with at least 212 citations each were selected for the review study. Key findings from each paper were taken from the abstracts of these papers and were discussed. Additionally, several brief conclusions were drawn and many relevant recommendations were made to enhance the future research landscape.

68.3 RESULTS

The brief information about the 25 most-cited papers with at least 212 citations each on the second generation industrial waste-based bioethanol fuels is given as follows. The primary research fronts are the pretreatment and hydrolysis of industrial wastes and production and evaluation of industrial waste-based bioethanol fuels with 11 and 14 highly cited papers (HCPs), respectively. Furthermore, there are two and nine HCPs for the pretreatment and hydrolysis of industrial wastes, respectively, while there are 13 and 1 HCPs for the production and evaluation of the bioethanol fuels, respectively.

68.3.1 Industrial Waste Pretreatment and Hydrolysis

The brief information about the 11 most-cited papers on the pretreatment and hydrolysis of industrial wastes with at least 212 citations each is given as follows (Table 68.1). There are two and nine HCPs for the pretreatment and hydrolysis of industrial wastes, respectively.

68.3.1.1 Industrial Waste Pretreatments

The key findings of two of these papers on the pretreatments of industrial wastes are given as follows.

Brandt et al. (2010) studied the effect of the ionic liquid (IL) anion in the pretreatment of pine wood chips in a paper with 269 citations. All ILs used contained the 1-butyl-3-methylimidazolium [Bmim] cation, while the anions were trifluoromethanesulfonate, methylsulfate, dimethylphosphate, dicyanamide, chloride, and acetate. They observed that this anion had a profound impact on the ability to promote both swelling and dissolution of biomass, while viscosity, temperature, and water content were also important parameters influencing the swelling process. Furthermore, the anion basicity described by the parameter β correlated with the ability to expand and dissolve pine lignocellulose. Finally, [Bmim] dicyanamide dissolved neither cellulose nor lignocellulosic material.

Chen et al. (2011) studied the effect of dilute sulfuric acid (H_2SO_4) pretreatment on sugarcane bagasse structure using microwave (MW) heating in a paper with 237 citations. They used three reaction temperatures of 130°C, 160°C, and 190°C with two heating times of 5 and 10 min. They found that an increase in reaction temperature destroyed the lignocellulosic structure of bagasse in a significant way. When the reaction temperature was as high as 190°C, the fragmentation of particles became fairly pronounced so that the specific surface area of the pretreated material grew substantially. Furthermore, almost all hemicellulose was removed from bagasse and the crystalline structure of cellulose disappeared. In contrast, the feature of lignin has remained clear. The effect of the heating time on the lignocellulosic structure was not significant, indicating that the pretreatment with 5 min was sufficiently long.

TABLE 68.1
Pretreatment and Hydrolysis of Industrial Wastes

No.	Papers	Wastes	Res. Fronts	Prts.	Parameters	Keywords	Lead Author	Affil.	Cits.
3	Aguilar et al. (2002)	Sugarcane bagasse	Hydrolysis	Acids	Acid hydrolysis, temperature, acid concentrations, optimization, sugar yield, fermentation inhibitors	Bagasse, hydrolysis	Vazquez Manuel 55188958100	Univ. Santiago Compostela Spain	385
4	Lavarack et al. (2002)	Sugarcane bagasse	Hydrolysis	Acids	Acid hydrolysis, temperature, solid and concentrations, acid type, sugar yield	Bagasse, hydrolysis	Griffin, Gregory J. 7201836626	RMIT Univ. Australia	368
5	Rezende et al. (2011)	Sugarcane bagasse	Hydrolysis	Acids, alkali, enzymes	Pretreatments, enzymatic digestibility, delignification, hydrolysis yield	Bagasse, delignification, digestibility	Polikarpov, Igor 7006220351	Univ. Sao Paulo Brazil	333
7	Binod et al. (2012)	Sugarcane bagasse	Hydrolysis	Microwave, enzymes, acids, alkali	Enzymatic hydrolysis, pretreatments, hydrolysis efficiency, delignification, sugar yield	Bagasse, saccharification, fermentable, pretreatment	Pandey, Ashok 7201771319	CSIR India	285
8	Brandt et al. (2010)	Wood chips	Pretreatment	IL	Pretreatment, anion effect and basicity, chip swelling, and dissolution	Wood chips, pretreatment, ionic liquids, pine	Welton, Tom 7003503272	Imperial Coll. UK	269
10	Dias et al. (2012)	Sugarcane bagasse	Evaluation	Na	Bioethanol fuel production from sugarcane bagasse, sugarcane, and both of them, technoeconomics	Bagasse, ethanol	Dias, Marina O. S.* 57226710236	Fed. Univ. Sao Paulo Brazil	263
12	Mussatto et al. (2008)	Brewer's spent grain	Hydrolysis	Acid, alkali, enzymes	Pretreatments, enzymatic hydrolysis, substrate concentration, hemicellulose, lignin, sugar yield	Spent grain, hydrolysis	Mussatto, Solange I.* 6602643634	Tech. Univ. Denmark Denmark	253
17	Chen et al. (2011)	Sugarcane bagasse	Pretreatment	Microwave	Pretreatment, cell disruption, temperature, specific surface area, heating time	Sugarcane, bagasse, pretreatment	Chen, Wei-Hsin 57200873137	Natl. Cheng Kung Univ. Taiwan	237
19	Gamez et al. (2006)	Sugarcane bagasse	Hydrolysis	Acids	Acid hydrolysis, sugars, fermentation inhibitors, optimization, sugar yield	Sugar cane, bagasse, hydrolysis	Vazquez Manuel 55188958100	Univ. Santiago Compostela Spain	235
21	Guerard et al. (2002)	Tuna waste	Hydrolysis	Enzymes	Enzymatic hydrolysis, enzyme/substrate ratio, hydrolysis degree, nitrogen recovery	Tuna, waste, hydrolysates	Guerard, Fabienne* 6603404306	Univ. Brest France	221
24	Martin et al. (2007)	Sugarcane bagasse	Hydrolysis	Wet oxidation, enzymes	Pretreatments, enzymatic hydrolysis, cellulose content, sugar yield, fermentation inhibitors	Sugarcane bagasse, pretreatment	Martin, Carlos 56484787200	Inland Norway Univ. Appl. Sci. Norway	219
25	Yu et al. (2009)	Rice hulls	Hydrolysis	Ultrasonic, H_2O_2, enzymes	Enzymatic hydrolysis, two-step pretreatment strategy, sugar yield	Rice hulls, pretreatment, hydrolysis	Yu, Ziniu 7404346720	Huazhong Agr. Univ. China	212

Prt., biomass pretreatments; Cits., number of citations received for each paper; *, female.

68.3.1.2 Industrial Waste Hydrolysis

The key findings about only nine HCPs on the hydrolysis of industrial waste biomass are given as follows.

Aguilar et al. (2002) produced xylose, glucose, and fermentation inhibitors from sugarcane bagasse in a paper with 385 citations. They performed H_2SO_4 hydrolysis at several temperatures (100°C, 122°C, and 128°C) and concentrations of acid (2%, 4%, and 6%). They developed kinetic models to explain the variation with time of xylose, glucose, acetic acid, and furfural generated in the hydrolysis. Optimal conditions were 2% H_2SO_4 at 122°C for 24 min, which yielded a solution with 21.6 g of xylose/L, 3 g of glucose/L, 0.5 g of furfural/L, and 3.65 g of acetic acid/L. In these conditions, they hydrolyzed ≈90% of the hemicelluloses.

Lavarack et al. (2002) produced sugars from sugarcane bagasse hemicellulose in a paper with 368 citations. They performed acid hydrolysis of the bagasse using a temperature-controlled digester. The reaction conditions varied were temperature (80°C–200°C), mass ratio of solid to liquid (1:5 to 1:20), type of bagasse material (i.e., bagasse or bagacillo), concentration of acid (0.25%–8% of liquid), type of acid (hydrochloric (HCl) or H_2SO_4), and reaction time (10–2,000). They found that the most accurate kinetic model of the global reaction for the decomposition of xylan was a simple series hydrolysis of xylan to xylose followed by xylose decomposition. They modeled the production of acid-soluble lignin (ASL) by a first-order decomposition of lignin to ASL followed by a reversible decomposition of ASL. They obtained yields of up to 220 mg xylose/g solid, i.e., about 80% of the theoretical xylose available from the bagasse. Furthermore, the bagasse particle size negligibly influenced the rate of hydrolysis, while HCl was less active for the degradation of xylose compared with H_2SO_4.

Rezende et al. (2011) investigated modifications in the morphology and chemical composition of sugarcane bagasse submitted to a two-step treatment, using diluted acid followed by a delignification process with increasing sodium hydroxide (NaOH) concentrations in a paper with 333 citations. They observed that up to 96% and 85% of hemicellulose and lignin fractions, respectively, were removed by this two-step method when NaOH concentrations of 1% (m/v) or higher were used. The efficient lignin removal resulted in an enhanced hydrolysis yield reaching values around 100%. Considering the cellulose loss due to the pretreatment (maximum of 30%, depending on the process), the total cellulose conversion increased significantly from 22.0 (value for the untreated bagasse) to 72.4%. They also observed the delignification process, with a consequent increase in the cellulose-to-lignin ratio. The morphological changes contributing to this remarkable improvement occurred as a consequence of lignin removal from the sample. Bagasse decomposition was favored by the loss of cohesion between neighboring cell walls, as well as by changes in the inner cell wall structure, such as damage, hole formation, and loss of mechanical resistance, facilitating liquid and enzyme access to crystalline cellulose. In conclusion, the proposed method improved the enzymatic digestibility of sugarcane bagasse.

Binod et al. (2012) studied the enzymatic hydrolysis of sugarcane bagasse using MW pretreatment in a paper with 285 citations. They compared three types of MW pretreatment such as MW acid, MW alkali, and combined MW alkali–acid. They found that the MW pretreatment of sugarcane bagasse with 1% NaOH at 600 W for 4 min followed by enzymatic hydrolysis gave a reducing sugar yield of 0.67 g/g dry biomass, while combined MW alkali–acid treatment with 1% NaOH followed by 1% H_2SO_4, the reducing sugar yield increased to 0.83 g/g dry biomass. Furthermore, MW alkali pretreatment at 450 W for 5 min resulted in almost 90% of lignin removal from the bagasse. In conclusion, combined MW alkali–acid treatment for a short duration enhanced the fermentable sugar yield of the biomass.

Mussatto et al. (2008) compared the enzymatic hydrolysis of brewer's spent grain in the forms of original (untreated), pretreated by dilute acid (cellulignin), and pretreated by a sequence of dilute acid and dilute alkali to verify the effect of hemicellulose and lignin on cellulose conversion into glucose in a paper with 253 citations. They performed the hydrolysis using celluclast 1.5 L in an enzyme/substrate ratio of 45 filter paper unit (FPU)/g, 2% (w/v) substrate concentration, and 45°C for 96h. They found that the cellulose hydrolysis was influenced by the presence of hemicellulose and/or lignin in the sample. Furthermore, the cellulose conversion ratio from cellulignin was 3.5 times higher than that from the untreated sample, while from cellulose pulp such value was four times higher, corresponding to 91.8% (glucose yield of 85.6%). This best result was probably due to the strong modification in the material structure caused by the hemicellulose and lignin removal from the sample. As a consequence, the cellulose fibers were separated being more susceptible to the enzymatic attack. In conclusion, the lower the hemicellulose and lignin contents in the sample, the higher the efficiency of cellulose hydrolysis.

Gamez et al. (2006) hydrolyzed samples of sugarcane bagasse with phosphoric acid under mild conditions (H_3PO_4 2%–6%, time 0–300min, and 122°C) in a paper with 235 citations. They developed kinetic models to describe the course of products of the acid hydrolysis and found that the course of xylose, glucose, arabinose, acetic acid, and furfural was satisfactorily described by these models. The optimal conditions were 122°C, 4% H_3PO_4, and 300min. Under these conditions, they obtained 17.6g of xylose/L, 2.6g of arabinose/L, 3.0g of glucose/L, 1.2g of furfural/L, and 4.0g of acetic acid/L. Furthermore, the efficiency in these conditions was 4.46g sugars/g inhibitors and the mass fraction of sugars in dissolved solids in liquid phase was superior to 55%.

Guerard et al. (2002) produced tuna waste hydrolysates by a commercial neutral protease preparation in a paper with 221 citations. They hydrolyzed a waste product from the canning industry using Umamizyme in a 1-l batch reactor at pH 7 and 45°C. They varied the enzyme/substrate ratio from 0.1% to 1.5% (w/w) protein. They obtained a degree of hydrolysis up to 22.5% with an enzyme/substrate ratio of 1.5%, after 4h of hydrolysis, and found a linear correlation between the hydrolysis degree and the nitrogen recovery. They then found that Umamizyme performed as effectively as Alcalase® 2, 4L for the tuna waste solubilization. However, the Umamizyme stability was lower than Alcalase® 2, 4L.

Martin et al. (2007) enhanced the enzymatic convertibility of sugarcane bagasse with the wet oxidation (WO) pretreatment method in a paper with 219 citations. They observed that WO resulted in an increase in the cellulose content of bagasse as a result of the solubilization of hemicelluloses and lignin. They obtained the highest cellulose content, nearly 70%, in the pretreatment at 195°C, 15min, and alkaline pH. Furthermore, pretreatments at 195°C and 15min solubilized 93%–94% of hemicelluloses and 40%–50% of lignin, while pretreatment at 185°C, 5min, and alkaline pH solubilized only 30% of hemicelluloses and 20% of lignin. They obtained the highest sugar yield in the liquid fraction, 16.1/100g, at 185°C, 5min, and acidic pH. The highest formation of carboxylic acids, phenols, and furaldehydes occurred at 195°C, 15min, and acidic pH as fermentation inhibitors. Alkaline pH reduced the formation of furaldehydes, which was irrelevant for most WO conditions. All pretreatment conditions improved the enzymatic convertibility of cellulose. They obtained the highest convertibility, 74.9% in the hydrolysis of the material obtained by pretreatment at 195°C, 15min, and alkaline pH.

Yu et al. (2009) used two two-step pretreatments for enzymatic hydrolysis of rice hulls to lower the severity requirement of fungal pretreatment time in a paper with 212 citations. These consisted of a mild physical or chemical step (ultrasonic and H_2O_2) and a subsequent biological treatment (*Pleurotus ostreatus (P. ostreatus)*). They observed that these combined pretreatments led to significant increases in lignin degradation than those of one-step pretreatments. After enzymatic hydrolysis of the pretreated hulls, the net yields of total soluble sugar and glucose increased greatly. Furthermore, the combined pretreatment of H_2O_2 (2%, 48h) and *P. ostreatus* (18d) was more effective than the sole pretreatment of *P. ostreatus* for 60d. It could remarkably shorten the residence time and reduce the losses of carbohydrates. This improvement was due to the structure disruption of the hulls during the first pretreatment step. In conclusion, two-step pretreatment was beneficial for the subsequent enzymatic hydrolysis of the biomass.

68.3.2 Industrial Waste-based Bioethanol Fuels

There are 14 HCPs for the production and evaluation of industrial waste-based bioethanol fuels with at least 220 citations each (Table 68.2). Only one of these papers relates to the evaluation of industrial waste-based bioethanol fuels. These papers also cover the fermentation of the hydrolysates of industrial wastes.

68.3.2.1 Bioethanol Fuel Production

Laser et al. (2002) compared liquid hot water (LHW) and steam explosion pretreatments of sugarcane bagasse in a paper with 480 citations. Solid concentration ranged from 1% to 8% for LHW pretreatment and was ≥50% for steam pretreatment, while reaction temperature and time ranged from 170°C to 230°C and 1 to 46 min, respectively. They found that LHW pretreatment achieved ≥80% conversion by simultaneous saccharification and fermentation (SSF), ≥80% xylan recovery, and no hydrolysate inhibition of glucose fermentation yield. Combined effectiveness was not as good for steam pretreatment due to low xylan recovery. SSF conversion increased and xylan recovery decreased as xylan dissolution increased for both modes. SSF conversion, xylan dissolution, hydrolysate furfural concentration, and hydrolysate inhibition increased, while xylan recovery and hydrolysate pH decreased, as a function of increasing LHW pretreatment solid concentration (1%–8%). Autohydrolysis played an important role in batch hydrothermal pretreatment.

Ito et al. (2005) produced biohydrogen (H_2) and bioethanol from biodiesel wastes and pure glycerol in a paper with 419 citations. They used *Enterobacter aerogenes* HU-101. They found that the addition of yeast extract and tryptone to the synthetic medium accelerated the production of H_2 and ethanol. However, the yields of H_2 and ethanol decreased with an increase in the concentrations of biodiesel wastes and pure glycerol. Furthermore, the rates of H_2 and ethanol production from biodiesel wastes were much lower than those at the same concentration of pure glycerol, partially due to the high salt content in the wastes. In a continuous culture with a packed-bed reactor using self-immobilized cells, the maximum rate of H_2 production from pure glycerol was 80 mmol/L/h yielding ethanol at 0.8 mol/mol-glycerol, while that from biodiesel wastes was only 30 mmol/L/h. However, using porous ceramics as a support material to fix cells in the reactor, the maximum H_2 production rate from biodiesel wastes reached 63 mmol/L/h obtaining an ethanol yield of 0.85 mol/mol-glycerol.

Chandel et al. (2007) detoxified sugarcane bagasse hydrolysate to improve ethanol production in a paper with 323 citations. They found that the sugarcane bagasse hydrolysis with 2.5% (v/v) HCl yielded 30.29 g/L total reducing sugars along with various fermentation inhibitors such as furans, phenolics, and acetic acid. The acid hydrolysate when treated with anion exchange resin brought about a maximum reduction in furans (63.4%) and total phenolics (75.8%). Furthermore, treatment of hydrolysate with activated charcoal caused 38.7% and 57.5% reduction in furans and total phenolics, respectively. Laccase reduced total phenolics (77.5%) without affecting furans and acetic acid content in the hydrolysate. Fermentation of these hydrolysates with *Candida shehatae* NCIM 3501 showed maximum ethanol yield (0.48 g/g) from ion-exchange-treated hydrolysate, followed by activated charcoal (0.42 g/g), laccase (0.37 g/g), overliming (0.30 g/g), and neutralized hydrolysate (0.22 g/g).

Da Silva et al. (2010) compared the effectiveness of ball milling (BM) and wet disk milling (WDM) on the enzymatic hydrolysis and fermentation of sugarcane bagasse and straw in a paper with 269 citations. They found that glucose and xylose hydrolysis yields at optimum conditions for BM-treated bagasse and straw were 78.7% and 72.1% and 77.6% and 56.8%, respectively. Furthermore, maximum glucose and xylose yields for bagasse and straw using WDM were 49.3% and 36.7% and 68.0% and 44.9%, respectively. BM improved the enzymatic hydrolysis by decreasing the crystallinity, while the defibrillation effect observed for WDM samples favored enzymatic conversion. They fermented bagasse and straw BM hydrolysates by *Saccharomyces cerevisiae* (*S. cerevisiae*) strains. Ethanol yields from total fermentable sugars using a C_6-fermenting strain reached 89.8% and 91.8% for bagasse and straw hydrolysates, respectively, and 82% and 78% when using a C_6/C_5-fermenting strain.

TABLE 68.2
Production and Evaluation of Industrial Waste-based Bioethanol Fuels

No.	Papers	Wastes	Res. Fronts	Prts.	Yeasts	Parameters	Keywords	Lead Author	Affil.	Cits.
1	Laser et al. (2002)	Sugarcane bagasse	Production	Enzymes, LHW, steam	Yeasts	Pretreatments, SSF conversion, xylan recovery, fermentation inhibition, slid concentration	Bagasse, ethanol, pretreatments	Lynd, Lee R. 35586183800	Dartmouth Coll. USA	480
2	Ito et al. (2005)	Glycerol	Production	Enzymes	*E. aerogenes*	H2 and ethanol production, glycerol, biodiesel waste, ethanol, and H_2 yield	Glycerol, ethanol	Nishio, Naomichi 7005754508	Hiroshima Univ. Japan	419
6	Chandel et al. (2007)	Sugarcane bagasse	Production	Acids	*C. shehatae*	Acid hydrolysis, fermentation inhibitors, fermentation, sugar and ethanol yield	Bagasse, ethanol, hydrolysate	Kuhad, Ramesh C. 55663451900	Central Univ. Haryana India	323
9	Da Silva et al. (2010)	Sugarcane bagasse	Production	Milling, enzymes	*S. cerevisiae*	Pretreatment, enzymatic hydrolysis, fermentation, sugar and ethanol yields	Bagasse, pretreatment, hydrolysis, fermentation, ethanol	Bon, Elba P. S.* 7007036976	Fed. Univ. Rio de Janeiro Brazil	269
10	Dias et al. (2012)	Sugarcane bagasse	Evaluation	Na	Na	Bioethanol fuel production from sugarcane bagasse, sugarcane, and both of them, technoeconomics	Bagasse, ethanol	Dias, Marina O. S.* 57226710236	Fed. Univ. Sao Paulo Brazil	263
11	Rabelo et al. (2011)	Sugarcane bagasse	Production	Alkali, enzymes	Yeasts	Pretreatments, enzymatic hydrolysis, energy recovery, biomethane production	Bagasse, bioethanol	Rabelo, Sarita C.* 22953880600	Sao Paulo State Univ. Brazil	260
13	Saha et al. (2005)	Rice hulls	Production	Acids, enzymes	*E. coli*	Enzymatic hydrolysis, pretreatment, sugar and ethanol yield, detoxification, SSF, SHF	Rice hulls, pretreatment, saccharification, ethanol, fermentation	Saha, Badal C. 7202946302	USDA Agr. Res. Serv. USA	253
14	Martinez et al. (2000)	Sugarcane bagasse	Production	Acids	*E. coli*	Fermentation inhibitors, detoxification, overliming, titration, fermentation	Bagasse, hydrolysates	Ingram, Lonnie O 7102962097	Univ. Florida	247
15	Martin et al. (2002)	Sugarcane bagasse	Production	Steam, enzymes	*S. cerevisiae*	Fermentation inhibitors, detoxification, ethanol yield, overliming, laccase	Bagasse, ethanol, hydrolysates	Jonsson, Leif J. 7102349315	Lund Univ. Sweden	245

(Continued)

TABLE 68.2 (*Continued*)
Production and Evaluation of Industrial Waste-based Bioethanol Fuels

No.	Papers	Wastes	Res. Fronts	Prts.	Yeasts	Parameters	Keywords	Lead Author	Affil.	Cits.
16	Kim et al. (2008)	Barley hulls	Production	Ammonia, enzymes	*E. coli*	Pretreatments, enzymatic hydrolysis, SSCF, ethanol yield, temperature, ammonia concentration	Barley, hulls, pretreatment, bioethanol	Kim, Tae Hyun 57210847338	Hanyang Univ. S. Korea	243
18	Yazdani and Gonzalez (2008)	Glycerol	Production	Strain eng.	*E. coli*	Bioethanol, H_2, formate production, strain engineering, ethanol yield	Glycerol, ethanol	Gonzalez, Roman 57192167471	Univ. S. Florida USA	235
20	Kadar et al. (2004)	Industrial wastes	Production	Enzymes	*K. marxianus*, baker's yeast, *S. cerevisiae*	Ethanol production, industrial wastes, ethanol yield, SSF, yeasts	Industrial wastes, saccharification, fermentation, ethanol, SSF	Reczey, Kati* 7004072336	Budapest Univ. Technol. Econ. Hungary	224
22	Dias et al. (2009)	Sugarcane bagasse	Production	Solvent, enzymes	Yeasts	Integrated bioethanol production, simulation, enzymatic hydrolysis, distillation, fermentation	Sugarcane bagasse, bioethanol	Dias, Marina O. S.* 57226710236	Fed. Univ. Sao Paulo Brazil	220
23	Ezeji and Blaschek (2008)	Distillers' grains	Production	Acids, LHW, ammonia	*C. beijerinckii*, *C. acetobutylicum*, *C. acetobutylicum*, *C. butylicum*	Fermentation, pretreatments, hydrolysis, yeasts, ethanol, ABE fermentation inhibitors	Distillers' grains, DDGS, fermentation	Blaschek, Hans P. 7006971390	Univ. Ill. U. C. USA	220

Prt., biomass pretreatments; Na, nonavailable; Cits., number of citations received for each paper. *, female.

Rabelo et al. (2011) produced bioethanol, biomethane, and heat from sugarcane bagasse in a biorefinery concept with lime or alkaline hydrogen peroxide pretreatments in a paper with 260 citations. They pretreated bagasse, enzymatically hydrolyzed, and used the wastes from pretreatment and hydrolysis to produce biogas. If pretreatment was carried out at a bagasse concentration of 4% dry matter (DM), they obtained the highest global methane production with the peroxide pretreatment: 72.1 L methane/kg bagasse. The recovery of lignin from the peroxide pretreatment liquor was also the highest, 112.7 g/kg of bagasse. Furthermore, 63%–65% of the energy that would be produced by bagasse incineration could be recovered by combining ethanol production with the combustion of lignin and hydrolysis residues, along with the anaerobic digestion of pretreatment liquors, while only 32%–33% of the energy was recovered by bioethanol production alone.

Saha et al. (2005) evaluated dilute H_2SO_4 pretreatment at varied temperature (120°C–190°C) and enzymatic hydrolysis (45°C, pH 5.0) for conversion of rice hull cellulose and hemicellulose to monomeric sugars in a paper with 253 citations. The maximum yield of monomeric sugars from rice hulls (15%, w/v) by dilute H_2SO_4 (1.0%, v/v) pretreatment and enzymatic hydrolysis (45°C, pH 5.0, 72 h) using cellulase, β-glucosidase, xylanase, esterase, and Tween 20 was 287 mg/g (60% yield based on total carbohydrate content). Under this condition, no furfural and hydroxymethylfurfural (HMF) were produced. The yield of ethanol per L by the mixed sugar utilizing recombinant *Escherichia coli* (*E. coli*) strain FBR 5 from rice hull hydrolysate containing 43.6 g fermentable sugars was 18.7 g (0.43 g/g sugars obtained and 0.13 g/g rice hulls) at pH 6.5 and 35°C. Furthermore, the detoxification of the acid- and enzyme-treated rice hull hydrolysate by overliming (pH 10.5, 90°C, 30 min) reduced the time required for maximum ethanol production (17 g from 42.0 g sugars per L) by the *E. coli* strain from 64 to 39 h in the case of separate hydrolysis and fermentation (SHF) and increased the maximum ethanol yield (per L) from 7.1 g in 140 h to 9.1 g in 112 h in the case of simultaneous SSF.

Martinez et al. (2000) studied the effects of overliming treatment on the composition and toxicity of sugarcane bagasse hemicellulose hydrolysates (primarily pentose sugars) in a paper with 247 citations. They optimized this overliming treatment for this hydrolysate using recombinant *E. coli* LY01 as the biocatalyst. They observed a substantial reduction in furfural and HMF, while organic acids (acetic, formic, and levulinic) were not affected. Furthermore, the extent of furan reduction correlated with increasing fermentability although furan reduction was not the sole cause for reduced toxicity. After optimal overliming, bagasse hydrolysate was rapidly and efficiently fermented (>90% yield) by LY01. Titration provided an estimate of total organic acids in the hydrolysate.

Martin et al. (2002) produced ethanol from the enzymatic hydrolysates of sugarcane bagasse in a paper with 245 citations. They pretreated the bagasse by steam explosion at 205°C and 215°C and hydrolyzed it with cellulolytic enzymes. They then detoxified these hydrolysates by treatment with the phenoloxidase laccase and with overliming. They removed ~80% of the phenolic compounds by the laccase treatment. Furthermore, overliming partially removed the phenolic compounds, but also other fermentation inhibitors such as acetic acid, furfural, and 5-HMF. They then fermented the detoxified hydrolysates with the recombinant xylose-utilizing *S. cerevisiae* TMB 3001. This strain was a CEN.PK derivative with overexpressed xylulokinase activity and expressing the xylose reductase and xylitol dehydrogenase of *Pichia stipitis* and the *S. cerevisiae* strain ATCC 96581. They found a nearly twofold increase in the specific productivity of the strain TMB 3001 in the detoxified hydrolysates compared with the un-detoxified hydrolysates. The ethanol yield in the fermentation of the hydrolysate detoxified by overliming was 0.18 g/g dry bagasse, whereas it reached only 0.13 g/g dry bagasse in the un-detoxified hydrolysate.

Kim et al. (2008) produced bioethanol from barley hulls using the soaking in aqueous ammonia (SAA) pretreatment in a paper with 243 citations. They soaked barley hull in 15 and 30 wt% aqueous ammonia at 30°C, 60°C, and 75°C for between 12 h and 11 weeks. They found that the best pretreatment conditions were 75°C, 48 h, 15 wt% aqueous ammonia, and 1:12 of solid:liquid ratio resulting in hydrolysis yields of 83% for glucan and 63% for xylan with 15 FPU/g-glucan enzyme loading. Pretreatment using 15 wt% ammonia for 24–72 h at 75°C removed 50%–66% of the original lignin from the solids, while it retained 65%–76% of the xylan without any glucan loss.

Furthermore, the addition of xylanase along with cellulase resulted in a synergetic effect on ethanol production in simultaneous saccharification and cofermentation (SSCF) using SAA-treated barley hull and recombinant *E. coli* (KO11). With 3% w/v glucan loading and 4 mL of xylanase enzyme loadings, the SSCF of the SAA-treated barley hull resulted in 24.1 g/L ethanol concentration at 15 FPU cellulase/g-glucan loading, which corresponds to 89.4% of the maximum theoretical yield based on glucan and xylan. The SAA pretreatment increased surface area and pore size, and these physical changes enhanced the enzymatic digestibility in the SAA-pretreated barley hull.

Yazdani and Gonzalez (2008) engineered *E. coli* for the efficient conversion of glycerol to ethanol and coproducts in a paper with 235 citations. They capitalized on the high degree of reduction in carbon in glycerol, thus enabling the production of not only ethanol but also coproducts such as hydrogen and formate. They created two strains for the coproduction of ethanol–hydrogen and ethanol–formate: SY03 and SY04, respectively. They obtained high ethanol yields in both strains by minimizing the synthesis of byproducts succinate and acetate through mutations that inactivated fumarate reductase (Δ*frdA*) and phosphate acetyltransferase (Δ*pta*), respectively. Similarly, strain SY04 also contained a mutation that inactivated formate–hydrogen lyase (Δ*fdhF*), thus preventing the conversion of formate to CO_2 and H_2. They obtained high rates of glycerol utilization and product synthesis by simultaneous overexpression of glycerol dehydrogenase (*gldA*) and dihydroxyacetone kinase (*dhaKLM*), which were the enzymes responsible for the conversion of glycerol to glycolytic intermediate dihydroxyacetone phosphate. The resulting strains, SY03 (pZSKLMgldA) and SY04 (pZSKLMgldA), produced ethanol–hydrogen and ethanol–formate from unrefined glycerol, respectively, at yields exceeding 95% of the theoretical maximum and specific rates in the order of 15–30 mmol/gcell/h.

Kadar et al. (2004) used SOLKA-FLOC, old corrugated container (OCC) waste, and paper sludge in the SSF process for the production of ethanol in a paper with 224 citations. They compared two yeast strains, baker's yeast and *Kluyveromyces marxianus (K. marxianus),* in two types of SSF experiments, i.e., isothermal SSF and non-isothermal SSF with temperature profiling. They found that OCC waste and paper sludge could be used as substrates for ethanol production in SSF. There was no significant difference observed between *S. cerevisiae* and *K. marxianus* when the results of SSF were compared. The ethanol yields were in the range of 0.31–0.34 g/g for both strains used, while the isothermal SSF resulted in higher ethanol yields compared with non-isothermal SSF.

Dias et al. (2009) produced bioethanol and other value-added products from sugarcane bagasse, integrated into the conventional bioethanol production process in a paper with 219 citations. They performed simulations of bioethanol production from sugarcane juice and bagasse. They considered a typical large-scale production plant of 1,000 m^3/day of ethanol using sugarcane juice as feedstock. Next, they considered a three-step hydrolysis process (prehydrolysis of hemicellulose, organosolv delignification, and cellulose hydrolysis) of sugarcane bagasse. They determined the minimum hot utility obtained with thermal integration of the plant, to find out the maximum availability of bagasse that could be used in the hydrolysis process, taking into consideration the use of 50% of generated sugarcane trash as fuel for bioelectricity and steam production. They found that the double-effect distillation system allowed 90% of generated bagasse to be used as raw material in the hydrolysis plant, which accounted for an increase of 26% in bioethanol production, considering exclusively the fermentation of hexoses obtained from the cellulosic fraction.

Ezeji and Blaschek (2008) fermented dried distillers' grains and solubles (DDGS) hydrolysates to acetone, butanol, and ethanol (ABE) in a paper with 220 citations. The pretreatment and hydrolysis of this lignocellulosic biomass using either dilute acid, LHW, or ammonium fiber expansion (AFEX) resulted in a complex mixture of sugars such as hexoses (glucose, galactose, and mannose) and pentoses (xylose and arabinose). They then fermented these sugars by the clostridia: *Clostridium (C.) beijerinckii* BA101, *C. acetobutylicum* 260, *C. acetobutylicum* 824, *C. saccharobutylicum* 262, and *C. butylicum* 592. They observed that all the sugars were utilized concurrently throughout the fermentation, although the rate of sugar utilization was sugar-specific. For all clostridia strains, the rate of glucose utilization was higher than for the other sugars in the mixture.

In addition, the availability of excess fermentable sugars in the bioreactor was necessary for both the onset and the maintenance of solvent production; otherwise, the fermentation would become acidogenic, leading to premature termination of the fermentation process. Furthermore, ferulic and p-coumaric acids were potent inhibitors of growth and ABE production. Strikingly, furfural and HMF were not inhibitory to the solventogenic clostridia as they had a stimulatory effect on growth and ABE production at concentrations up to 2.0 g/L.

68.3.2.2 Bioethanol Fuel Evaluation

Dias et al. (2012) compared stand-alone second generation ethanol production from surplus sugarcane bagasse and trash with first generation ethanol production from sugarcane and with integrated first- and second generation ethanol production in a paper with 263 citations. They developed simulations to represent the different technological scenarios, which provided data for economic and environmental analysis. They found that the integrated first- and second generation ethanol production process from sugarcane led to better economic results compared with the stand-alone second generation ethanol production, especially when advanced hydrolysis technologies and pentose fermentation were included.

68.4 DISCUSSION

68.4.1 Introduction

Crude oil-based gasoline fuels have been widely used in the transportation sector since the 1920s. However, there have been great public concerns over the adverse environmental and human impact of these fuels. Hence, biomass-based bioethanol fuels have increasingly been used in blending gasoline and petrodiesel fuels, in fuel cells, and in biochemical production in a biorefinery context.

However, it is necessary to pretreat the biomass to enhance the yield of the bioethanol before the bioethanol fuel production from the feedstocks through the hydrolysis and fermentation of the biomass and hydrolysates, respectively.

One of the most-studied feedstocks for bioethanol fuels has been industrial wastes such as milling, food processing, wood and paper, and other industrial wastes such as biodiesel industrial wastes. The research in the field of second generation industrial waste-based bioethanol fuels has intensified in this context in the key research fronts of the pretreatment and hydrolysis of industrial wastes, fermentation of industrial waste-based hydrolysates, and production and evaluation of the second generation industrial waste-based bioethanol fuels. Furthermore, milling, food processing, wood and paper, and other industrial wastes such as biodiesel industrial wastes have been studied intensively in this context.

However, it is essential to develop efficient incentive structures for the primary stakeholders to enhance the research in this field. Although there have been a limited number of review papers for this field, there has been no review of the 25 most-cited articles in this field.

Thus, this book chapter presents a review of the 25 most-cited articles on bioethanol fuel production and evaluation from the industrial wastes. Then, it discusses the key findings of these highly influential papers and comments on future research priorities in this field.

As a first step for the search of the relevant literature, the keywords were selected using the 300 most-cited first population papers. The selected keyword list was then optimized to obtain a representative sample of papers for the searched research field. This keyword list was provided in the appendix of Konur (2023) for future replicative studies.

As a second step, a sample dataset was used for this study. The first 25 articles with at least 212 citations each were selected for the review study. Key findings from each paper were taken from the abstracts of these papers and were discussed. Additionally, several brief conclusions were drawn and many relevant recommendations were made to enhance the future research landscape.

Information about the research fronts for the sample papers in the industrial waste-based bioethanol fuels with regard to the feedstocks used in these processes is given in Table 68.3. As this

TABLE 68.3
Most Prolific Research Fronts for the Second Generation Industrial Waste-based Bioethanol Fuels

No.	Research Fronts	N Paper % Review	N Paper % Sample	Surplus (%)
1	Milling industrial wastes	76.0	63.2	12.8
2	Biodiesel industrial wastes	8.0	2.8	5.2
3	Food processing industrial wastes	4.0	21.6	−17.6
4	Wood and paper industrial wastes	4.0	8.0	–4.0
5	Ethanol industrial wastes	4.0	1.6	2.4
6	Brewing industrial wastes	4.0	1.2	2.8
7	Other industrial wastes	0.0	7.6	−7.6
	Plastic industrial wastes	0.0	0.8	–0.8
	Textile industrial wastes	0.0	0.8	–0.8
	Leather industrial wastes	0.0	0.4	–0.4

N Paper (%) Review: the number of papers in the sample of 25 reviewed papers. N Paper (%) Sample: the number of papers in the population sample of 250 papers.

TABLE 68.4
Most Prolific Thematic Research Fronts for the Industrial Waste-based Bioethanol Fuels

No.	Research Fronts	N Paper % Review	N Paper % Sample	Surplus (%)
1	Biomass pretreatments	96.0	66.8	29.2
2	Biomass hydrolysis	72.0	66.4	5.6
3	Bioethanol production	36.0	37.6	−1.6
4	Hydrolysate fermentation	24.0	43.2	−19.2
5	Bioethanol fuel evaluation and utilization	8.0	4.4	3.6

N Paper (%) Review: the number of papers in the sample of 25 reviewed papers. N Paper (%) Sample: the number of papers in the population sample of 250 papers.

table shows, the most prolific research front for this field is the milling industrial wastes with 76% of the HCPs. Furthermore, the other prolific research fronts are biodiesel, ethanol, brewing, food processing, and wood and paper industrial wastes, with 4%–8% of the sample papers each.

However, the most influential feedstock is the milling industry wastes with 13% surplus, followed by the biodiesel, brewing, and ethanol industrial wastes with 2%–5% surplus each. Similarly, the least influential research fronts are food processing industrial wastes with 18% deficit, followed by other and wood and paper industrial wastes with 8% and 4% deficits, respectively.

Information about the thematic research fronts for the sample papers in the industrial waste-based bioethanol fuels is given in Table 68.4. As this table shows, the most prolific research fronts for this field are the pretreatment and hydrolysis of industrial wastes with 96% and 72% of the HCPs, respectively. The other research fronts are hydrolysate fermentation and bioethanol production with 24% and 36% of the sample papers, respectively. Furthermore, the minor research front is bioethanol fuel evaluation with 8% of the HCPs.

Furthermore, the most influential research fronts are pretreatment and hydrolysis of the food wastes with 29% and 6% surplus, respectively. Similarly, hydrolysate fermentation is the least influential research front with 17% deficit.

68.4.2 Industrial Waste Pretreatment and Hydrolysis

The brief information about the 11 most-cited papers on the pretreatment and hydrolysis of industrial wastes with at least 212 citations each is given as follows (Table 68.1). There are two and nine HCPs for the pretreatment and hydrolysis of industrial wastes, respectively. However, it should be noted that as Table 68.4 shows the number of HCPs related to the biomass pretreatments and hydrolysis is 24 and 18, respectively.

68.4.2.1 Industrial Waste Pretreatments

The most prolific pretreatments are the enzymatic and chemical pretreatments with 15 and 16 HCPs, respectively, followed by hydrothermal and mechanical pretreatments with five and four HCPs, respectively. On individual terms, however, acid pretreatments are the most prolific feedstocks with 10 HCPs. The other individual prolific pretreatments are the alkaline, LHW, ammonia, steam, and MW pretreatments with two to four HCPs each. The key findings about only two of these papers are given as follows briefly.

Brandt et al. (2010) studied the effect of the IL anion in the pretreatment of pine wood chips and observed that the anion had a profound impact on the ability to promote both swelling and dissolution of biomass. Furthermore, Chen et al. (2011) studied the effect of dilute H_2SO_4 pretreatment on sugarcane bagasse structure using MW heating and found that an increase in reaction temperature destroyed the lignocellulosic structure of bagasse in a significant way.

68.4.2.2 Industrial Waste Hydrolysis

As Table 68.4 shows, there are 18 HCPs related to the hydrolysis of industrial waste biomass. The key findings about only nine of these HCPs are given as follows. Thus, it appears that this research front is one of the most prolific research fronts for these HCPs.

Aguilar et al. (2002) produced xylose, glucose, and fermentation inhibitors from sugarcane bagasse and determined that the optimal conditions were 2% H_2SO_4 at 122°C for 24 min, which yielded a solution with 21.6 g of xylose/L, 3 g of glucose/L, 0.5 g of furfural/L, and 3.65 g of acetic acid/L. Furthermore, Lavarack et al. (2002) produced sugars from sugarcane bagasse hemicellulose and found that the most accurate kinetic model of the global reaction for the decomposition of xylan was a simple series hydrolysis of xylan to xylose followed by xylose decomposition.

Rezende et al. (2011) investigated modifications in the morphology and chemical composition of sugarcane bagasse submitted to a two-step treatment, using diluted acid followed by a delignification process with increasing NaOH concentrations, and observed that up to 96% and 85% of hemicellulose and lignin fractions, respectively, were removed by this two-step method. Furthermore, Binod et al. (2012) studied the enzymatic hydrolysis of sugarcane bagasse using MW pretreatment and found that the MW treatment of sugarcane bagasse with 1% NaOH at 600 W for 4 min followed by enzymatic hydrolysis gave a reducing sugar yield of 0.665 g/g dry biomass.

Mussatto et al. (2008) compared the enzymatic hydrolysis of brewer's spent grain in the forms of original (untreated), pretreated by dilute acid, and pretreated by a sequence of dilute acid and dilute alkali and found that the cellulose hydrolysis was influenced by the presence of hemicellulose and/or lignin in the sample. Furthermore, Gamez et al. (2006) hydrolyzed samples of sugarcane bagasse with phosphoric acid under mild conditions and found that the course of xylose, glucose, arabinose, acetic acid, and furfural was satisfactorily described by these models.

Guerard et al. (2002) produced tuna waste hydrolysates by a commercial neutral protease preparation and obtained a degree of hydrolysis up to 22.5% with an enzyme/substrate ratio of 1.5%, after 4 h of hydrolysis. Furthermore, Martin et al. (2007) enhanced the enzymatic convertibility of sugarcane bagasse with the WO pretreatment method and observed that WO resulted in an increase in cellulose content of bagasse as a result of the solubilization of hemicelluloses and lignin. Finally, Yu et al. (2009) used two two-step pretreatments for enzymatic hydrolysis of rice hulls and observed that these combined pretreatments led to significant increases in the lignin degradation than those of one-step pretreatments.

68.4.3 Industrial Waste-based Bioethanol Fuels

There are 14 HCPs for the production and evaluation of industrial waste-based bioethanol fuels with at least 220 citations each (Table 68.2). Only one of these papers relates to the evaluation of industrial waste-based bioethanol fuels. These papers also cover the fermentation of the hydrolysates of industrial wastes. As Table 68.4 shows, there are nine, six, and two HCPs for the production of bioethanol fuels, fermentation of the hydrolysates, and evaluation of the bioethanol fuels, respectively.

68.4.4 Bioethanol Fuel Production

Laser et al. (2002) compared LHW and steam pretreatments of sugarcane bagasse and found that LHW pretreatment achieved ≥80% conversion by SSF, ≥80% xylan recovery, and no hydrolysate inhibition of glucose fermentation yield. Furthermore, Ito et al. (2005) produced H_2 and ethanol from biodiesel wastes and pure glycerol and found that the yields of H_2 and ethanol decreased with an increase in the concentrations of biodiesel wastes and pure glycerol.

Chandel et al. (2007) detoxified sugarcane bagasse hydrolysate to improve ethanol production and found that the fermentation of these hydrolysates showed maximum ethanol yield from ion exchange-treated hydrolysate, followed by activated charcoal, laccase, overliming, and neutralized hydrolysate. Furthermore, da Silva et al. (2010) compared the effectiveness of BM and WDM on the enzymatic hydrolysis and fermentation of sugarcane bagasse and straw and found that the ethanol yields from total fermentable sugars using a C_6-fermenting strain reached 89.8% and 91.8% for bagasse and straw hydrolysates, respectively, and 82% and 78% when using a C_6/C_5-fermenting strain.

Rabelo et al. (2011) produced bioethanol, biomethane, and heat from sugarcane bagasse with lime or alkaline hydrogen peroxide and found that 63%–65% of the energy that would be produced by bagasse incineration could be recovered by combining ethanol production with the combustion of lignin and hydrolysis residues, along with the anaerobic digestion of pretreatment liquors, while only 32%–33% of the energy was recovered by bioethanol production alone. Furthermore, Saha et al. (2005) evaluated dilute H_2SO_4 pretreatment at varied temperature and enzymatic hydrolysis for the conversion of rice hull cellulose and hemicellulose to monomeric sugars and found that the yield of ethanol per L by the mixed sugar from rice hull hydrolysate containing 43.6 g fermentable sugars was 18.7 g (0.43 g/g sugars obtained and 0.13 g/g rice hulls) at pH 6.5 and 35°C.

Martinez et al. (2000) studied the effects of overliming treatment on the composition and toxicity of sugarcane bagasse hemicellulose hydrolysates and observed a substantial reduction in furfural and HMF, while organic acids (acetic, formic, and levulinic) were not affected. Furthermore, Martin et al. (2002) produced ethanol from the enzymatic hydrolysates of sugarcane bagasse and found that the ethanol yield in the fermentation of the hydrolysate detoxified by overliming was 0.18 g/g dry bagasse, whereas it reached only 0.13 g/g dry bagasse in the un-detoxified hydrolysate.

Kim et al. (2008) produced bioethanol from barley hulls using the SAA pretreatment and found that with 3% w/v glucan loading and 4 mL of xylanase enzyme loadings, the SSCF of the SAA-treated barley hull resulted in 24.1 g/L ethanol concentration at 15 FPU cellulase/g-glucan loading, which corresponded to 89.4% of the maximum theoretical yield based on glucan and xylan. Furthermore, Yazdani and Gonzalez (2008) engineered *E. coli* for the efficient conversion of glycerol to ethanol and coproducts and obtained high ethanol yields in both strains by minimizing the synthesis of byproducts succinate and acetate through mutations that inactivated Δ*frdA* and Δ*pta*, respectively.

Kadar et al. (2004) used SOLKA-FLOC, OCC waste, and paper sludge in the SSF process and found that OCC waste and paper sludge could be used as substrates for ethanol production in SSF. Furthermore, Dias et al. (2009) produced bioethanol and other value-added products from sugarcane bagasse, integrated into the conventional bioethanol production process and found that the double-effect distillation system allowed 90% of generated bagasse to be used as raw material in the hydrolysis plant, which accounted for an increase of 26% in bioethanol production. Finally, Ezeji and Blaschek (2008) fermented DDGS hydrolysates to ABE and found that ferulic and p-coumaric acids were potent inhibitors of growth and ABE production.

68.4.4.1 Bioethanol Fuel Evaluation

Dias et al. (2012) compared stand-alone second generation ethanol production from surplus sugarcane bagasse and trash with first generation ethanol production from sugarcane and with integrated first- and second generation ethanol production and found that the integrated first- and second generation ethanol production process from sugarcane led to better economic results compared with the stand-alone second generation ethanol production, especially when advanced hydrolysis technologies and pentose fermentation were included.

68.5 CONCLUSION AND FUTURE RESEARCH

The brief information about the key research fronts covered by the 25 most-cited papers with at least 212 citations each is given under two primary headings: the pretreatment and hydrolysis of industrial wastes and production and evaluation of the bioethanol fuels.

The usual characteristics of these HCPs are that the pretreatments and hydrolysis of industrial wastes and fermentation of the resulting hydrolysates are the primary processes for the bioethanol fuel production from industrial wastes to improve the ethanol yield as the industrial wastes are one of the most-studied feedstocks for the bioethanol production, especially for the countries with the large farmlands, forests, and crude oil deficiency.

The key findings on these research fronts should be read in light of the increasing public concerns about climate change, greenhouse gas (GHG) emissions, and global warming as these concerns have been certainly behind the boom in the research on the industrial waste-based bioethanol fuels as an alternative to crude oil-based gasoline and petrodiesel fuels in the last decades. It is also a sustainable alternative to first generation food crop-based bioethanol fuels such as corn grain-based bioethanol fuels. The recent supply shocks caused by the coronavirus disease 2019 (COVID-19) pandemic and the Russian invasion of Ukraine also highlight the importance of the production and utilization of bioethanol fuels as an alternative to crude oil-based gasoline and petrodiesel fuels.

As Table 68.3 shows, the most prolific research front for this field is the milling industrial wastes, while the other prolific research fronts are biodiesel, ethanol, brewing, food processing, and wood and paper industrial wastes.

Similarly, as Table 68.4 shows the most prolific research fronts for this field are the pretreatment and hydrolysis of industrial wastes, while the other research fronts are hydrolysate fermentation and bioethanol production together with the minor research front of bioethanol fuel evaluation.

These studies emphasize the importance of proper incentive structures for the efficient production of industrial waste-based bioethanol fuels in light of North's institutional framework (North, 1991). In this context, the major producers and users of bioethanol fuels such as the USA, Brazil, and Canada with vast forests and farmlands have developed strong incentive structures for efficient industrial waste-based bioethanol fuels. In light of the recent supply shocks caused primarily by the COVID-19 pandemic and the Russian invasion of Ukraine, it is expected that the incentive structures such as public funding would be enhanced to increase the share of bioethanol fuels in the global fuel portfolio as a strong alternative to crude oil-based gasoline and diesel fuels.

In this context, it is expected that the most prolific researchers, institutions, countries, funding bodies, and journals in this field would have a first-mover advantage to benefit from such potential incentives. This is especially true for the US and Brazilian stakeholders as the USA and Brazil have become the global leaders in both the production and utilization of second generation bioethanol fuels from the industrial wastes.

It is recommended that such review studies are performed for the primary research fronts of industrial waste-based bioethanol fuels.

ACKNOWLEDGMENTS

The contribution of the highly cited researchers in the field of industrial waste-based bioethanol fuels has been gratefully acknowledged.

REFERENCES

Aguilar, R., J. A. Ramirez, G. Garrote and M. Vazquez. 2002. Kinetic study of the acid hydrolysis of sugar cane bagasse. *Journal of Food Engineering* 55:309–318.

Alvira, P., E. Tomas-Pejo, M. Ballesteros and M. J. Negro. 2010. Pretreatment technologies for an efficient bioethanol production process based on enzymatic hydrolysis: A review. *Bioresource Technology* 101:4851–4861.

Angelici, C., B. M. Weckhuysen and P. C. A. Bruijnincx. 2013. Chemocatalytic conversion of ethanol into butadiene and other bulk chemicals. *ChemSusChem* 6:1595–1614.

Antolini, E. 2007. Catalysts for direct ethanol fuel cells. *Journal of Power Sources* 170:1–12.

Antolini, E. 2009. Palladium in fuel cell catalysis. *Energy and Environmental Science* 2:915–931.

Binod, P., K. Satyanagalakshmi and R. Sindhu, et al. 2012. Short duration microwave assisted pretreatment enhances the enzymatic saccharification and fermentable sugar yield from sugarcane bagasse. *Renewable Energy* 37:109–116.

Brandt, A., J. P. Hallett, D. J. Leak, R. J. Murphy and T. Welton. 2010. The effect of the ionic liquid anion in the pretreatment of pine wood chips. *Green Chemistry* 12:672–679.

Burnham, J. F. 2006. Scopus database: A review. *Biomedical Digital Libraries* 3:1–8.

Cardona, C. A., J. A. Quintero and I. C. Paz. 2010. Production of bioethanol from sugarcane bagasse: Status and perspectives. *Bioresource Technology* 101:4754–4766.

Chandel, A. K., R. K. Kapoor, A. Singh and R. C. Kuhad. 2007. Detoxification of sugarcane bagasse hydrolysate improves ethanol production by *Candida shehatae* NCIM 3501. *Bioresource Technology* 98:1947–1950.

Chen, W. H., Y. J. Tu and H. K. Sheen. 2011. Disruption of sugarcane bagasse lignocellulosic structure by means of dilute sulfuric acid pretreatment with microwave-assisted heating. *Applied Energy* 88:2726–2734.

da Silva, A. S., H. Inoue, T. Endo, S. Yano and E. P. S. Bon. 2010. Milling pretreatment of sugarcane bagasse and straw for enzymatic hydrolysis and ethanol fermentation. *Bioresource Technology* 101:7402–7409.

Dias, M. O. S., A. V. Ensinas and S. A. Nebra, et al. 2009. Production of bioethanol and other bio-based materials from sugarcane bagasse: Integration to conventional bioethanol production process. *Chemical Engineering Research and Design* 87:1206–1216.

Dias, M. O. S., T. L. Junqueira and O. Cavalett, et al. 2012. Integrated versus stand-alone second generation ethanol production from sugarcane bagasse and trash. *Bioresource Technology* 103:152–161.

Duff, S. J. B. and W. D. Murray. 1996. Bioconversion of forest products industry waste cellulosics to fuel ethanol: A review. *Bioresource Technology* 55:1–33.

Ezeji, T. and H. P. Blaschek. 2008. Fermentation of dried distillers' grains and solubles (DDGS) hydrolysates to solvents and value-added products by solventogenic clostridia. *Bioresource Technology* 99:5232–5242.

Fan, Z., C. South and K. Lyford, et al. 2003. Conversion of paper sludge to ethanol in a semicontinuous solids-fed reactor. *Bioprocess and Biosystems Engineering* 26:93–101.

Fernando, S., S. Adhikari, C. Chandrapal and M. Murali. 2006. Biorefineries: Current status, challenges, and future direction. *Energy & Fuels* 20:1727–1737.

Gamez, S., J. J. Gonzalez-Cabriales, J. A. Ramirez, G. Garrote and M. Vazquez. 2006. Study of the hydrolysis of sugar cane bagasse using phosphoric acid. *Journal of Food Engineering* 74:78–88.

Grohmann, K., R. G. Cameron and B. S. Buslig. 1995. Fractionation and pretreatment of orange peel by dilute acid hydrolysis. *Bioresource Technology* 54:129–141.

Guerard, F., L. Guimas and A. Binet. 2002. Production of tuna waste hydrolysates by a commercial neutral protease preparation. *Journal of Molecular Catalysis B: Enzymatic* 19:489–498.

Hahn-Hagerdal, B., M. Galbe, M. F. Gorwa-Grauslund, G. Liden and G. Zacchi. 2006. Bio-ethanol: The fuel of tomorrow from the residues of today. *Trends in Biotechnology* 24:549–556.

Hill, J., E. Nelson, D. Tilman, S. Polasky and D. Tiffany. 2006. Environmental, economic, and energetic costs and benefits of biodiesel and ethanol biofuels. *Proceedings of the National Academy of Sciences of the United States of America* 103:11206–11210.

Hill, J., S. Polasky and E. Nelson, et al. 2009. Climate change and health costs of air emissions from biofuels and gasoline. Proceedings of the National Academy of Sciences of the United States of America 106:2077–2082.

Hsieh, W. D., R. H. Chen, T. L. Wu and T. H. Lin. 2002. Engine performance and pollutant emission of an SI engine using ethanol-gasoline blended fuels. *Atmospheric Environment* 36:403–410.

Huang, H. J., S. Ramaswamy, U. W. Tschirner and B. V. Ramarao. 2008. A review of separation technologies in current and future biorefineries. *Separation and Purification Technology* 62:1–21.

Ito, T., Y. Nakashimada, K. Senba, T. Matsui and N. Nishio. 2005. Hydrogen and ethanol production from glycerol-containing wastes discharged after biodiesel manufacturing process. *Journal of Bioscience and Bioengineering* 100:260–265.

Kadar, Z., Z. Szengyel and K. Reczey. 2004. Simultaneous saccharification and fermentation (SSF) of industrial wastes for the production of ethanol. *Industrial Crops and Products* 20:103–110.

Kim, T. H., F. Taylor and K. B. Hicks. 2008. Bioethanol production from barley hull using SAA (soaking in aqueous ammonia) pretreatment. *Bioresource Technology* 99:5694–5702.

Konur, O. 2000. Creating enforceable civil rights for disabled students in higher education: An institutional theory perspective. *Disability & Society* 15:1041–1063.

Konur, O. 2002a. Access to nursing education by disabled students: Rights and duties of nursing programs. *Nurse Education Today* 22:364–374.

Konur, O. 2002b. Assessment of disabled students in higher education: Current public policy issues. *Assessment and Evaluation in Higher Education* 27:131–152.

Konur, O. 2002c. Access to employment by disabled people in the UK: Is the Disability Discrimination Act working? *International Journal of Discrimination and the Law* 5:247–279.

Konur, O. 2006a. Participation of children with dyslexia in compulsory education: Current public policy issues. *Dyslexia* 12:51–67.

Konur, O. 2006b. Teaching disabled students in higher education. *Teaching in Higher Education* 11:351–363.

Konur, O. 2007a. A judicial outcome analysis of the Disability Discrimination Act: A windfall for the employers? *Disability & Society* 22:187–204.

Konur, O. 2007b. Computer-assisted teaching and assessment of disabled students in higher education: The interface between academic standards and disability rights. *Journal of Computer Assisted Learning* 23:207–219.

Konur, O. 2012. The evaluation of the research on the bioethanol: A scientometric approach. *Energy Education Science and Technology Part A: Energy Science and Research* 28:1051–1064.

Konur, O. 2015. Current state of research on algal bioethanol. In *Marine Bioenergy: Trends and Developments*, Eds. S. K. Kim and C. G. Lee, pp. 217–244. Boca Raton, FL: CRC Press.

Konur, O. 2020. The scientometric analysis of the research on the bioethanol production from green macroalgae. In *Handbook of Algal Science*, Technology and Medicine, Ed. O. Konur, pp. 385–401. London: Academic Press.

Konur, O. 2023. Second generation Industrial waste-based bioethanol fuels: Scientometric study. In *Feedstock-based Bioethanol Fuels. II. Waste Feedstocks: Agricultural, Food, Industrial, Urban, Forestry, and Lignocellulosic Waste-based Bioethanol Fuels. Handbook of Bioethanol Fuels Volume 4*, Ed. O. Konur, pp. 191–216. Boca Raton, FL: CRC Press.

Koutinas, A. A., A. Vlysidis and D. Pleissner, et al. 2014. Valorization of industrial waste and by-product streams via fermentation for the production of chemicals and biopolymers. *Chemical Society Reviews* 43:2587–2627.

Laser, M., D. Schulman and S. G. Allen, et al. 2002. A comparison of liquid hot water and steam pretreatments of sugar cane bagasse for bioconversion to ethanol. *Bioresource Technology* 81:33–44.

Lavarack, B. P., G. J. Griffin and D. Rodman. 2002. The acid hydrolysis of sugarcane bagasse hemicellulose to produce xylose, arabinose, glucose and other products. *Biomass and Bioenergy* 23:367–380.

Lin, Y. and S. Tanaka. 2006. Ethanol fermentation from biomass resources: Current state and prospects. *Applied Microbiology and Biotechnology* 69:627–642.

Ma, X., L. Sun and C. Song. 2002. A new approach to deep desulfurization of gasoline, diesel fuel and jet fuel by selective adsorption for ultra-clean fuels and for fuel cell applications. *Catalysis Today* 77:107–116.

Martin, C., H. B. Klinke and A. B. Thomsen. 2007. Wet oxidation as a pretreatment method for enhancing the enzymatic convertibility of sugarcane bagasse. *Enzyme and Microbial Technology* 40:426–432.

Martin, C., M. Galbe, C. F. Wahlbom, B. Hahn-Hagerdal and L. J. Jonsson. 2002. Ethanol production from enzymatic hydrolysates of sugarcane bagasse using recombinant xylose-utilising *Saccharomyces cerevisiae*. *Enzyme and Microbial Technology* 31:274–282.

Martinez, A., M. E. Rodriguez, S. W. York, J. F. Preston and L. O. Ingram. 2000. Effects of $Ca(OH)_2$ treatments ('overliming') on the composition and toxicity of bagasse hemicellulose hydrolysates. *Biotechnology and Bioengineering* 69:526–536.

Morschbacker, A. 2009. Bio-ethanol based ethylene. *Polymer Reviews* 49:79–84.

Mussatto, S. I., E. M. S. Machado, L. M. Carneiro and T. A. Teixeira. 2012. Sugars metabolism and ethanol production by different yeast strains from coffee industry wastes hydrolysates. *Applied Energy* 92:763–768.

Mussatto, S. I., M. Fernandes, A. M. F. Milagres and I. C. Roberto. 2008. Effect of hemicellulose and lignin on enzymatic hydrolysis of cellulose from brewer's spent grain. *Enzyme and Microbial Technology* 43:124–129.

Najafi, G., B. Ghobadian and T. Tavakoli, et al. 2009. Performance and exhaust emissions of a gasoline engine with ethanol blended gasoline fuels using artificial neural network. *Applied Energy* 86:630–639.

Newman, P. W. G. and J. R. Kenworthy. 1989. Gasoline consumption and cities: A comparison of U.S. cities with a global survey. *Journal of the American Planning Association* 55:24–37.

North, D. C. 1991. Institutions. *Journal of Economic Perspectives* 5:97–112.

Olsson, L. and B. Hahn-Hagerdal. 1996. Fermentation of lignocellulosic hydrolysates for ethanol production. *Enzyme and Microbial Technology* 18:312–331.

Rabelo, S. C., H. Carrere, R. M. Filho, R. and A. C. Costa. 2011. Production of bioethanol, methane and heat from sugarcane bagasse in a biorefinery concept. *Bioresource Technology* 102:7887–7895.

Rezende, C. A., M. de Lima and P. Maziero, et al. 2011. Chemical and morphological characterization of sugarcane bagasse submitted to a delignification process for enhanced enzymatic digestibility. *Biotechnology for Biofuels* 4:54.

Saha, B. C., L. B. Iten, M. A. Cotta and Y. V. Wu. 2005. Dilute acid pretreatment, enzymatic saccharification, and fermentation of rice hulls to ethanol. *Biotechnology Progress* 21:816–822.

Sanchez, O. J. and C. A. Cardona. 2008. Trends in biotechnological production of fuel ethanol from different feedstocks. *Bioresource Technology* 99:5270–5295.

Sun, Y. and J. Cheng. 2002. Hydrolysis of lignocellulosic materials for ethanol production: A review. *Bioresource Technology* 83:1–11.

Taherzadeh, M. J. and K. Karimi. 2007. Enzyme-based hydrolysis processes for ethanol from lignocellulosic materials: A review. *Bioresources* 2:707–738.

Taherzadeh, M. J. and K. Karimi. 2008. Pretreatment of lignocellulosic wastes to improve ethanol and biogas production: A review. *International Journal of Molecular Sciences* 9:1621–1651.

Yazdani, S. S. and R. Gonzalez. 2008. Engineering *Escherichia coli* for the efficient conversion of glycerol to ethanol and co-products. *Metabolic Engineering* 10:340–351.

Yu, J., J. Zhang, J. He, Z. Liu, and Z. Yu. 2009. Combinations of mild physical or chemical pretreatment with biological pretreatment for enzymatic hydrolysis of rice hull. *Bioresource Technology* 100:903–908.

Part 21

Second Generation Urban Waste-based Bioethanol Fuels

69 Second Generation Urban Waste-based Bioethanol Fuels

Scientometric Study

Ozcan Konur
(Formerly) Ankara Yildirim Beyazit University

69.1 INTRODUCTION

Crude oil-based gasoline fuels (Ma et al., 2002; Newman and Kenworthy, 1989) have been widely used in the transportation sector since the 1920s. However, there have been great public concerns over the adverse environmental and human impact of these fuels (Hill et al., 2006, 2009). Hence, biomass-based bioethanol fuels (Hill et al., 2006; Konur, 2012e, 2015, 2019, 2020a) have increasingly been used in blending gasoline fuels (Hsieh et al., 2002; Najafi et al., 2009), in fuel cells (Antolini, 2007, 2009), and in biochemical production (Angelici et al., 2013; Morschbacker, 2009) in a biorefinery context (Fernando et al., 2006; Huang et al., 2008).

Bioethanol fuels also play a critical role in maintaining energy security (Kruyt et al., 2009; Winzer, 2012) in supply shocks (Kilian, 2008, 2009) related to oil price shocks (Hamilton, 1983, 2003), coronavirus disease 2019 (COVID-19) pandemics (Fauci et al., 2020; Li et al., 2020), or wars (Hamilton, 1983; Jones, 2012) in the aftermath of the Russian invasion of Ukraine (Reeves, 2014).

However, it is necessary to pretreat the biomass (Taherzadeh and Karimi, 2008; Yang and Wyman, 2008) to enhance the yield of the bioethanol (Hahn-Hagerdal et al., 2006; Sanchez and Cardona, 2008) before the bioethanol production through the hydrolysis (Sun and Cheng, 2002; Taherzadeh and Karimi, 2007) and fermentation (Lin and Tanaka, 2006; Olsson and Hahn-Hagerdal, 1996) of the biomass and hydrolysates, respectively.

One of the most-studied feedstocks for bioethanol fuels has been urban wastes such as food, municipal solid, paper, textile, plastic, and wood wastes. The research in the field of second generation urban waste-based bioethanol fuels has intensified in this context in the key research fronts of the pretreatment (Chang et al., 2001; Ravindran and Jaiswal, 2016) and hydrolysis (Rahman et al., 2007; Wilkins et al., 2007) of the urban wastes, fermentation (Guimaraes et al., 2010; Sen et al., 2016) of the urban waste-based hydrolysates, and production (Arapoglou et al., 2010; Mussatto et al., 2012) and evaluation (Kalogo et al., 2007; Wang et al., 2013) of the second generation urban waste-based bioethanol fuels. Furthermore, food wastes (Kim et al., 2011; Ravindran and Jaiswal, 2016), municipal solid wastes (Li et al., 2007; Liu et al., 2012), paper wastes (Castanon and Wilke, 1981; Chang et al. (2001), textile wastes (Agblevor et al., 2003; Jeihanipour and Taherzadeh, 2009), plastic wastes (Guclu et al., 2003; Yoshioka et al., 1994), and wood wastes (Okuda et al., 2007, 2008) have been studied intensively in this context.

However, it is essential to develop efficient incentive structures (North, 1991) for the primary stakeholders to enhance the research in this field (Konur, 2000, 2002a–c, 2006a,b, 2007a,b). The scientometric analysis has been used in this context to inform the primary stakeholders about the current state of the research in a selected research field (Garfield, 1955; Konur, 2011, 2012a–i, 2015, 2018b, 2019, 2020a).

As there have been no published scientometric studies in this field, this book chapter presents a scientometric study of the research in second generation urban waste-based bioethanol fuels. It

DOI: 10.1201/9781003226550-94

examines the scientometric characteristics of both the sample and population data presenting the scientometric characteristics of these both datasets in the order of documents, authors, publication years, institutions, funding bodies, source titles, countries, Scopus subject categories, Scopus keywords, and research fronts.

69.2 MATERIALS AND METHODS

The search for this study was carried out using the Scopus database (Burnham, 2006) in October 2022.

As a first step for the search of the relevant literature, the keywords were selected using the first most-cited 300 population papers. The selected keyword list was then optimized to obtain a representative sample of papers for the searched research field. This keyword list was provided in the Appendix for future replicative studies.

As a second step, two sets of data were used for this study. First, a population sample of 1,677 papers was used to examine the scientometric characteristics of the population data. Second, a sample of 168 most-cited papers, corresponding to 10% of the population papers, was used to examine the scientometric characteristics of these citation classics.

The scientometric characteristics of these both sample and population datasets were presented in the order of documents, authors, publication years, institutions, funding bodies, source titles, countries, Scopus subject categories, Scopus keywords, and research fronts.

Lastly, the key scientometric findings for both datasets were discussed to highlight the research landscape for second generation urban waste-based bioethanol fuels. Additionally, several brief conclusions were drawn and many relevant recommendations were made to enhance the future research landscape.

69.3 RESULTS

69.3.1 The Most Prolific Documents in the Second Generation Urban Waste-based Bioethanol Fuels

The information on the types of documents for both datasets is given in Table 69.1. The articles and conference papers, published in journals, dominate both the sample (94%) and population (96%) papers with 2% deficit. Furthermore, review papers and short surveys have a 4% surplus as they are overrepresented in the sample papers as they constitute 6% and 2% of the sample and population papers, respectively.

TABLE 69.1
Documents in Second Generation Urban Waste-based Bioethanol Fuels

Documents	Sample Dataset (%)	Population Dataset (%)	Surplus (%)
Article	91.7	92.5	–0.8
Review	6.0	2.4	3.6
Conference paper	1.8	3.4	–1.6
Note	0.6	0.6	0.0
Book chapter	0.0	1.1	–1.1
Book	0.0	0.0	0.0
Editorial	0.0	0.0	0.0
Letter	0.0	0.0	0.0
Short survey	0.0	0.0	0.0
Sample size	168	1,677	

Sample Dataset: the number of papers (%) in the set of 168 highly cited papers. Population Dataset: the number of papers (%) in the set of 1,677 population papers.

It is further notable that 97% of the population papers were published in journals, while 3% of them were published in books and book series. Similarly, 100% of the sample papers were published in the journals.

69.3.2 The Most Prolific Authors in the Second Generation Urban Waste-based Bioethanol Fuels

The information about the 16 most prolific authors with at least 1.8% of sample papers each is given in Table 69.2. The most prolific author is Karel Grohmann with 4.2% of the sample papers, followed by Mohammad J. Taherzadeh with 3.6% of the sample papers. The other prolific authors are Fikret Kargi, Serpil Ozmihci, and Jose A. Teixeira with 3% of the sample papers each.

However, the most influential author is Karel Grohmann with 3.6% surplus, followed by Mohammad J. Taherzadeh and Jose A. Teixeira with 2.9% and 2.6% surplus, respectively. The other influential authors are Fikret Kargi and Serpil Ozmihci with 2.5% surplus each.

The most prolific institutions for the sample dataset are the Dokuz Eylul University, Imperial College, and University of Minho with two authors each. However, the most prolific country for the sample dataset is the USA with three authors, followed by India, Portugal, Turkey, and the UK with two authors each. In total, ten countries house these top authors.

The most prolific research front for these top authors is the bioethanol fuels from food waste with 13 authors followed by the bioethanol fuels from paper wastes with three authors. The other research fronts are the bioethanol fuels from municipal solid wastes and textile wastes with one author each.

However, there is a significant gender deficit (Beaudry and Lariviere, 2016) for the sample dataset as surprisingly only three of these top researchers are female with a representation rate of 19%.

Additionally, there are other authors with a relatively low citation impact and with 0.4%–1.0% of the population papers each: Hongzhi Ma, Qunhui Wang, Suraini Abd-Aziz, Mohd Ali Hassan, Ekin Demiray, Gonul Donmez, Maria E. Russo, Noeli Sellin, Ozair Souca, M Zahangir Alam, Eulogio Castro, Asma Chaudhary, Chin H. Chia, Widya Fatrisiari, Abdel E. Ghaly, Dimitris Kekos, Soh K. Loh, Dimitrios Malamis, Cintia Marangomi, Antonio Marzocchella, Alessandra Procentese, Foster A. Agblevor, Satinder K. Brar, Manuel Cuevas, Ali Demirci, Haibo Huang, M. F. Ibrahim, Vinod K. Joshi, Kiat M. Lee, Carlos R. Soccol, Sebastian Sanchez, Arief Widjaja, Saranni Zakaria, and Wenyu Zhang.

69.3.3 The Most Prolific Research Output by Years in the Second Generation Urban Waste-based Bioethanol Fuels

Information about papers published between 1970 and 2022 is given in Figure 69.1. This figure clearly shows that the bulk of the research papers in the population dataset were published primarily in the 2010s and the early 2020s with 55% and 29% of the population datasets, respectively. Similarly, the publication rates for the 2000s, 1990s, 1980s, and 1970s were 9%, 4%, 3%, and 1% respectively. Additionally, 0.4% of the population papers were published in the pre-1970s.

Similarly, the bulk of the research papers in the sample dataset were published in the 2010s and 2000s with 61% and 27% of the sample datasets, respectively. Similarly, the publication rates for the early 2020s, 1990s, 1980s, and 1970s were 3%, 8%, 2%, and 0% of the sample papers, respectively.

The most prolific publication years for the population dataset were 2020, 2021, and 2022 with 9.7%, 9.6%, and 9.5% of the datasets, respectively, while 88% of the population papers were published between 2008 and 2022. Similarly, 84% of the sample papers were published between 2006 and 2020, while the most prolific publication years were 2007, 2010, 2013, and 2014 with 10.1%, 9.5%, 9.5%, and 8.9% of the sample papers, respectively. There was a clear rising trend for the population papers starting in 2005.

TABLE 69.2
Most Prolific Authors in Second Generation Urban Waste-based Bioethanol Fuels

No.	Author Name	Author Code	Sample Papers (%)	Population Papers (%)	Surplus (%)	Institution	Country	HI	N	Res. Front
1	Grohmann, Karel	7004503589	4.2	0.6	3.6	Renewable Spirits LLC	USA	49	111	F
2	Taherzadeh, Mohammad J.	6701407496	3.6	0.7	2.9	Univ. Boras	Sweden	66	419	F
3	Kargi, Fikret	57218389979	3.0	0.5	2.5	Dokuz Eylul Univ.	Turkey	54	207	F
4	Ozmihci, Serpil*	6506240734	3.0	0.5	2.5	Dokuz Eylul Univ.	Turkey	20	28	F
5	Teixeira, Jose A.	13402823200	3.0	0.4	2.6	Univ. Minho	Portugal	86	731	F
6	Karimi, Keikhosro	10046195700	2.4	0.5	1.9	Vrije Univ.	Netherlands	56	222	F, M, T
7	Oberoi, Harinder S.	6603479987	2.4	0.4	2.0	Indian Inst. Hort. Res.	India	27	66	F
8	Murphy, Richard J.	35556604100	2.4	0.3	2.1	Imperial Coll.	UK	36	128	P
9	Templer, Richard	7003885074	2.4	0.3	2.1	Imperial Coll.	UK	43	111	P
10	Wang, Lei	57070568200	2.4	0.3	2.1	Shenzhen Inst. Adv. Technol.	China	30	329	P
11	Bae, Hyeun-Jong	24280549300	1.8	0.5	1.3	Chonnam Natl. Univ.	S: Korea	33	117	F
12	Shirai, Yoshihito	7202788909	1.8	0.3	1.5	Kyushu Inst. Technol.	Japan	50	249	F
13	Wilkins, Mark R.	56492323200	1.8	0.3	1.5	Univ. Nebraska Lincoln	USA	28	97	F
14	Vadlani, Praveen V.	24075089500	1.8	0.2	1.6	Sri Sathya Sai Inst. High. Lear.	India	26	80	F
15	Baldwin, Elizabeth A.*	7006618084	1.8	0.2	1.6	USDA Agr. Res. Serv.	USA	42	80	F
16	Mussatto, Solange I.*	6602643634	1.8	0.2	1.6	Univ. Minho	Portugal	56	201	F

Author Code: the unique code given by Scopus to the authors. Sample Papers: the number of papers authored in the sample dataset. Population Papers: the number of papers authored in the population dataset.

*, female; F, food wastes; T, textile wastes; M, municipal solid wastes; and P, paper wastes; HI, Hirsch Index.

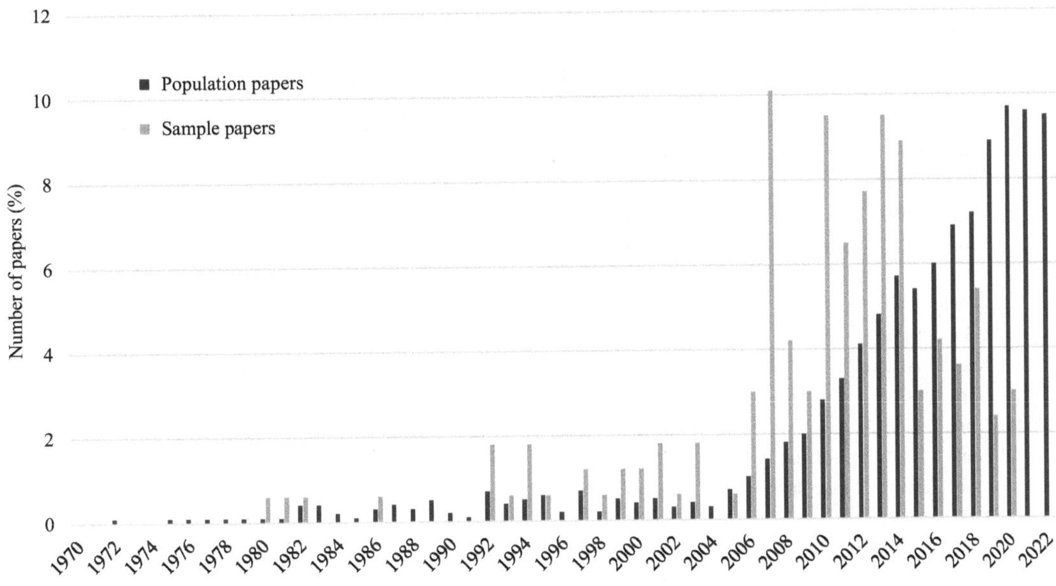

FIGURE 69.1 Research output by years regarding second generation urban waste-based bioethanol fuels.

69.3.4 The Most Prolific Institutions in the Second Generation Urban Waste-based Bioethanol Fuels

Information about the most prolific 22 institutions publishing papers on second generation urban waste-based bioethanol fuels with at least 1.8% of the sample papers each is given in Table 69.3.

The most prolific institutions are the Dokuz Eylul University and University of Boras with 3.6% of the sample papers each, followed by the University of Minho with 3% of the sample papers. Similarly, the top countries for these most prolific institutions are the USA and Malaysia with three institutions each, followed by Greece, South (S.) Korea, and Sweden with two institutions each. In total, 14 countries house these top institutions.

However, the institutions with the most citation impact are Dokuz Eylul University and University of Boras with 3.0% and 2.9% surplus, respectively, followed by University of Minho, Imperial College, Korea Institute of Bioscience and Biotechnology, and the US Department of Agriculture (USDA) with 1.9%–2.3% surplus each.

Additionally, there are other institutions with a relatively low citation impact and with 0.5%–1.3% of the population papers each: Chinese Academy of Sciences, University of Technology Petronas, University of Kebangsaan Malaysia, Prince of Songkla University, Tsinghua University, University of Jaen, Virginia Polytechnic Institute and State University, Federal University of Parana, National Research Council (NRC), Kasetsart University, Federal University of Technology, Chulalongkorn University, Indonesian Institute of Sciences, University of Malaysia Pahang, University of Sains Malaysia, Dalhousie University, Sichuan University, Ten November Institute of Technology, Kyushu University, Technical University of Denmark, City University of Hong Kong, International Islamic University Malaysia, Nanjing Forestry University, University of Vigo, University of the Punjab, and University of Indonesia.

69.3.5 The Most Prolific Funding Bodies in the Second Generation Urban Waste-based Bioethanol Fuels

Information about the 13 most prolific funding bodies funding at least 1.8% of the sample papers each is given in Table 69.4. Furthermore, only 48% and 44% of the sample and population papers each were funded, respectively.

TABLE 69.3
Most Prolific Institutions in Second Generation Urban Waste-based Bioethanol Fuels

No.	Institutions	Country	Sample Papers (%)	Population Papers (%)	Surplus (%)
1	Univ. Boras	Sweden	3.6	0.7	2.9
2	Dokuz Eylul Univ.	Turkey	3.6	0.6	3.0
3	Univ. Minho	Portugal	3.0	0.7	2.3
4	Univ. Putra Malaysia	Malaysia	2.4	2.0	0.4
5	Univ. Sao Paulo	Brazil	2.4	0.8	1.6
6	Isfahan Univ. Technol.	Iran	2.4	0.6	1.8
7	US Dept. Agr.	USA	2.4	0.5	1.9
8	Korea Inst. Biosci. Biotechnol.	S. Korea	2.4	0.5	1.9
9	Imperial Coll.	UK	2.4	0.4	2.0
10	Univ. Sci. Technol. China	China	1.8	1.5	0.3
11	Natl. Tech. Univ. Athens	Greece	1.8	1.1	0.7
12	Univ. Technol. Malaysia	Malaysia	1.8	1.1	0.7
14	Univ. Napoli	Italy	1.8	0.6	1.2
15	Univ. Sains Malaysia	Malaysia	1.8	0.6	1.2
16	Chonnam Natl. Univ.	S. Korea	1.8	0.5	1.3
17	Chalmers Univ. Technol.	Sweden	1.8	0.4	1.4
18	Kyushu Inst. Technol.	Japan	1.8	0.4	1.4
19	Florida Dept. Citrus	USA	1.8	0.3	1.5
20	Univ. Patras	Greece	1.8	0.3	1.5
21	Indian Inst. Hort. Res.	India	1.8	0.2	1.6
22	Kansas State Univ.	USA	1.8	0.2	1.6

TABLE 69.4
Most Prolific Funding Bodies in Second Generation Urban Waste-based Bioethanol Fuels

No.	Funding Bodies	Country	Sample Paper No. (%)	Population Paper No. (%)	Surplus (%)
1	European Commission	EU	5.4	1.7	3.7
2	National Council for Scientific and Technological Development	Brazil	4.2	2.6	1.6
3	Natural Resources Canada	Canada	3.6	2.5	1.1
4	Ministry of Education, Science and Technology	S. Korea	3.6	0.9	2.7
5	National Research Foundation of Korea	S. Korea	3.0	1.3	1.7
6	Ministry of Science, Technology and Innovation	Brazil	3.0	1.0	2.0
7	Horizon 2020 Framework Program	EU	3.0	0.7	2.3
8	Korea Institute of Energy Technology Evaluation and Planning	S. Korea	3.0	0.5	2.5
9	Ministry of Knowledge Economy	S. Korea	3.0	0.4	2.6
10	Dokuz Eylul University	Turkey	3.0	0.4	2.6
11	State Planning Organization	Turkey	2.4	0.3	2.1
12	European Regional Development Fund	EU	1.8	1.8	0.0
13	Ministry of Science and Technology of China	China	1.8	1.4	0.4

The most prolific funding body is the European Commission with 5.4% of the sample papers, followed by the National Council for Scientific and Technological Development with 4.2% of the sample papers. The other prolific funding bodies are Natural Resources Canada and the Ministry of Education, Science and Technology of S. Korea with 3.6% of the sample papers each. However, the most prolific countries for these top funding bodies are S. Korea and the European Union (EU) with four and three funding bodies, respectively. The other prolific funding bodies are Brazil and Turkey with two funding bodies each. In total, only five countries and the EU house these top funding bodies.

The funding body with the most citation impact is the European Commission with 3.7% surplus, followed by the Ministry of Education, Science and Technology with 2.7% surplus. The other influential funding bodies are the Dokuz Eylul University, Ministry of Knowledge Economy, and Korea Institute of Energy Technology Evaluation and Planning with 2.5%–2.6% surplus each. Furthermore, the funding body with the least citation impact is the European Regional Development Fund with 0.0% surplus.

The other funding bodies with a relatively low citation impact and with 0.5%–4.3% of the population papers each are the National Natural Science Foundation of China, National Key Research and Development Program of China, Ministry of Higher Education Malaysia, Fundamental Research Funds for the Central Universities, Ministry of Education of China, Ministry of Science and Innovation, Japan Society for the Promotion of Science, Ministry of Science and Technology India, Research Support Foundation of the State of Sao Paulo, Foundation for Science and Technology, Ministry of Finance, Ministry of Education, Culture, Sports, Science and Technology, Ministry of Science and Technology Taiwan, Ministry of Trade, Industry and Energy, USDA, Biotechnology and Biological Sciences Research Council, National Council of Science and Technology, Ministry of Science, ICT and Future Planning, Natural Sciences and Engineering Research Council of Canada, Thailand Research Fund, Council of Scientific and Industrial Research India, National Institute of Food and Agriculture, National Science Foundation, Prince of Songkla University, and University of Putra Malaysia.

69.3.6 The Most Prolific Source Titles in the Second Generation Urban Waste-based Bioethanol Fuels

Information about the 17 most prolific source titles publishing at least 1.8% of the sample papers each in the second generation urban waste-based bioethanol fuels is given in Table 69.5.

The most prolific source title is Bioresource Technology with 21% of the sample papers, followed by Biomass and Bioenergy and Waste Management with 6% of the sample papers each. The other prolific titles are Renewable Energy, Applied Energy, and Enzyme and Microbial Technology with 4% of the sample papers each.

However, the source title with the most citation impact is Bioresource Technology with 13% surplus, followed by Waste Management and Biomass and Bioenergy with 4% of the sample papers each. The other influential titles are Renewable Energy, Applied Energy, and Enzyme and Microbial Technology with 3% surplus each.

The other source titles with a relatively low citation impact with 0.5%–2.7% of the population papers each are Waste and Biomass Valorization, Biomass Conversion and Biorefinery, Fuel, Chemical Engineering Transactions, Bioprocess and Biosystems Engineering, Fermentation, Energy Sources Part A Recovery Utilization and Environmental Effects, Journal of Cleaner Production, Journal of Chemical Technology and Biotechnology, Advanced Materials Research, Biochemical Engineering Journal, Biocatalysis and Agricultural Biotechnology, International Journal of Biological Macromolecules, Energies, Processes, ACS Sustainable Chemistry and Engineering, African Journal of Biotechnology, Science of the Total Environment, Applied Sciences Switzerland, Bioenergy Research, Energy, 3 Biotech, Biofuels, Biotechnology Letters, Biotechnology Progress, International Journal of ChemTech Research, Materials Today Proceedings, and Preparative Biochemistry and Biotechnology.

TABLE 69.5
Most Prolific Source Titles in Second Generation Urban Waste-based Bioethanol Fuels

No.	Source Titles	Sample Papers (%)	Population Papers (%)	Surplus (%)
1	Bioresource Technology	21.4	8.1	13.3
2	Biomass and Bioenergy	6.0	1.9	4.1
3	Waste Management	6.0	1.8	4.2
4	Renewable Energy	4.2	1.1	3.1
5	Applied Energy	3.6	0.5	3.1
6	Enzyme and Microbial Technology	3.6	0.5	3.1
7	Applied Biochemistry and Biotechnology	3.0	2.6	0.4
8	Industrial Crops and Products	3.0	1.4	1.6
9	Process Biochemistry	2.4	0.9	1.5
10	Carbohydrate Polymers	2.4	0.6	1.8
11	Chemical Engineering Journal	2.4	0.5	1.9
12	Bioresources	1.8	2.6	−0.8
13	Biotechnology for Biofuels	1.8	0.7	1.1
14	Applied Microbiology and Biotechnology	1.8	0.6	1.2
15	Renewable and Sustainable Energy Reviews	1.8	0.4	1.4
16	Industrial and Engineering Chemistry Research	1.8	0.3	1.5
17	Journal of Agricultural and Food Chemistry	1.8	0.3	1.5

69.3.7 The Most Prolific Countries in the Second Generation Urban Waste-based Bioethanol Fuels

Information about the 16 most prolific countries publishing at least 3% of sample papers each in the second generation urban waste-based bioethanol fuels is given in Table 69.6.

The most prolific country is the USA with 15% of the sample papers, followed by S. Korea, Malaysia, India, and China with 11%, 10%, 8%, and 8% of the sample papers, respectively. The other prolific countries are Brazil, Spain, the UK, and Sweden with 6%–7% of the sample papers, respectively. Furthermore, seven European countries listed in Table 69.6 produce 35% and 20% of the sample and population papers, respectively, with 15% surplus. However, China is the largest producer of the population papers with a 13.8% publication rate, followed by India and Malaysia.

However, the country with the most citation impact is the USA with 7% surplus, followed by S. Korea and Sweden with 6% and 4% surplus, respectively. The other influential countries are the UK, Portugal, Greece, and Turkey with 2%–3% surplus each. Similarly, the country with the least citation impact is China with 6% deficit, followed by India with 3% deficit.

Additionally, there are other countries with relatively low citation impact and with 0.5%–6.4% of the sample papers each: Indonesia, Thailand, Nigeria, Taiwan, Mexico, Greece, Pakistan, France, Germany, South Africa, Denmark, Egypt, Australia, Colombia, Poland, Belgium, Netherlands, Saudi Arabia, Algeria, Bangladesh, Finland, Singapore, Ghana, Ireland, Israel, Tunisia, and Vietnam.

69.3.8 The Most Prolific Scopus Subject Categories in the Second Generation Urban Waste-based Bioethanol Fuels

Information about the nine most prolific Scopus subject categories indexing at least 4% of the sample papers each is given in Table 69.7.

TABLE 69.6
Most Prolific Countries in Second Generation Urban Waste-based Bioethanol Fuels

No.	Countries	Sample Papers (%)	Population Papers (%)	Surplus (%)
1	USA	14.9	8.0	6.9
2	S. Korea	10.7	5.2	5.5
3	Malaysia	9.5	9.7	–0.2
4	India	7.7	10.7	–3.0
5	China	7.7	13.8	–6.1
6	Brazil	6.5	6.3	0.2
7	Spain	6.0	4.3	1.7
8	UK	6.0	3.2	2.8
9	Sweden	6.0	1.6	4.4
10	Japan	5.4	5.1	0.3
11	Turkey	5.4	3.5	1.9
12	Italy	4.2	3.9	0.3
13	Portugal	4.2	1.7	2.5
14	Canada	3.6	2.7	0.9
15	Greece	3.6	1.6	2.0
16	Iran	3.0	2.2	0.8

TABLE 69.7
Most Prolific Scopus Subject Categories in Second Generation Urban Waste-based Bioethanol Fuels

No.	Scopus Subject Categories	Sample Papers (%)	Population Papers (%)	Surplus (%)
1	Chemical Engineering	50.6	36.9	13.7
2	Environmental Science	49.4	36.3	13.1
3	Energy	46.4	30.8	15.6
4	Biochemistry, Genetics and Molecular Biology	29.2	23.6	5.6
5	Immunology and Microbiology	22.6	13.8	8.8
6	Chemistry	14.3	14.7	–0.4
7	Agricultural and Biological Sciences	13.7	19.0	–5.3
8	Engineering	11.9	14.8	–2.9
9	Materials Science	4.2	7.1	–2.9

The most prolific Scopus subject category in the second generation urban waste-based bioethanol fuels is Chemical Engineering with 51% of the sample papers, followed by Environmental Science and Energy with 49% and 46% of the sample papers, respectively. The other prolific subject categories are Biochemistry, Genetics and Molecular Biology and Immunology and Microbiology with 29% and 23% of the sample papers, respectively. It is notable that Social Sciences including Economics and Business account for 2% and 5% of the sample and population studies, respectively.

However, the Scopus subject category with the most citation impact is Energy with 16% surplus, followed by Chemical Engineering and Environmental Science with 14% and 13% surplus, respectively. Similarly, the least influential subject category is Agricultural and Biological Sciences with 5% deficit, followed by Engineering and Materials Science with 3% deficit each.

69.3.9 The Most Prolific Keywords in the Second Generation Urban Waste-based Bioethanol Fuels

Information about the Scopus keywords used with at least 6% or 3.6% of the sample or population papers, respectively, is given in Table 69.8. For this purpose, keywords related to the keyword set given in the Appendix are selected from a list of the most prolific keyword set provided by the Scopus database.

TABLE 69.8
Most Prolific Keywords in Second Generation Urban Waste-based Bioethanol Fuels

No.	Keywords	Sample Papers (%)	Population Papers (%)	Surplus (%)
1		**Urban Waste**		
	Fruits	35.1	21.6	13.5
	Cellulose	34.5	18.8	15.7
	Biomass	19.6	13.5	6.1
	Lignin	15.5	14.4	1.1
	Waste	14.3	8.1	6.2
	Carbohydrate	14.3	6.9	7.4
	Empty fruit bunch	13.1	12.2	0.9
	Food waste	10.7	9.5	1.2
	Municipal solid waste	8.3	4.2	4.1
	Lignocellulose	7.7	5.4	2.3
	Lignocellulosic biomass	7.1	4.9	2.2
	Industrial waste	6.5	3.7	2.8
	Hemicellulose	6.0	4.8	1.2
	Citrus	6.0	1.7	4.3
	Starch	5.4	4.5	0.9
2		**Pretreatments**		
	Pre-treatment	21.4	10.7	10.7
	Enzymes	17.3	9.7	7.6
	Pretreatment	14.3	8.0	6.3
	pH	12.5	8.8	3.7
	Temperature	11.9	7.1	4.8
	Sulfuric acid	8.9	4.4	4.5
	Steam	6.0	0.0	6.0
	Delignification	5.4	4.7	0.7
	Sodium hydroxide	4.2	4.1	0.1
3		**Fermentation**		
	Fermentation	57.1	39.6	17.5
	Yeast	23.2	14.4	8.8
	Saccharomyces	22.0	11.1	10.9
	Bioreactors	15.5	7.3	8.2
	Kluyveromyces	11.9	4.9	7.0
	Fungi	10.7	5.2	5.5
	SSF	10.7	4.8	5.9
	Ethanol fermentation	7.1	4.3	2.8
	Acetic acids	4.8	3.7	1.1

(Continued)

TABLE 69.8 (*Continued*)
Most Prolific Keywords in Second Generation Urban Waste-based Bioethanol Fuels

No.	Keywords	Sample Papers (%)	Population Papers (%)	Surplus (%)
4		**Hydrolysis**		
	Hydrolysis	48.8	29.3	19.5
	Sugar	38.7	16.6	22.1
	Saccharification	28.0	14.5	13.5
	Enzymatic hydrolysis	26.8	17.9	8.9
	Glucose	22.6	13.9	8.7
	Enzyme activity	22.6	10.6	12.0
	Cellulases	15.5	7.6	7.9
	Fermentable sugars	4.8	4.1	0.7
	Xylose	4.8	3.8	1.0
	Enzymatic saccharification	4.8	3.6	1.2
	Reducing sugars	3.6	3.8	–0.2
5		**Waste Management**		
	Waste management	18.5	6.3	12.2
	Waste treatment	10.7	6.0	4.7
	Refuse disposal	10.7	4.7	6.0
	Waste disposal	8.3	4.9	3.4
	Waste products	7.7	2.7	5.0
6		**Products**		
	Ethanol	61.3	36.4	24.9
	Biofuels	35.1	16.1	19.0
	Bioethanol	28.0	21.6	6.4
	Bio-ethanol production	14.3	9.7	4.6
	Ethanol production	13.7	7.5	6.2
	Biofuel production	8.3	3.6	4.7

These keywords are grouped under six headings: urban waste, pretreatments, fermentation, hydrolysis and hydrolysates, waste management, and products.

The most prolific keywords related to the biomass and biomass constituents are cellulose and fruits with 35% of the sample papers each. The other prolific keywords are biomass, lignin, waste, carbohydrate, and empty fruit bunches with 13%–20% of the sample papers each.

Furthermore, the prolific keywords related to the pretreatments are pre-treatment and enzymes with 21% and 17% of the sample papers, respectively. The other prolific keywords are the pretreatment, pH, and temperature with 12%–14% of the sample papers.

The most prolific keyword related to fermentation is fermentation with 57% of the sample papers. The other prolific keywords are yeast, saccharomyces, and bioreactors with 16%–23% of the sample papers each.

Furthermore, the most prolific keywords related to hydrolysis and hydrolysates are hydrolysis and sugars with 49% and 39% of the sample papers, respectively. The other prolific keywords are saccharification, enzymatic hydrolysis, glucose, and enzyme activity with 23%–28% of the sample papers each.

The most prolific keywords related to waste management are the waste management, waste treatment, and refuse disposal with 11%–19% of the sample papers. Finally, the most prolific keyword related to the products is ethanol with 61% of the sample papers. The other prolific keywords are biofuels and bioethanol with 35% and 28% of the sample papers.

However, the most prolific keywords across all of the research fronts are ethanol, fermentation, hydrolysis, sugar, fruits, biofuels, cellulose, bioethanol, saccharification, enzymatic hydrolysis, yeast, glucose, enzyme activity, saccharomyces, pre-treatment, and biomass with 20%–61% of the sample papers each. Similarly, the most influential keywords are ethanol, sugar, hydrolysis, biofuels, fermentation, cellulose, fruits, saccharification, waste management, enzyme activity, saccharomyces, and pre-treatment with 10%–27% surplus each.

69.3.10 The Most Prolific Research Fronts in the Second Generation Urban Waste-based Bioethanol Fuels

Information about the research fronts for the sample papers in the second generation urban waste-based bioethanol fuels with regard to the urban waste feedstocks used for bioethanol production is given in Table 69.9.

As this table shows, the most prolific urban waste feedstock is food wastes with 79% of the sample papers, followed by other urban wastes with 25% of the sample papers. The latter research front includes primarily paper wastes and municipal solid wastes with 11% and 6% of the sample papers, respectively, while the other minor research fronts are plastic, textile, and wood wastes with 4%, 3%, and 2% of the sample papers, respectively.

Information about the thematic research fronts for the sample papers in the second generation urban waste-based bioethanol fuels is given in Table 69.10. As this table shows, the most prolific research fronts are pretreatments and hydrolysis of the urban waste feedstocks, fermentation of the urban waste hydrolysates, and bioethanol production with 66%, 53%, 54%, and 57% of the sample papers, respectively. The other minor research front is bioethanol fuel evaluation with 5% of the sample papers.

TABLE 69.9
Most Prolific Research Fronts for Second Generation Urban Waste-based Bioethanol Fuels

No.	Research Fronts	N Paper % Sample
1	Food waste	79.2
	Fruit wastes	39.9
	Food waste in general	15.5
	Vegetable wastes	6.5
	Dairy wastes	5.4
	Kitchen wastes	5.4
	Beverage wastes	4.2
	Other food wastes	2.4
2	Other urban wastes	25.1
	Paper wastes	10.7
	Municipal solid wastes	6.0
	Plastic wastes	3.6
	Textile wastes	3.0
	Wood wastes	1.8

N Paper (%) Sample: the number of papers in the population sample of 168 papers.

TABLE 69.10
Most Prolific Thematic Research Fronts for Second Generation Urban Waste-based Bioethanol Fuels

No.	Research Fronts	N Paper % Sample
1	Biomass hydrolysis	65.5
2	Bioethanol production	57.1
3	Hydrolysate fermentation	53.6
4	Biomass pretreatments	53.0
5	Bioethanol fuel evaluation	5.4

N Paper (%) Sample: the number of papers in the population sample of 168 papers.

69.4 DISCUSSION

69.4.1 Introduction

Crude oil-based gasoline fuels have been widely used in the transportation sector since the 1920s. However, there have been great public concerns over the adverse environmental and human impact of these fuels. Hence, biomass-based bioethanol fuels have increasingly been used in blending gasoline fuels, in fuel cells, and in biochemical production in a biorefinery context.

However, it is necessary to pretreat the biomass to enhance the yield of the bioethanol before the bioethanol production through hydrolysis and fermentation. One of the most-studied feedstocks for bioethanol fuels has been urban wastes such as food, municipal solid, paper, textile, plastic, and wood wastes. The research in the field of second generation urban waste-based bioethanol fuels has intensified in this context in the key research fronts of the pretreatment and hydrolysis of the urban wastes, fermentation of the urban waste-based hydrolysates, and production and evaluation of second generation urban waste-based bioethanol fuels. Furthermore, food, municipal solid, paper, textile, plastic, and wood wastes have been studied intensively in this context.

However, it is essential to develop efficient incentive structures for the primary stakeholders to enhance the research in this field. This is especially important to maintain energy security in the cases of supply shocks such as oil price shocks, war-related shocks as in the case of the Russian invasion of Ukraine, or COVID-19 shocks.

The scientometric analysis has been used in this context to inform the primary stakeholders about the current state of the research in a selected research field. As there has been no scientometric study in this field, this book chapter presents a scientometric study of the research in second generation urban waste-based bioethanol fuels. It examines the scientometric characteristics of both the sample and population data presenting the scientometric characteristics of these both datasets in the order of documents, authors, publication years, institutions, funding bodies, source titles, countries, Scopus subject categories, Scopus keywords, and research fronts.

As a first step for the search of the relevant literature, the keywords were selected using the first 300 most-cited papers. The selected keyword list was then optimized to obtain a representative sample of papers for the searched research field. A copy of this keyword list was provided in the appendix for future replicative studies. Furthermore, a selected list of the keywords is presented in Table 69.8.

As a second step, two sets of data were used for this study. First, a population sample of 1,677 papers was used to examine the scientometric characteristics of the population data. Second, a sample of 168 most-cited papers, corresponding to 10% of the population datasets, was used to examine the scientometric characteristics of these citation classics.

The scientometric characteristics of these sample and population datasets were presented in the order of documents, authors, publication years, institutions, funding bodies, source titles, countries, Scopus subject categories, Scopus keywords, and research fronts.

Lastly, the key scientometric findings for both datasets were discussed to highlight the research landscape for second generation urban waste-based bioethanol fuels. Additionally, several brief conclusions were drawn and many relevant recommendations were made to enhance the future research landscape.

69.4.2 The Most Prolific Documents in the Second Generation Urban Waste-based Bioethanol Fuels

Articles (together with conference papers) dominate both the sample (94%) and population (96%) papers (Table 69.1). Furthermore, review papers have a surplus (4%) and the representation of the reviews in the sample papers is relatively modest (6%).

Scopus differs from the Web of Science database in differentiating and showing articles (92%) and conference papers (2%) published in the journals separately. However, it should be noted that these conference papers are also published in journals as articles, compared with those published only in the conference proceedings. Hence, the total number of articles and review papers in the sample dataset is 94% and 6%, respectively.

It is observed during the search process that there has been inconsistency in the classification of the documents in Scopus and other databases such as Web of Science. This is especially relevant for the classification of papers as reviews or articles as the papers not involving a literature review may be erroneously classified as a review paper. There is also a case of review papers being classified as articles. For example, the total number of reviews in the sample dataset was manually found as nearly 7% compared with 6% as indexed by Scopus, decreasing the number of articles and conference papers to 93% for the sample dataset.

In this context, it would be helpful to provide a classification note for the published papers in the books and journals at the first instance. It would also be helpful to use the document types listed in Table 69.1 for this purpose. Book chapters may also be classified as articles or reviews as an additional classification to differentiate review chapters from experimental chapters as it is done by the Web of Science. It would be further helpful to additionally classify the conference papers as articles or review papers and it is done in the Web of Science database.

69.4.3 The Most Prolific Authors in the Second Generation Urban Waste-based Bioethanol Fuels

There have been 16 most prolific authors with at least 1.8% of the sample papers each as given in Table 69.2. These authors have shaped the development of the research in this field.

The most prolific authors are Karel Grohmann, Mohammad J. Taherzadeh, and to a lesser extent Fikret Kargi, Serpil Ozmihci, and Jose A. Teixeira.

It is important to note the inconsistencies in indexing of the author names in Scopus and other databases. It is especially an issue for names with more than two components such as 'Blake Sam de Hyun Kargi'. The probable outcomes are 'Kargi, B.S.D.H.', 'de Hyun Kargi, B.S.', or 'Hyun Kargi, B.S.D.'. The first choice is the gold standard of the publishing sector as the last word in the name is taken as the last name. In most of the academic databases such as PubMed and EBSCO databases, this version is used predominantly. The second choice is a strong alternative, while the last choice is an undesired outcome as two last words are taken as the last name. It is good practice to combine the words of the last name by a hyphen: 'Hyun-Kargi, B.S.D.'. It is notable that inconsistent indexing of the author names may cause substantial inefficiencies in the search process for the papers and allocating credit to the authors as there are different author entries for each outcome in the databases.

There are also inconsistencies in the shortening of Chinese names. For example, 'YangYing Wang' is often shortened as 'Wang, Y.', 'Wang, Y.-Y.', and 'Wang, Y.Y.' as it is done in the Web of Science database as well. However, the gold standard in this case is 'Wang, Y' where the last word is taken as the last name and the first word is taken as a single forename. In most of the academic databases such as PubMed and EBSCO, this first version is used predominantly. However, it makes sense to use the third option to differentiate Chinese names efficiently: 'Wang, Y.Y.'. Therefore, there have been difficulties in locating papers for Chinese authors. In such cases, the use of the unique author codes provided for each author by the Scopus database has been helpful.

There is also a difficulty in allowing credit for the authors, especially for the authors with common names such as 'Zhuang, X.' in conducting scientometric studies. These difficulties strongly influence the efficiency of the scientometric studies and allocating credit to the authors as there are the same author entries for different authors with the same name, e.g., 'Zhuang, X.' in the databases.

In this context, the coding of authors in the Scopus database is a welcome innovation compared with the other databases such as Web of Science. In this process, Scopus allocates a unique number to each author in the database (Aman, 2018). However, there might still be substantial inefficiencies in this coding system, especially for common names. For example, some of the papers for a certain author may be allocated to another researcher with a different author code. It is possible that Scopus uses many software programs to differentiate the author names and the program may not be false-proof (Kim, 2018).

In this context, it does not help that author names are not given in full in some journals and books. This makes it difficult to differentiate authors with common names and makes the scientometric studies further difficult in the author domain. Therefore, the author names should be given in all books and journals at the first instance. There is also a cultural issue where some authors do not use their full names in their papers. Instead, they use initials for their forenames: 'Kargi, H.J.', 'Kargi, H.', or 'Kargi, J.' instead of 'Kargi, Hyun Jae'.

There are also inconsistencies in the naming of the authors with more than two components by the authors themselves in journal papers and book chapters. For example, 'Grohmann, A.P.C.' might be given as 'Grohmann, A.' or 'Grohmann, A.C.' or 'Grohmann, A.P.' or 'Grohmann, C' in the journals and books. This also makes the scientometric studies difficult in the author domain. Hence, contributing authors should use their name consistently in their publications.

The other critical issue regarding the author names is the inconsistencies in the spelling of the author names in the national spellings (e.g., Şöğütçiğdem, Çağla) rather than in English spellings (e.g., Sogutcigdem, Cagla) in the Scopus database. Scopus differs from the Web of Science database and many other databases in this respect where the author names are given only in the English spellings. It is observed that national spellings of the author names do not help much in conducting scientometric studies and in allocating credits to the authors as sometimes there are different author entries for the English and national spellings in the Scopus database.

The most prolific institutions for the sample dataset are Dokuz Eylul University, Imperial College, and University of Minho. Furthermore, the most prolific countries for the sample dataset are the USA and to a lesser extent India, Portugal, Turkey, and the UK. These findings confirm the dominance of the USA, India, and Europe in this field. However, the primary research front is the bioethanol fuels from food waste, while the other research fronts are the bioethanol fuels from municipal solid wastes and textile wastes.

It is also notable that there is a significant gender deficit for the sample dataset as surprisingly with a representation rate of 19%. This finding is the most thought-provoking with strong public policy implications. Hence, institutions, funding bodies, and policymakers should take efficient measures to reduce the gender deficit in this field and other scientific fields with strong gender deficit. In this context, it is worth to note the level of representation of the researchers from minority groups in science based on race, sexuality, age, and disability, besides gender (Blankenship, 1993; Dirth and Branscombe, 2017; Konur, 2000, 2002a–c, 2006a,b, 2007a,b).

69.4.4 The Most Prolific Research Output by Years in the Second Generation Urban Waste-based Bioethanol Fuels

The research output observed between 1970 and 2022 is illustrated in Figure 69.1. This figure clearly shows that the bulk of the research papers in the population dataset were published primarily in the 2010s and early 2020s. Similarly, the bulk of the research papers in the sample dataset were published in the 2010s and 2000s. There was a clear rising trend for the population papers starting in 2005.

These findings suggest that the most prolific sample and population papers were primarily published in the 2010s. Furthermore, a significant portion of the sample and population papers were published in the early 2020s and 2000s, respectively.

These are the thought-provoking findings as there has been a significant research boom since 2008. In this context, the increasing public concerns about climate change (Change, 2007), greenhouse gas emissions (Carlson et al., 2017), and global warming (Kerr, 2007) have been certainly behind the boom in the research in this field in the last two decades. Furthermore, the recent supply shocks experienced due to the COVID-19 pandemic and the Ukrainian war might also be behind the research boom in this field since 2019.

Based on these findings, the size of the population papers is likely to more than double in the current decade, provided that the public concerns about climate change, greenhouse gas emissions, and global warming, as well as the supply shocks, are translated efficiently to the research funding in this field.

69.4.5 The Most Prolific Institutions in the Second Generation Urban Waste-based Bioethanol Fuels

The 22 most prolific institutions publishing papers on second generation urban waste-based bioethanol fuels with at least 1.8% of the sample papers each given in Table 69.3 have shaped the development of the research in this field.

The most prolific institutions are the Dokuz Eylul University, University of Boras, and University of Minho. Similarly, the top countries for these most prolific institutions are the USA, Malaysia, and to a lesser extent Greece, S. Korea, and Sweden. In total, 14 countries house these top institutions.

However, the institutions with the most citation impact are Dokuz Eylul University, University of Boras, and to a lesser extent University of Minho, Imperial College, Korea Institute of Bioscience and Biotechnology, and the USDA. These findings confirm the dominance of the institutions from the USA, Europe, and S. Korea.

69.4.6 The Most Prolific Funding Bodies in the Second Generation Urban Waste-based Bioethanol Fuels

The 13 most prolific funding bodies funding at least 1.8% of the sample papers each are given in Table 69.4. It is notable that only 48% and 44% of the sample and population papers were funded, respectively.

The most prolific funding bodies are the European Commission, the National Council for Scientific and Technological Development, and to a lesser extent the Natural Resources Canada and the Ministry of Education, Science and Technology of S. Korea. However, the most prolific countries for these top funding bodies are S. Korea, the EU, and to a lesser extent Brazil and Turkey. In total, only six countries and the EU house these top funding bodies.

The funding bodies with the most citation impact are European Commission, Ministry of Education, Science and Technology of S. Korea, and to a lesser extent Dokuz Eylul University, Ministry of Knowledge Economy, and Korea Institute of Energy Technology Evaluation and Planning.

These findings on the funding of the research in this field suggest that the level of funding, mostly since 2008, has been largely instrumental in enhancing the research in this field (Ebadi and Schiffauerova, 2016) in light of North's institutional framework (North, 1991). It is also notable that the funding rate in this field is relatively modest compared with those in the other research fronts of the bioethanol fuels such as algal bioethanol fuels. Furthermore, it is expected that this high funding rate would improve in light of the recent supply shocks. Furthermore, it emerges that Brazil, Europe, and S. Korea have heavily funded the research on urban waste-based bioethanol fuels.

69.4.7 The Most Prolific Source Titles in the Second Generation Urban Waste-based Bioethanol Fuels

The 17 most prolific source titles publishing at least 1.8% of the sample papers each in the second generation urban waste-based bioethanol fuels have shaped the development of the research in this field (Table 69.5).

The most prolific source titles are Bioresource Technology and to a lesser extent Biomass and Bioenergy, Waste Management, Renewable Energy, Applied Energy, and Enzyme and Microbial Technology. However, the source titles with the most citation impact are Bioresource Technology and to a lesser extent by Waste Management, Biomass and Bioenergy, Renewable Energy, Applied Energy, and Enzyme and Microbial Technology.

It is notable that these top source titles are primarily related to bioresources, wastes, energy, and microbial technology. This finding suggests that Bioresource Technology and the other prolific journals in these fields have significantly shaped the development of the research in this field as they focus primarily on second generation urban waste-based bioethanol fuels with a high yield. In this context, the influence of the top journal is quite extraordinary.

69.4.8 The Most Prolific Countries in the Second Generation Urban Waste-based Bioethanol Fuels

The 16 most prolific countries publishing at least 3% of the sample papers each have significantly shaped the development of the research in this field (Table 69.6).

The most prolific countries are the USA, S. Korea, Malaysia, and to a lesser extent India, China, Brazil, Spain, the UK, and Sweden. Furthermore, seven European countries listed in Table 69.6 pro duce 35% and 20% of the sample and population papers, respectively, with 15% surplus. However, China is the largest producer of the population papers with a 13.8% publication rate, followed by India and Malaysia.

However, the countries with the most citation impact are the USA, S. Korea, and to a lesser extent Sweden, the UK, Portugal, Greece, and Turkey. Similarly, the countries with the least impact are India and China.

A close examination of these findings suggests that Europe, the USA, S. Korea, Malaysia, China, and to a lesser extent Brazil are the major producers of the research in this field. It is a fact that the USA has been a major player in science (Leydesdorff and Wagner, 2009). The USA has further developed a strong research infrastructure to support its corn and grass-based bioethanol industry (Gillon, 2010).

However, China has been a rising mega star in scientific research in competition with the USA and Europe (Leydesdorff and Zhou, 2005). China is also a major player in this field as a major producer of bioethanol (Fang et al., 2010).

Next, Europe has been a persistent player in scientific research in competition with both the USA and China (Leydesdorff, 2000). Europe has also been a persistent producer of bioethanol along with the USA and Brazil (Gnansounou, 2010).

Furthermore, Brazil (Soccol et al., 2010), S. Korea (Leydesdorff and Zhou, 2005), Malaysia (Tan et al., 2015), and India (Basu and Kumar, 2000) are the other countries with substantial research activities in bioethanol fuels.

69.4.9 The Most Prolific Scopus Subject Categories in the Second Generation Urban Waste-based Bioethanol Fuels

The nine most prolific Scopus subject categories indexing at least 4% of the sample papers each, given in Table 69.7, have shaped the development of the research in this field.

The most prolific Scopus subject categories in the second generation urban waste-based bioethanol fuels are Chemical Engineering, Environmental Science, Energy, and to a lesser extent Biochemistry, Genetics and Molecular Biology, and Immunology and Microbiology. It is also notable that Social Sciences including Economics and Business have a minimal presence in both sample and population studies.

However, the Scopus subject categories with the most citation impact are Energy, Chemical Engineering, and Environmental Science. Similarly, the least influential subject categories are Agricultural and Biological Sciences, Engineering, and Materials Science.

These findings are thought-provoking suggesting that the primary subject categories are related to energy, chemical engineering, environmental sciences, and to a lesser extent genetics and microbiology as the core of the research in this field concerns with the production and utilization of the second generation urban waste-based bioethanol fuels. The other finding is that social sciences are not well represented in both the sample and population papers as in line with most fields in bioethanol fuels. Social, environmental, and economic studies account for the field of social sciences.

69.4.10 The Most Prolific Keywords in the Second Generation Urban Waste-based Bioethanol Fuels

A limited number of keywords have shaped the development of the research in this field as shown in Table 69.8 and the Appendix. These keywords are grouped under the five headings: urban waste, pretreatments, fermentation, hydrolysis and hydrolysates, and products.

The most prolific keywords across all of the research fronts are ethanol, fermentation, hydrolysis, sugar, fruits, biofuels, cellulose, bioethanol, saccharification, enzymatic hydrolysis, yeast, glucose, enzyme activity, saccharomyces, pre-treatment, and biomass. Similarly, the most influential keywords are ethanol, sugar, hydrolysis, biofuels, fermentation, cellulose, fruits, saccharification, waste management, enzyme activity, saccharomyces, and pre-treatment.

These findings suggest that it is necessary to determine the keyword set carefully to locate the relevant research in each of these research fronts. Additionally, the size of the samples for each keyword highlights the intensity of the research in the relevant research areas for both sample and population datasets. These findings also highlight different spelling of some strategic keywords such as pretreatment v. pre-treatment and ethanol v. bio-ethanol.

69.4.11 The Most Prolific Research Fronts in the Second Generation Urban Waste-based Bioethanol Fuels

Information about the research fronts for the sample papers in the second generation urban waste-based bioethanol fuels with regard to the urban waste feedstocks used for bioethanol production is given in Table 69.9. As this table shows, the most prolific urban waste feedstock is food wastes with 79% of the sample papers, followed by other urban wastes with 25% of the sample papers. The latter research front includes primarily paper wastes and municipal solid wastes, while the other minor research fronts are plastic, textile, and wood wastes.

Information about the thematic research fronts for the sample papers in the second generation urban waste-based bioethanol fuels is given in Table 69.10. As this table shows, the most prolific research fronts are the pretreatments and hydrolysis of the wood waste feedstocks, fermentation of the urban waste hydrolysates, and bioethanol production. The other minor research front is bioethanol fuel evaluation.

These findings are thought-provoking in seeking ways to increase urban waste feedstock-based bioethanol yield at the global scale. It is clear that all of these research fronts have public importance and merit substantial funding and other incentives. Furthermore, it is notable that second generation urban waste-based bioethanol fuels have become a core unit of bioethanol research to make it more competitive with crude oil-based gasoline and diesel fuels, especially for the USA, Europe, and China. It also solves the tremendous waste treatment (Morrissey and Browne, 2004; Wilson, 2007) problem of the huge amounts of urban wastes, avoiding ecological disasters.

In comparison with the other feedstock-based research fronts, it is notable that the pretreatment and hydrolysis of the wood wastes emerge as primary research fronts for this field. However, the research fronts of the fermentation of the urban waste hydrolysates and bioethanol production are also important. Furthermore, the field of the evaluation and utilization of bioethanol fuels is a neglected area. This suggests that the primary stakeholders have been primarily interested in these key processes of bioethanol production. It is also notable that evaluation of the second generation urban waste-based bioethanol fuels such as technoeconomics, life cycle, economics, social, land use, labor, and environment-related studies emerges as a case study for bioethanol fuels. Similarly, the utilization of these biofuels in gasoline or diesel engines is also an important research field from a societal perspective. In this context, the USA and Brazil have been the global leaders in the production and use of corn- and sugarcane-based bioethanol fuels since the 1970s in the aftermath of the global crude oil crisis in the early 1970s.

It is further notable that the research on second generation urban waste-based bioethanol fuels complements the research on the first generation bioethanol fuel research from starch and sugar feedstocks such as sugarcane or corn extracting further value from these primary feedstocks.

In the end, these most-cited papers in this field hint that the production of second generation urban waste-based bioethanol fuels could be optimized using the structure, processing, and property relationships of these urban wastes in the fronts of the feedstock pretreatment and hydrolysis, and hydrolysate fermentation (Formela et al., 2016; Konur, 2018a, 2020b, 2021a–d; Konur and Matthews, 1989).

69.5 CONCLUSION AND FUTURE RESEARCH

The research on second generation urban waste-based bioethanol fuels has been mapped through a scientometric study of both sample (168 papers) and population (1,677 papers) datasets.

The critical issue in this study has been to obtain a representative sample of the research as in any other scientometric study. Therefore, the keyword set has been carefully devised and optimized after several runs in the Scopus database. It is a representative sample of the wider population studies. This keyword set was provided in the Appendix, and the relevant keywords are presented in Table 69.8. However, it should be noted that it has been very difficult to compile a representative keyword set since this research field has been connected closely with many other fields. Therefore, it has been necessary to compile a keyword list to exclude papers concerned with other research fields. It covers food, plastic, wood, textile, and paper wastes as the key components of the urban wastes and the general field of municipal solid wastes.

The other issue has been the selection of a multidisciplinary database to carry out the scientometric study of the research in this field. For this purpose, the Scopus database has been selected. The journal coverage of this database has been notably wider than that of the Web of Science and other multi-subject databases.

The key scientometric properties of the research in this field have been determined and discussed in this book chapter. It is evident that a limited number of documents, authors, institutions, publication years, institutions, funding bodies, source titles, countries, Scopus subject categories, Scopus keywords, and research fronts have shaped the development of the research in this field.

There is ample scope to increase the efficiency of the scientometric studies in this field in the author and document domains by developing consistent policies and practices in both domains across all academic databases. In this respect, it seems that authors, journals, and academic databases have a lot to do. Furthermore, the significant gender deficit as in most scientific fields emerges as a public policy issue. The potential deficits based on age, race, disability, and sexuality need also to be explored in this field as in other scientific fields.

The research in this field has boomed since 2008, possibly promoted by public concerns about global warming, greenhouse gas emissions, and climate change. Furthermore, the recent COVID-19 pandemic and the Russian invasion of Ukraine have resulted in global supply shocks shifting the recent focus of the stakeholders from crude oil-based fuels to biomass-based fuels such as bioethanol fuels.

It is expected that there would be further incentives for the key stakeholders to carry out the research for the second generation urban waste-based bioethanol fuels to increase the ethanol yield and to make it more competitive with crude oil-based gasoline and petrodiesel fuels. This might be truer for the crude oil- and foreign exchange-deficient countries to maintain energy and food security in the face of the global supply shocks. It also solves the tremendous waste treatment problem of the huge amounts of urban wastes, avoiding ecological disasters.

The relatively modest funding rate of 48% and 44% for the sample and population papers, respectively, suggests that funding in this field significantly enhanced the research in this field primarily since 2008, possibly more than doubling in the current decade. However, it is evident that there is ample room for more funding and other incentives to enhance the research in this field further in light of the recent supply shocks.

The institutions from Malaysia and to a lesser extent Greece, S. Korea, and Sweden have mostly shaped the research in this field. Furthermore, Europe, the USA, S. Korea, Malaysia, and to a lesser extent India and China have been the major producers of the research in this field as the major producers and users of bioethanol fuels. It is evident that these countries have well-developed research infrastructure in bioethanol fuels and their derivatives.

It emerges that ethanol is more popular than bioethanol as a keyword with strong implications for the search strategy. In other words, the search strategy using only the bioethanol keyword would not be much helpful. The Scopus keywords are grouped under six headings: biomass, pretreatments, fermentation, hydrolysis and hydrolysates, waste management, and products.

As Table 69.9 shows, the most prolific urban waste feedstock is food wastes with 79% of the sample papers, followed by other urban wastes with 25% of the sample papers. The latter research front includes primarily paper and municipal solid wastes, while the other minor research fronts are plastic, textile, and wood wastes. Furthermore, food wastes include fruit wastes and to a lesser extent food waste, in general, vegetable, dairy, kitchen, beverage, and other food wastes.

However, Table 69.10 shows that the most prolific research fronts are pretreatments and hydrolysis of the urban waste feedstocks, fermentation of the urban waste hydrolysates, and bioethanol production. The other minor research front is bioethanol fuel evaluation. The first four research fronts dominate the research in this field, while the field of the utilization and evaluation of urban waste-based bioethanol fuels is relatively a neglected research field.

These findings are thought-provoking in seeking ways to increase urban waste feedstock-based bioethanol yield at the global scale. It is clear that all of these research fronts have public importance and merit substantial funding and other incentives. Furthermore, it is notable that second generation urban waste-based bioethanol fuels have become a core unit of bioethanol research to make it more competitive with crude oil-based gasoline and petrodiesel fuels, especially for the USA, Europe, Brazil, and China. It is further notable that the research on second generation urban waste-based bioethanol fuels complements the research on the first generation bioethanol fuels from starch and sugar feedstocks such as corn and sugarcane, extracting further value from these primary feedstocks.

Thus, the scientometric analysis has a great potential to gain valuable insights into the evolution of the research in this field as in other scientific fields, especially in the aftermath of the significant global supply shocks such as COVID-19 pandemic and the Russian invasion of Ukraine.

It is recommended that further scientometric studies are carried out for the primary research fronts. It is further recommended that reviews of the most-cited papers are carried out for each primary research front to complement these scientometric studies. Next, the scientometric studies of the hot papers in these primary fields are carried out.

ACKNOWLEDGMENTS

The contribution of the highly cited researchers in the field of second generation urban waste-based bioethanol fuels has been gratefully acknowledged.

APPENDIX: THE KEYWORD SET FOR SECOND GENERATION URBAN WASTE-BASED BIOETHANOL FUELS

(((((TITLE (kitchen OR "polyethylene terephthalate" OR pet OR urban OR municipal OR domestic OR newsp* OR paper* OR *food OR *wood OR msw OR cheese OR potato OR grape OR coffee OR fruit* OR citrus OR kinnow OR banana OR bread OR tea OR tuna OR vegetable OR mandarin OR lemon OR *apple OR pear OR mango OR *nut OR orange OR cassava OR sago OR taro OR date OR pomegranate OR cocoa OR textile* OR cotton OR pomelo) AND TITLE (*waste OR garbage OR wasted OR scrap OR residues OR hulls OR husks OR shells)) OR TITLE (peel* OR newsp* OR "spent coffee" OR "olive stone" OR "fruit bunch*" OR pomaces OR marc OR "cheese whey" OR "coffee husks" OR silverskin OR dregs OR msw OR "municipal solid waste" OR "kitchen *waste" OR "potato waste*" OR "food *waste" OR "kitchen garbage" OR "house* wood" OR "urban *waste" OR "domestic organic waste")) AND TITLE (ethanol OR bioethanol OR *saccharification OR *hydrolysis OR digestibili* OR ssf OR recalcitrance OR hydrolysate* OR hydrolyzate* OR ferment* OR fractionation OR delignification OR pretreat* OR "pre-treat*" OR "consolidated processing")) AND NOT (TITLE (mill OR "thermal pretreat" OR chromium OR *hydrogen OR methan* OR anaerobic OR {ns-2} OR "fatty acid" OR biomethan* OR sludge OR digestion OR struvite OR biogas OR combustion OR coal OR landfill* OR hydrothermal OR ash OR lactic OR compost* OR vfa OR h2 OR dots OR *butyrate OR xanthan OR "cellulase production" OR carboxylic OR *butanol OR agricultural OR "activated carbon" OR lignac* OR arsenic OR protein* OR ch4 OR succin* OR "solid state" OR anti* OR citric OR "cellulose production" OR furfural OR "solid substrate" OR *hythane OR biosorption OR pyrolysis OR wastewater OR binders OR aerogels OR "fuel cells" OR *alkanoates OR keratin* OR stem OR endoglucanase OR bromelain OR syngas OR *lipid OR beverage OR *oligosaccharides OR pectin OR amino OR cement OR collagen OR ester* OR *amylase OR lycopene OR biochar OR dark OR ions OR exopolysaccharide OR phenol* OR "agro waste" OR carbamate OR *worm OR supercrit* OR decant* OR volume* OR soil OR acetamin* OR caproate OR bioactive OR biodiesel OR trove OR polyphenol* OR "waste waters" OR acidogen* OR bakers OR isolation OR "*wood residues" OR "forest resid*" OR rats OR stillage OR plant OR chlorophenol OR vinegar OR cannery OR nanocr* OR nanofib* OR nanoce*) OR SRCTITLE (hazard* OR nutr* OR food* OR poultry OR carcin* OR water* OR meat OR lwt OR animal OR fluids OR aqua* OR dairy OR desal* OR peptides OR phyto* OR med* OR ruminant OR archa* OR amino OR hydrogen OR livestock OR pyrol* OR anal* OR hort* OR chromat* OR cellulose OR lebens* OR enol*) OR SUBJAREA (medi OR vete OR phar OR nurs OR heal OR neur OR dent)))) OR ((TITLE (ethanol OR bioethanol) AND TITLE (peel* OR newsp* OR "spent coffee" OR "olive stone" OR "fruit bunch*" OR pomaces OR marc OR "cheese whey" OR "coffee husks" OR silverskin OR dregs OR msw OR "municipal solid waste" OR "kitchen *waste" OR "potato waste*" OR "food waste" OR "kitchen garbage" OR "house* wood" OR "domestic organic waste" OR "urban waste")) AND NOT TITLE (anti* OR "fatty acid*" OR cannery OR *cyanins OR carotene OR {hydrogen prod*} OR *cyanidins OR {methane yield} OR glycoalkaloids OR caproate OR washing OR trove OR leaf OR eggplant OR pork OR foul* OR phenolic OR amylase OR sludge OR galactomannan OR supercritical OR "forest residues" OR anaerobic OR emissions OR solvent OR nanowh* OR "ethanol type")) AND (LIMIT-TO (DOCTYPE, "ar") OR LIMIT-TO (DOCTYPE, "cp") OR LIMIT-TO (DOCTYPE, "re") OR LIMIT-TO (DOCTYPE, "ch") OR LIMIT-TO (DOCTYPE, "no") OR LIMIT-TO (DOCTYPE, "sh")) AND (LIMIT-TO (LANGUAGE, "English")) AND (LIMIT-TO (SRCTYPE, "j") OR LIMIT-TO (SRCTYPE, "k") OR LIMIT-TO (SRCTYPE, "b")).

REFERENCES

Agblevor, F. A., S. Batz and J. Trumbo. 2003. Composition and ethanol production potential of cotton gin residues. *Applied Biochemistry and Biotechnology, Part A: Enzyme Engineering and Biotechnology* 105:219–230.

Aman, V. 2018. Does the Scopus author ID suffice to track scientific international mobility? A case study based on Leibniz laureates. *Scientometrics* 117:705–720.

Angelici, C., B. M. Weckhuysen and P. C. A. Bruijnincx. 2013. Chemocatalytic conversion of ethanol into butadiene and other bulk chemicals. *ChemSusChem* 6:1595–1614.

Antolini, E. 2007. Catalysts for direct ethanol fuel cells. *Journal of Power Sources* 170:1–12.

Antolini, E. 2009. Palladium in fuel cell catalysis. *Energy and Environmental Science* 2:915–931.

Arapoglou, D., T. Varzakas, A. Vlyssides and C. Israilides. 2010. Ethanol production from potato peel waste (PPW). *Waste Management* 30:1898–1902.

Basu, A. and B. V. Kumar. 2000. International collaboration in Indian scientific papers. *Scientometrics* 48:381–402.

Beaudry, C. and V. Lariviere. 2016. Which gender gap? Factors affecting researchers' scientific impact in science and medicine. *Research Policy* 45:1790–1817.

Blankenship, K. M. 1993. Bringing gender and race in: US employment discrimination policy. *Gender & Society* 7:204–226.

Burnham, J. F. 2006. Scopus database: A review. *Biomedical Digital Libraries* 3:1–8.

Carlson, K. M., J. S. Gerber and D. Mueller, et al. 2017. Greenhouse gas emissions intensity of global croplands. *Nature Climate Change* 7:63–68.

Castanon, M. and C. R. Wilke. 1981. Effects of the surfactant tween 80 on enzymatic hydrolysis of newspaper. *Biotechnology and Bioengineering* 23:1365–1372.

Chang, V. S., M. Nagwani, C. H. Kim and M. T. Holtzapple. 2001. Oxidative lime pretreatment of high-lignin biomass: Poplar wood and newspaper. *Applied Biochemistry and Biotechnology, Part A: Enzyme Engineering and Biotechnology* 94:1–28.

Change, C. 2007. Climate change impacts, adaptation and vulnerability. *Science of the Total Environment* 326:95–112.

Dirth, T. P. and N. R. Branscombe. 2017. Disability models affect disability policy support through awareness of structural discrimination. *Journal of Social Issues* 73:413–442.

Ebadi, A. and A. Schiffauerova. 2016. How to boost scientific production? A statistical analysis of research funding and other influencing factors. *Scientometrics* 106:1093–1116.

Fang, X., Y. Shen, J. Zhao, X. Bao and Y. Qu. 2010. Status and prospect of lignocellulosic bioethanol production in China. *Bioresource Technology* 101:4814–4819.

Fauci, A. S., H. C. Lane and R. R. Redfield. 2020. Covid-19-navigating the uncharted. *New England Journal of Medicine* 382:1268–1269.

Fernando, S., S. Adhikari, C. Chandrapal and M. Murali. 2006. Biorefineries: Current status, challenges, and future direction. *Energy & Fuels* 20:1727–1737.

Formela, K., A. Hejna, L. Piszczyk, M. R. Saeb and X. Colom. 2016. Processing and structure-property relationships of natural rubber/wheat bran biocomposites. *Cellulose* 23:3157–3175.

Garfield, E. 1955. Citation indexes for science. *Science* 122:108–111.

Gillon, S. 2010. Fields of dreams: Negotiating an ethanol agenda in the Midwest United States. *Journal of Peasant Studies* 37:723–748.

Gnansounou, E. 2010. Production and use of lignocellulosic bioethanol in Europe: Current situation and perspectives. *Bioresource Technology* 101:4842–4850.

Guclu, G., T. Yalçinyuva, S. Ozgumus and M. Orbay, M. 2003. Hydrolysis of waste polyethylene terephthalate and characterization of products by differential scanning calorimetry. *Thermochimica Acta* 404:193–205.

Guimaraes, P. M. R., J. A. Teixeira and L. Domingues. 2010. Fermentation of lactose to bio-ethanol by yeasts as part of integrated solutions for the valorisation of cheese whey. *Biotechnology Advances* 28:375–384.

Hahn-Hagerdal, B., M. Galbe, M. F. Gorwa-Grauslund, G. Liden and G. Zacchi. 2006. Bio-ethanol: The fuel of tomorrow from the residues of today. *Trends in Biotechnology* 24:549–556.

Hamilton, J. D. 1983. Oil and the macroeconomy since World War II. *Journal of Political Economy* 91:228–248.

Hamilton, J. D. 2003. What is an oil shock? *Journal of Econometrics* 113:363–398.

Hill, J., E. Nelson, D. Tilman, S. Polasky and D. Tiffany. 2006. Environmental, economic, and energetic costs and benefits of biodiesel and ethanol biofuels. *Proceedings of the National Academy of Sciences of the United States of America* 103:11206–11210.

Hill, J., S. Polasky and E. Nelson, et al. 2009. Climate change and health costs of air emissions from biofuels and gasoline. *Proceedings of the National Academy of Sciences of the United States of America* 106:2077–2082.

Hsieh, W. D., R. H. Chen, T. L. Wu and T. H. Lin. 2002. Engine performance and pollutant emission of an SI engine using ethanol-gasoline blended fuels. *Atmospheric Environment* 36:403–410.

Huang, H. J., S. Ramaswamy, U. W. Tschirner and B. V. Ramarao. 2008. A review of separation technologies in current and future biorefineries. *Separation and Purification Technology* 62:1–21.

Jeihanipour, A. and M. J. Taherzadeh. 2009. Ethanol production from cotton-based waste textiles. *Bioresource Technology* 100:1007–1010.

Jones, T. C. 2012. America, oil, and war in the Middle East. *Journal of American History* 99:208–218.

Kalogo, Y., S. Habibi, H. L. Maclean and S. V. Joshi. 2007. Environmental implications of municipal solid waste-derived ethanol. *Environmental Science and Technology* 41:35–41.

Kerr, R. A. 2007. Global warming is changing the world. *Science* 316:188–190.

Kilian, L. 2008. Exogenous oil supply shocks: How big are they and how much do they matter for the US economy? *Review of Economics and Statistics* 90:216–240.

Kilian, L. 2009. Not all oil price shocks are alike: Disentangling demand and supply shocks in the crude oil market. *American Economic Review* 99:1053–1069.

Kim, J. 2018. Evaluating author name disambiguation for digital libraries: A case of DBLP. *Scientometrics* 116:1867–1886.

Kim, J. H., J. C. Lee and D. Pak. 2011. Feasibility of producing ethanol from food waste. *Waste Management* 31:2121–2125.

Konur, O. 2000. Creating enforceable civil rights for disabled students in higher education: An institutional theory perspective. *Disability & Society* 15:1041–1063.

Konur, O. 2002a. Access to nursing education by disabled students: Rights and duties of nursing programs. *Nurse Education Today* 22:364–374.

Konur, O. 2002b. Assessment of disabled students in higher education: Current public policy issues. *Assessment and Evaluation in Higher Education* 27:131–152.

Konur, O. 2002c. Access to employment by disabled people in the UK: Is the Disability Discrimination Act working? *International Journal of Discrimination and the Law* 5:247–279.

Konur, O. 2006a. Participation of children with dyslexia in compulsory education: Current public policy issues. *Dyslexia* 12:51–67.

Konur, O. 2006b. Teaching disabled students in higher education. *Teaching in Higher Education* 11:351–363.

Konur, O. 2007a. A judicial outcome analysis of the Disability Discrimination Act: A windfall for the employers? *Disability & Society* 22:187–204.

Konur, O. 2007b. Computer-assisted teaching and assessment of disabled students in higher education: The interface between academic standards and disability rights. *Journal of Computer Assisted Learning* 23:207–219.

Konur, O. 2011. The scientometric evaluation of the research on the algae and bio-energy. *Applied Energy* 88:3532–3540.

Konur, O. 2012a. The evaluation of the biogas research: A scientometric approach. *Energy Education Science and Technology, Part A: Energy Science and Research* 29:1277–1292.

Konur, O. 2012b. The evaluation of the educational research: A scientometric approach. *Energy Education Science and Technology, Part B: Social and Educational Studies* 4:1935–1948.

Konur, O. 2012c. The evaluation of the global energy and fuels research: A scientometric approach. *Energy Education Science and Technology, Part A: Energy Science and Research* 30:613–628.

Konur, O. 2012d. The evaluation of the research on the biodiesel: A scientometric approach. *Energy Education Science and Technology, Part A: Energy Science and Research* 28:1003–1014.

Konur, O. 2012e. The evaluation of the research on the bioethanol: A scientometric approach. *Energy Education Science and Technology, Part A: Energy Science and Research* 28:1051–1064.

Konur, O. 2012f. The evaluation of the research on the biofuels: A scientometric approach. *Energy Education Science and Technology, Part A: Energy Science and Research* 28:903–916.

Konur, O. 2012g. The evaluation of the research on the biohydrogen: A scientometric approach. *Energy Education Science and Technology, Part A: Energy Science and Research* 29:323–338.

Konur, O. 2012h. The evaluation of the research on the microbial fuel cells: A scientometric approach. *Energy Education Science and Technology, Part A: Energy Science and Research* 29:309–322.

Konur, O. 2012i. The scientometric evaluation of the research on the production of bioenergy from biomass. *Biomass and Bioenergy* 47:504–515.

Konur, O. 2015. Current state of research on algal bioethanol. In *Marine Bioenergy: Trends and Developments*, Eds. S. K. Kim and C. G. Lee, pp. 217–244. Boca Raton, FL: CRC Press.

Konur, O., Ed. 2018a. *Bioenergy and Biofuels*. Boca Raton, FL: CRC Press.

Konur, O. 2018b. Bioenergy and biofuels science and technology: Scientometric overview and citation classics. In *Bioenergy and Biofuels*, Ed. O. Konur, pp. 3–63. Boca Raton, FL: CRC Press.

Konur, O. 2019. Cyanobacterial bioenergy and biofuels science and technology: A scientometric overview. In *Cyanobacteria: From Basic Science to Applications*, Eds. A. K. Mishra, D. N. Tiwari and A. N. Rai, pp. 419–442. Amsterdam: Elsevier.

Konur, O. 2020a. The scientometric analysis of the research on the bioethanol production from green macroalgae. In *Handbook of Algal Science, Technology and Medicine,* Ed. O. Konur, pp. 385–401. London: Academic Press.

Konur, O., Ed. 2020b. *Handbook of Algal Science, Technology and Medicine.* London: Academic Press.

Konur, O., Ed. 2021a. *Handbook of Biodiesel and Petrodiesel Fuels: Science, Technology, Health, and Environment*. Boca Raton, FL: CRC Press.

Konur, O., Ed. 2021b. *Handbook of Biodiesel and Petrodiesel Fuels: Science, Technology, Health, and Environment. Volume 1. Biodiesel Fuels: Science, Technology, Health, and Environment.* Boca Raton, FL: CRC Press.

Konur, O., Ed. 2021c. *Handbook of Biodiesel and Petrodiesel Fuels: Science, Technology, Health, and Environment. Volume 2. Biodiesel Fuels Based on the Edible and Nonedible Feedstocks, Wastes, and Algae: Science, Technology, Health, and Environment.* Boca Raton, FL: CRC Press.

Konur, O., Ed. 2021d. *Handbook of Biodiesel and Petrodiesel Fuels: Science, Technology, Health, and Environment. Volume 3. Petrodiesel Fuels: Science, Technology, Health, and Environment.* Boca Raton, FL: CRC Press.

Konur, O. and F. L. Matthews. 1989. Effect of the properties of the constituents on the fatigue performance of composites: A review. *Composites* 20:317–328.

Kruyt, B., D. P. van Vuuren, H. J. de Vries and H. Groenenberg. 2009. Indicators for energy security. *Energy Policy* 37:2166–2181.

Leydesdorff, L. 2000. Is the European Union becoming a single publication system? *Scientometrics* 47:265–280.

Leydesdorff, L. and C. Wagner. 2009. Is the United States losing ground in science? A global perspective on the world science system. *Scientometrics* 78:23–36.

Leydesdorff, L. and P. Zhou. 2005. Are the contributions of China and Korea upsetting the world system of science? *Scientometrics* 63:617–630.

Li, A., V. Antizar-Ladislao and M. Khraisheh. 2007. Bioconversion of municipal solid waste to glucose for bioethanol production. *Bioprocess and Biosystems Engineering* 30:189–196.

Li, H., S. M. Liu, X. H. Yu, S. L. Tang and C. K. Tang. 2020. Coronavirus disease 2019 (COVID-19): Current status and future perspectives. *International Journal of Antimicrobial Agents* 55:105951.

Lin, Y. and S. Tanaka. 2006. Ethanol fermentation from biomass resources: Current state and prospects. *Applied Microbiology and Biotechnology* 69:627–642.

Liu, X., W. Wang, X. Gao, Y. Zhou and R. Shen. 2012. Effect of thermal pretreatment on the physical and chemical properties of municipal biomass waste. *Waste Management* 32:249–255.

Ma, X., L. Sun and C. Song. 2002. A new approach to deep desulfurization of gasoline, diesel fuel and jet fuel by selective adsorption for ultra-clean fuels and for fuel cell applications. *Catalysis Today* 77:107–116.

Morrissey, A. J. and J. Browne. 2004. Waste management models and their application to sustainable waste management. *Waste Management* 24:297–308.

Morschbacker, A. 2009. Bio-ethanol based ethylene. *Polymer Reviews* 49:79–84.

Mussatto, S. I., E. M. S. Machado, L. M. Carneiro and J. A. Teixeira. 2012. Sugars metabolism and ethanol production by different yeast strains from coffee industry wastes hydrolysates. *Applied Energy* 92:763–768.

Najafi, G., B. Ghobadian and T. Tavakoli, et al. 2009. Performance and exhaust emissions of a gasoline engine with ethanol blended gasoline fuels using artificial neural network. *Applied Energy* 86:630–639.

Newman, P. W. G. and J. R. Kenworthy. 1989. Gasoline consumption and cities: A comparison of U.S. cities with a global survey. *Journal of the American Planning Association* 55:24–37.

North, D. C. 1991. Institutions. *Journal of Economic Perspectives* 5:97–112.

Okuda, N., K. Ninomiya, M. Takao, Y. Katakura and S. Shioya. 2007. Microaeration enhances productivity of bioethanol from hydrolysate of waste house wood using ethanologenic *Escherichia coli* KO11. *Journal of Bioscience and Bioengineering* 103:350–357.

Okuda, N., M. Soneura, K. Ninomiya, Y. Katakura and S. Shioya. 2008. Biological detoxification of waste house wood hydrolysate using *Ureibacillus thermosphaericus* for bioethanol production. *Journal of Bioscience and Bioengineering* 106:128–133.

Olsson, L. and B. Hahn-Hagerdal. 1996. Fermentation of lignocellulosic hydrolysates for ethanol production. *Enzyme and Microbial Technology* 18:312–331.

Rahman, S. H. A., J. P. Choudhury, A. L. Ahmad and A. H. Kamaruddin. 2007. Optimization studies on acid hydrolysis of oil palm empty fruit bunch fiber for production of xylose. *Bioresource Technology* 98:554–559.

Ravindran, R. and A. K. Jaiswal. 2016. A comprehensive review on pre-treatment strategy for lignocellulosic food industry waste: Challenges and opportunities. *Bioresource Technology* 199:92–102.

Reeves, S. 2014. To Russia with love: How moral arguments for a humanitarian intervention in Syria opened the door for an invasion of the Ukraine. *Michigan State University International Law Review* 23:199.

Sanchez, O. J. and C. A. Cardona. 2008. Trends in biotechnological production of fuel ethanol from different feedstocks. *Bioresource Technology* 99:5270–5295.

Sen, B., J. Aravind, P. Kanmani and C. H. Lay. 2016. State of the art and future concept of food waste fermentation to bioenergy. *Renewable and Sustainable Energy Reviews* 53:547–557.

Soccol, C. R., L. P. de Souza Vandenberghe and A. B. P. Medeiros, et al. 2010. Bioethanol from lignocelluloses: Status and perspectives in Brazil. *Bioresource Technology* 101:4820–4825.

Sun, Y. and J. Cheng. 2002. Hydrolysis of lignocellulosic materials for ethanol production: A review. *Bioresource Technology* 83:1–11.

Taherzadeh, M. J. and K. Karimi. 2007. Enzyme-based hydrolysis processes for ethanol from lignocellulosic materials: A review. *Bioresources* 2:707–738.

Taherzadeh, M. J. and K. Karimi. 2008. Pretreatment of lignocellulosic wastes to improve ethanol and biogas production: A review. *International Journal of Molecular Sciences* 9:1621–1651.

Tan, H. X., E. A. Ujum, K. F. Choong and K. Ratnavelu. 2015. Impact analysis of domestic and international research collaborations: A Malaysian case study. *Scientometrics* 102:885–904.

Wang, L., M. Sharifzadeh, R. Templer and R. J. Murphy. 2013. Bioethanol production from various waste papers: Economic feasibility and sensitivity analysis. *Applied Energy* 111:1172–1182.

Wilkins, M. R., W. W. Widmer, K. Grohmann and R. G. Cameron. 2007. Hydrolysis of grapefruit peel waste with cellulase and pectinase enzymes. *Bioresource Technology* 98:1596–1601.

Wilson, D. C. 2007. Development drivers for waste management. *Waste Management & Research* 25:198–207.

Winzer, C. 2012. Conceptualizing energy security. *Energy Policy* 46:36–48.

Yang, B. and C. E. Wyman. 2008. Pretreatment: The key to unlocking low-cost cellulosic ethanol. *Biofuels, Bioproducts and Biorefining* 2:26–40.

Yoshioka, T., T. Sato and A. Okuwaki. 1994. Hydrolysis of waste PET by sulfuric acid at 150°C for a chemical recycling. *Journal of Applied Polymer Science* 52:1353–1355.

70 Second Generation Urban Waste-based Bioethanol Fuels
Review

Ozcan Konur
(Formerly) Ankara Yildirim Beyazit University

70.1 INTRODUCTION

The crude oil-based gasoline fuels (Ma et al., 2002; Newman and Kenworthy, 1989) have been widely used in the transportation sector since the 1920s. However, there have been great public concerns over the adverse environmental and human impact of these fuels (Hill et al., 2006, 2009). Hence, biomass-based bioethanol fuels (Hill et al., 2006; Konur, 2012, 2015, 2019, 2020) have increasingly been used in blending gasoline fuels (Hsieh et al., 2002; Najafi et al., 2009), in the fuel cells (Antolini, 2007, 2009), and in the biochemical production (Angelici et al., 2013; Morschbacker, 2009) in a biorefinery context (Fernando et al., 2006; Huang et al., 2008).

However, it is necessary to pretreat the biomass (Alvira et al., 2010; Taherzadeh and Karimi, 2008) to enhance the yield of the bioethanol (Hahn-Hagerdal et al., 2006; Sanchez and Cardona, 2008) prior to the bioethanol fuel production from the feedstocks through the hydrolysis (Sun and Cheng, 2002; Taherzadeh and Karimi, 2007) and fermentation (Lin and Tanaka, 2006; Olsson and Hahn-Hagerdal, 1996) of the biomass and hydrolysates, respectively.

One of the most-studied feedstocks for the bioethanol fuels has been the urban wastes such as food, paper, textile, plastic, and wood wastes. The research in the field of second generation urban waste-based bioethanol fuels has intensified in this context in the key research fronts of the pretreatment (Arapoglou et al., 2010; Chang et al., 2001) and hydrolysis (Chang et al., 2001; Wilkins et al., 2007a) of the urban wastes, fermentation (Mussatto et al., 2012; Wilkins et al., 2007b) of the urban waste-based hydrolysates, and production (Arapoglou et al., 2010; Mussatto et al., 2012) and evaluation (Kalogo et al., 2007; Wang et al., 2013) of second generation urban waste-based bioethanol fuels. Further, food wastes (Arapoglou et al., 2010; Wilkins et al., 2007a), municipal solid wastes (MSWs) (Holtzapple et al., 1992; Li et al., 2007), paper wastes (Castanon and Wilke, 1981; Chang et al. (2001), textile wastes (Agblevor et al., 2003; Jeihanipour and Taherzadeh, 2009), plastic wastes (Guclu et al., 2003; Yoshioka et al., 1994), and wood wastes (Okuda et al., 2007, 2008) have been studied intensively in this context.

However, it is essential to develop efficient incentive structures (North, 1991) for the primary stakeholders to enhance the research in this field (Konur, 2000, 2002a–c, 2006a,b, 2007a,b). Although there have been a number of review papers on the urban waste-based bioethanol fuels (Guimaraes et al., 2010; Maitan-Alfenas et al., 2015; Ravindran and Jaiswal, 2016), there has been no review of the most-cited 25 papers in this field.

Thus, this book chapter presents a review of the most-cited 25 articles in the field of the urban waste-based bioethanol fuels. Then, it discusses the key findings of these highly influential papers and comments on the future research priorities in this field.

70.2 MATERIALS AND METHODS

The search for this study was carried out using Scopus database (Burnham, 2006) in September 2022.

DOI: 10.1201/9781003226550-95

As a first step for the search of the relevant literature, the keywords were selected using the most-cited first 300 population papers. The selected keyword list was then optimized to obtain a representative sample of papers for the searched research field. This final keyword set was provided in the appendix of Konur (2023) for future replication studies.

As a second step, a sample dataset was used for this study. The first 25 articles with at least 104 citations each were selected for the review study. Key findings from each paper were taken from the abstracts of these papers and were discussed. Additionally, a number of brief conclusions were drawn, and a number of relevant recommendations were made to enhance the future research landscape.

70.3 RESULTS

The brief information about 25 most-cited papers with at least 104 citations each on second generation urban waste-based bioethanol fuels is given below. The primary research fronts are the hydrolysis of the urban wastes and production of the urban waste-based bioethanol fuels with 11 and 14 highly cited papers (HCPs), respectively. Further, there are 2 and 19 HCPs for the pretreatment and hydrolysis of the urban wastes, respectively.

70.3.1 Urban Waste Hydrolysis

The brief information about 11 most-cited papers on the hydrolysis of urban wastes with at least 104 citations each is given below (Table 70.1). These papers also cover the research on the pretreatment of the urban wastes.

Guerard et al. (2002) produced tuna waste hydrolysates in a paper with 221 citations. They performed the enzymatic hydrolysis of the biomass using Umamizyme in a 1-l batch reactor at pH 7 and 45°C varying enzyme/substrate ratio ranging from 0.1% to 1.5% (w/w) protein). They obtained a degree of hydrolysis up to 22.5% with an enzyme/substrate ratio of 1.5%, after 4h of hydrolysis. They observed a linear correlation between the hydrolysis degree and the nitrogen recovery. The enzyme performed as effectively as Alcalase® 2, 4L for the biomass solubilization. However, this enyzme's stability was lower than Alcalase® 2, 4L.

Chang et al. (2001) used lime ($Ca(OH)_2$) and oxygen (O_2) to enhance the enzymatic digestibility of poplar wood and newspaper in a paper with 190 citations. Under the conditions of 140°C, 3h, 0.3g of $Ca(OH)_2$/g of dry biomass, 16mL of water/g of dry biomass, and 7.1 bar absolute oxygen, they observed that the reducing sugar yield using a cellulase loading of 5 filter paper unit (FPU)/g of raw dry biomass increased from 240 to 565mg of eq. glucose/g of raw dry biomass.

Wilkins et al. (2007a) tested different loadings of commercial cellulase and pectinase enzymes and pH levels to hydrolyze grapefruit peel waste to produce sugars in a paper with 174 citations. They used pectinase and cellulase loadings of 0, 1, 2, 5, and 10mg protein/g peel dry matter at 45°C. They supplemented them with 2.1mg β-glucosidase protein/g peel dry matter. They observed that 5mg pectinase/g peel dry matter and 2mg cellulase/g peel dry matter were the lowest loadings to yield the most glucose, while the optimum pH was 4.8.

Rahman et al. (2007) optimized the acid hydrolysis of oil palm empty fruit bunch fiber for the production of xylose in a paper with 157 citations. They performed batch reactions under various reaction temperature, reaction time, and acid concentrations to optimize the hydrolysis process in order to obtain high xylose yield. They observed that the optimum reaction temperature, reaction time, and acid concentration were 119°C, 60min, and 2%, respectively. Under these conditions, xylose yield and selectivity were 91.27% and 17.97 g/g, respectively.

Castanon and Wilke (1981) evaluated the effects of Tween 80 surfactant on the enzymatic hydrolysis of newspaper in a paper with 140 citations. They observed that the surfactant (0.1%) increased the rate and extent of cellulose hydrolysis. Consequently, the rate of enzyme usage in the hydrolysis reactor was improved by 33%. In addition, in the presence of surfactant, the recovery of enzymes

TABLE 70.1
The Hydrolysis of Urban Wastes

No.	Papers	Wastes	Prts.	Parameters	Keywords	Lead Author	Affil.	Cits.
1	Guerard et al. (2002)	Food waste Tuna wastes	Enzymes	Enzymatic hydrolysis, enzyme efficiency, enzyme/substrate ratio, hydrolysis efficiency	Tuna, waste, hydrolysates	Guerard, Fabienne* 6603404306	Univ. Brest France	221
3	Chang et al. (2001)	Paper waste Newspapers	Enzymes, alkali	Enzymatic hydrolysis, enzymatic digestibility, sugar yield	Wood newspaper, pretreatment	Holtzapple, Mark T. 7004167004	Texas A&M Univ. USA	190
4	Wilkins et al. (2007a)	Food waste Grapefruit peels	Enzymes	Enzymatic hydrolysis, enzyme loadings, glucose yield	Grapefruit, waste, peels, hydrolysis	Wilkins, Mark R. 56492323200	Univ. Nebraska Lincoln USA	174
5	Rahman et al. (2007)	Food waste Fruit bunches	Acids	Acid hydrolysis, optimization, temperature, reaction time, acid concentrations, xylose yield, selectivity	Fruit, bunch, pretreatment	Choudhury, J. P: 57198146735	Univ. Sains Malaysia Malaysia	157
8	Castanon and Wilke (1981)	Paper waste Newspapers	Surfactants, enzymes	Enzymatic hydrolysis, surfactants, hydrolysis efficiency, enzyme use and recovery	Newspaper, hydrolysis	Castanon, Marisi* 56857335700	Univ. Calif. Berkeley USA	140
10	Park et al. (1992)	Paper waste Newspapers	Surfactants, enzymes	Enzymatic hydrolysis, surfactants, enzyme quantity and the conversion	Newspaper, hydrolysis	Park, Jin Won 56471045700	Yonsei Univ. S. Korea	133
12	Fernandez-Bolanos et al. (2001)	Food waste Olive stones	Steam, acids	Enzymatic hydrolysis, pretreatments, sugar yield	Olive stones, hydrolysis	Heredia, Antonio 35560421500	CSIC Spain	129
13	Grohmann et al. (1995)	Food waste Orange peels	Acids, enzymes	Acid and enzymatic hydrolysis, pretreatments, carbohydrates	Orange, peel, hydrolysis, fractionation, pretreatment	Grohmann, Karel 7004503589	Renewable Spirits PLC USA	126
17	Li et al. (2007)	Municipal waste	Acids, steam, MW, enzymes	Enzymatic pretreatment, pretreatments, sugar yield, enzyme loading, acid concentration	Municipal, waste, bioethanol	Antizar-Ladislao, Blanca* 6602212294	Univ. Coll. London UK	118
22	Holtzapple et al. (1992)	Paper waste	Enzymes, ammonia	Enzymatic hydrolysis, pretreatments, newspapers	Municipal waste, pretreatment	Holtzapple, Mark T. 7004167004	Texas A&M Univ. USA	110
25	Yunus et al. (2010)	Food waste Fruit bunch	Ultrasonics, acids	Acid hydrolysis, pretreatments, sugar yield	Fruit bunch, hydrolysis, pre-treatment	Yunus, Robiah* 6603243672	Univ. Putra Malaysia Malaysia	104

Prt., biomass pretreatments; Cits., number of citations received for each paper; *, female.

was higher. Thus, the surfactant hindered the immobilization of the enzymes on the substrates by reducing the strength of adsorption.

Park et al. (1992) evaluated the effects of surfactants with a polyoxyethylene glycol (POG) group on enzymatic hydrolysis of used newspaper in a paper with 133 citations. They found that the surfactants enhanced the hydrolysis of the biomass. The optimum surfactant concentration was 0.05% (wt/substrate wt) in the case of POG(21) sorbitane oleic ester. Among the surfactants, POG phenyl ether types showed the highest enhancement effect, for example, with two times higher conversion at 80h than that without surfactant. Further, as the hydrophile–lypophile balance (HLB) value increased, both the free enzyme quantity and the conversion increased. In conclusion, surfactants helped the enzyme to desorb from the binding site on the substrate surface after the completion of hydrolysis at that site.

Fernandez-Bolanos et al. (2001) processed the olive stones (whole stones and seed husks in fragments) by steam explosion under different experimental conditions of temperature and time, 200°C–236°C for 2–4min, with or without previous acid impregnation with 0.1% H_2SO_4 (w/w) in a paper with 129 citations. They observed that the maximum yield of the pentosan recovered in the water solution was 63% pentose in the starting material for seed husk treated at 200°C for 2min (log R_0=3.24) prior to acid-impregnation, or at 215°C for 2min (log R_0=3.69) without acid, compared to 39% of the potential yield for whole stones preimpregnated with acid under more severe conditions (at log R_0=4.07). Thus, the autohydrolysis of hemicellulose in seed husks was enhanced compared to whole stones. The depolymerization of hemicelluloses was a function of the severity of the treatment. Steam explosion improved the accessibility of the cellulose and increased the enzymatic hydrolysis yield after steam explosion with respect to biomass without steam explosion. When the biomass was post-treated with Na-chlorite, the enzymatic hydrolysis improved as the water-insoluble residue being almost completely hydrolyzed in 8h of incubation.

Grohmann et al. (1995) fractionated and pretreated orange peels by dilute acid hydrolysis in a paper with 126 citations. They solubilized and depolymerized carbohydrates by pretreatment of orange peel with dilute (0.06% and 0.5%) sulfuric acid at 100°C, 120°C, and 140°C. They found that the acid pretreatments solubilized a large portion of total carbohydrates in the biomass. However, only soluble sugars and sugars derived from hydrolysis of hemicelluloses were efficiently released by the pretreatment with hot dilute sulfuric acid. Further, cellulose and segments of pectin-containing galacturonic acid units were very resistant to acid-catalyzed hydrolysis. The treatment with dilute sulfuric acid had a positive effect on the rate of subsequent enzymatic hydrolysis of orange peel by a mixture of cellulolytic and pectinolytic enzymes.

Li et al. (2007) converted MSW to glucose in a paper with 118 citations. They first pretreated the biomass with different prehydrolysis treatments. These pretreatments included dilute acid (H_2SO_4, H_NO_3, or HCl, 1% and 4%, 180min, 60°C), steam explosion pretreatment (121°C and 134°C, 15min), microwave pretreatment (700 W, 2min), or a combination of two of them. They performed the enzymatic hydrolysis of the biomass with *Trichoderma reesei* and *T. viride* (10 and 60 FPU/g substrate). They obtained the highest glucose yield (72.80%) with a pretreatment consisting of H_2SO_4 at 1% concentration, followed by steam pretreatment at 121°C and enzymatic hydrolysis with *T. viride* at 60 FPU/g substrate. The contribution of enzyme loading and acid concentration was significantly higher (49.39% and 47.70%, respectively) than the contribution of temperature during steam treatment (0.13%) to the glucose yield.

Holtzapple et al. (1992) retreated lignocellulosic MSW by ammonia fiber explosion (AFEX) in a paper with 110 citations. They used mixed MSW and individual components such as softwood newspaper, kenaf newspaper, copy paper, paper towels, cereal boxes, paper bags, corrugated boxes, magazines, and waxed paper. They found that softwood newspaper was the most difficult component to digest because of its high lignin content. A combination of oxidative lignin cleavage and AFEX was required to increase softwood newspaper digestibility substantially, whereas AFEX alone made kenaf newspaper digestible. Because most MSW components were substantially delignified in the paper-making process, AFEX only marginally increased their digestibility.

Yunus et al. (2010) studied the effect of ultrasonic pretreatment on low-temperature acid hydrolysis of oil palm empty fruit bunch prior to the acid hydrolysis of the biomass in a paper with 104 citations. They used 2% sulfuric acid, 1:25 solid liquid ratio, and 100°C operating temperature. They obtained a maximum xylose yield of 58% when the biomass was ultrasonicated at 90% amplitude for 45 min. In the absence of ultrasonic pretreatment, they obtained only 22% of xylose. However, they observed no substantial increase of xylose formation for acid hydrolysis at higher temperatures of 120°C and 140°C on ultrasonicated biomass. Further, there were some morphological changes within the biomass for different acid hydrolysis conditions.

70.3.2 Urban Waste-based Bioethanol Production

There are 14 HCPs for the production the urban waste-based bioethanol fuels with at least 106 citations each (Table 70.2). These papers also cover the fermentation of the hydrolysates of the urban wastes. As the pretreatment and hydrolysis are the fundamental parts of the bioethanol production, these narrated papers often cover these processes too.

Arapoglou et al. (2010) produced ethanol from potato peel waste in a paper with 199 citations. They hydrolyzed a number of batches of potato peel wastes with various enzymes and/or acid and fermented by *Saccharomyces cerevisiae* var. bayanus to determine fermentability and ethanol production. They observed that the enzymatic hydrolysis with a combination of three enzymes released 18.5 g/L reducing sugar and produced 7.6 g/L of ethanol after fermentation.

Mussatto et al. (2012) evaluated the production of ethanol by *S. cerevisiae*, *Pichia stipites*, and *Kluyveromyces fragilis* in the hydrolysates produced by acid hydrolysis of coffee silverskin (CS) and spent coffee grounds (SCG) in a paper with 149 citations. They observed that *S. cerevisiae* provided the best ethanol production from SCG hydrolysate (11.7 g/L, 50.2% efficiency), while they obtained insignificant (≤1.0 g/L) ethanol production from CS hydrolysate, for all the evaluated yeast strains, probably due to the low sugar concentration present in this medium (~22 g/L). In conclusion, SCG was a potential biomass for the bioethanol production.

Wilkins et al. (2007b) performed the simultaneous saccharification and fermentation (SSF) of citrus peel waste by *S. cerevisiae* at 37°C to produce ethanol in a paper with 142 citations. They first performed the steam explosion pretreatment of the biomass to remove more than 90% of the initial d-limonene, a fermentation inhibitor. They observed that ethanol concentrations after 24 h were reduced in fermentations with initial d-limonene concentrations greater than or equal to 0.33% (v/v) and final (24 h) d-limonene concentrations greater than or equal to 0.14% (v/v). Further, ethanol production was reduced when enzyme loadings were (international units (IU) or FPU/g peel dry solids) <25, pectinase, 0.02, cellulase, and 13, β-glucosidase. However, ethanol production was greatest when the initial pH of the peel waste was adjusted to 6.0.

Kwon et al. (2013) produced bioethanol and biodiesel from spent coffee grounds in a paper with 139 citations. They converted the crude lipids extracted from the spent coffee grounds into fatty acid methyl ester (FAME) and fatty acid ethyl ester (FAEE) via the noncatalytic biodiesel transesterification reaction, and then converted the de-fatted biomass into bioethanol fuels. They observed that the yields of bioethanol and biodiesel were 0.46 g/g and 97.5%, which were calculated based on the consumed sugar and lipids extracted from spent coffee grounds, respectively.

Kim et al. (2011) produced ethanol fuels from food waste in a lab-scale fermentor in a paper with 129 citations. They pretreated the biomass with carbohydrase, glucoamylase, cellulase, and protease for hydrolysis of food waste. They observed that the carbohydrase hydrolyzed and produced glucose with a glucose yield of 0.63 g glucose/g total solid. They then fermented the resulting hydrolysate with *S. cerevisiae* in the batch mode. For separate hydrolysis and fermentation (SHF), ethanol concentration reached at the level corresponding to an ethanol yield of 0.43 g ethanol/g total solids, while for SSF, the ethanol yield was 0.31 g ethanol/g total solids. During the continuous operation of SHF, the volumetric ethanol production rate was 1.18 g/L h with an ethanol yield of 0.3 g ethanol/g total solids, while for SSF process, the volumetric ethanol production rate was 0.8 g/L h with an ethanol yield of 0.2 g ethanol/g total solids.

TABLE 70.2
The Production of Urban Waste-based Bioethanol Fuels

No.	Papers	Wastes	Prts.	Yeasts	Parameters	Keywords	Lead Author	Affil.	Cits.
2	Arapoglou et al. (2010)	Food waste Potato peels	Enzymes, acids	*S. cerevisiae*	Enzymatic hydrolysis, sugar and ethanol production and yields	Potato, peel, waste, ethanol	Varzakas, Theodoros 6603098855	Univ. Peloponnese Greece	199
6	Mussatto et al. (2012)	Food waste Spent coffee grounds	Acids	*S. cerevisiae, P. stipitis, K. fragilis*	Ethanol production, fermentation, yeasts, coffee wastes, ethanol yield	Coffee, waste, hydrolysates, ethanol	Mussatto, Solange I.* 6602643634	Univ. Minho Portugal	149
7	Wilkins et al. (2007b)	Food waste Citrus peels	Enzymes, steam	*S. cerevisiae*	Ethanol production, fermentation, SSF, d-limonene concentration, enzyme loading, pH, ethanol yield	Citrus, peel, waste, saccharification, fermentation, ethanol	Wilkins, Mark R. 56492323200	Univ. Nebraska Lincoln USA	142
9	Kwon et al. (2013)	Food waste Spent coffee grounds	Enzymes	Yeasts	Bioethanol and biodiesel production, lipid conversion, sugar conversion	Spent, coffee, bioethanol	Jeon, Yeong Jae 7201888480	Pukyong Natl. Univ. S. Korea	139
11	Kim et al. (2011)	Food waste	Enzymes	*S. cerevisiae*	Enzymatic hydrolysis, enzymes, glucose and ethanol yield, SHF, SSF, fermentation, volumetric ethanol production rate	Food, waste, ethanol	Pak, Daewon 7005142765	Seoul Natl. Univ. Sci. Technol. S. Korea	129
14	Piarpuzan et al. (2011)	Food waste Fruit bunches	Alkali, enzymes, autoclaving	*S. cerevisiae*	Ethanol production, enzymatic hydrolysis, sugar and ethanol yield, pentoses, energy consumption	Fruit, bunches, ethanol	Cardona, Carlos A. 57214443163	Colombia Natl. Univ. Colombia	125
15	Le Man et al. (2010)	Food waste	Enzymes	Yeasts	Ethanol production, optimization, temperature, pH, reducing sugar concentration, ethanol yield	Food waste, ethanol	Park, Hung-Suck 7601570293	Univ. Ulsan S. Korea	124
16	Choi et al. (2015)	Food waste Fruit peels	Enzymes	Yeasts	Enzymatic hydrolysis, fruit wastes, fermentation inhibitors, d-limonene, ethanol yield	Fruit, citrus, peel, waste, bioethanol	Bae, Hyeun-Jong 24280549300	Chonnam Natl. Univ. S. Korea	123

(Continued)

TABLE 70.2 (*Continued*)
The Production of Urban Waste-based Bioethanol Fuels

No.	Papers	Wastes	Prts.	Yeasts	Parameters	Keywords	Lead Author	Affil.	Cits.
18	Oberoi et al. (2011)	Food waste Banana peels	Hydrothermal, enzymes	Yeasts	Ethanol production, enzymatic hydrolysis, fermentation, optimization, ethanol yield	Banana peels, ethanol, saccharification, fermentation	Oberoi, Harinder S. 6603479987	Indian Inst. Hort. Cult. India	115
19	Jeihanipour and Taherzadeh (2009)	Textile waste	Alkali, enzymes	*S. cerevisiae*	Ethanol production, enzymatic hydrolysis, fermentation, SSF, pretreatments, alkali content, temperature	Textiles, waste, ethanol	Jeihanipour, Azam* 24597740600	Karlsruhe Inst. Technol. Germany	113
20	Silveira et al. (2005)	Food waste Cheese whey	Enzymes	*K. marxianus*	Ethanol production, hydrolysis, fermentation, ethanol yield and productivity, lactose concentration, oxygen level, culture	Cheese whey, ethanol	Passos, Flavia M. L.* 7004639512	Fed. Univ. Vicosa Brazil	113
21	Wilkins et al. (2007c)	Food waste Orange peels	Enzymes	*K. marxianus, S. cerevisiae*	Ethanol production, hydrolysis, fermentation, peel oil concentration	Orange peels, ethanol	Wilkins, Mark R. 56492323200	Univ. Nebraska Lincoln USA	110
23	Widmer et al. (2010)	Food waste Orange peels	Steam, acids, alkali, enzymes	*S. cerevisiae*	Ethanol production, hydrolysis, fermentation, limonene removal, sugar and ethanol yield, T, pH, time	Orange, waste, pretreatment, ethanol, saccharification, fermentation	Widmer, Wilbur 7006097359	USDA Agr. Res. Serv. USA	106
24	Kim et al. (2008)	Food waste	Enzymes	Yeasts	Ethanol production, hydrolysis, fermentation, optimization, T, pH, enzyme content, sugar and ethanol yield	Food waste, saccharification, fermentation, ethanol	Kim, Si Wouk 56689100900	Chosun Univ. S. Korea	106

Prt., biomass pretreatments; Cits., number of citations received for each paper; *, female.

Piarpuzan et al. (2011) produced ethanol from empty fruit bunches using alkaline pretreatment and enzymatic hydrolysis in a paper with 125 citations. They fermented the hydrolysates with a native *S. cerevisiae* strain and concentrated the ethanol concentration using a glass bench-scale distillation column. They found that coupling alkaline pretreatment with a later autoclaving improved the sugar yield in enzymatic hydrolysis. Ethanol yield obtained from both experiments and simulation was very similar (66.50 and 65.84 dm^3 of ethanol per each t of empty fruit bunches, respectively). These low ethanol yields were obtained since the native *S. cerevisiae* did not assimilate all reducing sugars, suggesting that those sugars were pentoses. Simulated alkaline and autoclaving pretreatment contributed only with 2% of the total energy consumption (198.4 GJ/m ethanol), while product recovery represented 57% of the total energy.

Le Man et al. (2010) optimized the ethanol production from food waste leachate in a paper with 124 citations. They obtained maximum ethanol concentration of 24.17 g/L at the optimum condition of temperature (38°C), pH (5.45), and reducing sugar concentration (75 g/L). The experimental value agreed very well with the predicted one (23.66 g/L). In conclusion, with an ethanol yield of 0.32 g ethanol/g reducing sugar, food waste leachate was a promising biomass resource for the production of ethanol.

Choi et al. (2015) converted single-source citrus peels such as orange, mandarin, grapefruit, lemon, or lime or citrus peels in combination with other fruit wastes such as banana peel, apple pomace, and pear waste to produce bioethanol in a paper with 123 citations. They produced two in-house enzymes from Avicel and citrus peels and tested them with fruit waste at 12%–15% (w/v) solid loading. They observed that the rates of enzymatic conversion of fruit wastes to fermentable sugars were ~90% for all feedstocks after 48 h. They also designed a d-limonene removal column (LRC) that successfully removed this inhibitor from the fruit waste. When the LRC was coupled with an immobilized cell reactor (ICR), yeast fermentation resulted in ethanol concentrations (14.4–29.5 g/L) and yields (90.2%–93.1%) that were 12-fold greater than products from ICR fermentation alone.

Oberoi et al. (2011) produced ethanol from banana peels using statistically optimized SSF process in a paper with 115 citations. They optimized concentrations of cellulase and pectinase, temperature, and time. They used a laboratory batch fermenter with cellulase, pectinase, temperature, and time of nine cellulase FPU/g-cellulose, 72 IU/g-pectin, 37°C, and 15 h, respectively. They obtained higher ethanol concentration than the one predicted by the model equation but also saved fermentation time. Both hydrothermal pretreatment and SSF could be successfully carried out in a single vessel, and the use of optimized process parameters helped to achieve significant ethanol productivity. They obtained ethanol concentration and ethanol productivity of 28.2 g/L and 2.3 g/L/h, respectively, from banana peels.

Jeihanipour and Taherzadeh (2009) produced ethanol from cotton-based waste textiles in a paper with 113 citations. They found that alkaline pretreatment followed by enzymatic hydrolysis resulted in almost complete conversion of the cotton and jeans to glucose, which was then fermented by *S. cerevisiae* to ethanol. If no pretreatment applied, hydrolyses of the textiles by cellulase and β-glucosidase for 24 h followed by SSF in 4 days resulted in 0.140–0.145 g ethanol/g textiles, which was 25%–26% of the corresponding theoretical yield. A pretreatment with concentrated phosphoric acid prior to the hydrolysis improved ethanol production from the textiles up to 66% of the theoretical yield. However, the best results were obtained from alkaline pretreatment of the materials by NaOH. They performed the alkaline pretreatment of cotton fibers with 0%–20% NaOH at 0°C, 23°C, and 100°C, followed by enzymatic hydrolysis up to 4 days. In general, higher concentration of NaOH resulted in a better yield of the hydrolysis, whereas temperature had a reverse effect and better results were obtained at lower temperature. They obtained the best conditions for the alkaline pretreatment of the cotton at 12% NaOH and 0°C and 3 h. In this condition, they enzymatically hydrolyzed the biomass with 3% solid content at 85.1% of the theoretical yield in 24 h and 99.1% in 4 days. The alkaline pretreatment of the waste textiles at these conditions and subsequent SSF resulted in 0.48 g ethanol/g pretreated textiles used.

Silveira et al. (2005) produced ethanol from cheese whey permeate in a paper with 113 citations. They evaluated the effect of lactose concentration and oxygen level on the growth and metabolism of *K. marxianus* UFV-3 in cheese whey permeate and performed batch cultures under aerobic, hypoxic, and anoxic conditions, with lactose at initial concentration ranging from 1 to 240 g/L. They found that the increase in lactose concentration increased ethanol yield and ethanol volumetric productivity, and reduced cell yield. When lactose concentration was equal or above 50 g/L and the oxygen levels were low, the ethanol yield was close to its theoretical value. They obtained maximum ethanol concentrations of 76 and 80 g/L in hipoxia and anoxia, respectively. The lactose consumption rate in anoxia was greater than in aerobiosis and hypoxia. All oxygen levels showed a tendency for saturation of the ethanol production rate above 65 g/L lactose. Ethanol production rate was also higher on anoxia.

Wilkins et al. (2007c) produced ethanol from orange peel oil by *S. cerevisiae* and *K. marxianus* at 37°C in a paper with 110 citations. They observed that minimum inhibitory peel oil concentrations for ethanol production were 0.05% at 24 h, 0.10% at 48 h, and 0.15% at 72 h for both yeasts. Further, *S. cerevisiae* produced more ethanol than *K. marxianus* at each time point.

Widmer et al. (2010) pretreated orange processing waste under different times, pH, and temperature in a paper with 106 citations. They found that pretreatments at 160°C for longer than 4 min with steam purging were needed to remove limonene to below 0.1%. While hemicelluloses were solubilized well following all pretreatments at 160°C, just 70% of the pectin was solubilized in natural biomass compared to over 80% after pretreatments using acid-modified biomass (pH 2.8). Pretreatments at 160°C on base-modified biomass (initial pH 6.8) quickly destroyed pectin, had significantly lower dissolved solids, and were excessively viscous. Total sugars fermentable by *S. cerevisiae* were not changed after pretreatment at 160°C for up to 8 min in biomass between pH 2.2 and 8.2. Ethanol yields based on sugar content after enzymatic hydrolysis after 48 h SSF ranged from 76% to 94%. Further, ethanol yields were slightly lower but not statistically different using base-modified pretreatments.

Kim et al. (2008) optimized enzymatic hydrolysis and ethanol fermentation using food waste in a paper with 106 citations. They found optimum conditions such as hydrolysis pH of 5.20, enzyme reaction temperature of 46.3°C, enzyme concentration of 0.16% (v/v), fermentation pH of 6.85, fermentation temperature of 35.3°C, and fermentation time of 14 h. The model predicted that maximum concentrations of reducing sugar and ethanol under these optimum conditions were 117.0 g reducing sugar/L and 57.6 g ethanol/L, respectively. Experimental results were in close agreement with model prediction with 120.1 g reducing sugar/L and 57.5 g ethanol/L, respectively.

70.4 DISCUSSION

70.4.1 INTRODUCTION

The crude oil-based gasoline fuels have been widely used in the transportation sector since the 1920s. However, there have been great public concerns over the adverse environmental and human impact of these fuels. Hence, biomass-based bioethanol fuels have increasingly been used in blending gasoline and petrodiesel fuels, in the fuel cells, and in the biochemical production in a biorefinery context.

However, it is necessary to pretreat the biomass to enhance the yield of the bioethanol prior to the bioethanol fuel production from the feedstocks through the hydrolysis and fermentation of the biomass and hydrolysates, respectively.

One of the most-studied feedstocks for the bioethanol fuels has been the urban wastes such as food, municipal solid, paper, textile, plastic, and wood wastes. The research in the field of second generation urban waste-based bioethanol fuels has intensified in this context in the key research fronts of the pretreatment and hydrolysis of the urban wastes, fermentation of the urban waste-based hydrolysates, and production and evaluation of second generation urban waste-based bioethanol fuels. Further, food, municipal solid, and to a lesser extent paper, textile, plastic, and wood wastes have been studied intensively in this context.

However, it is essential to develop efficient incentive structures for the primary stakeholders to enhance the research in this field. Although there have been a number of review papers for this field, there has been no review of the most-cited 25 articles in this field.

Thus, this book chapter presents a review of the most-cited 25 articles on the bioethanol fuel production and evaluation from the urban wastes. Then, it discusses the key findings of these highly influential papers and comments on the future research priorities in this field.

As a first step for the search of the relevant literature, the keywords were selected using the most-cited first 300 population papers. The selected keyword list was then optimized to obtain a representative sample of papers for the searched research field. This keyword list was provided in the appendix of Konur (2023) for future replicative studies.

As a second step, a sample dataset was used for this study. The first 25 articles with at least 104 citations each were selected for the review study. Key findings from each paper were taken from the abstracts of these papers and were discussed. Additionally, a number of brief conclusions were drawn, and a number of relevant recommendations were made to enhance the future research landscape.

Information about the research fronts for the sample papers in the urban waste-based bioethanol fuels with regard to the feedstocks used in these processes is given in Table 70.3. As this table shows, the most prolific research front for this field is the food wastes with 76% of the HCPs, followed by the other urban wastes with 24% of the sample papers. The latter front includes paper, municipal solid, and textile wastes with 16%, 4%, and 4% of the sample papers, respectively.

On the other hand, the most influential feedstocks are the other food, paper, fruit, and beverage wastes with 4%–6% surplus each. Similarly, the least influential research fronts are food waste in general, kitchen wastes, and plastic wastes with 4%–8% deficit each.

TABLE 70.3
The Most-Prolific Research Fronts for the Second Generation Urban Waste-based Bioethanol Fuels

No.	Research Fronts	N Paper % Review	N Paper % Sample	Surplus (%)
1	Food waste	76.0	79.2	–3.2
	Fruit wastes	44.0	39.9	4.1
	Food waste in general	8.0	15.5	–7.5
	Beverage wastes	8.0	4.2	3.8
	Other food wastes	8.0	2.4	5.6
	Vegetable wastes	4.0	6.5	–2.5
	Dairy wastes	4.0	5.4	–1.4
	Kitchen wastes	0.0	5.4	–5.4
2	Other urban wastes	24.0	25.1	–1.1
	Paper wastes	16.0	10.7	5.3
	Municipal solid wastes	4.0	6.0	–2.0
	Textile wastes	4.0	3.0	1.0
	Plastic wastes	0.0	3.6	–3.6
	Wood wastes	0.0	1.8	–1.8

N Paper (%) review: the number of papers in the sample of 25 reviewed papers. N paper (%) sample: the number of papers in the population sample of 168 papers.

Information about the thematic research fronts for the sample papers in the urban waste-based bioethanol fuels is given in Table 70.4. As this table shows, the most prolific research fronts for this field are the pretreatment and hydrolysis of the urban wastes with 100% of the HCPs each. The other research fronts are the hydrolysate fermentation and bioethanol production with 56% of the sample papers each.

Further, the most influential research fronts are pretreatment and hydrolysis of the food wastes with 47% and 35% surplus, respectively. Similarly, the evaluation and production of bioethanol fuels are the least influential research fronts with 5% and 1% deficits, respectively.

70.4.2 Urban Waste Hydrolysis

The brief information about 11 most-cited papers on the hydrolysis of urban wastes with at least 104 citations each is given below (Table 70.1). However, it should be noted that as the Table 70.4 shows, the number of HCPs related to the biomass hydrolysis is 25. These papers also cover the research on the pretreatment of the urban wastes.

These narrated studies highlight the importance of the pretreatment and hydrolysis processes for the production of the bioethanol fuels form the urban wastes with a high ethanol yield. These pretreatments, primarily enzymatic and chemical pretreatments, fractionate the urban waste biomass and enhance the enzymatic digestibility of the biomass.

Guerard et al. (2002) produced tuna waste hydrolysates and obtained a degree of hydrolysis up to 22.5% with an enzyme/substrate ratio of 1.5%, after 4h of hydrolysis. Further, Chang et al. (2001) used lime and oxygen to enhance the enzymatic digestibility of poplar wood and newspaper and observed that the reducing sugar yield using a cellulase increased from 240 to 565 mg of eq. glucose/g of raw dry biomass upon the pretreatments.

Wilkins et al. (2007a) tested different loadings of commercial cellulase and pectinase enzymes and pH levels to hydrolyze grapefruit peel waste and observed that five mg pectinase/g peel dry matter and two mg cellulase/g peel dry matter were the lowest loadings to yield the most glucose, while the optimum pH was 4.8. Further, Rahman et al. (2007) optimized the acid hydrolysis of oil palm empty fruit bunch fiber and observed that the optimum reaction temperature, reaction time, and acid concentration were 119°C, 60 min, and 2%, respectively.

Castanon and Wilke (1981) evaluated the effects of Tween 80 on the enzymatic hydrolysis of newspaper and observed that this surfactant (0.1%) increased the rate and extent of cellulose hydrolysis. Further, Park et al. (1992) evaluated the effects of surfactants with a POG group on enzymatic hydrolysis of used newspaper and found that the surfactants enhanced the hydrolysis of the biomass.

Fernandez-Bolanos et al. (2001) processed the olive stones (whole stones and seed husks in fragments) by steam explosion under different experimental conditions of temperature and time with

TABLE 70.4
The Most-Prolific Thematic Research Fronts for the Urban Waste-based Bioethanol Fuels

No.	Research Fronts	N Paper % Review	N Paper 2% Sample	Surplus (%)
1	Biomass hydrolysis	100.0	65.5	34.5
2	Biomass pretreatments	100.0	53.0	47.0
3	Bioethanol production	56.0	57.1	–1.1
4	Hydrolysate fermentation	56.0	53.6	2.4
5	Bioethanol fuel evaluation	0.0	5.4	–5.4

N Paper (%) review: the number of papers in the sample of 25 reviewed papers. N paper (%) sample: the number of papers in the population sample of 168 papers.

or without previous acid impregnation and observed that autohydrolysis of hemicellulose in seed husks was enhanced compared to whole stones. Further, Grohmann et al. (1995) fractionated and pretreated orange peels by dilute acid hydrolysis and found that the acid pretreatments solubilized a large portion of total carbohydrates in the biomass.

Li et al. (2007) converted MSW to glucose using a number of pretreatments and obtained the highest glucose yield (72.80%) with a pretreatment consisting of H_2SO_4 at 1% concentration, followed by steam explosion pretreatment at 121°C, and enzymatic hydrolysis with *T. viride* at 60 FPU/g substrate. Further, Holtzapple et al. (1992) retreated lignocellulosic MSW by AFEX and found that softwood newspaper was the most difficult component to digest because of its high lignin content. Finally, Yunus et al. (2010) studied the effect of ultrasonic pretreatment on low-temperature acid hydrolysis of oil palm empty fruit bunch prior to the acid hydrolysis of the biomass and obtained a maximum xylose yield of 58% when the biomass was ultrasonicated at 90% amplitude for 45 min.

70.4.3 Urban Waste-based Bioethanol Production

There are only 14 HCPs for the production the urban waste-based bioethanol fuels with at least 206 citations each (Table 70.2). These papers also cover the fermentation of the hydrolysates of the urban wastes. As the pretreatment and hydrolysis are the fundamental parts of the bioethanol production, these narrated papers often cover these processes too.

These narrated studies highlight the importance of the pretreatment (primarily chemical, enzymatic, or hydrothermal) and hydrolysis (primarily enzymatic or acid) processes as well as of the fermentation processes (SSF or SHF) on the production of the bioethanol fuels from the urban wastes with a high ethanol yield. Thus, the studies solely focusing on the fermentation of the urban waste hydrolysates are rare.

Arapoglou et al. (2010) produced ethanol from potato peel waste and observed that the enzymatic hydrolysis with a combination of three enzymes released 18.5 g/L reducing sugar and produced 7.6 g/L of ethanol after fermentation. Further, Mussatto et al. (2012) evaluated the production of ethanol from coffee waste by three yeasts in the hydrolysates produced by acid hydrolysis of CS and SCG and observed that *S. cerevisiae* provided the best ethanol production from SCG hydrolysate (11.7 g/L, 50.2% efficiency), while they obtained insignificant (≤1.0 g/L) ethanol production from CS hydrolysate, for all the evaluated yeast strains.

Wilkins et al. (2007b) performed the SSF of citrus peel waste by *S. cerevisiae* at 37°C to produce ethanol and observed that ethanol concentrations after 24 h were reduced in fermentations with initial d-limonene concentrations greater than or equal to 0.33% (v/v) and final (24 h) d-limonene concentrations greater than or equal to 0.14% (v/v). Further, Kwon et al. (2013) produced bioethanol and biodiesel from SCG and observed that the yields of bioethanol and biodiesel were 0.46 g/g and 97.5%. Further, Kim et al. (2011) produced ethanol fuels from food waste and observed that for SHF process, ethanol concentration reached at the level corresponding to an ethanol yield of 0.43 g ethanol/g total solids, while for SSF, the ethanol yield was 0.31 g ethanol/g total solids.

Piarpuzan et al. (2011) produced ethanol from empty fruit bunches using alkaline pretreatment and enzymatic hydrolysis and observed that ethanol yield obtained from both experiments and simulation were very similar (66.50 and 65.84 dm^3 of ethanol per each t of empty fruit bunches, respectively). Further, Le Man et al. (2010) optimized the ethanol production from food waste leachate and obtained maximum ethanol concentration of 24.17 g/L at the optimum condition of temperature (38°C), pH (5.45), and reducing sugar concentration (75 g/L).

Choi et al. (2015) converted single-source citrus peels in combination with other fruit wastes to produce bioethanol and observed that the yeast fermentation resulted in ethanol concentrations (14.4–29.5 g/L) and yields (90.2%–93.1%) that were 12-fold greater than products from ICR fermentation alone. Further, Oberoi et al. (2011) produced ethanol from banana peels using statistically optimized SSF process and obtained higher ethanol concentration than the one predicted by the model equation but also saved fermentation time.

Jeihanipour and Taherzadeh (2009) produced ethanol from cotton-based waste textiles and found that a pretreatment with concentrated phosphoric acid prior to the hydrolysis improved ethanol production from the textiles up to 66% of the theoretical yield. Further, Silveira et al. (2005) produced ethanol from cheese whey permeate and found that the increase in lactose concentration increased ethanol yield and ethanol volumetric productivity, and reduced cell yield.

Wilkins et al. (2007c) produced ethanol from orange peel oil by two yeast strains and observed that the minimum inhibitory peel oil concentrations for ethanol production were 0.05% at 24 h, 0.10% at 48 h, and 0.15% at 72 h for both yeasts, while *S. cerevisiae* produced more ethanol than *K. marxianus* at each time point. Further, Widmer et al. (2010) pretreated orange processing waste under different times, pH, and temperature and found that the ethanol yields based on sugar content after enzymatic hydrolysis after 48 h SSF ranged from 76% to 94%. Further, Kim et al. (2008) optimized enzymatic hydrolysis and ethanol fermentation using food waste and found that the experimental results were in close agreement with model prediction with 120.1 g reducing sugar/L and 57.5 g ethanol/L, respectively.

70.5 CONCLUSION AND FUTURE RESEARCH

The brief information about the key research fronts covered by the 25 most-cited papers with at least 104 citations each is given under two primary headings: the hydrolysis of the urban wastes and production of the bioethanol fuels.

The usual characteristics of these HCPs are that the pretreatments and hydrolysis of the urban wastes and fermentation of the resulting hydrolysates are the primary processes for the bioethanol fuel production from urban wastes to improve the ethanol yield as the urban wastes are one of the most studied feedstocks for the bioethanol production especially for the countries with the large farmlands, forests, and crude oil deficiency.

The key findings on these research fronts should be read in the light of the increasing public concerns about climate change, GHG emissions, and global warming as these concerns have been certainly behind the boom in the research on the urban waste-based bioethanol fuels as an alternative to crude oil-based gasoline and petrodiesel fuels in the last decades. It is also a sustainable alternative to the first generation food crop-based bioethanol fuels such as corn grain- and sugarcane-based bioethanol fuels. The recent supply shocks caused by the COVID-19 pandemics and the Russian invasion of Ukraine also highlight the importance of the production and utilization of the bioethanol fuels from the urban wastes as an alternative to the crude oil-based gasoline and petrodiesel fuels.

As Table 70.3 shows, the most-prolific research front for this field is the food wastes with 76% of the HCPs, followed by the other urban wastes with 24% of the sample papers. The latter front includes paper and to a lesser extent municipal solid and textile wastes. Thus, these HCPs cover most-prolific research fronts for the urban waste-based bioethanol fuels. However, they do not cover household wood wastes.

Further, the most-prolific food wastes have been fruit wastes such as orange peel waste and to a lesser extent food waste in general, beverage wastes such as spent coffee grounds, other food wastes, vegetable wastes such as potato peels, dairy wastes such as cheese whey, and kitchen wastes in general such as waste meals and bread.

Similarly, as Table 70.4 shows, the most-prolific thematic research fronts for this field are the pretreatment and hydrolysis of the urban wastes with 100% of the HCPs each. The other research fronts are the hydrolysate fermentation and bioethanol production with 56% of the sample papers each. It is notable that there are no HCPs on the utilization and evaluation of the urban waste-based bioethanol fuels.

These studies emphasize the importance of proper incentive structures for the efficient production of urban waste-based bioethanol fuels in the light of North's institutional framework (North, 1991). In this context, the major producers and users of bioethanol fuels such as the USA and

Brazil with vast forests and farmlands have developed strong incentive structures for the efficient urban waste-based bioethanol fuels. In the light of the recent supply shocks caused primarily by the COVID-19 pandemics and Russian invasion of Ukraine, it is expected that the incentive structures such as public funding would be enhanced to increase the share of bioethanol fuels in the global fuel portfolio as a strong alternative to crude oil-based gasoline and petrodiesel fuels.

In this context, it is expected that the most-prolific researchers, institutions, countries, funding bodies, and journals in this field would have a first-mover advantage to benefit from such potential incentives. This is especially true for the US stakeholders as the USA has become the global leader in both the production and utilization of second generation bioethanol fuels from the urban wastes.

It is recommended that such review studies are performed for the primary research fronts of the urban waste-based bioethanol fuels.

ACKNOWLEDGMENTS

The contribution of the highly cited researchers in the field of the urban waste-based bioethanol fuels has been gratefully acknowledged.

REFERENCES

Agblevor, F. A., S. Batz and J. Trumbo. 2003. Composition and ethanol production potential of cotton gin residues. *Applied Biochemistry and Biotechnology, Part A: Enzyme Engineering and Biotechnology* 105:219–230.

Alvira, P., E. Tomas-Pejo, M. Ballesteros and M. J. Negro. 2010. Pretreatment technologies for an efficient bioethanol production process based on enzymatic hydrolysis: A review. *Bioresource Technology* 101:4851–4861.

Angelici, C., B. M. Weckhuysen and P. C. A. Bruijnincx. 2013. Chemocatalytic conversion of ethanol into butadiene and other bulk chemicals. *ChemSusChem* 6:1595–1614.

Antolini, E. 2007. Catalysts for direct ethanol fuel cells. *Journal of Power Sources* 170:1–12.

Antolini, E. 2009. Palladium in fuel cell catalysis. *Energy and Environmental Science* 2:915–931.

Arapoglou, D., T. Varzakas, A. Vlyssides and C. Israilides. 2010. Ethanol production from potato peel waste (PPW). *Waste Management* 30:1898–1902.

Burnham, J. F. 2006. Scopus database: A review. *Biomedical Digital Libraries* 3:1–8.

Castanon, M. and C. R. Wilke. 1981. Effects of the surfactant tween 80 on enzymatic hydrolysis of newspaper. *Biotechnology and Bioengineering* 23:1365–1372.

Chang, V. S., M. Nagwani, C. H. Kim and M. T. Holtzapple. 2001. Oxidative lime pretreatment of high-lignin biomass: Poplar wood and newspaper. *Applied Biochemistry and Biotechnology, Part A, Enzyme Engineering and Biotechnology* 94:1–28.

Choi, I. S., Y. G. Lee, S. K. Khanal, B. J. Park and H. J. Bae. 2015. A low-energy, cost-effective approach to fruit and citrus peel waste processing for bioethanol production. *Applied Energy* 140:65–74.

Fernandez-Bolanos, J., B. Felizon and A. Heredia, et al. 2001. Steam-explosion of olive stones: Hemicellulose solubilization and enhancement of enzymatic hydrolysis of cellulose. *Bioresource Technology* 79:53–61.

Fernando, S., S. Adhikari, C. Chandrapal and M. Murali. 2006. Biorefineries: Current status, challenges, and future direction. *Energy & Fuels* 20:1727–1737.

Grohmann, K., R. G. Cameron and B. S. Buslig. 1995. Fractionation and pretreatment of orange peel by dilute acid hydrolysis. *Bioresource Technology* 54:129–141.

Guclu, G., Yalcinyuva, T., S. Ozgumus and M. Orbay. 2003. Hydrolysis of waste polyethylene terephthalate and characterization of products by differential scanning calorimetry. *Thermochimica Acta* 404:193–205.

Guerard, F., L. Guimas and A. Binet. 2002. Production of tuna waste hydrolysates by a commercial neutral protease preparation. *Journal of Molecular Catalysis B: Enzymatic* 19:489–498.

Guimaraes, P. M. R., J. A. Teixeira and L. Domingues. 2010. Fermentation of lactose to bio-ethanol by yeasts as part of integrated solutions for the valorisation of cheese whey. *Biotechnology Advances* 28:375–384.

Hahn-Hagerdal, B., M. Galbe, M. F. Gorwa-Grauslund, G. Liden and G. Zacchi. 2006. Bio-ethanol: The fuel of tomorrow from the residues of today. *Trends in Biotechnology* 24:549–556.

Hill, J., E. Nelson, D. Tilman, S. Polasky and D. Tiffany. 2006. Environmental, economic, and energetic costs and benefits of biodiesel and ethanol biofuels. *Proceedings of the National Academy of Sciences of the United States of America* 103:11206–11210.

Hill, J., S. Polasky and E. Nelson, et al. 2009. Climate change and health costs of air emissions from biofuels and gasoline. *Proceedings of the National Academy of Sciences of the United States of America* 106:2077–2082.

Holtzapple, M. T., J. E. Lundeen, R. Sturgis, J. E. Lewis and B. E. Dale. 1992. Pretreatment of lignocellulosic municipal solid waste by ammonia fiber explosion (AFEX). *Applied Biochemistry and Biotechnology* 34–35:5–21.

Hsieh, W. D., R. H. Chen, T. L. Wu and T. H. Lin. 2002. Engine performance and pollutant emission of an SI engine using ethanol-gasoline blended fuels. *Atmospheric Environment* 36:403–410.

Huang, H. J., S. Ramaswamy, U. W. Tschirner and B. V. Ramarao. 2008. A review of separation technologies in current and future biorefineries. *Separation and Purification Technology* 62:1–21.

Jeihanipour, A. and M. J. Taherzadeh. 2009. Ethanol production from cotton-based waste textiles. *Bioresource Technology* 100:1007–1010.

Kalogo, Y., S. Habibi, H. L. Maclean and S. V. Joshi. 2007. Environmental implications of municipal solid waste-derived ethanol. *Environmental Science and Technology* 41:35–41.

Kim, J. H., J. C. Lee and D. Pak. 2011. Feasibility of producing ethanol from food waste. *Waste Management* 31:2121–2125.

Kim, J. K., B. R. Oh, H. J. Shin, C. Y. Eom and S. W. Kim. 2008. Statistical optimization of enzymatic saccharification and ethanol fermentation using food waste. *Process Biochemistry* 43:1308–1312.

Konur, O. 2000. Creating enforceable civil rights for disabled students in higher education: An institutional theory perspective. *Disability & Society* 15:1041–1063.

Konur, O. 2002a. Access to nursing education by disabled students: Rights and duties of nursing programs. *Nurse Education Today* 22:364–374.

Konur, O. 2002b. Assessment of disabled students in higher education: Current public policy issues. *Assessment and Evaluation in Higher Education* 27:131–152.

Konur, O. 2002c. Access to employment by disabled people in the UK: Is the Disability Discrimination Act working? *International Journal of Discrimination and the Law* 5:247–279.

Konur, O. 2006a. Participation of children with dyslexia in compulsory education: Current public policy issues. *Dyslexia* 12:51–67.

Konur, O. 2006b. Teaching disabled students in higher education. *Teaching in Higher Education* 11:351–363.

Konur, O. 2007a. A judicial outcome analysis of the Disability Discrimination Act: A windfall for the employers? *Disability & Society* 22:187–204.

Konur, O. 2007b. Computer-assisted teaching and assessment of disabled students in higher education: The interface between academic standards and disability rights. *Journal of Computer Assisted Learning* 23:207–219.

Konur, O. 2012. The evaluation of the research on the bioethanol: A scientometric approach. *Energy Education Science and Technology Part A: Energy Science and Research* 28:1051–1064.

Konur, O. 2015. Current state of research on algal bioethanol. In *Marine Bioenergy: Trends and Developments*, Eds. S. K. Kim and C. G. Lee, pp. 217–244. Boca Raton, FL: CRC Press.

Konur, O. 2019. Cyanobacterial bioenergy and biofuels science and technology: A scientometric overview. In *Cyanobacteria: From Basic Science to Applications*, Eds. A. K. Mishra, D. N. Tiwari and A. N. Rai, pp. 419–442. Amsterdam: Elsevier.

Konur, O. 2020. The scientometric analysis of the research on the bioethanol production from green macroalgae. In *Handbook of Algal Science, Technology and Medicine,* Ed. O. Konur, pp. 385–401. London: Academic Press.

Konur, O. 2023. Second generation urban waste-based bioethanol fuels: Scientometric study. In *Feedstock-based Bioethanol Fuels. II. Waste Feedstocks: Agricultural, Food, Industrial, Urban, Forestry, and Lignocellulosic Waste-based Bioethanol Fuels. Handbook of Bioethanol Fuels Volume 4*, Ed. O. Konur, pp. 237–261. Boca Raton, FL: CRC Press.

Kwon, E. E., H. Yi and Y. J. Jeon. 2013. Sequential co-production of biodiesel and bioethanol with spent coffee grounds. *Bioresource Technology* 136:475–480.

Le Man, H., S. K. Behera and H. S. Park. 2010. Optimization of operational parameters for ethanol production from Korean food waste leachate. *International Journal of Environmental Science and Technology* 7:157–164.

Li, A., B. Antizar-Ladislao and M. Khraisheh. 2007. Bioconversion of municipal solid waste to glucose for bio-ethanol production. *Bioprocess and Biosystems Engineering* 30:189–196.

Lin, Y. and S. Tanaka. 2006. Ethanol fermentation from biomass resources: Current state and prospects. *Applied Microbiology and Biotechnology* 69:627–642.

Ma, X., L. Sun and C. Song. 2002. A new approach to deep desulfurization of gasoline, diesel fuel and jet fuel by selective adsorption for ultra-clean fuels and for fuel cell applications. *Catalysis Today* 77:107–116.

Maitan-Alfenas, G. P., E. M. Visser and V. M. Guimaraes. 2015. Enzymatic hydrolysis of lignocellulosic biomass: Converting food waste in valuable products. *Current Opinion in Food Science* 1:44–49.

Morschbacker, A. 2009. Bio-ethanol based ethylene. *Polymer Reviews* 49:79–84.

Mussatto, S. I., E. M. S. Machado, L. M. Carneiro and J. A. Teixeira. 2012. Sugars metabolism and ethanol production by different yeast strains from coffee industry wastes hydrolysates. *Applied Energy* 92:763–768.

Najafi, G., B. Ghobadian and T. Tavakoli, et al. 2009. Performance and exhaust emissions of a gasoline engine with ethanol blended gasoline fuels using artificial neural network. *Applied Energy* 86:630–639.

Newman, P. W. G. and J. R. Kenworthy. 1989. Gasoline consumption and cities: A comparison of U.S. cities with a global survey. *Journal of the American Planning Association* 55:24–37.

North, D. C. 1991. Institutions. *Journal of Economic Perspectives* 5:97–112.

Oberoi, H. S., P. V. Vadlani, L. Saida, S. Bansal and J. D. Hughes. 2011. Ethanol production from banana peels using statistically optimized simultaneous saccharification and fermentation process. *Waste Management* 31:1576–1584.

Okuda, N., K. Ninomiya, M. Takao, Y. Katakura and S. Shioya. 2007. Microaeration enhances productivity of bioethanol from hydrolysate of waste house wood using ethanologenic *Escherichia coli* KO11. *Journal of Bioscience and Bioengineering* 103:350–357.

Okuda, N., M. Soneura, K. Ninomiya, Y. Katakura and S. Shioya. 2008. Biological detoxification of waste house wood hydrolysate using *Ureibacillus thermosphaericus* for bioethanol production. *Journal of Bioscience and Bioengineering* 106:128–133.

Olsson, L. and B. Hahn-Hagerdal. 1996. Fermentation of lignocellulosic hydrolysates for ethanol production. *Enzyme and Microbial Technology* 18:312–331.

Park, J. W., Y. Takahata, T. Kajiuchi and T. Akehata. 1992. Effects of nonionic surfactant on enzymatic hydrolysis of used newspaper. *Biotechnology and Bioengineering* 39:117–120.

Piarpuzan, D., J. A. Quintero and C. A. Cardona. 2011. Empty fruit bunches from oil palm as a potential raw material for fuel ethanol production. *Biomass and Bioenergy* 35:1130–1137.

Rahman, S. H. A., J. P. Choudhury, A. L. Ahmad and A. H. Kamaruddin. 2007. Optimization studies on acid hydrolysis of oil palm empty fruit bunch fiber for production of xylose. *Bioresource Technology* 98:554–559.

Ravindran, R. and A. K. Jaiswal. 2016. A comprehensive review on pre-treatment strategy for lignocellulosic food industry waste: Challenges and opportunities. *Bioresource Technology* 199:92–102.

Sanchez, O. J. and C. A. Cardona. 2008. Trends in biotechnological production of fuel ethanol from different feedstocks. *Bioresource Technology* 99:5270–5295.

Silveira, W. B., F. J. V. Passos, H. C. Mantovani and F. M. L. Passos. 2005. Ethanol production from cheese whey permeate by *Kluyveromyces marxianus* UFV-3: A flux analysis of oxido-reductive metabolism as a function of lactose concentration and oxygen levels. *Enzyme and Microbial Technology* 36:930–936.

Sun, Y. and J. Cheng. 2002. Hydrolysis of lignocellulosic materials for ethanol production: A review. *Bioresource Technology* 83:1–11.

Taherzadeh, M. J. and K. Karimi. 2007. Enzyme-based hydrolysis processes for ethanol from lignocellulosic materials: A review. *Bioresources* 2:707–738.

Taherzadeh, M. J. and K. Karimi. 2008. Pretreatment of lignocellulosic wastes to improve ethanol and biogas production: A review. *International Journal of Molecular Sciences* 9:1621–1651.

Wang, L., M. Sharifzadeh, R. Templer and M. J. Murphy. 2013. Bioethanol production from various waste papers: Economic feasibility and sensitivity analysis. *Applied Energy* 111:1172–1182.

Widmer, W., W. Zhou and K. Grohmann. 2010. Pretreatment effects on orange processing waste for making ethanol by simultaneous saccharification and fermentation. *Bioresource Technology* 101:5242–5249.

Wilkins, M. R., L. Suryawati, N. O. Maness and D. Chrz. 2007c. Ethanol production by *Saccharomyces cerevisiae* and *Kluyveromyces marxianus* in the presence of orange-peel oil. *World Journal of Microbiology and Biotechnology* 23:1161–1168.

Wilkins, M. R., W. W. Widmer and K. Grohmann. 2007b. Simultaneous saccharification and fermentation of citrus peel waste by *Saccharomyces cerevisiae* to produce ethanol. *Process Biochemistry* 42:1614–1619.

Wilkins, M. R., W. W. Widmer, K. Grohmann and R. G. Cameron. 2007a. Hydrolysis of grapefruit peel waste with cellulase and pectinase enzymes. *Bioresource Technology* 98:1596–1601.

Yoshioka, T., T. Sato and A. Okuwaki. 1994. Hydrolysis of waste PET by sulfuric acid at 150°C for a chemical recycling. *Journal of Applied Polymer Science* 52:1353–1355.

Yunus, R., S. F. Salleh, N. Abdullah and D. R. A. Biak. 2010. Effect of ultrasonic pre-treatment on low temperature acid hydrolysis of oil palm empty fruit bunch. *Bioresource Technology* 101:9792–9796.

Part 22

Second Generation Forestry Waste-based Bioethanol Fuels

71 Second Generation Forestry Waste-based Bioethanol Fuels

Scientometric Study

Ozcan Konur
(Formerly) Ankara Yildirim Beyazit University

71.1 INTRODUCTION

The crude oil-based gasoline fuels (Ma et al., 2002; Newman and Kenworthy, 1989) have been widely used in the transportation sector since the 1920s. However, there have been great public concerns over the adverse environmental and human impact of these fuels (Hill et al., 2006, 2009). Hence, biomass-based bioethanol fuels (Hill et al., 2006; Konur, 2012e, 2015, 2019, 2020a) have increasingly been used in blending gasoline fuels (Hsieh et al., 2002; Najafi et al., 2009), in the fuel cells (Antolini, 2007, 2009), and in the biochemical production (Angelici et al., 2013; Morschbacker, 2009) in a biorefinery context (Fernando et al., 2006; Huang et al., 2008).

Bioethanol fuels also play a critical role in maintaining the energy security (Kruyt et al., 2009; Winzer, 2012) in the supply shocks (Kilian, 2008, 2009) related to oil price shocks (Hamilton, 1983, 2003), COVID-19 pandemics (Fauci et al., 2020; Li et al., 2020), or wars (Hamilton, 1983; Jones, 2012) in the aftermath of the Russian invasion of Ukraine (Reeves, 2014).

However, it is necessary to pretreat the biomass (Taherzadeh and Karimi, 2008; Yang and Wyman, 2008) to enhance the yield of the bioethanol (Hahn-Hagerdal et al., 2006; Sanchez and Cardona, 2008) prior to the bioethanol production through the hydrolysis (Sun and Cheng, 2002; Taherzadeh and Karimi, 2007) and fermentation (Lin and Tanaka, 2006; Olsson and Hahn-Hagerdal, 1996) of the biomass and hydrolysates, respectively.

One of the most-studied feedstocks for the bioethanol fuels has been the forestry wastes such as wood chips, bark, sawdust, forestry residues, tree prunings, spent sulfite liquor, and wood wastes. The research in the field of second generation forestry waste-based bioethanol fuels has intensified in this context in the key research fronts of the pretreatment (Brandt et al., 2010; Duff and Murray, 1996) and hydrolysis (Duff and Murray, 1996; Lu et al., 2002) of the forestry wastes, fermentation (Cara et al., 2008; Duff and Murray, 1996) of the forestry waste-based hydrolysates, and production (Cara et al., 2008; Duff and Murray, 1996) and evaluation (Franko et al., 2016; Jonker et al., 2016) of second generation forestry waste-based bioethanol fuels. Further, wood chips (Brandt et al., 2010; Brownell et al., 1986), forest residues (Huang et al., 2019; Lu et al., 2002), sawdust (Vidal and Molinier, 1988, Weil et al., 1997), tree prunings (Cara et al., 2007, 2008), bark (Hu et al., 2008; Lima et al., 2013), wood waste (Ayeni et al., 2013; Salehian and Karimi, 2013), and spent sulfite liquor (Nigam, 2001; Survase et al., 2011) have been studied intensively in this context.

However, it is essential to develop efficient incentive structures (North, 1991) for the primary stakeholders to enhance the research in this field (Konur, 2000, 2002a–c, 2006a,b, 2007a,b). The scientometric analysis has been used in this context to inform the primary stakeholders about the current state of the research in a selected research field (Garfield, 1955; Konur, 2011, 2012a–i, 2015, 2018b, 2019, 2020a).

As there have been no published scientometric studies in this field, this book chapter presents a scientometric study of the research in second generation forestry waste-based bioethanol fuels. It examines

DOI: 10.1201/9781003226550-97

the scientometric characteristics of both the sample and population data presenting scientometric characteristics of both these datasets in the order of documents, authors, publication years, institutions, funding bodies, source titles, countries, Scopus subject categories, Scopus keywords, and research fronts.

71.2 MATERIALS AND METHODS

The search for this study was carried out using Scopus database (Burnham, 2006) in July 2022.

As a first step for the search of the relevant literature, the keywords were selected using the first most-cited 100 population papers. The selected keyword list was then optimized to obtain a representative sample of papers for the searched research field. This keyword list was provided in the appendix for future replicative studies.

As a second step, two sets of data were used for this study. First, a population sample of 535 papers was used to examine the scientometric characteristics of the population data. Second, a sample of 100 most-cited papers, corresponding to 19% of the population papers, was used to examine the scientometric characteristics of these citation classics.

The scientometric characteristics of these both sample and population datasets were presented in the order of documents, authors, publication years, institutions, funding bodies, source titles, countries, Scopus subject categories, Scopus keywords, and research fronts.

Lastly, the key scientometric findings for both datasets were discussed to highlight the research landscape for second generation forestry waste-based bioethanol fuels. Additionally, a number of brief conclusions were drawn, and a number of relevant recommendations were made to enhance the future research landscape.

71.3 RESULTS

71.3.1 The Most-Prolific Documents in Second Generation Forestry Waste-based Bioethanol Fuels

The information on the types of documents for both datasets is given in Table 71.1. The articles and conference papers, published in journals, dominate both the sample (96%) and population (97%) papers with 1% deficit. Further, review papers and short surveys have a 2% surplus as they are over-represented in the sample papers as they constitute 4% and 2% of the sample and population papers, respectively.

TABLE 71.1
Documents in Second Generation Forestry Waste-based Bioethanol Fuels

Documents	Sample Dataset (%)	Population Dataset (%)	Surplus (%)
Article	88.0	93.6	–5.6
Review	4.0	2.2	1.8
Conference paper	8.0	3.4	4.6
Book chapter	0.0	0.4	–0.4
Note	0.0	0.4	–0.4
Letter	0.0	0.0	0.0
Short survey	0.0	0.0	0.0
Book	0.0	0.0	0.0
Editorial	0.0	0.0	0.0
Sample size	100	535	

Sample dataset: the number of papers (%) in the set of 100 highly cited papers. Population dataset: the number of papers (%) in the set of the 535 population papers.

It is further notable that 99% of the population papers were published in journals, while 1% of them were published in books and book series. Similarly, 100% of the sample papers were published in the journals.

71.3.2 The Most-Prolific Authors in Second Generation Forestry Waste-based Bioethanol Fuels

The information about the most-prolific 20 authors with at least 3% of sample papers each is given in Table 71.2. The most-prolific authors are John N. Saddler, Qiang Yong, and Caoxing Huang with 8% of the sample papers, followed by Eulogio Castro with 7% of the sample papers. The other prolific authors are Ignacio Ballesteros, Mercedes Ballesteros, Maria J. Negro, Jose M. Oliva, Inmaculada Romero, and Encarnacion Ruiz with 5% of the sample papers each.

On the other hand, the most influential author is Caoxing Huang with 5.9% surplus, followed by Qiang Yong with 5.8% surplus. The other influential authors are Eulogio Castro, John N. Saddler, Jose M. Oliva, Inmaculada Romero, and Encarnacion Ruiz with 3.7%–4.9% surplus each.

TABLE 71.2
The Most-Prolific Authors in Second Generation Forestry Waste-based Bioethanol Fuels

No.	Author Name	Author Code	Sample Papers (%)	Population Papers (%)	Surplus (%)	Institution	Country	HI	N	Res. Front
1	Saddler, John N.	57202481615 7005297559	8.0	3.2	4.8	Univ. British Columbia	Canada	89	408	C
2	Yong, Qiang	7003718716	8.0	2.2	5.8	Nanjing Forest. Univ.	China	37	270	R
3	Huang, Caoxing	56495726800	8.0	2.1	5.9	Nanjing Forest. Univ.	China	35	149	R
4	Castro, Eulogio	7102441948	7.0	2.1	4.9	Univ. Jaen	Spain	44	189	T
5	Ballesteros, Ignacio	6602732963	5.0	1.5	3.5	CIEMAT	Spain	38	70	T
6	Ballesteros, Mercedes*	7006110611	5.0	1.5	3.5	CIEMAT	Spain	51	135	T
7	Negro, Maria J.*	6701512649	5.0	1.5	3.5	CIEMAT	Spain	38	74	T
8	Oliva, Jose M.	57194220606	5.0	1.3	3.7	CIEMAT	Spain	34	54	T
9	Romero, Inmaculada*	25032007200	5.0	1.3	3.7	Univ. Jaen	Spain	30	76	T
10	Ruiz, Encarnacion*	25646493300	5.0	1.3	3.7	Univ. Jaen	Spain	34	75	T
11	Sanchez, Sebastian	7202745657	3.0	1.7	1.3	Univ. Jaen	Spain	29	125	T
12	Zhu, Junyong	7405692678	3.0	1.5	1.5	USDA Forest Serv.	USA	63	309	R
13	Gleisner, Roland	55928749700	3.0	1.3	1.7	USDA Forest Serv.	USA	28	70	R
14	Grohmann, Karel	7004503589	3.0	0.7	2.3	Renewable Spirits LLC	USA	49	111	C
15	Cara, Cristobal	22949914200	3.0	0.6	2.4	Univ. Jaen	Spain	33	53	T
16	Jin, Yongcan	9846535500	3.0	0.6	2.4	Nanjing Forest. Univ.	China	32	177	R
17	Ladisch, Michael R.	7005670397	3.0	0.6	2.4	Purdue Univ.	USA	59	290	S
18	Li, Xin	55951836300	3.0	0.6	2.4	S. China Agr. Univ.	China	63	136	R
19	Moya, Manuel	35262479500	3.0	0.6	2.4	Univ. Jaen	Spain	26	63	T
20	Ramos, Luiz P.	7202180586	3.0	0.6	2.4	Fed. Univ. Parana	Brazil	43	172	C

Author code: the unique code given by Scopus to the authors. Sample papers: the number of papers authored in the sample dataset. Population papers: the number of papers authored in the population dataset.
*, Female; R, Wood residues; T, Tree prunings; C, Wood chips; S, Sawdust.

The most-prolific institution for the sample dataset is the University of Jaen with six authors, followed by CIEMAT with four authors. The other prolific institutions are Nanjing Forestry University and the U.S. Forest Service with two authors each. On the other hand, the most-prolific country for the sample dataset is Spain with ten authors, followed by China and the USA with four authors each. In total, only five countries house these top authors.

The most-prolific research front for these top authors is the bioethanol fuels from tree prunings with 10 authors followed by the bioethanol fuels from forest residues with six authors. The other research fronts are the bioethanol fuels from the wood chips and sawdust with three and one authors, respectively.

On the other hand, there is significant gender deficit (Beaudry and Lariviere, 2016) for the sample dataset as surprisingly only five of these top researchers are female with a representation rate of 25%.

Additionally, there are other authors with the relatively low citation impact and with 0.7%–1.5% of the population papers each: Soledad Mateo, Alberto J. Moya, Maria C. Area, Raimo Alen, Fernando E. Felissia, Paloma Manzanares, Maria E. Eugenio, Qingxi Hou, Mikhail Iakovlev, Joni Lehto, Lucian A. Lucia, Raquel Martín-Sampedro, Ines Conceicao Roberto, Juan Miguel Romero-Garcia, Weiping Ban, Ursula Fillat, Mats Galbe, Francisco Girio, Juan He, David Ibarra, Hasan Jameel, Julia Kruyeniski, Chenhuan Lai, Teixeira Mendonca, Jyri Pekka Mikkola, Islam S. M. Rafiqul, Alejandro Rodriguez, Tsutomu Suzuki, Qian Yang, Adriaan van Heiningen, and Birgitte Ahring.

71.3.3 The Most-Prolific Research Output by Years in Second Generation Forestry Waste-based Bioethanol Fuels

Information about papers published between 1970 and 2022 is given in Figure 71.1. This figure clearly shows that the bulk of the research papers in the population dataset were published primarily in the 2010s and the early 2020s with 54% and 23% of the population dataset, respectively. Similarly, the publication rates for the 2000s, 1990s, 1980s, and 1970s were 11%, 6%, 4%, and 1% respectively. Additionally, 1% of the population papers were published in the pre-1970s.

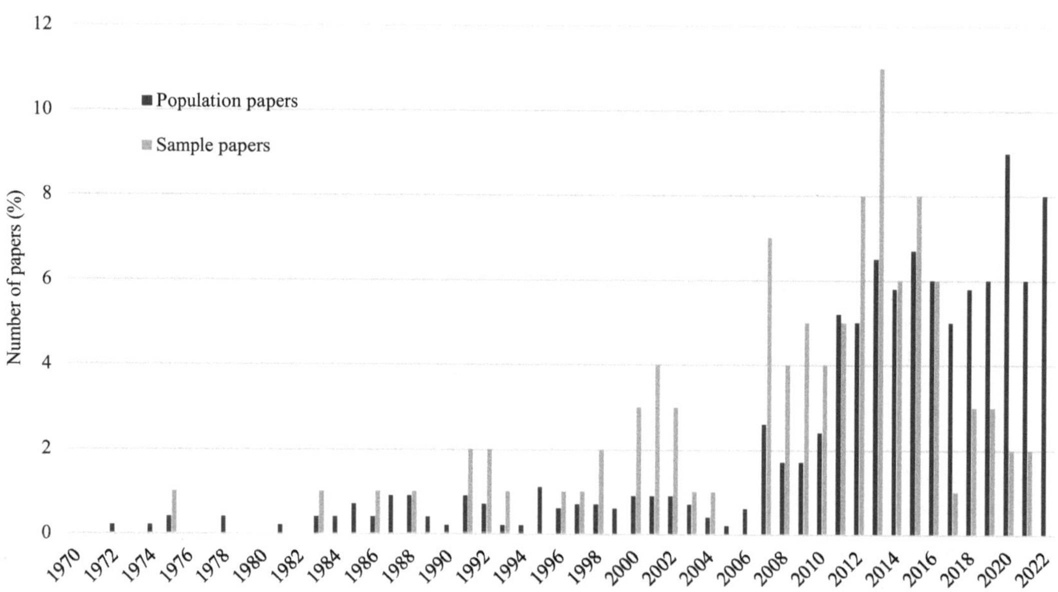

FIGURE 71.1 The research output by years regarding second generation forestry waste-based bioethanol fuels.

Similarly, the bulk of the research papers in the sample dataset were published in the 2010s and 2000s with 55% and 28% of the sample dataset, respectively. Similarly, the publication rates for the early 2020s, 1990s, 1980s, and 1970s were 4%, 9%, 3%, and 1% of the sample papers, respectively.

The most-prolific publication years for the population dataset were 2020 and 2022 with 9% and 8% of the dataset, respectively, while 83% of the population papers were published between 2007 and 2022. Similarly, 71% of the sample papers were published between 2007 and 2019, while the most-prolific publication years were 2013, 2012, and 2015 with 11%, 8%, and 8% of the sample papers, respectively.

Further, the number of population papers first rose between 2007 and 2015, and thereafter, it steadied with spikes in 2020 and 2022.

71.3.4 The Most-Prolific Institutions in Second Generation Forestry Waste-based Bioethanol Fuels

Information about the most-prolific 12 institutions publishing papers on second generation forestry waste-based bioethanol fuels with at least 3% of the sample papers each is given in Table 71.3.

The most-prolific institution is the Nanjing Forestry University with 10% of the sample papers, followed by the Center for Energy, Environmental and Technological Research (CIEMAT), University of Jaen, and University of British Columbia with 9%, 8%, and 7% of the sample papers, respectively. The other prolific institutions are the National Renewable Energy Laboratory (NREL) and NC State University with 5% and 4% of the sample papers, respectively.

Similarly, the top country for these most-prolific institutions is the USA with six institutions, followed by China and Spain with two institutions each. In total, only five countries house these top institutions.

On the other hand, the institution with the most citation impact is the CIEMAT with 6.2% surplus, followed by Nanjing Forestry University and University of Jaen with 4.4% and 4.1% surplus, respectively. The other influential institutions are the University of British Columbia and National Renewable Energy Laboratory with 3.6% and 3.5% surplus, respectively.

Additionally, there are other institutions with the relatively low citation impact and with 0.9%–2.4% of the population papers each: University of Jyvaskyla, Chinese Academy of Forestry, South China University of Technology, Umea University, Tianjin University of Science & Technology, Georgia Institute of Technology, Aalto University, University of Concepcion, Abo Academy University, Lund University, University of the Republic, State Key Laboratory of Pulp and Paper Engineering, National

TABLE 71.3
The Most-Prolific Institutions in Second Generation Forestry Waste-based Bioethanol Fuels

No.	Institutions	Country	Sample Papers (%)	Population Papers (%)	Surplus (%)
1	Nanjing Forestry Univ.	China	10.0	5.6	4.4
2	CIEMAT	Spain	9.0	2.8	6.2
3	Univ. Jaen	Spain	8.0	3.9	4.1
4	Univ. British Columbia	Canada	7.0	3.4	3.6
5	Natl. Renew. Ener. Lab.	USA	5.0	1.5	3.5
6	NC State Univ.	USA	4.0	1.9	2.1
7	Chinese Acad. Sci.	China	3.0	2.4	0.6
8	Univ. Sao Paulo	Brazil	3.0	2.2	0.8
9	Univ. Maine	USA	3.0	2.1	0.9
10	USDA Forest Serv.	USA	3.0	2.8	0.2
11	Oregon State Univ.	USA	3.0	0.7	2.3
12	Purdue Univ.	USA	3.0	0.6	2.4

Council for Scientific and Technical Research, University of New Brunswick, National Institute of Agricultural and Food Research and Technology, National Center for Scientific Research (CNRS), Kyoto University, University of Oulu, Seoul National University, Lakehead University, University of Vigo, United States Department of Agriculture, University Malaysia Pahang,

71.3.5 The Most-Prolific Funding Bodies in Second Generation Forestry Waste-based Bioethanol Fuels

Information about the most-prolific 14 funding bodies funding at least 3% of the sample papers each is given in Table 71.4. Further, only 50% of the sample and population papers each were funded.

The most-prolific funding body is the National Natural Science Foundation of China with 11% of the sample papers, followed by the European Regional Development Fund and Natural Science Foundation of Jiangsu Province with 8% and 6% of the sample papers, respectively. The other prolific funding bodies are the European Commission, National Institute of Food and Agriculture, Government of Canada, and National Council for Scientific and Technological Development with 4% of the sample papers each. The National Natural Science Foundation of China is also the largest funder of the population papers with 11% funding rate.

On the other hand, the most-prolific countries for these top funding bodies are China and Canada with four and three funding bodies, respectively. The other prolific funding bodies are the European Commission and the USA with two funding bodies each. In total, only six countries and the EU house these top funding bodies.

TABLE 71.4
The Most-Prolific Funding Bodies in Second Generation Forestry Waste-based Bioethanol Fuels

No.	Funding Bodies	Country	Sample Paper No. (%)	Population Paper No. (%)	Surplus (%)
1	National Natural Science Foundation of China	China	11.0	9.5	1.5
2	European Regional Development Fund	EU	8.0	4.7	3.3
3	Natural Science Foundation of Jiangsu Province	China	6.0	1.7	4.3
4	European Commission	EU	4.0	2.6	1.4
5	National Institute of Food and Agriculture	USA	4.0	2.4	1.6
6	Government of Canada	Canada	4.0	1.9	2.1
7	National Council for Scientific and Technological Development	Brazil	4.0	1.9	2.1
8	Natural Sciences and Engineering Research Council of Canada	Canada	3.0	2.8	0.2
9	Northwest Advanced Renewables Alliance	USA	3.0	2.1	0.9
10	Priority Academic Program Development of Jiangsu Higher Education Institutions	China	3.0	2.1	0.9
11	Ministry of Economy and Competitiveness	Spain	3.0	1.5	1.5
12	Nanjing Forestry University	China	3.0	1.3	1.7
13	Tekes – Finnish Funding Agency for Technology and Innovation	Finland	3.0	0.9	2.1
14	Natural Resources Canada	Canada	3.0	0.7	2.3

The funding body with the most citation impact is the Natural Science Foundation of Jiangsu Province with 4.3% surplus, followed by the European Regional Development Fund and Natural Resources Canada with 3.3% and 2.3% surplus, respectively. The other influential funding bodies are the Government of Canada, National Council for Scientific and Technological Development, and Finnish Funding Agency for Technology and Innovation (Tekes) with 2.1% surplus each. Further, the funding body with the least citation impact is the Natural Sciences and Engineering Research Council of Canada with 0.2% surplus.

The other funding bodies with the relatively low citation impact and with 0.7%–2.2% of the population papers each are the National Key Research and Development Program of China, Higher Education Personnel Improvement Coordination, U.S. Department of Agriculture, Ministry of Science and Technology of China, National Council for Scientific and Technical Research, Foundation for Science and Technology, National Research Foundation of Korea, U.S. Department of Energy, National Agency for Research and Innovation, Japan Society for the Promotion of Science, Ministry of Science, Technology and Innovation, Thematic Operational Program Competitiveness Factors, China Scholarship Council, Energy Agency, Fundamental Research Funds for the Central Universities, Research Support Foundation of the State of Sao Paulo, Korea Institute of Energy Technology Evaluation and Planning, Ministry of Science and Innovation, Ministry of Education of China, National Basic Research Program of China (973 Program), National University of Misiones, and University Grants Commission.

71.3.6 The Most-Prolific Source Titles in Second Generation Forestry Waste-based Bioethanol Fuels

Information about the most-prolific 15 source titles publishing at least 2% of the sample papers each in second generation forestry waste-based bioethanol fuels is given in Table 71.5.

The most-prolific source title is the Bioresource Technology with 30% of the sample papers, followed by Applied Biochemistry and Biotechnology Part A Enzyme Engineering and Biotechnology

TABLE 71.5
The Most-Prolific Source Titles in Second Generation Forestry Waste-based Bioethanol Fuels

No.	Source Titles	Sample Papers (%)	Population Papers (%)	Surplus (%)
1	Bioresource Technology	30.0	14.4	15.6
2	Applied Biochemistry and Biotechnology, Part A: Enzyme Engineering and Biotechnology	7.0	1.7	5.3
3	Industrial Crops and Products	5.0	4.7	0.3
4	Applied Biochemistry and Biotechnology	5.0	3.6	1.4
5	Industrial and Engineering Chemistry Research	5.0	1.9	3.1
6	Biotechnology and Bioengineering	5.0	1.3	3.7
7	Biomass and Bioenergy	4.0	3.0	1.0
8	Green Chemistry	3.0	1.1	1.9
9	Journal of Wood Chemistry and Technology	3.0	1.1	1.9
10	Journal of Bioscience and Bioengineering	3.0	0.9	2.1
11	Applied Energy	3.0	0.7	2.3
12	Chemical Engineering Science	3.0	0.6	2.4
13	Holzforschung	2.0	3.0	−1.0
14	Biotechnology for Biofuels	2.0	1.7	0.3
15	Fuel	2.0	1.3	0.7

with 7% of the sample papers. The other prolific titles are the Industrial Crops and Products, Applied Biochemistry and Biotechnology, Industrial and Engineering Chemistry Research, and Biotechnology and Bioengineering with 5% of the sample papers each.

On the other hand, the source title with the most citation impact is the Bioresource Technology with 15.6% surplus, followed by Applied Biochemistry and Biotechnology Part A Enzyme Engineering and Biotechnology with 5.3% Surplus. The other influential titles are the Biotechnology and Bioengineering, Industrial and Engineering Chemistry Research, Chemical Engineering Science, and Applied Energy with 2.3%–3.7% surplus each.

The other source titles with the relatively low citation impact with 0.6%–6.4% of the population papers each are Bioresources, Biomass Conversion and Biorefinery, Wood Science and Technology, Cellulose Chemistry and Technology, Renewable Energy, ACS Sustainable Chemistry and Engineering, Energies, Cellulose, Nordic Pulp and Paper Research Journal, Waste and Biomass Valorization, Bioenergy Research, Energy and Fuels, Journal of Industrial Microbiology and Biotechnology, RSC Advances, Enzyme and Microbial Technology, Fermentation, Industrial Biotechnology, Journal of Chemical Technology and Biotechnology, Journal of Pulp and Paper Science, and Separation and Purification Technology.

71.3.7 The Most-Prolific Countries in Second Generation Forestry Waste-based Bioethanol Fuels

Information about the most-prolific 11 countries publishing at least 3% of sample papers each in second generation forestry waste-based bioethanol fuels is given in Table 71.6.

The most-prolific country is the USA with 27% of the sample papers, followed by China, Spain, and Canada with 19%, 14%, and 12% of the sample papers, respectively. The other prolific countries are Brazil, Finland, Japan, and Portugal with 5%–6% of the sample papers, respectively. Further, five European countries listed in Table 71.6 produce 31% and 25% of the sample and population papers, respectively, and China is the largest producer of the population papers with a publication rate of 18.6%.

On the other hand, the country with the most citation impact is the USA with 10.4% surplus, followed by Spain with 5.2% surplus. The other influential countries are Canada and Portugal with 3% and 2.2% surplus, respectively. Similarly, the country with the least citation impact is India with 2.6% deficit, followed by Finland and Sweden with 1.2% deficit each.

Additionally, there are other countries with relatively low citation impact and with 0.7%–4.5% of the sample papers each: South Korea, Nigeria, France, Chile, Malaysia, Argentina, Australia,

TABLE 71.6
The Most-Prolific Countries in Second Generation Forestry Waste-based Bioethanol Fuels

No.	Countries	Sample Papers (%)	Population Papers (%)	Surplus (%)
1	USA	27.0	16.6	10.4
2	China	20.0	18.6	1.4
3	Spain	14.0	8.8	5.2
4	Canada	12.0	9.0	3.0
5	Brazil	6.0	4.5	1.5
6	Finland	5.0	6.2	−1.2
7	Japan	5.0	5.6	−0.6
8	Portugal	5.0	2.8	2.2
9	Sweden	4.0	5.2	−1.2
10	India	3.0	5.6	−2.6
11	UK	3.0	1.7	1.3

Italy, Poland, Belgium, Iran, Uruguay, Pakistan, Austria, Colombia, Mexico, Norway, Russian Federation, Turkey, Bangladesh, Germany, Hong Kong, Taiwan, and Thailand.

71.3.8 The Most-Prolific Scopus Subject Categories in Second Generation Forestry Waste-based Bioethanol Fuels

Information about the most-prolific nine Scopus subject categories indexing at least 7% of the sample papers each is given in Table 71.7.

The most-prolific Scopus subject category in second generation forestry waste-based bioethanol fuels is the Chemical Engineering with 74% of the sample papers, followed by Environmental Science and Energy with 48% of the sample papers each. The other prolific subject categories are Biochemistry, Genetics and Molecular Biology, Immunology and Microbiology, and Chemistry with 18%–29% of the sample papers each. It is notable that Social Sciences including Economics and Business account for 1% and 2% of the sample and population studies, respectively.

On the other hand, the Scopus subject category with the most citation impact is the Chemical Engineering with 23% surplus, followed by Energy, Environmental Science, and Immunology and Microbiology with 11%–13% surplus each. Similarly, the least influential subject categories are Agricultural and Biological Sciences and Materials Science with 9% deficit each.

71.3.9 The Most-Prolific Keywords in Second Generation Forestry Waste-based Bioethanol Fuels

Information about the Scopus keywords used with at least 6% or 3.9% of the sample or population papers, respectively, is given in Table 71.8. For this purpose, keywords related to the keyword set given in the appendix are selected from a list of the most-prolific keyword set provided by Scopus database.

These keywords are grouped under the five headings: forestry waste, pretreatments, fermentation, hydrolysis and hydrolysates, and products.

The most-prolific keywords related to the biomass and biomass constituents are cellulose, lignin, and wood with 51%, 48%, and 43% of the sample papers, respectively. The other prolific keywords are biomass, hemicellulose, carbohydrate, lignocellulose, and forestry with 16%–33% of the sample papers each.

Further, the prolific keywords related to the pretreatments are pretreatments, pretreatments, and steam with 32%, 21%, and 21% of the sample papers, respectively. The other prolific keywords are the acids, sulfuric acid, water, and delignification with 11%–15% of the sample papers.

TABLE 71.7
The Most-Prolific Scopus Subject Categories in Second Generation Forestry Waste-based Bioethanol Fuels

No.	Scopus Subject Categories	Sample Papers (%)	Population Papers (%)	Surplus (%)
1	Chemical Engineering	74.0	50.8	23.2
2	Environmental Science	48.0	35.9	12.1
3	Energy	48.0	35.0	13.0
4	Biochemistry, Genetics and Molecular Biology	29.0	20.9	8.1
5	Immunology and Microbiology	25.0	14.4	10.6
6	Chemistry	18.0	17.9	0.1
7	Engineering	14.0	13.3	0.7
8	Agricultural and Biological Sciences	10.0	18.5	–8.5
9	Materials Science	7.0	15.5	–8.5

TABLE 71.8
The Most-Prolific Keywords in Second Generation Forestry Waste-based Bioethanol Fuels

No.	Keywords	Sample Papers (%)	Population Papers (%)	Surplus (%)
1		**Forestry Waste**		
	Cellulose	51.0	31.0	20.0
	Lignin	48.0	38.5	9.5
	Wood	43.0	27.7	15.3
	Biomass	33.0	23.9	9.1
	Hemicellulose	29.0	16.3	12.7
	Carbohydrate	20.0	13.3	6.7
	Lignocellulose	17.0	9.2	7.8
	Forestry	16.0	15.1	0.9
	Eucalyptus	13.0	7.9	5.1
	Softwood	12.0	9.2	2.8
	Pseudotsuga	12.0	4.9	7.1
	Xylan	12.0	3.7	8.3
	Wood chips	10.0	7.5	2.5
	Bamboo	10.0	6.0	4.0
	Hardwood	10.0	5.6	4.4
	Bark	8.0	5.6	2.4
	Chips	8.0	4.3	3.7
	Pruning	6.0	3.4	2.6
	Sawdust	5.0	8.6	–3.6
	Lignocellulosic biomass	4.0	7.5	–3.5
2		**Pretreatments**		
	Pretreatment	23.0	17.6	5.4
	Pre-treatment	21.0	16.3	4.7
	Steam	21.0	9.2	11.8
	Acids	15.0	4.9	10.1
	Sulfuric acid	13.0	8.2	4.8
	Water	13.0	6.0	7.0
	Delignification	11.0	13.8	–2.8
	Temperature	9.0	6.2	2.8
	Water vapor	9.0	3.0	6.0
	Sulfur dioxide	8.0	2.2	5.8
	Enzymes	7.0	10.5	–3.5
	Fractionation	6.0	5.2	0.8
	Ionic liquids	5.0	3.9	1.1
	pH	4.0	4.5	–0.5
	Sodium hydroxide		4.3	–4.3
3		**Fermentation**		
	Fermentation	35.0	25.8	9.2
	Furfural	11.0	4.7	6.3
	Acetic acids	10.0	4.9	5.1
	Saccharomyces	10.0	4.7	5.3
	Yeast	8.0	7.1	0.9
	Detoxification	6.0	3.9	2.1

(Continued)

TABLE 71.8 (*Continued*)
The Most-Prolific Keywords in Second Generation Forestry Waste-based Bioethanol Fuels

No.	Keywords	Sample Papers (%)	Population Papers (%)	Surplus (%)
4	**Hydrolysis**			
	Hydrolysis	61.0	38.5	22.5
	Sugar	47.0	20.2	26.8
	Enzyme activity	34.0	20.2	13.8
	Enzymatic hydrolysis	33.0	26.5	6.5
	Glucose	28.0	17.9	10.1
	Saccharification	21.0	17.6	3.4
	Xylose	15.0	6.9	8.1
	Cellulases	14.0	7.3	6.7
	Enzymatic digestibility	13.0	4.1	8.9
	Enzymolysis	11.0	8.0	3.0
	Enzymatic saccharification	8.0	7.1	0.9
	Acid hydrolysis	6.0	3.6	2.4
5	**Products**			
	Ethanol	48.0	33.3	14.7
	Biofuels	22.0	12.7	9.3
	Bioethanol	17.0	16.8	0.2
	Ethanol production	11.0	5.2	5.8
	Bioethanol production	6.0	6.7	−0.7
	Ethanol yield	6.0	2.2	3.8
	Biorefinery	5.0	13.5	−8.5

The most-prolific keyword related to the fermentation is fermentation with 35% of the sample papers. The other prolific keywords are furfural, acetic acids, and saccharomyces with 10%–11% of the sample papers each.

Further, the most-prolific keywords related to the hydrolysis and hydrolysates are hydrolysis and sugars with 61% and 47% of the sample papers, respectively. The other prolific keywords are enzyme activity, enzymatic hydrolysis, glucose, and saccharification with 21%–34% of the sample papers each.

Finally, the most-prolific keyword related to the products is ethanol with 48% of the sample papers. The other prolific keywords are biofuels, bioethanol, and ethanol production with 11%–22% of the sample papers.

On the other hand, the most-prolific keywords across all of the research fronts are hydrolysis, cellulose, lignin, ethanol, sugar, wood, fermentation, enzyme activity, biomass, enzymatic hydrolysis, hemicellulose, and glucose with 28%–61% of the sample papers each. Similarly, the most influential keywords are sugar, hydrolysis, cellulose, wood, ethanol, enzyme activity, hemicellulose, steam, glucose, and acids with 10%–27% surplus each.

71.3.10 The Most-Prolific Research Fronts in Second Generation Forestry Waste-based Bioethanol Fuels

Information about the research fronts for the sample papers in second generation forestry waste-based bioethanol fuels with regard to the forestry waste feedstock used for the bioethanol production is given in Table 71.9.

TABLE 71.9
The Most-Prolific Research Fronts for Second Generation Forestry Waste-based Bioethanol Fuels

No.	Research Fronts	N Paper % Sample
1	Wood chips	31.0
2	Forest residues	27.0
3	Sawdust	13.0
4	Tree prunings	10.0
5	Bark	11.0
6	Wood waste	8.0
7	Spent sulfite liquor	5.0

N paper (%) sample: the number of papers in the population sample of 100 papers.

TABLE 71.10
The Most-Prolific Thematic Research Fronts for Second Generation Forestry Waste-based Bioethanol Fuels

No.	Research Fronts	N Paper % Sample
1	Biomass pretreatments	92.0
2	Biomass hydrolysis	79.0
3	Bioethanol production	26.0
4	Hydrolysate fermentation	25.0
5	Bioethanol fuel evaluation	5.0

N paper (%) sample: the number of papers in the population sample of 100 papers.

As this table shows, the most-prolific forestry waste feedstock is the wood chips with 31% of the sample papers, followed by forest residues with 27% of the sample papers. The other research fronts are the sawdust, tree prunings, bark, wood waste, and spent sulfite liquor with 5%–13% of the sample papers.

Information about the thematic research fronts for the sample papers in second generation forestry waste-based bioethanol fuels is given in Table 71.10. As this table shows, the most-prolific research fronts are the pretreatments and hydrolysis of the wood waste feedstocks with 92% and 79% of the sample papers, respectively. The other research fronts are the bioethanol fuel production, hydrolysate fermentation, and bioethanol fuel evaluation with 26%, 25%, and 5% of the sample papers, respectively.

71.4 DISCUSSION

71.4.1 Introduction

The crude oil-based gasoline fuels have been widely used in the transportation sector since the 1920s. However, there have been great public concerns over the adverse environmental and human impact of these fuels. Hence, biomass-based bioethanol fuels have increasingly been used in blending gasoline fuels, in the fuel cells, and in the biochemical production in a biorefinery context.

However, it is necessary to pretreat the biomass to enhance the yield of the bioethanol prior to the bioethanol production through the hydrolysis and fermentation. One of the most-studied feedstocks for the bioethanol fuels has been the forestry wastes such as wood chips, bark, sawdust, forestry residues, tree prunings, spent sulfite liquor, and wood wastes. The research in the field of second generation forestry waste-based bioethanol fuels has intensified in this context in the key research fronts of the pretreatment and hydrolysis of the forestry wastes, fermentation of the forestry waste-based hydrolysates, and production and evaluation of second generation forestry waste-based bioethanol fuels.

However, it is essential to develop efficient incentive structures for the primary stakeholders to enhance the research in this field. This is especially important to maintain energy security in the cases of supply shocks such as oil price shocks, war-related shocks as in the case of Russian invasion of Ukraine, or COVID-19 shocks.

The scientometric analysis has been used in this context to inform the primary stakeholders about the current state of the research in a selected research field. As there has been no scientometric study in this field, this book chapter presents a scientometric study of the research in second generation forestry waste-based bioethanol fuels. It examines the scientometric characteristics of both the sample and population data presenting scientometric characteristics of both these datasets in the order of documents, authors, publication years, institutions, funding bodies, source titles, countries, Scopus subject categories, Scopus keywords, and research fronts.

As a first step for the search of the relevant literature, the keywords were selected using the first most-cited 100 papers. The selected keyword list was then optimized to obtain a representative sample of papers for the searched research field. A copy of this keyword list was provided in the appendix for future replicative studies. Further, a selected list of the keywords was presented in Table 71.8.

As a second step, two sets of data were used for this study. First, a population sample of 535 papers was used to examine the scientometric characteristics of the population data. Second, a sample of 100 most-cited papers, corresponding to 19% of the population dataset, was used to examine the scientometric characteristics of these citation classics.

The scientometric characteristics of these sample and population datasets were presented in the order of documents, authors, publication years, institutions, funding bodies, source titles, countries, Scopus subject categories, Scopus keywords, and research fronts.

Lastly, the key scientometric findings for both datasets were discussed to highlight the research landscape for second generation forestry waste-based bioethanol fuels. Additionally, a number of brief conclusions were drawn, and a number of relevant recommendations were made to enhance the future research landscape.

71.4.2 The Most-Prolific Documents in Second Generation Forestry Waste-based Bioethanol Fuels

Articles (together with conference papers) dominate both the sample (96%) and population (97%) papers (Table 71.1). Further, review papers have a surplus (2%). The representation of the reviews in the sample papers is relatively modest (4%).

Scopus differs from the Web of Science database in differentiating and showing articles (88%) and conference papers (8%) published in the journals separately. However, it should be noted that these conference papers are also published in journals as articles, compared to those published only in the conference proceedings. Hence, the total number of articles and review papers in the sample dataset are 96% and 4%, respectively.

It is observed during the search process that there has been inconsistency in the classification of the documents in Scopus as well as in other databases such as Web of Science. This is especially relevant for the classification of papers as reviews or articles as the papers not involving a literature review may be erroneously classified as a review paper. There is also a case of review papers being classified as articles. For example, the total number of the reviews in the sample data set was manually found as nearly 5% compared to 4% as indexed by Scopus, decreasing the number of articles and conference papers to 95% for the sample dataset.

In this context, it would be helpful to provide a classification note for the published papers in the books and journals at the first instance. It would also be helpful to use the document types listed in Table 71.1 for this purpose. Book chapters may also be classified as articles or reviews as an additional classification to differentiate review chapters from the experimental chapters as it is done by the Web of Science. It would be further helpful to additionally classify the conference papers as articles or review papers as well as it is done in the Web of Science database.

71.4.3 The Most-Prolific Authors in Second Generation Forestry Waste-based Bioethanol Fuels

There have been most-prolific 20 authors with at least 3% of the sample papers each as given in Table 71.2. These authors have shaped the development of the research in this field.

The most-prolific authors are John N. Saddler, Qiang Yong, Caoxing Huang, Eulogio Castro and to a lesser extent Ignacio Ballesteros, Mercedes Ballesteros, Maria J. Negro, Jose M. Oliva, Inmaculada Romero, and Encarnacion Ruiz.

It is important to note the inconsistencies in indexing of the author names in Scopus and other databases. It is especially an issue for the names with more than two components such as 'Blake Sam de Hyun Castro'. The probable outcomes are 'Castro, B.S.D.H.', 'de Hyun Castro, B.S.', or 'Hyun Castro, B.S.D.'. The first choice is the gold standard of the publishing sector as the last word in the name is taken as the last name. In most of the academic databases such as PUBMED and EBSCO databases, this version is used predominantly. The second choice is a strong alternative, while the last choice is an undesired outcome as two last words are taken as the last name. It is a good practice to combine the words of the last name by a hyphen: 'Hyun-Castro, B.S.D.'. It is notable that inconsistent indexing of the author names may cause substantial inefficiencies in the search process for the papers as well as allocating credit to the authors as there are different author entries for each outcome in the databases.

There are also inconsistencies in the shortening Chinese names. For example, 'YangYing Yong' is often shortened as 'Yong, Y.', 'Yong, Y.-Y.', and 'Yong, Y.Y.', as it is done in the Web of Science database as well. However, the gold standard in this case is 'Yong, Y', where the last word is taken as the last name and the first word is taken as a single forename. In most of the academic databases such as PUBMED and EBSCO, this first version is used predominantly. However, it makes sense to use the third option to differentiate Chinese names efficiently: 'Yong, Y.Y.' Therefore, there have been difficulties in locating papers for the Chinese authors. In such cases, the use of the unique author codes provided for each author by the Scopus database has been helpful.

There is also a difficulty in allowing credit for the authors especially for the authors with common names such as 'Yong, X.' in conducting scientometric studies. These difficulties strongly influence the efficiency of the scientometric studies as well as allocating credit to the authors as there are the same author entries for different authors with the same name, e.g., 'Yong, X.' in the databases.

In this context, the coding of authors in Scopus database is a welcome innovation compared to the other databases such as Web of Science. In this process, Scopus allocates a unique number to each author in the database (Aman, 2018). However, there might still be substantial inefficiencies in this coding system especially for common names. For example, some of the papers for a certain author maybe allocated to another researcher with a different author code. It is possible that Scopus uses a number of software programs to differentiate the author names and the program may not be false-proof (Kim, 2018).

In this context, it does not help that author names are not given in full in some journals and books. This makes difficult to differentiate authors with common names and makes the scientometric studies further difficult in the author domain. Therefore, the author names should be given in all books and journals at the first instance. There is also a cultural issue where some authors do not use their full names in their papers. Instead, they use initials for their forenames: 'Saddler, H.J.', 'Saddler, H.', or 'Saddler, J.' instead of 'Saddler, Hyun Jae'.

There are also inconsistencies in naming of the authors with more than two components by the authors themselves in journal papers and book chapters. For example, 'Saddler, A.P.C.' might be given as 'Saddler, A.' or 'Saddler, A.C.' or 'Saddler, A.P.' or 'Saddler, C' in the journals and books. This also makes the scientometric studies difficult in the author domain. Hence, contributing authors should use their name consistently in their publications.

The other critical issue regarding the author names is the inconsistencies in the spelling of the author names in the national spellings (e.g., Şöğütçığlık, Gökçe) rather than in the English spellings (e.g., Sogutciglik, Gokce) in the Scopus database. Scopus differs from the Web of Science database and many other databases in this respect where the author names are given only in the English spellings. It is observed that national spellings of the author names do not help much in conducting scientometric studies as well in allocating credits to the authors as sometimes there are the different author entries for the English and National spellings in the Scopus database.

The most-prolific institutions for the sample dataset are University of Jaen, CIEMAT, and to a lesser extent Nanjing Forestry University and USA Forest Service. Further, the most-prolific countries for the sample dataset are Spain and to a lesser extent China and the USA. These findings confirm the dominance of Spain and to a lesser extent the USA and China in this field. On the other hand, the primary research fronts are the bioethanol fuels from tree prunings, forest residues, and to a lesser extent wood chips and sawdust.

It is also notable that there is a significant gender deficit for the sample dataset as surprisingly with a representation rate of 25%. This finding is the most thought-provoking with strong public policy implications. Hence, institutions, funding bodies, and policymakers should take efficient measures to reduce the gender deficit in this field as well as other scientific fields with strong gender deficit. In this context, it is worth to note the level of representation of the researchers from the minority groups in science on the basis of race, sexuality, age, and disability, besides the gender (Blankenship, 1993; Dirth and Branscombe, 2017; Konur, 2000, 2002a–c, 2006a,b, 2007a,b).

71.4.4 The Most-Prolific Research Output by Years in Second Generation Forestry Waste-based Bioethanol Fuels

The research output observed between 1970 and 2022 is illustrated in Figure 71.1. This figure clearly shows that the bulk of the research papers in the population dataset were published primarily in the 2010s and early 2020s. Similarly, the bulk of the research papers in the sample dataset were published in the 2010s and 2000s.

These findings suggest that the most-prolific sample and population papers were primarily published in the 2010s. Further, a significant portion of the sample and population papers were published in the early 2020s and 2000s, respectively.

These are the thought-provoking findings as there has been significant research boom in since 2007. In this context, the increasing public concerns about climate change (Change, 2007), greenhouse gas emissions (Carlson et al., 2017), and global warming (Kerr, 2007) have been certainly behind the boom in the research in this field in the last two decades. Furthermore, the recent supply shocks experiences due to the COVID-19 pandemics and the Ukrainian war might also be behind the research boom in this field since 2019.

Based on these findings, the size of the population papers is likely to more than double in the current decade, provided that the public concerns about climate change, greenhouse gas emissions, and global warming, as well as the supply shocks are translated efficiently to the research funding in this field.

71.4.5 The Most-Prolific Institutions in Second Generation Forestry Waste-based Bioethanol Fuels

The most-prolific 12 institutions publishing papers on second generation forestry waste-based bioethanol fuels with at least 3% of the sample papers each given in Table 71.3 have shaped the development of the research in this field.

The most-prolific institutions are the Nanjing Forestry University, CIEMAT, University of Jaen, University of British Columbia, and to a lesser extent National Renewable Energy Laboratory and NC State University. Similarly, the top countries for these most-prolific institutions are the USA and to a lesser extent Spain and China. In total, only five countries house these top institutions.

On the other hand, the institutions with the most citation impact are the CIEMAT and to a lesser extent Nanjing Forestry University, University of Jaen, University of British Columbia, and NREL.

These findings confirm the dominance of the institutions from the USA, Spain, and China and to a lesser extent from Canada in this field.

71.4.6 The Most-Prolific Funding Bodies in Second Generation Forestry Waste-based Bioethanol Fuels

The most-prolific 14 funding bodies funding at least 3% of the sample papers each is given in Table 71.4. It is notable that only 50% of the sample and population papers each were funded.

The most-prolific funding bodies are the National Natural Science Foundation of China, European Regional Development Fund and to a lesser extent the Natural Science Foundation of Jiangsu Province, European Commission, National Institute of Food and Agriculture, Government of Canada, and National Council for Scientific and Technological Development. The most-prolific countries for these top funding bodies are China, Canada, and to a lesser extent the European Commission and the USA. In total, only six countries and the EU house these top funding bodies.

The funding bodies with the most citation impact are the Natural Science Foundation of Jiangsu Province, European Regional Development Fund, and to a lesser extent Natural Resources Canada, Government of Canada, National Council for Scientific and Technological Development, and Finnish Funding Agency for Technology and Innovation (Tekes).

These findings on the funding of the research in this field suggest that the level of the funding, mostly since 2007, has been largely instrumental in enhancing the research in this field (Ebadi and Schiffauerova, 2016) in the light of North's institutional framework (North, 1991). It is also notable that the funding rate in this field is relatively modest compared to those in the other research fronts of the bioethanol fuels such as algal bioethanol fuels. Further, it is expected that this high funding rate would improve in the light of the recent supply shocks. Further, it emerges that China, the USA, Canada, and Europe have heavily funded the research on the forestry waste-based bioethanol fuels.

71.4.7 The Most-Prolific Source Titles in Second Generation Forestry Waste-based Bioethanol Fuels

The most-prolific 15 source titles publishing at least 2% of the sample papers each in second generation forestry waste-based bioethanol fuels have shaped the development of the research in this field (Table 71.5).

The most-prolific source titles are the Bioresource Technology and to a lesser extent Applied Biochemistry and Biotechnology Part A Enzyme Engineering and Biotechnology, Industrial Crops and Products, Applied Biochemistry and Biotechnology, Industrial and Engineering Chemistry Research, and Biotechnology and Bioengineering.

On the other hand, the source titles with the most citation impact are the Bioresource Technology and to a lesser extent Applied Biochemistry and Biotechnology Part A Enzyme Engineering and Biotechnology, Biotechnology and Bioengineering, Industrial and Engineering Chemistry Research, Chemical Engineering Science, and Applied Energy.

It is notable that these top source titles are primarily related to the bioresources, biotechnology, chemical engineering, energy, and wood science. This finding suggests that Bioresource Technology and the other prolific journals in these fields have significantly shaped the development of the research in this field as they focus primarily on second generation forestry waste-based bioethanol fuels with a high yield. In this context, the influence of the top journal is quite extraordinary.

71.4.8 The Most-Prolific Countries in Second Generation Forestry Waste-based Bioethanol Fuels

The most-prolific 11 countries publishing at least 3% of the sample papers each have significantly shaped the development of the research in this field (Table 71.6).

The most-prolific countries are the USA, China, Spain, Canada, and to a lesser extent Brazil, Finland, Japan, and Portugal. Further, five European countries listed in Table 71.6 produce 31 and 25% of the sample and population papers, respectively, and China is the largest producer of the population papers.

On the other hand, the countries with the most citation impact are the USA and to a lesser extent Spain, Canada, and Portugal. Similarly, the countries with the least impact are India, Finland, and Sweden.

The close examination of these findings suggests that the USA, China, Europe, Canada, and to a lesser extent Brazil and Japan are the major producers of the research in this field. It is a fact that the USA has been a major player in science (Leydesdorff and Wagner, 2009). The USA has further developed a strong research infrastructure to support its corn- and grass-based bioethanol industry (Gillon, 2010).

However, China has been a rising mega star in scientific research in competition with the USA and Europe (Leydesdorff and Zhou, 2005). China is also a major player in this field as a major producer of bioethanol (Fang et al., 2010).

Next, Europe has been a persistent player in the scientific research in competition with both the USA and China (Leydesdorff, 2000). Europe has also been a persistent producer of bioethanol along with the USA and Brazil (Gnansounou, 2010).

Further, Brazil (Soccol et al., 2010), Japan (Fukuzawa and Ida, 2016), and Canada (Lariviere et al., 2006) are the other countries with substantial research activities in bioethanol fuels.

71.4.9 The Most-Prolific Scopus Subject Categories in Second Generation Forestry Waste-based Bioethanol Fuels

The most-prolific nine Scopus subject categories indexing at least 7% of the sample papers each, given in Table 71.7, have shaped the development of the research in this field.

The most-prolific Scopus subject categories in second generation forestry waste-based bioethanol fuels are Chemical Engineering and to a lesser extent Environmental Science, Energy, Biochemistry, Genetics and Molecular Biology, Immunology and Microbiology, and Chemistry. It is also notable that Social Sciences including Economics and Business have a minimal presence in both sample and population studies.

On the other hand, the Scopus subject categories with the most citation impact are Chemical Engineering and to a lesser extent Energy, Environmental Science, and Immunology and Microbiology. Similarly, the least influential subject categories are Agricultural and Biological Sciences and Materials Science.

These findings are thought-provoking, suggesting that the primary subject categories are related to energy, chemical engineering, environmental sciences, genetics, and microbiology as the core of the research in this field concerns with the production and utilization of second generation forestry waste-based bioethanol fuels. The other finding is that social sciences are not well represented in both the sample and population papers as in line with the most fields in bioethanol fuels. The social, environmental, and economics studies account for the field of social sciences.

71.4.10 The Most-Prolific Keywords in Second Generation Forestry Waste-based Bioethanol Fuels

A limited number of keywords have shaped the development of the research in this field as shown in Table 71.8 and in Appendix. These keywords are grouped under five headings: forestry waste, pretreatments, fermentation, hydrolysis and hydrolysates, and products.

The most-prolific keywords across all of the research fronts are hydrolysis, cellulose, lignin, ethanol, sugar, wood, fermentation, enzyme activity, biomass, enzymatic hydrolysis, hemicellulose, and glucose. Similarly, the most influential keywords are sugar, hydrolysis, cellulose, wood, ethanol, enzyme activity, hemicellulose, steam, glucose, and acids.

These findings suggest that it is necessary to determine the keyword set carefully to locate the relevant research in each of these research fronts. Additionally, the size of the samples for each keyword highlights the intensity of the research in the relevant research areas for both sample and population datasets. These findings also highlight different spelling of some strategic keywords such as pretreatment v. pre-treatment and ethanol v. bio-ethanol, etc.

71.4.11 The Most-Prolific Research Fronts in Second Generation Forestry Waste-based Bioethanol Fuels

Information about the research fronts for the sample papers in second generation forestry waste-based bioethanol fuels with regard to the forestry waste feedstocks used for the bioethanol production is given in Table 71.9. As this table shows, the most-prolific forestry waste feedstocks are the wood chips, forest residues, and to a lesser extent the sawdust, tree prunings, bark, wood waste, and spent sulfite liquor.

Information about the thematic research fronts for the sample papers in second generation forestry waste-based bioethanol fuels is given in Table 71.10. As this table shows, the most-prolific research fronts are the pretreatments and hydrolysis of the wood waste feedstocks and to a lesser extent bioethanol fuel production, hydrolysate fermentation, and bioethanol fuel evaluation. The first two research fronts dominate the research in this field.

These findings are thought-provoking in seeking ways to increase forestry waste feedstock-based bioethanol yield at the global scale. It is clear that all of these research fronts have public importance and merit substantial funding and other incentives. Further, it is notable that second generation forestry waste-based bioethanol fuels have become a core unit of the bioethanol research to make it more competitive with the crude oil-based gasoline and petrodiesel fuels, especially for the USA, Europe, and China.

In comparison to the other feedstock-based research fronts, it is notable that the pretreatment and hydrolysis of the wood wastes emerge as primary research fronts for this field. This suggests that the primary stakeholders have been primarily interested in these key processes of the bioethanol production. It is also notable that evaluation of second generation forestry waste-based bioethanol fuels such as technoeconomics, life cycle, economics, social, land use, labor, and environment-related studies emerges as a case study for the bioethanol fuels. In this context, the USA has been the global leader in the production and the use of the corn-based bioethanol fuels since the 1970s in the aftermath of the global crude oil crisis in the early 1970s.

It is further notable that the research on the forest waste-based bioethanol fuels forms only 10% of the first generation wood-based bioethanol fuel research and complements it, extracting further value from the wood feedstocks.

In the end, these most-cited papers in this field hint that the production of second generation forestry waste-based bioethanol fuels could be optimized using the structure, processing, and property relationships of these forestry wastes in the fronts of the feedstock pretreatment and hydrolysis, and hydrolysate fermentation (Formela et al., 2016; Konur, 2018a, 2020b, 2021a–d; Konur and Matthews, 1989).

71.5 CONCLUSION AND FUTURE RESEARCH

The research on second generation forestry waste-based bioethanol fuels has been mapped through a scientometric study of both sample (100 papers) and population (535 papers) datasets.

The critical issue in this study has been to obtain a representative sample of the research as in any other scientometric study. Therefore, the keyword set has been carefully devised and optimized after a number of runs in the Scopus database. It is a representative sample of the wider population studies. This keyword set was provided in the appendix, and the relevant keywords are presented in Table 71.8. However, it should be noted that it has been very difficult to compile a representative keyword set since this research field has been connected closely with many other fields. Therefore, it has been necessary to compile a keyword list to exclude papers concerned with the other research fields.

The other issue has been the selection of a multidisciplinary database to carry out the scientometric study of the research in this field. For this purpose, Scopus database has been selected. The journal coverage of this database has been notably wider than that of Web of Science and other multisubject databases.

The key scientometric properties of the research in this field have been determined and discussed in this book chapter. It is evident that a limited number of documents, authors, institutions, publication years, institutions, funding bodies, source titles, countries, Scopus subject categories, Scopus keywords, and research fronts have shaped the development of the research in this field.

There is ample scope to increase the efficiency of the scientometric studies in this field in the author and document domains by developing consistent policies and practices in both domains across all the academic databases. In this respect, it seems that authors, journals, and academic databases have a lot to do. Furthermore, the significant gender deficit as in most scientific fields emerges as a public policy issue. The potential deficits on the basis of age, race, disability, and sexuality need also to be explored in this field as in other scientific fields.

The research in this field has boomed since 2007, possibly promoted by the public concerns on global warming, greenhouse gas emissions, and climate change. Furthermore, the recent COVID-19 pandemics and Russian invasion of Ukraine have resulted in a global supply shocks shifting the focus of the stakeholders from the crude oil-based fuels to biomass-based fuels such as bioethanol fuels. It is expected that there would be further incentives for the key stakeholders to carry out the research for second generation forestry waste-based bioethanol fuels to increase the ethanol yield and to make it more competitive with the crude oil-based gasoline and diesel fuels. This might be truer for the crude oil- and foreign exchange-deficient countries to maintain the energy and food security at the face of the global supply shocks.

The relatively modest funding rate of 50% for the sample and population papers each suggests that funding in this field significantly enhanced the research in this field primarily since 2007, possibly more than doubling in the current decade. However, it is evident that there is ample room for more funding and other incentives to enhance the research in this field further.

The institutions from the USA and to a lesser extent Spain and China have mostly shaped the research in this field. Further, the USA, China, Europe, and to a lesser extent Canada, Brazil, and Japan have been the major producers of the research in this field as the major producers and users of bioethanol fuels. It is evident that these countries have well-developed research infrastructure in bioethanol fuels and their derivatives.

It emerges that ethanol is more popular than bioethanol as a keyword with strong implications for the search strategy. In other words, the search strategy using only bioethanol keyword would not be much helpful. The Scopus keywords are grouped under five headings: biomass, pretreatments, fermentation, hydrolysis and hydrolysates, and products. It is also important to use the Latin terms for the forestry waste in addition to the English terms.

As Table 71.9 shows, the most-prolific forestry waste feedstocks are the wood chips, forest residues, and to a lesser extent the sawdust, tree prunings, bark, wood waste, and spent sulfite liquor. On the other hand, Table 71.10 shows that the most-prolific research fronts are the pretreatments and hydrolysis of the wood waste feedstocks and to a lesser extent bioethanol fuel production, hydrolysate fermentation, and bioethanol fuel evaluation. The first two research fronts dominate the research in this field.

These findings are thought-provoking in seeking ways to increase forestry waste feedstock-based bioethanol yield at the global scale. It is clear that all of these research fronts have public importance and merit substantial funding and other incentives. Further, it is notable that second generation forestry waste-based bioethanol fuels have become a core unit of the bioethanol research to make it more competitive with the crude oil-based gasoline and petrodiesel fuels, especially for the USA, Europe, and China. It is further notable that the research on the second generation forest waste-based bioethanol forms only 10% of the wood-based bioethanol fuel research and complements it extracting further value from the wood feedstocks.

Thus, the scientometric analysis has a great potential to gain valuable insights into the evolution of the research in this field as in other scientific fields especially in the aftermath of the significant global supply shocks such as COVID-19 pandemics and the Russian invasion of Ukraine.

It is recommended that further scientometric studies are carried out for the primary research fronts. It is further recommended that reviews of the most-cited papers are carried out for each primary research front to complement these scientometric studies. Next, the scientometric studies of the hot papers in these primary fields are carried out.

ACKNOWLEDGMENTS

The contribution of the highly cited researchers in the field of second generation forestry waste-based bioethanol fuels has been gratefully acknowledged.

APPENDIX: THE KEYWORD SET FOR SECOND GENERATION FORESTRY WASTE-BASED BIOETHANOL FUELS

((((TITLE (forest* OR *wood OR woody OR "olive tree*" OR eucalyptus OR pine OR pinus OR poplar* OR aspen* OR populus OR cedar OR spruce OR sweetgum OR beech OR oak OR willow* OR salix OR birch OR bamboo* OR maple* OR fir OR mahogany OR "palm tree" OR timber) AND TITLE (waste OR spent OR residues OR "residual biomass" OR wasted OR chips OR grounds OR "second generation" OR bark)) OR TITLE ("Saw dust" OR sawdust OR "tree leaves" OR "tree pruning*" OR "grapevine pruning*" OR "tree branches")) AND TITLE (*saccharification OR *hydrolysis OR digestibili* OR ssf OR recalcitrance OR hydrolysate* OR hydrolyzate* OR ferment* OR coferment* OR delignification OR depolymerization OR fractionation OR pretreat* OR "pre-treat*" OR "ionic liquid*" OR "consolidated processing" OR "compressed water" OR "wet oxidation")) AND NOT (TITLE (lignac* OR biochars OR 13c OR nano* OR isotopic OR hydrochar OR {enzymes production} OR pyrolysis OR biogas OR biomethan* OR soil* OR {Cellulase production} OR tannin OR suberin OR phenol* OR anaerobic OR *oil OR lead OR methane OR *cyanidins OR beetle OR torref* OR liquefaction OR refiner* OR *alkanoates OR drought OR combustion OR *hydrogen OR *cane OR rubber OR xylitol OR lactic OR rumen OR adhesive OR reforming OR remediat* OR *butanol OR "corn resid*" OR *protein OR ficus OR "sunflower stalk*" OR "kraft pulp*" OR "bark ratio" OR "cellulose fib*" OR sinensis OR silicon) OR SRCTITLE (animal* OR entomol* OR food* OR pyrolysis OR materials OR livestock OR ruminant OR polymer* OR plant* OR water OR macromol* OR fluids OR lwt OR adhes* OR medi* OR biomed* OR anal* OR phar*) OR SUBJAREA (medi OR vete OR phar OR nurs OR eart))) OR ((TITLE (ethanol OR bioethanol) AND (TITLE (forest* OR *wood OR woody OR "olive tree*" OR eucalyptus OR pine OR pinus OR poplar* OR aspen* OR populus OR cedar OR spruce OR sweetgum OR beech OR oak OR willow* OR salix OR birch OR bamboo* OR maple* OR fir OR mahogany OR "palm tree" OR timber) AND TITLE (waste OR spent OR residues OR "residual biomass" OR wasted OR chips OR grounds OR "second generation" OR bark)) OR TITLE ("saw dust" OR sawdust OR "tree leaves" OR "tree pruning*" OR "grapevine pruning*" OR "tree branches")) AND NOT TITLE (nanowh* OR bagasse OR beetle OR algal OR liquefaction OR sugarcane OR extraction OR solvent)) AND (LIMIT-TO (DOCTYPE, "ar")

OR LIMIT-TO (DOCTYPE, "cp") OR LIMIT-TO (DOCTYPE, "re") OR LIMIT-TO (DOCTYPE, "sh") OR LIMIT-TO (DOCTYPE, "ch") OR LIMIT-TO (DOCTYPE, "no")) AND (LIMIT-TO (LANGUAGE, "English")) AND (LIMIT-TO (SRCTYPE, "j") OR LIMIT-TO (SRCTYPE, "k") OR LIMIT-TO (SRCTYPE, "b"))

REFERENCES

Aman, V. 2018. Does the Scopus author ID suffice to track scientific international mobility? A case study based on Leibniz laureates. *Scientometrics* 117:705–720.

Angelici, C., B. M. Weckhuysen and P. C. A. Bruijnincx. 2013. Chemocatalytic conversion of ethanol into butadiene and other bulk chemicals. *ChemSusChem* 6:1595–1614.

Antolini, E. 2007. Catalysts for direct ethanol fuel cells. *Journal of Power Sources* 170:1–12.

Antolini, E. 2009. Palladium in fuel cell catalysis. *Energy and Environmental Science* 2:915–931.

Ayeni, A. O., F. K. Hymore and S. N. Mudliar, et al. 2013. Hydrogen peroxide and lime based oxidative pretreatment of wood waste to enhance enzymatic hydrolysis for a biorefinery: Process parameters optimization using response surface methodology. *Fuel* 106:187–194.

Beaudry, C. and V. Lariviere. 2016. Which gender gap? Factors affecting researchers' scientific impact in science and medicine. *Research Policy* 45:1790–1817.

Blankenship, K. M. 1993. Bringing gender and race in: US employment discrimination policy. *Gender & Society* 7:204–226.

Brandt, A., J. P. Hallett, D. J. Leak, R. J. Murphy and T. Welton. 2010. The effect of the ionic liquid anion in the pretreatment of pine wood chips. *Green Chemistry* 12:672–679.

Brownell, H. H., E. K. C. Yu and J. N. Saddler. 1986. Steam-explosion pretreatment of wood: Effect of chip size, acid, moisture content and pressure drop. *Biotechnology and Bioengineering* 28:792–801.

Burnham, J. F. 2006. Scopus database: A review. *Biomedical Digital Libraries* 3:1–8.

Cara, C., E. Ruiz, M. Ballesteros, P. Manzanares, M. J. Negro and E. Castro. 2008. Production of fuel ethanol from steam-explosion pretreated olive tree pruning. *Fuel* 87:692–700.

Cara, C., I. Romero, J. M. Oliva, F. Saez and E. Castro. 2007. Liquid hot water pretreatment of olive tree pruning residues. *Applied Biochemistry and Biotechnology* 137–140:379–394.

Carlson, K. M., J. S. Gerber and D. Mueller, et al. 2017. Greenhouse gas emissions intensity of global croplands. *Nature Climate Change* 7:63–68.

Change, C. 2007. Climate change impacts, adaptation and vulnerability. *Science of the Total Environment* 326:95–112.

Dirth, T. P. and N. R. Branscombe. 2017. Disability models affect disability policy support through awareness of structural discrimination. *Journal of Social Issues* 73:413–442.

Duff, S. J. B. and W. D. Murray. 1996. Bioconversion of forest products industry waste cellulosics to fuel ethanol: A review. *Bioresource Technology* 55:1–33.

Ebadi, A. and A. Schiffauerova. 2016. How to boost scientific production? A statistical analysis of research funding and other influencing factors. *Scientometrics* 106:1093–1116.

Fang, X., Y. Shen, J. Zhao, X. Bao and Y. Qu. 2010. Status and prospect of lignocellulosic bioethanol production in China. *Bioresource Technology* 101:4814–4819.

Fauci, A. S., H. C. Lane and R. R. Redfield. 2020. Covid-19-navigating the uncharted. *New England Journal of Medicine* 382:1268–1269.

Fernando, S., S. Adhikari, C. Chandrapal and M. Murali. 2006. Biorefineries: Current status, challenges, and future direction. *Energy & Fuels* 20:1727–1737.

Formela, K., A. Hejna, L. Piszczyk, M. R. Saeb and X. Colom. 2016. Processing and structure-property relationships of natural rubber/wheat bran biocomposites. *Cellulose* 23:3157–3175.

Franko, B., M. Galbe and O. Wallberg. 2016. Bioethanol production from forestry residues: A comparative techno-economic analysis. *Applied Energy* 184:727–736.

Fukuzawa, N. and T. Ida. 2016. Science linkages between scientific articles and patents for leading scientists in the life and medical sciences field: The case of Japan. *Scientometrics* 106:629–644.

Garfield, E. 1955. Citation indexes for science. *Science* 122:108–111.

Gillon, S. 2010. Fields of dreams: Negotiating an ethanol agenda in the Midwest United States. *Journal of Peasant Studies* 37:723–748.

Gnansounou, E. 2010. Production and use of lignocellulosic bioethanol in Europe: Current situation and perspectives. *Bioresource Technology* 101:4842–4850.

Hahn-Hagerdal, B., M. Galbe, M. F. Gorwa-Grauslund, G. Liden and G. Zacchi. 2006. Bio-ethanol: The fuel of tomorrow from the residues of today. *Trends in Biotechnology* 24:549–556.

Hamilton, J. D. 1983. Oil and the macroeconomy since World War II. *Journal of Political Economy* 91:228–248.
Hamilton, J. D. 2003. What is an oil shock? *Journal of Econometrics* 113:363–398.
Hill, J., E. Nelson, D. Tilman, S. Polasky and D. Tiffany. 2006. Environmental, economic, and energetic costs and benefits of biodiesel and ethanol biofuels. *Proceedings of the National Academy of Sciences of the United States of America* 103:11206–11210.
Hill, J., S. Polasky and E. Nelson, et al. 2009. Climate change and health costs of air emissions from biofuels and gasoline. *Proceedings of the National Academy of Sciences of the United States of America* 106:2077–2082.
Hsieh, W. D., R. H. Chen, T. L. Wu and T. H. Lin. 2002. Engine performance and pollutant emission of an SI engine using ethanol-gasoline blended fuels. *Atmospheric Environment* 36:403–410.
Hu, G., J. A. Heitmann, O. J. Rojas. 2008. Feedstock pretreatment strategies for producing ethanol from wood, bark, and forest residues. *BioResources* 3:270–294.
Huang, C., W. Lin and C. Lai, et al. 2019. Coupling the post-extraction process to remove residual lignin and alter the recalcitrant structures for improving the enzymatic digestibility of acid-pretreated bamboo residues. *Bioresource Technology* 285:121355.
Huang, H. J., S. Ramaswamy, U. W. Tschirner and B. V. Ramarao. 2008. A review of separation technologies in current and future biorefineries. *Separation and Purification Technology* 62:1–21.
Jones, T. C. 2012. America, oil, and war in the Middle East. *Journal of American History* 99:208–218.
Jonker, J. G. G., M. M. Junginger and J. A. Verstegen, et al. 2016. Supply chain optimization of sugarcane first generation and eucalyptus second generation ethanol production in Brazil. *Applied Energy* 173:494–510.
Kerr, R. A. 2007. Global warming is changing the world. *Science* 316:188–190.
Kilian, L. 2008. Exogenous oil supply shocks: How big are they and how much do they matter for the US economy? *Review of Economics and Statistics* 90:216–240.
Kilian, L. 2009. Not all oil price shocks are alike: Disentangling demand and supply shocks in the crude oil market. *American Economic Review*, 99:1053–69.
Kim, J. 2018. Evaluating author name disambiguation for digital libraries: A case of DBLP. *Scientometrics* 116:1867–1886.
Konur, O. 2000. Creating enforceable civil rights for disabled students in higher education: An institutional theory perspective. *Disability & Society* 15:1041–1063.
Konur, O. 2002a. Access to nursing education by disabled students: Rights and duties of nursing programs. *Nurse Education Today* 22:364–374.
Konur, O. 2002b. Assessment of disabled students in higher education: Current public policy issues. *Assessment and Evaluation in Higher Education* 27:131–152.
Konur, O. 2002c. Access to employment by disabled people in the UK: Is the Disability Discrimination Act working? *International Journal of Discrimination and the Law* 5:247–279.
Konur, O. 2006a. Participation of children with dyslexia in compulsory education: Current public policy issues. *Dyslexia* 12:51–67.
Konur, O. 2006b. Teaching disabled students in higher education. *Teaching in Higher Education* 11:351–363.
Konur, O. 2007a. A judicial outcome analysis of the Disability Discrimination Act: A windfall for the employers? *Disability & Society* 22:187–204.
Konur, O. 2007b. Computer-assisted teaching and assessment of disabled students in higher education: The interface between academic standards and disability rights. *Journal of Computer Assisted Learning* 23:207–219.
Konur, O. 2011. The scientometric evaluation of the research on the algae and bio-energy. *Applied Energy* 88:3532–3540.
Konur, O. 2012a. The evaluation of the biogas research: A scientometric approach. *Energy Education Science and Technology, Part A: Energy Science and Research* 29:1277–1292.
Konur, O. 2012b. The evaluation of the educational research: A scientometric approach. *Energy Education Science and Technology, Part B: Social and Educational Studies* 4:1935–1948.
Konur, O. 2012c. The evaluation of the global energy and fuels research: A scientometric approach. *Energy Education Science and Technology, Part A: Energy Science and Research* 30:613–628.
Konur, O. 2012d. The evaluation of the research on the biodiesel: A scientometric approach. *Energy Education Science and Technology, Part A: Energy Science and Research* 28:1003–1014.
Konur, O. 2012e. The evaluation of the research on the bioethanol: A scientometric approach. *Energy Education Science and Technology, Part A: Energy Science and Research* 28:1051–1064.
Konur, O. 2012f. The evaluation of the research on the biofuels: A scientometric approach. *Energy Education Science and Technology, Part A: Energy Science and Research* 28:903–916.
Konur, O. 2012g. The evaluation of the research on the biohydrogen: A scientometric approach. *Energy Education Science and Technology, Part A: Energy Science and Research* 29:323–338.

Konur, O. 2012h. The evaluation of the research on the microbial fuel cells: A scientometric approach. *Energy Education Science and Technology, Part A: Energy Science and Research* 29:309–322.

Konur, O. 2012i. The scientometric evaluation of the research on the production of bioenergy from biomass. *Biomass and Bioenergy* 47:504–515.

Konur, O. 2015. Current state of research on algal bioethanol. In *Marine Bioenergy: Trends and Developments*, Eds. S. K. Kim and C. G. Lee, pp. 217–244. Boca Raton, FL: CRC Press.

Konur, O., Ed. 2018a. *Bioenergy and Biofuels*. Boca Raton, FL: CRC Press.

Konur, O. 2018b. Bioenergy and biofuels science and technology: Scientometric overview and citation classics. In Bioenergy and Biofuels, Ed. O. Konur, pp. 3–63. Boca Raton, FL: CRC Press.

Konur, O. 2019. Cyanobacterial bioenergy and biofuels science and technology: A scientometric overview. In *Cyanobacteria: From Basic Science to Applications*, Eds. A. K. Mishra, D. N. Tiwari and A. N. Rai, pp. 419–442. Amsterdam: Elsevier.

Konur, O. 2020a. The scientometric analysis of the research on the bioethanol production from green macroalgae. In *Handbook of Algal Science, Technology and Medicine,* Ed. O. Konur, pp. 385–401. London: Academic Press.

Konur, O., Ed. 2020b. *Handbook of Algal Science, Technology and Medicine.* London: Academic Press.

Konur, O., Ed. 2021a. *Handbook of Biodiesel and Petrodiesel Fuels: Science, Technology, Health, and Environment.* Boca Raton, FL: CRC Press.

Konur, O., Ed. 2021b. *Handbook of Biodiesel and Petrodiesel Fuels: Science, Technology, Health, and Environment. Volume 1. Biodiesel Fuels: Science, Technology, Health, and Environment.* Boca Raton, FL: CRC Press.

Konur, O., Ed. 2021c. *Handbook of Biodiesel and Petrodiesel Fuels: Science, Technology, Health, and Environment. Volume 2. Biodiesel Fuels Based on the Edible and Nonedible Feedstocks, Wastes, and Algae: Science, Technology, Health, and Environment.* Boca Raton, FL: CRC Press.

Konur, O., Ed. 2021d. *Handbook of Biodiesel and Petrodiesel Fuels: Science, Technology, Health, and Environment. Volume 3. Petrodiesel Fuels: Science, Technology, Health, and Environment.* Boca Raton, FL: CRC Press.

Konur, O. and F. L. Matthews. 1989. Effect of the properties of the constituents on the fatigue performance of composites: A review. *Composites* 20:317–328.

Kruyt, B., D. P. van Vuuren, H. J. de Vries and H. Groenenberg. 2009. Indicators for energy security. *Energy Policy* 37:2166–2181.

Lariviere, V., Y. Gingras and E. Archambault. 2006. Canadian collaboration networks: A comparative analysis of the natural sciences, social sciences and the humanities. *Scientometrics* 68:519–533.

Leydesdorff, L. 2000. Is the European Union becoming a single publication system? *Scientometrics* 47:265–280.

Leydesdorff, L. and C. Wagner. 2009. Is the United States losing ground in science? A global perspective on the world science system. *Scientometrics* 78:23–36.

Leydesdorff, L. and P. Zhou. 2005. Are the contributions of China and Korea upsetting the world system of science? *Scientometrics* 63:617–630.

Li, H., S. M. Liu, X. H. Yu, S. L. Tang and C. K. Tang. 2020. Coronavirus disease 2019 (COVID-19): Current status and future perspectives. *International Journal of Antimicrobial Agents* 55:105951.

Lima, M. A., G. B. Lavorente and H. K. P. da Silva, et al. 2013. Effects of pretreatment on morphology, chemical composition and enzymatic digestibility of eucalyptus bark: A potentially valuable source of fermentable sugars for biofuel production - Part 1. *Biotechnology for Biofuels* 6:75.

Lin, Y. and S. Tanaka. 2006. Ethanol fermentation from biomass resources: Current state and prospects. *Applied Microbiology and Biotechnology* 69:627–642.

Lu, Y., B. Yang, G. Gregg, J. N. Saddler and S. D. Mansfield. 2002. Cellulase adsorption and an evaluation of enzyme recycle during hydrolysis of steam-exploded softwood residues. *Applied Biochemistry and Biotechnology, Part A: Enzyme Engineering and Biotechnology* 98–100:641–654.

Ma, X., L. Sun and C. Song. 2002. A new approach to deep desulfurization of gasoline, diesel fuel and jet fuel by selective adsorption for ultra-clean fuels and for fuel cell applications. *Catalysis Today* 77:107–116.

Morschbacker, A. 2009. Bio-ethanol based ethylene. *Polymer Reviews* 49:79–84.

Najafi, G., B. Ghobadian and T. Tavakoli, et al. 2009. Performance and exhaust emissions of a gasoline engine with ethanol blended gasoline fuels using artificial neural network. *Applied Energy* 86:630–639.

Newman, P. W. G. and J. R. Kenworthy. 1989. Gasoline consumption and cities: A comparison of U.S. cities with a global survey. *Journal of the American Planning Association* 55:24–37.

Nigam, J. N. 2001. Ethanol production from hardwood spent sulfite liquor using an adapted strain of Pichia stipites. *Journal of Industrial Microbiology and Biotechnology* 26:145–150.

North, D. C. 1991. Institutions. *Journal of Economic Perspectives* 5:97–112.

Olsson, L. and B. Hahn-Hagerdal. 1996. Fermentation of lignocellulosic hydrolysates for ethanol production. *Enzyme and Microbial Technology* 18:312–331.

Reeves, S. 2014. To Russia with love: How moral arguments for a humanitarian intervention in Syria opened the door for an invasion of the Ukraine. *Michigan State University International Law Review* 23:199.

Salehian, P. and K. Karimi. 2013. Alkali pretreatment for improvement of biogas and ethanol production from different waste parts of pine tree. *Industrial and Engineering Chemistry Research* 52:972–978.

Sanchez, O. J. and C. A. Cardona. 2008. Trends in biotechnological production of fuel ethanol from different feedstocks. *Bioresource Technology* 99:5270–5295.

Soccol, C. R., L. P. de Souza Vandenberghe and A. B.P. Medeiros, et al. 2010. Bioethanol from lignocelluloses: Status and perspectives in Brazil. *Bioresource Technology* 101:4820–4825.

Sun, Y. and J. Cheng. 2002. Hydrolysis of lignocellulosic materials for ethanol production: A review. *Bioresource Technology* 83:1–11.

Survase, S. A., E. Sklavounos, G. Jurgens, A. van Heiningen and T. Granstrom. 2011. Continuous acetone-butanol-ethanol fermentation using SO_2-ethanol-water spent liquor from spruce. *Bioresource Technology* 102:10996–11002.

Taherzadeh, M. J. and K. Karimi. 2007. Enzyme-based hydrolysis processes for ethanol from lignocellulosic materials: A review. *Bioresources* 2:707–738.

Taherzadeh, M. J. and K. Karimi. 2008. Pretreatment of lignocellulosic wastes to improve ethanol and biogas production: A review. *International Journal of Molecular Sciences* 9:1621–1651.

Vidal, P. F. and J. Molinier. 1988. Ozonolysis of lignin: Improvement of in vitro digestibility of poplar sawdust. *Biomass* 16:1–17.

Weil, J., A. Sarikaya and S. L. Rau, et al. 1997. Pretreatment of yellow poplar sawdust by pressure cooking in water. *Applied Biochemistry and Biotechnology, Part A: Enzyme Engineering and Biotechnology* 68:21–40.

Winzer, C. 2012. Conceptualizing energy security. *Energy Policy* 46:36–48.

Yang, B. and C. E. Wyman. 2008. Pretreatment: The key to unlocking low-cost cellulosic ethanol. *Biofuels, Bioproducts and Biorefining* 2:26–40.

72 Second Generation Forestry Waste-based Bioethanol Fuels

Review

Ozcan Konur
(Formerly) Ankara Yildirim Beyazit University

72.1 INTRODUCTION

The crude oil-based gasoline fuels (Ma et al., 2002; Newman and Kenworthy, 1989) have been widely used in the transportation sector since the 1920s. However, there have been great public concerns over the adverse environmental and human impact of these fuels (Hill et al., 2006, 2009). Hence, biomass-based bioethanol fuels (Hill et al., 2006; Konur, 2012, 2015, 2020) have increasingly been used in blending gasoline fuels (Hsieh et al., 2002; Najafi et al., 2009), in the fuel cells (Antolini, 2007, 2009), and in the biochemical production (Angelici et al., 2013; Morschbacker, 2009) in a biorefinery context (Fernando et al., 2006; Huang et al., 2008).

However, it is necessary to pretreat the biomass (Alvira et al., 2010; Taherzadeh and Karimi, 2008) to enhance the yield of the bioethanol (Hahn-Hagerdal et al., 2006; Sanchez and Cardona, 2008) prior to the bioethanol fuel production from the feedstocks through the hydrolysis (Sun and Cheng, 2002; Taherzadeh and Karimi, 2007) and fermentation (Lin and Tanaka, 2006; Olsson and Hahn-Hagerdal, 1996) of the biomass and hydrolysates, respectively.

One of the most-studied feedstocks for the bioethanol fuels has been the forestry wastes such as wood chips, bark, sawdust, forestry residues, tree prunings, spent sulfite liquor, and wood wastes. The research in the field of the second generation forestry waste-based bioethanol fuels has intensified in this context in the key research fronts of the pretreatment (Brandt et al., 2010; Brownell et al., 1986; Lu et al., 2002) and hydrolysis (Brownell et al., 1986; Huang et al., 2019; Lu et al., 2002) of the forestry wastes, fermentation (Allen et al., 2001; Cara et al., 2008; Okuda et al., 2008) of the forestry waste-based hydrolysates, and production (Allen et al., 2001; Cara et al., 2008; Zhu et al., 2015) and evaluation (Franko et al., 2016; Jonker et al., 2016; Schell et al., 1991) of the second generation forestry waste-based bioethanol fuels. Further, wood chips (Brandt et al., 2010; Brownell et al., 1986; Iranmahboob et al., 2002), forest residues (Huang et al., 2019; Lin et al., 2020; Lu et al., 2002), sawdust (Weil et al., 1997, 1998), tree prunings (Cara et al., 2007, 2008), bark (Kim et al., 2005; Lima et al., 2013), wood waste (Ayeni et al., 2013; Salehian and Karimi, 2013), and spent sulfite liquor (Nigam, 2001; Survase et al., 2011) have been studied intensively in this context.

However, it is essential to develop efficient incentive structures (North, 1991) for the primary stakeholders to enhance the research in this field (Konur, 2000, 2002a–c, 2006a,b, 2007a,b). Although there have been a number of review papers on the forestry waste-based bioethanol fuels (Duff and Murray, 1996; Hu et al., 2008; Pereira et al., 2013), there has been no review of the most-cited 25 papers in this field.

Thus, this book chapter presents a review of the most-cited 25 articles in the field of the forestry waste-based bioethanol fuels. Then, it discusses the key findings of these highly influential papers and comments on the future research priorities in this field.

DOI: 10.1201/9781003226550-98

72.2 MATERIALS AND METHODS

The search for this study was carried out using Scopus database (Burnham, 2006) in September 2022.

As a first step for the search of the relevant literature, the keywords were selected using the most-cited first 200 population papers. The selected keyword list was then optimized to obtain a representative sample of papers for the searched research field. This final keyword set was provided in the appendix of Konur (2023) for future replication studies.

As a second step, a sample dataset was used for this study. The first 25 articles with at least 76 citations each were selected for the review study. Key findings from each paper were taken from the abstracts of these papers and were discussed. Additionally, a number of brief conclusions were drawn and a number of relevant recommendations were made to enhance the future research landscape.

72.3 RESULTS

The brief information about 25 most-cited papers with at least 76 citations each on the second generation forestry waste-based bioethanol fuels is given below. The primary research fronts are the pretreatment and hydrolysis of the forestry wastes and production of the forestry waste-based bioethanol fuels with 21 and 4 highly cited papers (HCPs), respectively. Further, there are 2 and 19 HCPs for the pretreatment and hydrolysis of the forestry wastes, respectively.

72.3.1 Forestry Waste Pretreatment and Hydrolysis

The brief information about 21 most-cited papers on the pretreatment and hydrolysis of forestry wastes with at least 79 citations each is given below (Table 72.1). There are 2 and 19 HCPs for the pretreatment and hydrolysis of the forestry wastes, respectively.

72.3.1.1 Forestry Waste Pretreatments

The key findings about two these papers on the pretreatments of the forestry wastes are given below. Brandt et al. (2010) studied the effect of the ionic liquid (IL) anion in the pretreatment of pine wood chips in a paper with 269 citations. All ILs contained the 1-butyl-3-methylimidazolium (Bmim) cation; the anions were trifluoromethanesulfonate, methylsulfate, dimethylphosphate, dicyanamide, chloride and acetate. Using a protocol for assessing the ability to swell small wood blocks ($10 \times 10 \times 5\,mm^3$), they observed that the anion had a profound impact on the ability to promote both swelling and dissolution of biomass. Viscosity, temperature, and water content were also important parameters influencing the swelling process. Further, the anion basicity described by the parameter β correlated with the ability to expand and dissolve pine lignocellulose. BMIM dicyanamide dissolved neither cellulose nor lignocellulosic material.

Wang et al. (2011) extracted cellulose from wood chips in an IL 1-allyl-3-methylimidazolium chloride ([Amim]Cl) in a paper with 141 citations. They observed that pine was the most suitable wood species for cellulose extraction with ILs. Its cellulose extraction rate could reach as high as 62% under optimized conditions, and its cellulose content was as high as 85% when dimethyl sulfoxide (DMSO)/water is used as the precipitant. The reaction time could be significantly reduced by microwave irradiation. IL dissolved pine wood by destroying inter- and intramolecular hydrogen bonds between lignocelluloses. The major component of pine extract was cellulose with a homogeneous and dense structure. Further, after extraction, the IL could be easily recycled and reused.

72.3.1.2 Forestry Waste Hydrolysis

The key findings about only 19 HCPS on the hydrolysis of the forestry waste biomass are given below. Thus, it appears that this research front is the most prolific research front for these HCPs.

TABLE 72.1
The Pretreatment and Hydrolysis of Forestry Wastes

No.	Papers	Forestry Waste	Res. Fronts	Prts.	Yeasts	Parameters	Keywords	Lead Author	Affil.	Cits.
1	Brandt et al. (2010)	Wood chips	Pretreatment	Ionic liquids	Na	IL pretreatment, anion effect, wood dissolution, anion basicity	Wood, chips, pretreatment, ionic liquids	Welton, Tom 7003503272	Imperial Coll. UK	269
2	Lu et al. (2002)	Softwood residues	Hydrolysis	Cellulases, H_2O_2, steam	Na	Enzymatic hydrolysis, pretreatments, sugar yield, enzyme recycling, and adsorption	Softwood, hydrolysis, residues	Mansfield, Shawn D. 7006421766	Univ. British Columbia Canada	207
3	Brownell et al. (1986)	Wood chips	Hydrolysis	Steam, acid, *T. harzianum*	*K. pneumoniae*	Enzymatic hydrolysis, pretreatments, sugar yield, ethanol	Wood, pretreatment, chips	Brownell, Harold H. 7003584800	Forintek Canada Corp. Canada	207
4	Huang et al. (2019)	Bamboo residues	Hydrolysis	Acids, solvents, enzymes	Na	Enzymatic hydrolysis, pretreatments, post-extraction, enzymatic digestibility	Bamboo, residues, digestibility	Yong, Qiang 7003718716	Nanjing Forestry Univ. China	195
5	Iranmahboob et al. (2002)	Wood chips	Hydrolysis	Acids	Na	Acid hydrolysis, acid concentration, heating period, dextrose yields	Wood chips, hydrolysis, ethanol	Nadim, Farhad 7007057957	Univ. Connecticut USA	193
7	Li et al. (2010)	Hardwood chips	Hydrolysis	Acids	Na	Acid hydrolysis, hemicellulose removal, pretreatment temperature, xylose yield	Hardwood chips, hydrolysis	Ni, Yonghao 7402909909	Univ. New Brunswick Canada	157
8	Lin et al. (2020)	Bamboo residues	Hydrolysis	Enzymes	Na	Enzymatic hydrolysis, pretreatments, enzymatic digestibility	Bamboo residues, pretreatment, digestibility	Huang, Caoxing 56495726800	Nanjing Forestry Univ. China	148

(*Continued*)

TABLE 72.1 (*Continued*)
The Pretreatment and Hydrolysis of Forestry Wastes

No.	Papers	Forestry Waste	Res. Fronts	Prts.	Yeasts	Parameters	Keywords	Lead Author	Affil.	Cits.
9	Emmel et al. (2003)	Eucalyptus chips	Hydrolysis	Acids, steam, enzymes	Na	Enzymatic hydrolysis, pretreatments, pretreatment temperature	Eucalyptus chips, fractionation,	Ramos, Luiz P. 7202180586	Fed. Univ. Parana Brazil	144
10	Wang et al. (2011)	Wood chips	Pretreatment	Ionic liquids	Na	Pretreatments, ILs, wood type, cellulose extraction, IL reusage	Wood chips, ionic liquids,	Li, Huiquan 35201433300	Chinese Acad. Sci. China	141
11	Canettieri et al. (2007)	Eucalyptus residues	Hydrolysis	Acids	Na	Acid hydrolysis, xylose removal, acid concentration, temperature, residue/acid solution ratio	Eucalyptus residues, hydrolysis	De Carvalho, Joao A. 57212048688	State Univ. Paulista Brazil	131
13	Ramos et al. (1992)	Wood chips	Hydrolysis	Steam, SO_2, enzymes, H_2O_2	Na	Enzymatic hydrolysis, pretreatment, chip type, substrate concentrations, enzyme loadings	Eucalyptus, aspen, spruce wood, chips, pretreatment, hydrolysis	Ramos, Luiz P. 7202180586	Fed. Univ. Parana Brazil	124
14	Teramoto et al. (2008)	Eucalyptus wood chips	Hydrolysis	Solvents, milling, enzymes	Na	Enzymatic hydrolysis, pretreatment, milling scale, cellulose conversion	eucalyptus wood chips, saccharification	Teramoto, Yoshikuni 7006223067	Kyoto Univ. Japan	122
15	Nitsos et al. (2016)	Poplar branches, grapevine pruning, sawdust	Hydrolysis	Hydrothermal, enzymes	Na	Enzymatic hydrolysis, pretreatment, milling scale, cellulose conversion	Hardwood, softwood, residues, hydrolysis	Triantafyllidis, Kostas S. 8406363900	Aristotle Univ. Thessaloniki Greece	111
16	Lin et al. (2019)	Bamboo residues	Hydrolysis	Acids, surfactants, enzymes	Na	Enzymatic hydrolysis, surfactants, enzymatic digestibility, hydrophobicity, hydrolysis yield	Bamboo residues, hydrolysis, pretreated	Huang, Caoxing 56495726800	Nanjing Forestry Univ. China	98

(*Continued*)

TABLE 72.1 (*Continued*)
The Pretreatment and Hydrolysis of Forestry Wastes

No.	Papers	Forestry Waste	Res. Fronts	Prts.	Yeasts	Parameters	Keywords	Lead Author	Affil.	Cits.
17	Zhang et al. (2007)	Bamboo residues	Hydrolysis	Fungi enzymes	Na	Enzymatic hydrolysis, hydrolysis rate, sugar yield	Bamboo residues, hydrolysis	Zhang, Xiaoyu 56176446300	Huazhong Univ. Sci. Technol. China	96
18	Lima et al. (2013)	Eucalyptus bark	Hydrolysis	Acids, alkali, enzymes	Na	Enzymatic hydrolysis, enzymatic digestibility, cellulose conversion, lignin and hemicellulose removal	Eucalyptus bark, digestibility	Polikarpov, Igor 7006220351	Univ. Sao Paulo Brazil	94
19	Shevchenko et al. (2000)	Softwood chips	Hydrolysis	Steam, SO_2, enzymes	Na	Enzymatic hydrolysis, hemicellulose conversion, sugar recovery	Softwood chips, hydrolysis	Shevchenko, Sergey M. 7101791686	Shevchenko Cons. USA	91
20	Boussaid et al. (2000)	Wood chips	Hydrolysis	Steam, SO_2, enzymes	Na	Enzymatic hydrolysis, pretreatment severity, sugar recovery, wood type	Fir, wood chips, hydrolysis, pretreatment	Saddler, John N. 7005297559	Univ. British Columbia Canada	90
21	Kim et al. (2014)	Hardwood chips	Hydrolysis	Hydrothermal, enzymes	Na	Enzymatic hydrolysis, pretreatment severity, fermentation inhibitors, sugar yield	Hardwood chips, pretreatment	Ladisch, Michael R. 7005670397	Purdue Univ. USA	89
22	Castoldi et al. (2014)	Eucalyptus sawdust	Hydrolysis	Fungi	Na	Enzymatic hydrolysis, fungi, hydrolyzable cellulose	Eucalyptus sawdust, pretreatment	Peralta, Rosane M.* 7007044821	State Univ. Maringa Brazil	85
24	Monavari et al. (2009)	Softwood chips	Hydrolysis	Steam, SO_2, acids, enzymes	Na	Enzymatic hydrolysis, impregnation time, chip size, sugar yields	Softwood chips, ethanol, pretreatment	Monavari, Sanam* 24076966600	Lund Univ. Sweden	79

Prt., Biomass pretreatments; Na, Non available; Cits., Number of citations received for each paper; *, Female.

Lu et al. (2002) determined sugar yield and enzyme adsorption profile during hydrolysis of sulfur dioxide (SO_2)-catalyzed and steam-exploded Douglas fir residues in a paper with 207 citations. After hot alkali peroxide (H_2O_2) post-pretreatment, they observed that the rates and yield of hydrolysis attained from the post-pretreated Douglas fir were significantly higher, even at lower enzyme loadings, than those obtained with the corresponding steam-exploded Douglas fir. The enzymatic adsorption profiles observed during hydrolysis of the two substrates were significantly different. They employed ultrafiltration to recover enzyme in solution and reused in subsequent hydrolysis reactions with added, fresh substrate. The enzyme remained relatively active for three rounds of recycle. In conclusion, enzyme recovery and reuse during the hydrolysis of post-treated softwood substrates could lead to reductions in the need for the addition of fresh enzyme during softwood-based bioconversion processes.

Brownell et al. (1986) performed the steam explosion pretreatment of aspen wood chips in a paper with 207 citations. Material balances for pentosan, lignin, and hexosan, during steam explosion pretreatment of aspen wood, showed almost quantitative recovery of cellulose in the water-insoluble fraction. Dilute acid impregnation resulted in more selective hydrolysis of pentosan relative to undesirable pyrolysis, and gave a more accessible substrate for enzymatic hydrolysis. Small chips gave approximately equal rates of pentosan destruction and solubilization and similar yields of glucose and of total reducing sugars on enzymatic hydrolysis with *Trichoderma harzianum*. Partial pyrolysis, destroying one third of the pentosan of aspen wood at atmospheric pressure by dry steam at 276°C, gave little increase in yield of reducing sugars on enzymatic hydrolysis. Pretreatment with saturated steam at 240°C gave essentially the same yields of glucose and of total reducing sugars, and the same yields of butanediol and ethanol on fermentation with *Klebsiella pneumoniae,* whether or not 80% of the steam was bled off before explosion and even if the chips remained intact, showing that explosion was unnecessary.

Huang et al. (2019) improved the enzymatic digestibility of acid-pretreated bamboo residues in a paper with 195 citations. They evaluated a mild and facile post-extraction using different reagents to overcome the recalcitrance by removing the lignin and disrupting its inhibitory properties. They observed that the enzymatic digestibility of acid-pretreated bamboo residues could be improved from 15.4% to 61.4%, 59.7%, and 42.8% by room temperature post-extraction with phosphoric acid, urea, and ethanol, respectively. They observed compelling correlations between enzymatic digestibility and structural changes, including delignification, reducing of substrate hydrophobicity, altering cellulose crystallinity, and elevations to the residual lignin syringyl-to-guaiacyl (S/G) ratio and functional groups. In conclusion, coupling a post-extraction process with acid pretreatment of bamboo residues resulted in greater fermentable sugar production.

Iranmahboob et al. (2002) optimized acid hydrolysis of wood chips for production of ethanol biofuels in a paper with 193 citations. They explored the sugar recovery rates through three sets of acid hydrolysis experiments. They sorted wood chips to include equal ratios (by weight) of softwood and hardwood. They found that the acid concentration and the heating period were the two main factors affecting dextrose yields. With the use of 26% by weight sulfuric acid (H_2SO4), highest dextrose yields could be reached within 2 h of heating time. This corresponded to overall conversion efficiency of mixed wood chip cellulose to dextrose in the range of 78%–82% based on theoretical values.

Li et al. (2010) removed hemicellulose from hardwood chips in the prehydrolysis step of the kraft-based dissolving pulp production process in a paper with 157 citations. These chips consisted of maple, aspen, and birch with a ratio of 7:2:1. They observed that the prehydrolysis was a dynamic process, in which the removal of hemicelluloses increased with time while the conversion of extracted hemicelluloses to monosaccharides due to acid hydrolysis increased and part of the xylose was converted to furfural, fermentation inhibitor. The maximum temperature was the most critical parameter for hemicelluloses extraction and conversion, and a temperature of 170°C was the optimum for hemicelluloses extraction with relatively low conversion of xylose to furfural. Further, about 11% of the xylan was removed at 170°C. Due to the presence of a high amount of xylan, birch produced the highest amount of xylose, followed by maple, and then aspen.

Lin et al. (2020) explored the performance of deep eutectic solvent (DES) pretreatment on improving enzymatic digestibility of bamboo residues in a paper with 148 citations. They used DES with different molar ratios of choline chloride/lactic acid. They observed that enzymatic digestibility of DES-pretreated bamboo residues was enhanced with the increasing molar ratio of choline chloride/lactic acid. DES pretreatment had the ability to remove lignin and xylan, reduce the degree of polymerization of cellulose, enhance the crystallite size of cellulose, and improve cellulose accessibility. Further, they observed linear correlations between enzymatic digestibility and these changes of physicochemical properties. In conclusion, DES pretreatment improved the enzymatic digestibility of the wood waste.

Emmel et al. (2003) fractionated eucalyptus chips by dilute acid-catalyzed steam explosion in a paper with 144 citations. They performed the steam explosion of the chips under various pretreatment conditions (200°C–210°C, 2–5 min) after impregnation of the wood chips with 0.087% and 0.175% (w/w) H_2SO_4. The pretreatment temperature was the most critical variable affecting the yield of steam-treated fractions and they observed that pretreatment of 0.175% (w/w) H_2SO_4-impregnated chips at 210°C for 2 min was the best condition for hemicellulose recovery (mostly as xylose) in the water-soluble fraction, reaching almost 70% of the corresponding xylose theoretical yield. By contrast, lower pretreatment temperatures of 200°C were enough to yield steam-treated substrates from which a 90% cellulose conversion was obtained in 48 h, using low enzyme loadings of a Celluclast 1.5 l plus Novozym 188 mixture. Further, the concentration of water-soluble chromophores increased with pretreatment severity, while the yield of alkali-soluble lignin increased at higher levels of acid impregnation and pretreatment temperatures. Finally, there was a pattern of lignin fragmentation toward greater pretreatment severities, but lignin condensation prevailed at the most drastic pretreatment conditions.

Canettieri et al. (2007) optimized acid hydrolysis from the hemicellulosic fraction of eucalyptus residues in a 1.4-l pilot-scale reactor in a paper with 131 citations. They used a model composition corresponding to a 23-orthogonal factorial design and employed the response surface methodology (RSM) to optimize the hydrolysis conditions to attain maximum xylose extraction from hemicellulose of residue. The optimal conditions were H_2SO_4 concentration of 0.65%, temperature of 157°C, and residue/acid solution ratio of 1/8.6 with a reaction time of 20 min. Under these conditions, they removed 79.6% of the total xylose and the hydrolysate contained 1.65 g/L glucose, 13.65 g/L xylose, and 1.55 g/L arabinose as sugars 3.10 g/L acetic acid, 1.23 g/L furfural, and 0.20 g/L 5-hydroxymethylfurfural (HMF) as fermentation inhibitors.

Ramos et al. (1992) compared the steam explosion pretreatment of eucalyptus, aspen, and spruce wood chips and their enzymatic hydrolysis in a paper with 124 citations. They performed the enzymatic hydrolysis of SO_2-impregnated, steam-exploded eucalyptus at increasing substrate concentrations and enzyme loadings. When low enzyme loadings were used, they observed that the H_2O_2-pretreated fraction derived from eucalyptus chips was more readily hydrolyzed than the water-insoluble and the alkali-insoluble fractions. The various cellulosic fractions derived from steam-exploded eucalyptus chips exhibited a greater susceptibility to hydrolysis than the equivalent aspen and spruce substrates, particularly at high substrate concentrations (10%, w/v).

Teramoto et al. (2008) performed the pretreatment of eucalyptus wood chips for enzymatic hydrolysis using combined H_2SO_4-free ethanol solvent and ball milling in a paper with 122 citations. They exposed chips to an ethanol/water/acetic acid mixed solvent in an autoclave which could cause the fibrillation of wood chips. During the process, they observed that the production of furfural was extremely low and delignification was insignificant. Subsequently, they pulverized the activated solid products by ball milling. Under the enzymatic hydrolysis experiments, they obtained the 100% conversion of the cellulosic components into glucose under optimal conditions. The scale affecting the improvement of enzymatic digestibility ranged from 10 nm to 1 μm. The solvent pretreatment induced a pore formation by the removal of part of the lignin and hemicellulose fractions in the size range from a few of tens nanometers to several hundred nanometers.

Nitsos et al. (2016) optimized hydrothermal pretreatment of residual poplar branches, grapevine pruning, and pine sawdust for selective hemicellulose recovery and improved cellulose enzymatic hydrolysis in a paper with 111 citations. They performed this pretreatment in a batch-mode, high-pressure reactor under autogenous pressure at varying temperature (170°C–220°C) and time (15–180 min) regimes and at liquid-to-solid ratio of 15. They observed that the maximum hemicellulose recovery for poplar, grapevine, and pine was around 60% at ~70%–85% hemicellulose removal, based on initial hemicellulose content of each biomass type, and at relatively moderate pretreatment severities (log R_0=3.8–4.1). Formation of the fermentation inhibitors such as acids (i.e., formic and levulinic acid) and furans (i.e., furfural and HMF) was relatively low and below ca. 1 mg/mL for the whole range of pretreatment severities. Enzymatic hydrolysis of the parent lignocellulosic materials toward glucose was very low (i.e., 10%) and remained low for the pretreated pine biomass (16%) but was substantially improved for poplar (49%) and especially for grapevine (77%) as a result of hydrothermal pretreatment at the highest severity (log R_0=4.7). The significant improvement of enzymatic hydrolysis of grapevine was due to the nearly complete removal of hemicellulose and to the changes in the morphological and textural characteristics of biomass particles, with the most pronounced one being the 9-fold increase in surface and pore volume.

Lin et al. (2019) improved enzymatic hydrolysis of acid-pretreated bamboo residues (APBR) using amphiphilic surfactant in a paper with 98 citations. They obtained surfactant using dehydroabietic acid from pine rosin and then pre-adsorbed with APBR to block the residual lignin adsorption site, which was expected to improve its enzymatic digestibility. They observed that the surfactant with polyethylene glycol (PEG) with polymerization degree of 34 (D-34) aggregated to form worm-like micelles, which improved enzymatic hydrolysis yield of APBR from 24.3% to 71.9% by pre-adsorbing with 0.8 g/L. Surfactants pre-adsorbed on APBR could reduce hydrophobicity of APBR, adsorption affinity, and adsorption capacity of lignin for cellulase from 0.51 L/g to 0.48–0.32 L/g, from 2.9 mL/mg to 1.8–1.4 mL/mg, and from 122.3 mg/g to 101.9–21.4 mg/g, respectively.

Zhang et al. (2007) pretreated bamboo residues with *Coriolus versicolor* B1 for enzymatic hydrolysis in a paper with 96 citations. They observed that the hydrolysis rate was significantly enhanced with a maximum hydrolysis rate of 37.0% after pretreatment. Further, reducing sugars yield was 223.2 mg/g of bamboo residues, which was 2.34 times that of the raw material. In conclusion, this pretreatment was feasible for the hydrolysis of bamboo residues.

Lima et al. (2013) studied the effects of pretreatment on morphology, chemical composition, and enzymatic digestibility of eucalyptus bark in paper with 94 citations. They evaluated effects of a delignification process with increasing sodium hydroxide (NaOH) concentrations, preceded or not by diluted acid, on the bark of two eucalyptus clones: *Eucalyptus grandis* (EG) and the hybrid, *E. grandis* x urophylla (HGU). They observed an increase in the cellulose content, reaching around 81% and 76% of glucose for HGU and EG, respectively, using a two-step treatment with HCl 1%, followed by 4% NaOH. Lignin removal was 84% (HGU) and 79% (EG), while the hemicellulose removal was 95% and 97% for HGU and EG, respectively. However, when they applied a one-step treatment, with 4% NaOH, they observed higher hydrolysis efficiencies after 48 h for both clones, reaching almost 100% for HGU and 80% for EG, in spite of the lower lignin and hemicellulose removal. Total cellulose conversion increased from 5% and 7% to around 65% for HGU and 59% for EG. In conclusion, the single-step alkaline pretreatment improved the enzymatic digestibility of this bark.

Shevchenko et al. (2000) optimized monosaccharide recovery by post-hydrolysis of the water-soluble hemicellulose component after steam explosion of softwood chips in a paper with 91 citations. They observed that the SO_2-catalyzed steam explosion effectively fractionated softwood carbohydrates, releasing 80%–90% of the hemicellulose into solution, and enhanced the enzymatic hydrolysis of the water-insoluble cellulose fraction. When they modified the steam pretreatment conditions to ensure maximum hemicellulose recovery, a significant part of hemicellulose-derived carbohydrates were solubilized in an oligomeric form. When Douglas fir wood chips were used as the substrate, sulfuric acid-catalyzed post-hydrolysis at 120°C for 1 h allowed most of the original hemicellulose to be recovered in the monomeric form. Although higher catalyst concentrations (1%–3%) resulted in complete depolymerization of the carbohydrates, they observed noticeable losses due to sugar degradation. Lower concentrations of the added catalyst increased the monomer yields, and resulted in less secondary degradation. Partial oxidation of the added SO_2 to sulfuric acid during steam explosion resulted in the depolymerization of the hemicelluloses during post-hydrolysis, even without further acid addition.

Boussaid et al. (2000) performed the steam explosion pretreatment of Douglas fir wood chips for efficient enzymatic hydrolysis in a paper with 90 citations. They impregnated the wood with SO_2 and steam exploded at three severity levels, and enzymatically hydrolyzed the cellulose-rich, water-insoluble component. They observed that the high-severity conditions resulted in near complete solubilization and some degradation of hemicelluloses and a significant improvement in the efficiency of enzymatic digestibility of the cellulose component. At lower severity, some of the hemicellulose remained unhydrolyzed, and the cellulose present in the pretreated solids was not readily hydrolyzed. The medium-severity pretreatment conditions were a good compromise because they improved the enzymatic hydrolyzability of the solids and resulted in the recovery of the majority of hemicellulose in a monomeric form within the water-soluble stream. Sapwood-derived wood chips exhibited a higher susceptibility to both pretreatment and hydrolysis and, on steam explosion, formed smaller particles as compared to heartwood-derived wood chips.

Kim et al. (2014) performed the subcritical liquid hot water (LHW) pretreatment of hardwood chips at various severities (log R_0=3.65–4.81) to assess the efficiencies of the pretreatments with respect to achieving high pentose sugar yields and improved enzymatic digestibility of pretreated cellulose in a paper with 89 citations. In multi-stage pretreatment, the first low-severity pretreatment was optimized for solubilizing fast-hydrolyzing hemicellulose while minimizing formation of furans. The subsequent pretreatment was carried out at over 200°C to recover the difficult-to-hydrolyze hemicellulose fraction as well as to increase susceptibility of pretreated cellulose to enzymes. They obtained high recovery (>92%) of hemicellulose-derived pentose sugars and enhanced enzymatic hydrolysis of pretreated cellulose (where >80% glucose yield resulted with 20 FPU=32 mg protein/g glucan or 10–13 mg/g initial hardwood) by applying a multi-stage pretreatment. In conclusion, they showed how the severity equation might be used to obtain a single characteristic curve that correlates xylan solubilization and enzymatic cellulose hydrolysis as a function of severity at pretreatment temperatures up to 230°C.

Castoldi et al. (2014) performed the biological pretreatment of eucalyptus sawdust with white-rot fungi in a paper with 85 citations. They used *Ganoderma lucidum, Phanerochaete chrysosporium, Pleurotus ostreatus, Pleurotus pulmonarius*, and *Trametes* sp. They observed that *P. ostreatus* and *P. pulmonarius* promoted more extensive selective modifications in the lignin content. They analyzed enzymatic hydrolysis of *E. grandis* sawdust cellulose with and without biological pretreatment. In general terms, the biological pretreatments diminished the initial delay, increased the easily hydrolyzable fraction, and generated a second hydrolyzable fraction. The generation of an easily hydrolyzable cellulose fraction obeyed the following decreasing sequence: *P. ostreatus* (16.7% of total cellulose)>*P. pulmonarius* (15.4%)>*Trametes* sp. (10.1%)≫*P. chrysosporium* (2.8%)≈no treatment (2.8%). The generation of the second hydrolyzable fraction was more efficient in the case of *P. pulmonarius* and *P. ostreatus* pretreatments. For the latter, the total amount of reducing sugars released after 48 h of saccharification of sawdust was increased from 2.5 to 48.0 μmol/mL.

Monavari et al. (2009) studied the effect of impregnation time and chip size on sugar yield in pretreatment of softwood for ethanol production in a paper with 79 citations. They impregnated chips of different sizes with SO_2 and steam-pretreated. Dilute acid pretreatment together with subsequent enzymatic hydrolysis resulted in solubilization of between 69% and 73% of the fermentable sugars (glucose and mannose) in the raw material for the combinations of impregnation times and chip sizes investigated. They observed that shorter impregnation times resulted in slightly lower mannose yields for the larger chips, probably due to poor diffusion of the catalyst. Small differences in glucose yield after enzymatic hydrolysis showed that the overall glucose yield was slightly higher for the smaller chips.

72.3.2 Forestry Waste-based Bioethanol Production

There are only four HCPs for the production the forestry waste-based bioethanol fuels with at least 76 citations each (Table 72.2). These papers also cover the fermentation of the hydrolysates of the wood wastes.

Cara et al. (2008) produced ethanol from steam-explosion-pretreated olive tree prunings in a paper with 190 citations. They performed steam explosion pretreatment in the temperature range 190°C–240°C, with or without previous impregnation by water or sulfuric acid solutions. They explored the effect of both pretreatment temperature and impregnation conditions on sugar and ethanol yields by the simultaneous saccharification and fermentation (SSF) process on the pretreated solids. They obtained the maximum ethanol yield (7.2 g ethanol/100 g raw material) from water-impregnated, steam-pretreated residue at 240°C. Further, if all sugars solubilized during pretreatment were taken into account, up to 15.9 g ethanol/100 g raw material could be obtained (pretreatment conditions: 230°C and impregnation with 1% w/w sulfuric acid concentration), assuming theoretical conversion of these sugars to ethanol.

Weil et al. (1997) pretreated yellow poplar sawdust by pressure cooking in water at temperatures ranging from 220°C to 260°C in a paper with 124 citations. They pretreated the wood sawdust at a 6%–6.6% solid/liquid slurry in a 2 L, 304 SS, Parr reactor. They obtained heat-up times to the final temperatures of 220°C, 240°C, or 260°C in 60–70 min, while the hold time at the final temperature was <1 min. A serpentine cooling coil cooled the reactor's contents within 3 min after the maximum

TABLE 72.2
The Production of Forestry Waste-based Bioethanol Fuels

No.	Papers	Forestry Waste	Res. Fronts	Prts.	Yeasts	Parameters	Keywords	Lead Author	Affil.	Cits.
6	Cara et al. (2008)	Tree prunings	Production	Steam, acids, enzymes	Yeasts	Enzymatic hydrolysis, fermentation, SSF, sugar and ethanol yields, pretreatment temperature	Tree prunings, ethanol, pretreated	Castro, Eulogio 7102441948	Univ. Jaen Spain	190
12	Weil et al. (1997)	Poplar sawdust	Production	Hydrothermal, enzymes	Yeasts	Ethanol production, SSF, pretreatment, sugar and ethanol yield	Poplar sawdust, pretreatment	Ladisch, Michael R. 7005670397	Purdue Univ. USA	124
23	Allen et al. (2001)	Poplar sawdust	Production	Hydrothermal, acids, steam, enzymes	S. cerevisiae	Enzymatic hydrolysis, fermentation SSF, ethanol yield, xylan recovery and solubilization	Poplar sawdust, pretreatment	Allen, Stephen G. 7403049682	Univ. Hawaii USA	84
25	Zhu et al. (2015)	Douglas fir residues	Production	Sulfite, SO_2	Yeasts	Fermentation, pretreatments, ethanol yield and titer, lignosulfonate	Fir, forest, residues, bioethanol	Zhu, Junyong 7405692678	USDA Forest Serv. USA	76

Prt., Biomass pretreatments; Na, Non available; Cits., Number of citations received for each paper.

temperature was attained. They found that enzymatic hydrolysis gave 80%–90% conversion of cellulose in the pretreated wood to glucose. The SSF of washed, pretreated lignocellulose gave an ethanol yield that was 55% of theoretical. Untreated wood sawdust gave <5% hydrolysis under the same conditions.

Allen et al. (2001) compared the pretreatment of the sawdust (<6 mm, 25–30 wt% moisture) with LHW (190°C–220°C, 2.0–7.5 min, 2–7 wt% solids loading) or steam explosion (220°C, 2.0 min) with dilute acid pretreatment (0.8% w/v H_2SO_4, 175°C, 10 min, 20–25 wt% solids loading) in a paper with 84 citations. They determined the most favorable conditions for LHW pretreatment (220°C, 2.0 min, 5 wt% solids loading) and dilute acid pretreatment (175°C, 10.0 min, 0.8% sulfuric acid, 20 wt% solids loading). They performed the SSF process using *Saccharomyces cerevisiae* and 25 IFPU cellulase/g of cellulose. They obtained the ethanol yields of 97 and 92 wt% of the theoretical yield and xylan recoveries of 85 and 70 wt% (92 and 96 wt% solubilization) with the LHW and dilute acid pretreatments, respectively. Steam explosion pretreatment (220°C, 2 min) was not as effective as the dilute acid pretreatment, resulting in an ethanol yield of 70 wt% of the theoretical yield, with 88 wt% xylan recovery (54 wt% solubilization). There was a positive correlation between xylan removal and the ethanol yield obtained by SSF. Xylan recovery, however, was inversely proportional to xylan solubilization. Further, removal of up to 57 wt% of the acid-insoluble lignin in yellow poplar was also possible with LHW pretreatment, with 75 wt% of this acid-insoluble lignin recoverable by simple filtration of the liquid pretreatment products.

Zhu et al. (2015) produced ethanol from Douglas fir forest harvest residues at high titer and yield in a paper with 76 citations. They used sulfite chemistry without solid–liquor separation and detoxification. They applied the sulfite pretreatment to overcome the recalcitrance of lignocelluloses (SPORL) directly to the ground forest harvest residue with no further mechanical size reduction, at a low temperature of 145°C and calcium bisulfite or total SO_2 loadings of only 6.5 or 6.6 wt% on oven dry forest residue, respectively. They observed that the low temperature pretreatment facilitated high solids fermentation of the un-detoxified pretreated whole slurry. They obtained an ethanol yield of 282 L/tonne, equivalent to 70% theoretical, with a titer of 42 g/L. SPORL solubilized ~45% of the wood lignin as lignosulfonate with properties equivalent to or better than a commercial lignosulfonate, important to improve the economics of biofuel production.

72.4 DISCUSSION

72.4.1 INTRODUCTION

The crude oil-based gasoline fuels have been widely used in the transportation sector since the 1920s. However, there have been great public concerns over the adverse environmental and human impact of these fuels. Hence, biomass-based bioethanol fuels have increasingly been used in blending gasoline and petrodiesel fuels, in the fuel cells, and in the biochemical production in a biorefinery context.

However, it is necessary to pretreat the biomass to enhance the yield of the bioethanol prior to the bioethanol fuel production from the feedstocks through the hydrolysis and fermentation of the biomass and hydrolysates, respectively.

One of the most-studied feedstocks for the bioethanol fuels has been the forestry wastes such as wood chips, bark, sawdust, forest residues, tree prunings, spent sulfite liquor, and wood wastes. The research in the field of the second generation forestry waste-based bioethanol fuels has intensified in this context in the key research fronts of the pretreatment and hydrolysis of the forestry wastes, fermentation of the forestry waste-based hydrolysates, and production and evaluation of the second generation forestry waste-based bioethanol fuels.

However, it is essential to develop efficient incentive structures for the primary stakeholders to enhance the research in this field. Although there have been a limited number of review papers for this field, there has been no review of the most-cited 25 articles in this field.

Thus, this book chapter presents a review of the most-cited 25 articles on the bioethanol fuel production and evaluation from the forestry wastes. Then, it discusses the key findings of these highly influential papers and comments on the future research priorities in this field.

As a first step for the search of the relevant literature, the keywords were selected using the most-cited first 200 population papers. The selected keyword list was then optimized to obtain a representative sample of papers for the searched research field. This keyword list was provided in the appendix of Konur (2023) for future replicative studies.

As a second step, a sample dataset was used for this study. The first 25 articles with at least 76 citations each were selected for the review study. Key findings from each paper were taken from the abstracts of these papers and were discussed. Additionally, a number of brief conclusions were drawn and a number of relevant recommendations were made to enhance the future research landscape.

Information about the research fronts for the sample papers in the forestry waste-based bioethanol fuels with regard to the feedstocks used in these processes is given in Table 72.3. As this table shows, the most prolific research fronts for this field are the wood chips and forest residues with 52% and 32% of the HCPs. Further, the other prolific research fronts are sawdust, tree prunings, and bark with 16%, 12%, and 4% of the sample papers, respectively.

On the other hand, the most influential feedstock is the wood chips with 21% surplus, followed by the forest residues with 5% surplus. Similarly, the least influential research fronts are wood waste, bark, and spent sulfite liquor with 8%, 7%, and 5% deficit, respectively.

Information about the thematic research fronts for the sample papers in the forestry waste-based bioethanol fuels is given in Table 72.4. As this table shows, the most prolific research fronts for this field are the pretreatment and hydrolysis of the forestry wastes with 100% and 88% of the HCPs, respectively. The other research fronts are the hydrolysate fermentation and bioethanol production with 16% of the sample papers each.

Further, the most influential research fronts are pretreatment and hydrolysis of the food wastes with 9% and 8% surplus, respectively. Similarly, hydrolysate fermentation and bioethanol production are the least influential research fronts with 10% and 9% deficits, respectively, followed by the bioethanol fuel evaluation with 5% deficit.

72.4.2 Forestry Waste Pretreatment and Hydrolysis

The brief information about 21 most-cited papers on the pretreatment and hydrolysis of forestry wastes with at least 79 citations each is given below (Table 72.1). There are 2 and 19 HCPs for the pretreatment and hydrolysis of the forestry wastes, respectively. However, it should be noted that as the Table 72.4 shows, the number of HCPs related to the biomass pretreatments and hydrolysis is 25 and 22, respectively.

TABLE 72.3
The Most Prolific Research Fronts for the Second Generation Forestry Waste-based Bioethanol Fuels

No.	Research Fronts	N Paper % Review	N Paper % Sample	Surplus (%)
1	Wood chips	52.0	31.0	21.0
2	Forest residues	32.0	27.0	5.0
3	Sawdust	16.0	13.0	3.0
4	Tree prunings	12.0	10.0	2.0
5	Bark	4.0	11.0	−7.0
6	Wood waste	0.0	8.0	−8.0
7	Spent sulfite liquor	0.0	5.0	−5.0

N Paper (%) review: The number of papers in the sample of 25 reviewed papers. N paper (%) sample: The number of papers in the population sample of 100 papers

TABLE 72.4
The Most Prolific Thematic Research Fronts for the Forestry Waste-based Bioethanol Fuels

No.	Research Fronts	N Paper % Review	N Paper % Sample	Surplus (%)
1	Biomass pretreatments	100.0	92.0	8.0
2	Biomass hydrolysis	88.0	79.0	9.0
3	Hydrolysate fermentation	16.0	26.0	−10.0
4	Bioethanol production	16.0	25.0	−9.0
5	Bioethanol fuel evaluation	0.0	5.0	−5.0

N Paper (%) review: The number of papers in the sample of 25 reviewed papers. N paper (%) sample: The number of papers in the population sample of 100 papers.

72.4.2.1 Forestry Waste Pretreatments

As the Table 72.3 shows, the most prolific pretreatments are the enzymatic and chemical pretreatments with 19 HCPs each, followed by hydrothermal pretreatments with 9 HCPs. On the individual terms, however, the most prolific pretreatments are acid and steam explosion pretreatments with 11 and 9 HCPs, respectively, followed by the hydrothermal and SO_2 pretreatments with five and four HCPs, respectively. The key findings about only two of these papers are given below.

Brandt et al. (2010) studied the effect of the IL anion in the pretreatment of pine wood chips and observed that the anion had a profound impact on the ability to promote both swelling and dissolution of biomass. Further, Wang et al. (2011) extracted cellulose from wood chips in [Amim] and the cellulose extraction rate could reach as high as 62% under optimized conditions and its cellulose content was as high as 85% when DMSO/water is used as the precipitant.

72.4.2.2 Forestry Waste Hydrolysis

As the Table 72.4 shows, there are 22 HCPs related to the hydrolysis of the forest waste biomass. The key findings about only 19 of these HCPS are given above. Thus, it appears that this research front is the most prolific research front for these HCPs.

Lu et al. (2002) determined sugar yield and enzyme adsorption profile during hydrolysis of SO_2-catalyzed and steam-exploded Douglas fir residues and observed that the rates and yield of hydrolysis attained from the post-treated Douglas fir were significantly higher, even at lower enzyme loadings, than those obtained with the corresponding steam-exploded Douglas fir. Further, Brownell et al. (1986) performed the steam explosion pretreatment of aspen wood chips and observed that the pretreatment with saturated steam at 240°C gave essentially the same yields of glucose and of total reducing sugars.

Huang et al. (2019) improved the enzymatic digestibility of acid-pretreated bamboo residues and observed that the enzymatic digestibility of acid-pretreated bamboo residues could be improved from 15.4% to 61.4%, 59.7%, and 42.8% by room temperature post-extraction with phosphoric acid, urea, and ethanol, respectively. Further, Iranmahboob et al. (2002) optimized acid hydrolysis of wood chips for production of ethanol biofuels and found that the acid concentration and the heating period were the two main factors affecting dextrose yields.

Li et al. (2010) removed hemicellulose from hardwood chips in the prehydrolysis step of the kraft-based dissolving pulp production process and observed that the prehydrolysis was a dynamic process, in which the removal of hemicelluloses increased with time while the conversion of extracted hemicelluloses to monosaccharides due to acid hydrolysis increased. Further, Lin et al. (2020) explored the performance of DES pretreatment on improving enzymatic digestibility of bamboo residues and observed that enzymatic digestibility of DES-pretreated bamboo residues was enhanced with the increasing molar ratio of choline chloride/lactic acid.

Emmel et al. (2003) fractionate eucalyptus chips by dilute acid-catalyzed steam explosion and found that pretreatment of 0.175% (w/w) H_2SO_4-impregnated chips at 210°C for 2 min was the best condition for hemicellulose recovery (mostly as xylose) in the water-soluble fraction, reaching almost 70% of the corresponding xylose theoretical yield. Further, Canettieri et al. (2007) optimized acid hydrolysis from the hemicellulosic fraction of eucalyptus residues in a 1.4-l pilot-scale reactor and determined that the optimal conditions were H_2SO_4 concentration of 0.65%, temperature of 157°C, and residue/acid solution ratio of 1/8.6 with a reaction time of 20 min.

Ramos et al. (1992) compared the steam pretreatment of eucalyptus, aspen, and spruce wood chips and their enzymatic hydrolysis and observed that the various cellulosic fractions derived from steam-exploded eucalyptus chips exhibited a greater susceptibility to hydrolysis than the equivalent aspen and spruce substrates, particularly at high substrate concentrations (10%, w/v). Further, Teramoto et al. (2008) performed the pretreatment of eucalyptus wood chips for enzymatic hydrolysis using combined H_2SO_4-free ethanol solvent and ball milling and obtained the 100% conversion of the cellulosic components into glucose under optimal conditions.

Nitsos et al. (2016) optimized hydrothermal pretreatment of residual poplar branches, grapevine pruning, and pine sawdust for selective hemicellulose recovery and improved cellulose enzymatic hydrolysis and observed that the maximum hemicellulose recovery for poplar, grapevine, and pine was around 60% at ~70%–85% hemicellulose removal, based on initial hemicellulose content of each biomass type, and at relatively moderate pretreatment severities. Further, Lin et al. (2019) improved enzymatic hydrolysis of acid-pretreated bamboo residues using amphiphilic surfactant and observed that the surfactant with PEG improved enzymatic hydrolysis yield of these residues from 24.3% to 71.9% by pre-adsorbing with 0.8 g/L.

Zhang et al. (2007) pretreated bamboo residues with *Coriolus versicolor* B1 for enzymatic hydrolysis and observed that the hydrolysis rate was significantly enhanced with a maximum hydrolysis rate of 37.0% after pretreatment. Further, Lima et al. (2013) studied the effects of pretreatment on morphology, chemical composition, and enzymatic digestibility of eucalyptus bark and observed an increase in the cellulose content, reaching around 81% and 76% of glucose for HGU and EG, respectively, using a two-step treatment with HCl 1%, followed by 4% NaOH.

Shevchenko et al. (2000) optimized monosaccharide recovery by post-hydrolysis of the water-soluble hemicellulose component after steam explosion of softwood chips and observed that the SO_2-catalyzed steam explosion enhanced the enzymatic hydrolysis of the water-insoluble cellulose fraction. Further, Boussaid et al. (2000) performed the steam pretreatment of Douglas fir wood chips for efficient enzymatic hydrolysis and observed that the high-severity conditions resulted in a significant improvement in the efficiency of enzymatic digestibility of the cellulose component.

Kim et al. (2014) performed the subcritical LHW pretreatment of hardwood chips at various severities and obtained high recovery (>92%) of hemicellulose-derived pentose sugars and enhanced enzymatic hydrolysis of pretreated cellulose by applying a multi-stage pretreatment. Further, Castoldi et al. (2014) performed the biological pretreatment of eucalyptus sawdust with white-rot fungi and found that the biological pretreatments diminished the initial delay, increased the easily hydrolyzable fraction, and generated a second hydrolyzable fraction. Finally, Monavari et al. (2009) studied the effect of impregnation time and chip size on sugar yield in pretreatment of softwood for ethanol production and found that the dilute acid pretreatment together with subsequent enzymatic hydrolysis resulted in solubilization of between 69% and 73% of the fermentable sugars (glucose and mannose) in the raw material for the combinations of impregnation times and chip sizes investigated.

72.4.3 Forestry Waste-based Bioethanol Production

There are only four HCPs for the production the forestry waste-based bioethanol fuels with at least 76 citations each (Table 72.2). These papers also cover the fermentation of the hydrolysates of the forestry wastes.

Cara et al. (2008) produced ethanol from steam explosion-pretreated olive tree prunings and obtained the maximum ethanol yield (7.2 g ethanol/100 g raw material) from water-impregnated, steam-pretreated residue at 240°C. Further, Weil et al. (1997) pretreated yellow poplar sawdust by pressure cooking in water at temperatures ranging from 220°C to 260°C and observed that the SSF of washed, pretreated lignocellulose gave an ethanol yield that was 55% of theoretical.

Allen et al. (2001) compared the pretreatment of the sawdust with LHW or steam explosion with dilute acid pretreatment and obtained the ethanol yields of 97 and 92 wt% of the theoretical yield and xylan recoveries of 85 and 70 wt% (92 and 96 wt% solubilization) with the LHW and dilute acid pretreatments, respectively. Further, Zhu et al. (2015) produced ethanol from Douglas fir forest harvest residues and obtained an ethanol yield of 282 L/tonne, equivalent to 70% theoretical, with a titer of 42 g/L.

72.5 CONCLUSION AND FUTURE RESEARCH

The brief information about the key research fronts covered by the 25 most-cited papers with at least 76 citations each is given under two primary headings: the pretreatment and hydrolysis of the forestry wastes and production of the bioethanol fuels.

The usual characteristics of these HCPs are that the pretreatments and hydrolysis of the forestry wastes and fermentation of the resulting hydrolysates are the primary processes for the bioethanol fuel production from forestry wastes to improve the ethanol yield as the forestry wastes are one of the most-studied feedstocks for the bioethanol production especially for the countries with the large forests and crude oil deficiency.

The key findings on these research fronts should be read in light of the increasing public concerns about climate change, GHG emissions, and global warming as these concerns have been certainly behind the boom in the research on the forestry waste-based bioethanol fuels as an alternative to crude oil-based gasoline and petrodiesel fuels in recent decades. It is also a sustainable alternative to first generation food crop-based bioethanol fuels such as corn grain-based bioethanol fuels. The recent supply shocks caused by the COVID-19 pandemics and the Russian invasion of Ukraine also highlight the importance of the production and utilization of the bioethanol fuels as an alternative to the crude oil-based gasoline and petrodiesel fuels.

As Table 72.3 shows, the most prolific feedstocks for this field are the wood chips and forest residues and to a lesser extent sawdust, tree prunings, and bark. Similarly, as Table 72.4 shows the most prolific research fronts for this field are the pretreatment and hydrolysis of the forestry wastes and to a lesser extent the hydrolysate fermentation and bioethanol production.

These studies emphasize the importance of proper incentive structures for the efficient production of forestry waste-based bioethanol fuels in light of North's institutional framework (North, 1991). In this context, the major producers and users of bioethanol fuels such as the USA and Canada with vast forests have developed strong incentive structures for the efficient forestry waste-based bioethanol fuels.

In light of the recent supply shocks caused primarily by the COVID-19 pandemics and Russian invasion of Ukraine, it is expected that the incentive structures such as public funding would be enhanced to increase the share of bioethanol fuels from forestry wastes in the global fuel portfolio as a strong alternative to crude oil-based gasoline and petrodiesel fuels.

In this context, it is expected that the most prolific researchers, institutions, countries, funding bodies, and journals in this field would have a first-mover advantage to benefit from such potential incentives. This is especially true for the US stakeholders as the USA has become the global leader in both the production and utilization of second generation bioethanol fuels from the forestry wastes.

It is recommended that such review studies are performed for the primary research fronts of the forestry waste-based bioethanol fuels.

ACKNOWLEDGMENTS

The contribution of the highly cited researchers in the field of the forestry waste-based bioethanol fuels has been gratefully acknowledged.

REFERENCES

Allen, S. G., D. Schulman and J. Lichwa, et al. 2001. A comparison of aqueous and dilute-acid single-temperature pretreatment of yellow poplar sawdust. *Industrial and Engineering Chemistry Research* 40:2352–2361.

Alvira, P., E. Tomas-Pejo, M. Ballesteros and M. J. Negro. 2010. Pretreatment technologies for an efficient bioethanol production process based on enzymatic hydrolysis: A review. *Bioresource Technology* 101:4851–4861.

Angelici, C., B. M. Weckhuysen and P. C. A. Bruijnincx. 2013. Chemocatalytic conversion of ethanol into butadiene and other bulk chemicals. *ChemSusChem* 6:1595–1614.

Antolini, E. 2007. Catalysts for direct ethanol fuel cells. *Journal of Power Sources* 170:1–12.

Antolini, E. 2009. Palladium in fuel cell catalysis. *Energy and Environmental Science* 2:915–931.

Ayeni, A. O., F. K. Hymore and S. N. Mudliar, et al. 2013. Hydrogen peroxide and lime based oxidative pretreatment of wood waste to enhance enzymatic hydrolysis for a biorefinery: Process parameters optimization using response surface methodology. *Fuel* 106:187–194.

Boussaid, A. L., A. R. Esteghlalian, D. J. Gregg, K. H. Lee and J. N. Saddler. 2000. Steam pretreatment of Douglas-fir wood chips. Can conditions for optimum hemicellulose recovery still provide adequate access for efficient enzymatic hydrolysis? Applied Biochemistry and Biotechnology, *Part A: Enzyme Engineering and Biotechnology* 84–86:693–705.

Brandt, A., J. P. Hallett, D. J. Leak, R. J. Murphy and T. Welton. 2010. The effect of the ionic liquid anion in the pretreatment of pine wood chips. *Green Chemistry* 12:672–679.

Brownell, H. H., E. K. C. Yu and J. N. Saddler. 1986. Steam-explosion pretreatment of wood: Effect of chip size, acid, moisture content and pressure drop. *Biotechnology and Bioengineering* 28:792–801.

Burnham, J. F. 2006. Scopus database: A review. *Biomedical Digital Libraries* 3:1–8.

Canettieri, E. V., de Moraes Rocha, J. A. de Carvalho and J. B. de Almeida e Silva. 2007. Optimization of acid hydrolysis from the hemicellulosic fraction of *Eucalyptus grandis* residue using response surface methodology. *Bioresource Technology* 98:422–428.

Cara, C., E. Ruiz and M. Ballesteros, et al. 2008. Production of fuel ethanol from steam-explosion pretreated olive tree pruning. *Fuel* 87:692–700.

Cara, C., I. Romero, J. M. Oliva, F. Saez and E. Castro. 2007. Liquid hot water pretreatment of olive tree pruning residues. *Applied Biochemistry and Biotechnology* 137–140:379–394.

Castoldi, R., A. Bracht and G. R. de Morais, et al. 2014. Biological pretreatment of *Eucalyptus grandis* sawdust with white-rot fungi: Study of degradation patterns and saccharification kinetics. *Chemical Engineering Journal* 258:240–246.

Duff, S. J. B. and W. D. Murray. 1996. Bioconversion of forest products industry waste cellulosics to fuel ethanol: A review. *Bioresource Technology* 55:1–33.

Emmel, A., A. L. Mathias, F. Wypych and L. P. Ramos. 2003. Fractionation of *Eucalyptus grandis* chips by dilute acid-catalysed steam explosion. *Bioresource Technology* 86:105–115.

Fernando, S., S. Adhikari, C. Chandrapal and M. Murali. 2006. Biorefineries: Current status, challenges, and future direction. *Energy & Fuels* 20:1727–1737.

Franko, B., M. Galbe and O. Wallberg. 2016. Bioethanol production from forestry residues: A comparative techno-economic analysis. *Applied Energy* 184:727–736.

Hahn-Hagerdal, B., M. Galbe, M. F. Gorwa-Grauslund, G. Liden and G. Zacchi. 2006. Bio-ethanol: The fuel of tomorrow from the residues of today. *Trends in Biotechnology* 24:549–556.

Hill, J., E. Nelson, D. Tilman, S. Polasky and D. Tiffany. 2006. Environmental, economic, and energetic costs and benefits of biodiesel and ethanol biofuels. *Proceedings of the National Academy of Sciences of the United States of America* 103:11206–11210.

Hill, J., S. Polasky and E. Nelson, et al. 2009. Climate change and health costs of air emissions from biofuels and gasoline. *Proceedings of the National Academy of Sciences of the United States of America* 106:2077–2082.

Hsieh, W. D., R. H. Chen, T. L. Wu and T. H. Lin. 2002. Engine performance and pollutant emission of an SI engine using ethanol-gasoline blended fuels. *Atmospheric Environment* 36:403–410.

Hu, G., J. A. Heitmann and O. J. Rojas. 2008. Feedstock pretreatment strategies for producing ethanol from wood, bark, and forest residues. *BioResources* 3:270–294.

Huang, C., W. Lin and X. Lai, et al. 2019. Coupling the post-extraction process to remove residual lignin and alter the recalcitrant structures for improving the enzymatic digestibility of acid-pretreated bamboo residues. *Bioresource Technology* 285:121355.

Huang, H. J., S. Ramaswamy, U. W. Tschirner and B. V. Ramarao. 2008. A review of separation technologies in current and future biorefineries. *Separation and Purification Technology* 62:1–21.

Iranmahboob, J., F. Nadim and S. Monemi. 2002. Optimizing acid-hydrolysis: A critical step for production of ethanol from mixed wood chips. *Biomass and Bioenergy* 22:401–404.

Jonker, J. G. G., H. M. Junginger and J. A. Verstegen, et al. 2016. Supply chain optimization of sugarcane first generation and eucalyptus second generation ethanol production in Brazil. *Applied Energy* 173:494–510.

Kim, K. H., M. Tucker and Q. Nguyen. 2005. Conversion of bark-rich biomass mixture into fermentable sugar by two-stage dilute acid-catalyzed hydrolysis. *Bioresource Technology* 96:1249–1255.

Kim, Y., T. Kreke, T. N. S. Mosier and M. R. Ladisch. 2014. Severity factor coefficients for subcritical liquid hot water pretreatment of hardwood chips. *Biotechnology and Bioengineering* 111:254–263.

Konur, O. 2000. Creating enforceable civil rights for disabled students in higher education: An institutional theory perspective. *Disability & Society* 15:1041–1063.

Konur, O. 2002a. Access to nursing education by disabled students: Rights and duties of nursing programs. *Nurse Education Today* 22:364–374.

Konur, O. 2002b. Assessment of disabled students in higher education: Current public policy issues. *Assessment and Evaluation in Higher Education* 27:131–152.

Konur, O. 2002c. Access to employment by disabled people in the UK: Is the Disability Discrimination Act working? *International Journal of Discrimination and the Law* 5:247–279.

Konur, O. 2006a. Participation of children with dyslexia in compulsory education: Current public policy issues. *Dyslexia* 12:51–67.

Konur, O. 2006b. Teaching disabled students in higher education. *Teaching in Higher Education* 11:351–363.

Konur, O. 2007a. A judicial outcome analysis of the Disability Discrimination Act: A windfall for the employers? *Disability & Society* 22:187–204.

Konur, O. 2007b. Computer-assisted teaching and assessment of disabled students in higher education: The interface between academic standards and disability rights. *Journal of Computer Assisted Learning* 23:207–219.

Konur, O. 2012. The evaluation of the research on the bioethanol: A scientometric approach. *Energy Education Science and Technology Part A: Energy Science and Research* 28:1051–1064.

Konur, O. 2015. Current state of research on algal bioethanol. In *Marine Bioenergy: Trends and Developments*, Eds. S. K. Kim and C. G. Lee, pp. 217–244. Boca Raton, FL: CRC Press.

Konur, O. 2020. The scientometric analysis of the research on the bioethanol production from green macroalgae. In *Handbook of Algal Science, Technology and Medicine*, Ed. O. Konur, pp. 385–401. London: Academic Press.

Konur, O. 2023. Second generation forestry waste-based bioethanol fuels: Scientometric study. In *Feedstock-based Bioethanol Fuels. II. Waste Feedstocks: Agricultural, Food, Industrial, Urban, Forestry, and Lignocellulosic Waste-based Bioethanol Fuels. Handbook of Bioethanol Fuels Volume 4*, Ed. O. Konur, pp. 281–304. Boca Raton, FL: CRC Press.

Li, H., A. Saeed, M. S. Jahan, Y. Ni and A. van Heiningen. 2010. Hemicellulose removal from hardwood chips in the pre-hydrolysis step of the kraft-based dissolving pulp production process. *Journal of Wood Chemistry and Technology* 30:48–60.

Lima, M. A., G. B. Lavorente and H. K. P. da Silva, et al. 2013. Effects of pretreatment on morphology, chemical composition and enzymatic digestibility of eucalyptus bark: A potentially valuable source of fermentable sugars for biofuel production: Part 1. *Biotechnology for Biofuels* 6:75.

Lin, W., D. Chen, Q. Yong, C. Huang and S. Huang. 2019. Improving enzymatic hydrolysis of acid-pretreated bamboo residues using amphiphilic surfactant derived from dehydroabietic acid. *Bioresource Technology* 293:122055.

Lin, W., S. Xing and Y. Jin, et al. 2020. Insight into understanding the performance of deep eutectic solvent pretreatment on improving enzymatic digestibility of bamboo residues. *Bioresource Technology* 306:123163.

Lin, Y. and S. Tanaka. 2006. Ethanol fermentation from biomass resources: Current state and prospects. *Applied Microbiology and Biotechnology* 69:627–642.

Lu, Y., B. Yang, D. Gregg, J. N. Saddler and S. D. Mansfield. 2002. Cellulase adsorption and an evaluation of enzyme recycle during hydrolysis of steam-exploded softwood residues. *Applied Biochemistry and Biotechnology, Part A: Enzyme Engineering and Biotechnology* 98–100:641–654.

Ma, X., L. Sun and C. Song. 2002. A new approach to deep desulfurization of gasoline, diesel fuel and jet fuel by selective adsorption for ultra-clean fuels and for fuel cell applications. *Catalysis Today* 77:107–116.

Monavari, S., M. Galbe and G. Zacchi. 2009. Impact of impregnation time and chip size on sugar yield in pretreatment of softwood for ethanol production. *Bioresource Technology* 100:6312–6316.

Morschbacker, A. 2009. Bio-ethanol based ethylene. *Polymer Reviews* 49:79–84.

Najafi, G., B. Ghobadian and T. Tavakoli, et al. 2009. Performance and exhaust emissions of a gasoline engine with ethanol blended gasoline fuels using artificial neural network. *Applied Energy* 86:630–639.

Newman, P. W. G. and J. R. Kenworthy. 1989. Gasoline consumption and cities: A comparison of U.S. cities with a global survey. *Journal of the American Planning Association* 55:24–37.

Nigam, J. N. 2001. Ethanol production from hardwood spent sulfite liquor using an adapted strain of *Pichia stipites*. *Journal of Industrial Microbiology and Biotechnology* 26:145–150.

Nitsos, C. K., T. Choli-Papadopoulou, M. A. Matis and K. S. Triantafyllidis. 2016. Optimization of hydrothermal pretreatment of hardwood and softwood lignocellulosic residues for selective hemicellulose recovery and improved cellulose enzymatic hydrolysis. *ACS Sustainable Chemistry and Engineering* 4:4529–4544.

North, D. C. 1991. Institutions. *Journal of Economic Perspectives* 5:97–112.

Okuda, N., M. Soneura, K. Ninomiya, Y. Katakura and S. Shioya. 2008. Biological detoxification of waste house wood hydrolysate using *Ureibacillus thermosphaericus* for bioethanol production. *Journal of Bioscience and Bioengineering* 106:128–133.

Olsson, L. and B. Hahn-Hagerdal. 1996. Fermentation of lignocellulosic hydrolysates for ethanol production. *Enzyme and Microbial Technology* 18:312–331.

Pereira, S. R., D. J. Portugal-Nunes, D. V. Evtuguin, L. S. Serafim and A. M. R. B. Xavier. 2013. Advances in ethanol production from hardwood spent sulphite liquors. *Process Biochemistry* 48:272–282.

Ramos, L. P., C. Breuil and J. N. Saddler. 1992. Comparison of steam pretreatment of eucalyptus, aspen, and spruce wood chips and their enzymatic hydrolysis. *Applied Biochemistry and Biotechnology* 34–35:37–48.

Salehian, P. and K. Karimi. 2013. Alkali pretreatment for improvement of biogas and ethanol production from different waste parts of pine tree. *Industrial and Engineering Chemistry Research* 52:972–978.

Sanchez, O. J. and C. A. Cardona. 2008. Trends in biotechnological production of fuel ethanol from different feedstocks. *Bioresource Technology* 99:5270–5295.

Schell, D. J., R. Torget and A. Power, et al. 1991. A technical and economic analysis of acid-catalyzed steam explosion and dilute sulfuric acid pretreatments using wheat straw or aspen wood chips. *Applied Biochemistry and Biotechnology* 28–29:87–97.

Shevchenko, S. M., K. Chang, J. Robinson and J. N. Saddler. 2000. Optimization of monosaccharide recovery by post-hydrolysis of the water-soluble hemicellulose component after steam explosion of softwood chips. *Bioresource Technology* 72:207–211.

Sun, Y. and J. Cheng. 2002. Hydrolysis of lignocellulosic materials for ethanol production: A review. *Bioresource Technology* 83:1–11.

Survase, S. A., E. Sklavounos, G. Jurgens, A. van Heiningen and T. Granstrom. 2011. Continuous acetone-butanol-ethanol fermentation using SO_2-ethanol-water spent liquor from spruce. *Bioresource Technology* 102:10996–11002.

Taherzadeh, M. J. and K. Karimi. 2007. Enzyme-based hydrolysis processes for ethanol from lignocellulosic materials: A review. *Bioresources* 2:707–738.

Taherzadeh, M. J. and K. Karimi. 2008. Pretreatment of lignocellulosic wastes to improve ethanol and biogas production: A review. *International Journal of Molecular Sciences* 9:1621–1651.

Teramoto, Y., N. Tanaka, S. H. Lee and T. Endo. 2008. Pretreatment of eucalyptus wood chips for enzymatic saccharification using combined sulfuric acid-free ethanol cooking and ball milling. *Biotechnology and Bioengineering* 99:75–85.

Wang, X., H. Li, Y. Cao and Q. Tang. 2011. Cellulose extraction from wood chip in an ionic liquid 1-allyl-3-methylimidazolium chloride (AmimCl). *Bioresource Technology* 102:7959–7965.

Weil, J., A. Sarikaya and S. L. Rau, et al. 1997. Pretreatment of yellow poplar sawdust by pressure cooking in water. *Applied Biochemistry and Biotechnology, Part A: Enzyme Engineering and Biotechnology* 68:21–40.

Weil, J., M. Brewer, R. Hendrickson, A. Sarikaya and M. R. Ladisch. 1998. Continuous pH monitoring during pretreatment of yellow poplar wood sawdust pressure cooking in water. *Applied Biochemistry and Biotechnology, Part A: Enzyme Engineering and Biotechnology* 70–72:99–111.

Zhang, X., C. Xu and H. Wang. 2007. Pretreatment of bamboo residues with *Coriolus versicolor* for enzymatic hydrolysis. *Journal of Bioscience and Bioengineering* 104:149–151.

Zhu, J. Y., M. S. Chandra and F. Gu, et al. 2015. Using sulfite chemistry for robust bioconversion of Douglas-fir forest residue to bioethanol at high titer and lignosulfonate: A pilot-scale evaluation. *Bioresource Technology* 179:390–397.

Part 23

Lignocellulosic Biomass-based Bioethanol Fuels

73 Lignocellulosic Biomass-based Bioethanol Fuels

Scientometric Study

Ozcan Konur
(Formerly) Ankara Yildirim Beyazit University

73.1 INTRODUCTION

The crude oil-based gasoline fuels (Ma et al., 2002; Newman and Kenworthy, 1989) have been widely used in the transportation sector since the 1920s. However, there have been great public concerns over the adverse environmental and human impact of these fuels (Hill et al., 2006, 2009). Hence, biomass-based bioethanol fuels (Hill et al., 2006; Konur, 2012e, 2015, 2019, 2020a) have increasingly been used in blending gasoline fuels (Hsieh et al., 2002; Najafi et al., 2009), in the fuel cells (Antolini, 2007, 2009), and in the biochemical production (Angelici et al., 2013; Morschbacker, 2009) in a biorefinery context (Fernando et al., 2006; Huang et al., 2008).

Bioethanol fuels also play a critical role in maintaining the energy security (Kruyt et al., 2009; Winzer, 2012) in the supply shocks (Kilian, 2008, 2009) related to oil price shocks (Hamilton, 1983, 2003), COVID-19 pandemics (Fauci et al., 2020; Li et al., 2020), or wars (Hamilton, 1983; Jones, 2012) in the aftermath of the Russian invasion of Ukraine (Reeves, 2014).

However, it is necessary to pretreat the biomass (Taherzadeh and Karimi, 2008; Yang and Wyman, 2008) to enhance the yield of the bioethanol (Hahn-Hagerdal et al., 2006; Sanchez and Cardona, 2008) prior to the bioethanol production through the hydrolysis (Sun and Cheng, 2002; Taherzadeh and Karimi, 2007) and fermentation (Lin and Tanaka, 2006; Olsson and Hahn-Hagerdal, 1996) of the biomass.

One of the most-studied feedstocks for the bioethanol fuels has been the lignocellulosic biomass. The research in the field of the lignocellulosic biomass-based bioethanol fuels has intensified in this context in the key research fronts of the pretreatment (Hendriks and Zeeman, 2009; Mosier et al., 2005) and hydrolysis (Jorgensen et al., 2007; Sun and Cheng, 2002) of the lignocellulosic biomass, fermentation (Palmqvist and Hahn-Hagerdal, 2000a,b) of the lignocellulosic biomass-based hydrolysates, and production (Limayem and Ricke, 2012; Zaldivar et al., 2001) and evaluation (Hamelinck et al., 2005; Sassner et al., 2008) of the lignocellulosic biomass-based bioethanol fuels. Thus, it emerges as a distinctive research field, complementing primarily the research on the second generation bioethanol fuels from the agricultural residues as well as food, industrial, urban, and forestry wastes among others.

However, it is essential to develop efficient incentive structures (North, 1991) for the primary stakeholders to enhance the research in this field (Konur, 2000, 2002a–c, 2006a,b, 2007a,b). The scientometric analysis has been used in this context to inform the primary stakeholders about the current state of the research in a selected research field (Garfield, 1955; Konur, 2011, 2012a–i, 2015, 2018b, 2019, 2020a).

As there have been no published scientometric studies in this field, this book chapter presents a scientometric study of the research in the lignocellulosic biomass-based bioethanol fuels. It examines the scientometric characteristics of both the sample and population data presenting scientometric characteristics of these both datasets in the order of documents, authors, publication years, institutions, funding bodies, source titles, countries, Scopus subject categories, Scopus keywords, and research fronts.

DOI: 10.1201/9781003226550-100

73.2 MATERIALS AND METHODS

The search for this study was carried out using Scopus database (Burnham, 2006) in October 2022.

As a first step for the search of the relevant literature, the keywords were selected using the first most-cited 400 population papers. The selected keyword list was then optimized to obtain a representative sample of papers for the searched research field. This keyword list was provided in the Appendix for future replicative studies.

As a second step, two sets of data were used for this study. First, a population sample of 3,644 papers was used to examine the scientometric characteristics of the population data. Secondly, a sample of 182 most-cited papers, corresponding to 5% of the population papers, was used to examine the scientometric characteristics of these citation classics.

The scientometric characteristics of these both sample and population datasets were presented in the order of documents, authors, publication years, institutions, funding bodies, source titles, countries, Scopus subject categories, Scopus keywords, and research fronts.

Lastly, the key scientometric findings for both datasets were discussed to highlight the research landscape for the lignocellulosic biomass-based bioethanol fuels. Additionally, a number of brief conclusions were drawn and a number of relevant recommendations were made to enhance the future research landscape.

73.3 RESULTS

73.3.1 The Most Prolific Documents in the Lignocellulosic Biomass-based Bioethanol Fuels

The information on the types of documents for both datasets is given in Table 73.1. The articles and conference papers, published in journals, dominate both the sample (51%) and population (79%) papers with 28% deficit. Further, review papers and short surveys have a 35% surplus as they are over-represented in the sample papers as they constitute 49% and 14% of the sample and population papers, respectively.

It is further notable that 93%, 5%, and 2% of the population papers were published in journals, books, and book series, respectively. Similarly, 98% and 2% of the sample papers were published in the journals and book series, respectively.

73.3.2 The Most Prolific Authors in the Lignocellulosic Biomass-based Bioethanol Fuels

The information about the most prolific 35 authors with at least 1.6% of sample papers each is given in Table 73.2. The most prolific author is Bruce E. Dale with 4.4% of the sample papers, followed by John N. Saddler and Leif J. Jonsson with 3.8% and 3.3% of the sample papers, respectively. The other prolific authors are Arthur J. Ragauskas, Venkatesh Balan, Barbel Hahn-Hagerdal, and Shishir P. S. Chundawat with 2.7% of the sample papers each.

On the other hand, the most influential author is Bruce E. Dale with 3.7% of the sample papers, followed by John N. Saddler and Leif J. Jonsson with 3.0% and 2.8% of the sample papers, respectively. The other influential authors Barbel Hahn-Hagerdal, Shishir P. S. Chundawat, Venkatesh Balan, Arthur J. Ragauskas, and Nils-Olof Nilvebrant with 2.0%–2.3% surplus each.

The most prolific institution for the sample dataset is the Lund University with four authors, followed by Imperial College and the Lawrence Berkeley National Laboratory with three authors each. The other prolific institutions are National Renewable Energy Laboratory (NREL), Purdue University, and Tsinghua University with two authors each. On the other hand, the most prolific country for the sample dataset is the USA with 17 authors, followed by Sweden and the UK with seven and three authors, respectively. The other prolific countries are China and Denmark with two authors each. In total, only nine countries house these top authors.

TABLE 73.1
Documents in the Lignocellulosic Biomass-Based Bioethanol Fuels

Documents	Sample Dataset (%)	Population Dataset (%)	Surplus (%)
Article	48.4	76.7	–28.3
Review	46.2	13.2	33.0
Conference paper	2.7	2.4	0.3
Short survey	2.7	0.3	2.4
Book chapter	0.0	6.2	–6.2
Editorial	0.0	0.5	–0.5
Letter	0.0	0.3	–0.3
Book	0.0	0.2	–0.2
Note	0.0	0.2	–0.2
Sample size	182	3,644	

Sample dataset: The number of papers (%) in the set of 182 highly cited papers. Population dataset: The number of papers (%) in the set of the 3,644 population papers.

The most prolific research front for these top authors is the pretreatments of the lignocellulosic biomass with 30 authors followed by the hydrolysis of the lignocellulosic biomass with 18 authors. The other research fronts are the fermentation of the lignocellulosic biomass-based hydrolysates and the bioethanol production with 11 authors each.

On the other hand, there is significant gender deficit (Beaudry and Lariviere, 2016) for the sample dataset as surprisingly only four of these top researchers are female with a representation rate of 11%.

Additionally, there are other authors with the relatively low citation impact and with 0.3%–0.6% of the population papers each: Je Bao, Jinghuang Hu, Liangcai Peng, Mercedes Ballesteros, Hongming Lou, Ashok Pandey, Xueqing Qiu, Yinbo Qu, Vincenza Faraco, Kyoung Heon Kim, Yanting Wang, Verawat Champreda, Xianzhi Meng, S. Singh, Anuj K. Chandel, Pletschke, Muhammad Bilal, Cheng Cai, Shulin Chen, Lucilia Domingues, Hasunuma, Tomohisa, Yuxia Pang, Ben-Guang Rong, Yogendra Shashri, Reeta R. Singhania, Yi-Heng P. Zhang, Junyong Zhu, Shengdong Zhu, Silvio S. da Silva, Parameswaran Binod, B. S. Chadha, Claus Felby, Krist V. Gernaev, Johann F. Gorgens, Rajeev Kumar, Shao-Yuan Leu, Guodong Liu, Jeremy S. Luterbacher, Yunqiao Pu, Wei Qi, Noppadon Sathitsuksanoh, and Shen, Fei, Yuanyuan Tu.

73.3.3 The Most Prolific Research Output by Years in the Lignocellulosic Biomass-based Bioethanol Fuels

Information about papers published between 1970 and 2022 is given in Figure 73.1. This figure clearly shows that the bulk of the research papers in the population dataset were published primarily in the 2010s and the early 2020s with 60% and 24% of the population dataset, respectively. Similarly, the publication rates for the 2000s, 1990s, 1980s, and 1970s were 8%, 4%, 3%, and 0% respectively.

Similarly, the bulk of the research papers in the sample dataset were published in the 2010s and 2000s with 59% and 32% of the sample dataset, respectively. Similarly, the publication rates for the early 2020s, 1990s, 1980s, and 1970s 2%, 4%, 2%, and 1% of the sample papers, respectively.

The most prolific publication years for the population dataset were 2021, 2020, and 2017 with 8.3%, 7.9%, and 7.9% of the dataset, respectively, while 86% of the population papers were published between 2009 and 2022. Similarly, 81% of the sample papers were published between 2007 and 2020 while the most prolific publication years were 2011, 2010, 2009, and 2016 with 11.5%, 9.9%, 8.2%, and 8.2% of the sample papers, respectively.

TABLE 73.2
Most Prolific Authors in the Lignocellulosic Biomass-Based Bioethanol Fuels

No.	Author Name	Author Code	Sample Papers (%)	Population Papers (%)	Surplus	Institution	Country	HI	N	Res. Front
1	Dale, Bruce E.	7201511969	4.4	0.7	3.7	Michigan State Univ.	USA	92	430	P, H, F, R
2	Saddler, John N.	7005297559	3.8	0.8	3.0	Univ. British Columbia	Canada	99	420	P, H, F, R
3	Jonsson, Leif J.	7102349315	3.3	0.5	2.8	Umea Univ.	Sweden	41	148	F
4	Ragauskas, Arthur J.	7006265204	2.7	0.7	2.0	Univ. Tennessee	USA	93	762	P, H
5	Balan, Venkatesh	15757087100	2.7	0.6	2.1	Univ. Houston	USA	56	213	P, H
6	Hahn-Hagerdal, Barbel*	7005389381	2.7	0.4	2.3	Lund Univ.	Sweden	76	258	F, R
7	Chundawat, Shishir P.S.	12803763300	2.7	0.4	2.3	Rutgers	USA	32	90	P, H
8	Taherzadeh, Mohammad J.	6701407496	2.2	0.9	1.3	Univ. Boras	Sweden	66	419	P, H, F, R
9	Wyman, Charles E.	7004396809	2.2	0.5	1.7	Univ. Calf. Riverside	USA	80	287	P, H, F, R
10	Himmel, Michael E.	7007125552	2.2	0.4	1.8	Natl. Renew. Ener. Lab.	USA	73	423	P, H
11	Nilvebrant, Nils-Olof	57209815309	2.2	0.2	2.0	Borregaard	Norway	22	43	F
12	Olsson, Lisbeth*	7203077540	1.6	0.7	0.9	Chalmers Univ. Technol.	Sweden	60	243	F, R
13	Simmons, Blake A.	7102183263	1.6	0.6	1.0	Lawrence Berkeley Natl. Lab.	USA	76	444	P
14	Karimi, Keikhosro	10046195700	1.6	0.5	1.1	Vrije Univ.	Netherlands	56	222	P, H, F, R
15	Kondo, Akihiko	57203868143	1.6	0.5	1.1	Kobe Univ.	Japan	79	801	P, H, F, R
16	Zhao, Xuebing	8961267200	1.6	0.4	1.2	Tsinghua Univ.	China	40	119	P, H
17	Galbe, Mats	7003788758	1.6	0.4	1.2	Lund Univ.	Sweden	51	131	P, R
18	Liden, Gunnar	7004458708	1.6	0.3	1.3	Lund Univ.	Sweden	48	144	P
19	Liu, Dehua	35233867100	1.6	0.3	1.3	Tsinghua Univ.	China	56	310	P, H
20	Jorgensen, Henning	7202554496	1.6	0.3	1.3	Univ. Copenhagen	Denmark	43	96	P, H

(Continued)

TABLE 73.2 (*Continued*)
Most Prolific Authors in the Lignocellulosic Biomass-Based Bioethanol Fuels

No.	Author Name	Author Code	Sample Papers (%)	Population Papers (%)	Surplus	Institution	Country	HI	N	Res. Front
21	Meyer, Anne S.*	57210905309	1.6	0.3	1.3	Tech. Univ. Denmark	Denmark	76	400	P, H, R
22	Zacchi, Guido	7006727748	1.6	0.3	1.3	Lund Univ.	Sweden	68	204	P, H, F, R
23	Mosier, Nathan S.	6602426392	1.6	0.3	1.3	Purdue Univ.	USA	43	117	P
24	Ladisch, Michael R.	7005670397	1.6	0.2	1.4	Purdue Univ.	USA	75	334	P
25	Lee, Yoon Y	8948274900	1.6	0.2	1.4	Auburn Univ.	USA	45	102	P, H
26	Beckham, Gregg T.	16240926200	1.6	0.2	1.4	Natl. Renew. Ener. Lab.	USA	71	266	P
27	Berlin, Alex	8639650700	1.6	0.2	1.4	Novozymes Biotech PLC	USA	19	34	H
28	Brandt, Agnieszka*	35785816800	1.6	0.2	1.4	Imperial Coll.	UK	23	41	P
29	Chandra, Richard P.	7401681548	1.6	0.2	1.4	Trinity W Univ.	USA	31	77	P, H
30	Hallett, Jason P.	7102331746	1.6	0.2	1.4	Imperial Coll.	UK	47	181	P
31	Holtzapple, Mark T.	7004167004	1.6	0.2	1.4	Texas A&M Univ.	USA	47	199	P
32	Klein-Marcuschamer, Daniel	22940762900	1.6	0.2	1.4	Lawrence Berkeley Natl. Lab.	USA	19	23	P
33	Rogers, Robin D.	35474829200	1.6	0.2	1.4	Univ. Alabama	USA	118	891	P
34	Sun, Ning	14038069100	1.6	0.2	1.4	Lawrence Berkeley Natl. Lab.	USA	21	28	P
35	Welton, Tom	7003503272	1.6	0.2	1.4	Imperial Coll.	UK	74	191	P

Author code: the unique code given by Scopus to the authors. Sample papers: the number of papers authored in the sample dataset. Population papers: the number of papers authored in the population dataset.

*, Female; P, Pretreatment of the lignocellulosic biomass; H, Hydrolysis of the lignocellulosic biomass; F, Fermentation of the lignocellulosic biomass-based hydrolysates; R, Bioethanol fuel production; E, Bioethanol fuel evaluation.

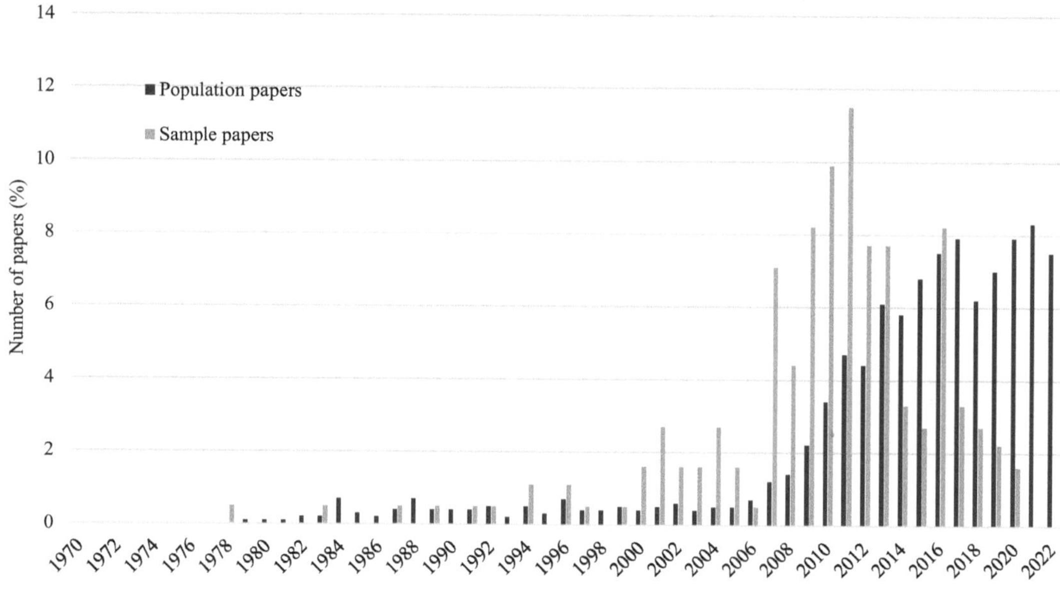

FIGURE 73.1 The research output by years regarding the lignocellulosic biomass-based bioethanol fuels.

The number of publications for the population papers increased between 2009 and 2017 and thereafter it steadied around 8% of the population papers for each year. However, there was a sharp rise in the research output for the population papers in 2020 and 2021 possibly due to the supply shocks.

73.3.4 The Most Prolific Institutions in the Lignocellulosic Biomass-based Bioethanol Fuels

Information about the most prolific 29 institutions publishing papers on the lignocellulosic biomass-based bioethanol fuels with at least 1.6% of the sample papers each is given in Table 73.3.

The most prolific institution is the Lund University with 8.2% of the sample papers, followed by the NREL, Michigan State University, and University of British Columbia with 7.1%, 4.4%, and 3.8% of the sample papers, respectively. The other prolific institutions are Chinese Academy of Sciences, Technical University of Denmark, Oak Ridge National Laboratory, and USDA Forest Service with 3.3% of the sample papers each. Similarly, the top country for these most prolific institutions is the USA with 13 institutions. The other prolific institutions are Sweden, China, and Denmark with four, three, and two institutions, respectively. In total, ten countries house these top institutions.

On the other hand, the institutions with the most citation impact are the Lund University and NREL with 6.5% surplus each. The other influential institutions are University of British Columbia, USDA Forest Service, Imperial College, Oak Ridge National Laboratory, and Technical University of Denmark with 2.0%–2.6% surplus each.

Additionally, there are other institutions with the relatively low citation impact and with 0.5%–1.5% of the population papers each: University of Sao Paulo, State Key Laboratory of Pulp and Paper Engineering, Nanjing Forestry University, Chalmers University of Technology, Korea University, NC State University, Beijing Forestry University, East China University of Science and Technology, State University of Campinas, Shandong University, National Scientific Research Center (CNRS), Tianjin University, University of Georgia, State Key Laboratory of Bioreactor Engineering, Ministry of Agriculture of China, Kyoto University, China Agricultural University, Rhenish-Westphalian Technical University, Stellenbosch University, USDA Agricultural Research Service, National Institute for Interdisciplinary Science and Technology, Thailand National Center for Genetic Engineering and Biotechnology, Sandia National Laboratories, Beijing University of Chemical Technology, Huazhong Agricultural University, Tianjin University of Science &

TABLE 73.3
The Most Prolific Institutions in the Lignocellulosic Biomass-Based Bioethanol Fuels

No.	Institutions	Country	Sample Papers (%)	Population Papers (%)	Surplus (%)
1	Lund Univ.	Sweden	8.2	1.7	6.5
2	Natl. Renew. Ener. Lab.	USA	7.1	0.6	6.5
3	Michigan State Univ.	USA	4.4	1.2	3.2
4	Univ. British Columbia	Canada	3.8	1.2	2.6
5	Chinese Acad. Sci.	China	3.3	2.4	0.9
6	Tech. Univ. Denmark	Denmark	3.3	1.3	2.0
7	Oak Ridge Natl. lab.	USA	3.3	1.2	2.1
8	USDA Forest Serv.	USA	3.3	0.9	2.4
9	Univ. Wisconsin Madison	USA	2.7	1.1	1.6
10	Imperial Coll.	UK	2.7	0.4	2.3
11	Lawrence Livermore Natl. Lab.	USA	2.2	0.9	1.3
12	Purdue Univ.	USA	2.2	0.9	1.3
14	Univ. Boras	Sweden	2.2	0.9	1.3
15	Umea Univ.	Sweden	2.2	0.6	1.6
16	LU Leuven	Belgium	2.2	0.5	1.7
17	Lund Univ. Technol.	Sweden	2.2	0.3	1.9
18	S. China Univ. Technol.	China	1.6	1.5	0.1
19	Univ. Tennessee	USA	1.6	0.9	0.7
20	Univ. Copenhagen	Denmark	1.6	0.9	0.7
21	Tsinghua Univ.	China	1.6	0.8	0.8
22	Univ. Helsinki	Finland	1.6	0.7	0.9
23	Kobe Univ.	Japan	1.6	0.6	1.0
24	Isfahan Univ. Technol.	Iran	1.6	0.5	1.1
25	Auburn Univ.	USA	1.6	0.5	1.1
26	Univ. Calif. Davis	USA	1.6	0.4	1.2
27	Virginia Polytech. Inst. State Univ.	USA	1.6	0.4	1.2
28	Georgia Inst. Technol.	USA	1.6	0.4	1.2
29	Univ. Alabama	USA	1.6	0.3	1.3

Technology, Zhejiang University, Qilu University of Technology, University of California Berkeley, and University of California Riverside.

73.3.5 The Most Prolific Funding Bodies in the Lignocellulosic Biomass-based Bioethanol Fuels

Information about the most prolific 18 funding bodies funding at least 1.6% of the sample papers each is given in Table 73.4. Further, only 38% and 43% of the sample and population papers each were funded, respectively

The most prolific funding body is the U.S. Department of Energy (DOE) with 6% of the sample papers, followed by the European Commission and National Natural Science Foundation of China with 4.4% and 3.3% of the sample papers, respectively. The other funding bodies are the Office of Science and National Science Foundation with 2.7% of the sample papers each. It is also notable that the National Natural Science Foundation of China is also the largest funder of the population papers with 11% funding rate.

On the other hand, the most prolific countries for these top funding bodies are China, Japan, Sweden, and the USA with three funding bodies each. The other prolific countries are India and the EU with two funding bodies each. In total, only seven countries and the EU house these top funding bodies.

TABLE 73.4
The Most Prolific Funding Bodies in the Lignocellulosic Biomass-Based Bioethanol Fuels

No.	Funding Bodies	Country	Sample Paper No. (%)	Population Paper No. (%)	Surplus (%)
1	U.S. Department of Energy	USA	6.0	3.3	2.7
2	European Commission	EU	4.4	1.6	2.8
3	National Natural Science Foundation of China	China	3.3	11.0	−7.7
4	Office of Science	USA	2.7	1.8	0.9
5	National Science Foundation	USA	2.7	1.6	1.1
6	Ministry of Science and Technology of China	China	1.6	2.1	−0.5
7	Ministry of Education of China	China	1.6	1.3	0.3
8	Japan Society for the Promotion of Science	Japan	1.6	1.1	0.5
9	Ministry of Education, Culture, Sports, Science and Technology	Japan	1.6	1.0	0.6
10	Natural Resources Canada	Canada	1.6	1.0	0.6
11	Seventh Framework Program	EU	1.6	0.9	0.7
12	University Grants Commission	India	1.6	0.6	1.0
13	Engineering and Physical Sciences Research Council	UK	1.6	0.5	1.1
14	New Energy and Industrial Technology Development Organization	Japan	1.6	0.5	1.1
15	Government of West Bengal	India	1.6	0.4	1.2
16	Swedish Research Council	Sweden	1.6	0.4	1.2
17	Knut and Alice Wallenberg Foundation	Sweden	1.6	0.3	1.3
18	Swedish National Board for Industrial and Technical Development	Sweden	1.6	0.3	1.3

The funding bodies with the most citation impact are the European Commission and the US DOE with 2.8% and 2.7% surplus, respectively. The other influential funding bodies are the Knut and Alice Wallenberg Foundation, Swedish National Board for Industrial and Technical Development, Government of West Bengal, and Swedish Research Council with 1.1%–1.3% surplus each. Further, the funding body with the least citation impact is the National Natural Science Foundation of China with 7.7% deficit.

The other funding bodies with the relatively low citation impact and with 0.5%–2.4% of the population papers each are the National Council for Scientific and Technological Development, National Key Research and Development Program of China, Higher Education Personnel Improvement Coordination, Fundamental Research Funds for the Central Universities, National Research Foundation of Korea, Research Support Foundation of the State of Sao Paulo, European Regional Development Fund, China Postdoctoral Science Foundation, Ministry of Science and Technology India, Energy Agency Sweden, Biotechnology and Biological Sciences Research Council, National Council of Science and Technology, Ministry of Science, Technology and Innovation, Horizon 2020 Framework Program, Ministry of Finance, Natural Science Foundation of Jiangsu Province, Chinese Academy of Sciences, National Basic Research Program of China (973 Program), Ministry of Science, ICT and Future Planning, National Institute of Food and Agriculture, Ministry of Education, Science and Technology, National High-tech Research and Development Program,

National Research Foundation, Priority Academic Program Development of Jiangsu Higher Education Institutions, China Scholarship Council, and Government of Canada.

73.3.6 The Most Prolific Source Titles in the Lignocellulosic Biomass-based Bioethanol Fuels

Information about the most prolific 15 source titles publishing at least 1.6% of the sample papers each in the lignocellulosic biomass-based bioethanol fuels is given in Table 73.5.

The most prolific source title is the Bioresource Technology with 20% of the sample papers, followed by Green Chemistry with 8% of the sample papers. The other prolific titles are the Biotechnology for Biofuels, Biotechnology and Bioengineering, and Biotechnology Advances with 4.4% of the sample papers each.

On the other hand, the source title with the most citation impact is the Bioresource Technology with 9% surplus, followed by Green Chemistry, Biotechnology Advances, and Renewable and Sustainable Energy Reviews with 6%, 4%, and 3% surplus, respectively. The other influential titles are the Biotechnology and Bioengineering, Enzyme and Microbial Technology, and Current Opinion in Biotechnology with 2% surplus each.

The other source titles with the relatively low citation impact with 0.5%–1.8% of the population papers each are Applied Biochemistry and Biotechnology, Industrial Crops and Products, Biomass Conversion and Biorefinery, Renewable Energy, Cellulose, ACS Sustainable Chemistry and Engineering, Process Biochemistry, Industrial and Engineering Chemistry Research, Biochemical Engineering Journal, Bioenergy Research, Applied and Environmental Microbiology, Biotechnology Letters, Chemical Engineering Transactions, RSC Advances, Waste and Biomass Valorization, Journal of Cleaner Production, Energies, Biomass Fractionation Technologies for a Lignocellulosic Feedstock based Biorefinery, Plos One, Chemical Engineering Journal, International Biodeterioration and Biodegradation, 3 Biotech, Fuel, Scientific Reports, and Journal of Chemical Technology and Biotechnology.

TABLE 73.5
The Most Prolific Source Titles in the Lignocellulosic Biomass-based Bioethanol Fuels

No.	Source Titles	Sample Papers (%)	Population Papers (%)	Surplus (%)
1	Bioresource Technology	20.3	11.7	8.6
2	Green Chemistry	7.7	1.5	6.2
3	Biotechnology for Biofuels	4.4	4.1	0.3
4	Biotechnology and Bioengineering	4.4	1.9	2.5
5	Biotechnology Advances	4.4	0.5	3.9
6	Renewable and Sustainable Energy Reviews	3.8	0.9	2.9
7	Applied Microbiology and Biotechnology	3.3	1.9	1.4
8	Biomass and Bioenergy	3.3	1.8	1.5
9	Enzyme and Microbial Technology	3.3	1.2	2.1
10	Biofuels Bioproducts and Biorefining	2.7	0.9	1.8
11	Applied Biochemistry and Biotechnology Part A	2.2	0.9	1.3
12	Current Opinion in Biotechnology	2.2	0.2	2.0
13	Bioresources	1.6	2.2	–0.6
14	ChemSusChem	1.6	0.7	0.9
15	Advances in Biochemical Engineering Biotechnology	1.6	0.2	1.4

73.3.7 The Most Prolific Countries in the Lignocellulosic Biomass-based Bioethanol Fuels

Information about the most prolific 17 countries publishing at least 2.2% of sample papers each in the lignocellulosic biomass-based bioethanol fuels is given in Table 73.6.

The most prolific country is the USA with 32% of the sample papers, followed by Sweden and China with 15% and 12% of the sample papers, respectively. The other prolific countries are India, Denmark, Canada, and the UK with 5%–9% of the sample papers each. It is also notable that China is the largest producer of the population papers with a 21.3% publication rate. Further, seven European countries listed in Table 73.6 produce 39% and 20% of the sample and population papers, respectively, with a 19% surplus.

On the other hand, the country with the most citation impact is the USA with 14% surplus, followed by Sweden and Denmark with 10% and 5% surplus, respectively. The other influential countries are Belgium, Canada, and the UK with 1.4%–1.8% surplus each. Similarly, the country with the least citation impact is China with 10% deficit, followed by India and Brazil with 3% deficit each.

Additionally, there are other countries with relatively low citation impact and with 0.5%–3.3% of the sample papers each: Spain, Germany, Italy, Australia, Mexico, Thailand, Pakistan, Taiwan, Indonesia, Turkey, Portugal, Austria, Nigeria, Colombia, Greece, Russia, Switzerland, Poland, Argentina, Chile, Czech Republic, and Egypt.

73.3.8 The Most Prolific Scopus Subject Categories in the Lignocellulosic Biomass-based Bioethanol Fuels

Information about the most prolific nine Scopus subject categories indexing at least 4% of the sample papers each is given in Table 73.7.

The most prolific Scopus subject category in the lignocellulosic biomass-based bioethanol fuels is the Chemical Engineering with 57% of the sample papers, followed by Environmental Science,

TABLE 73.6
The Most Prolific Countries in the Lignocellulosic Biomass-based Bioethanol Fuels

No.	Countries	Sample Papers (%)	Population Papers (%)	Surplus (%)
1	USA	32.4	18.9	13.5
2	Sweden	15.4	5.2	10.2
3	China	11.5	21.3	−9.8
4	India	8.8	11.8	−3.0
5	Denmark	7.7	2.7	5.0
6	Canada	6.0	4.3	1.7
7	UK	4.9	3.5	1.4
8	Japan	4.4	3.9	0.5
9	S. Korea	3.8	3.9	−0.1
10	Belgium	3.3	1.5	1.8
11	France	2.7	2.9	−0.2
12	Netherlands	2.7	1.9	0.8
13	Brazil	2.2	5.1	−2.9
14	Malaysia	2.2	2.5	−0.3
15	S. Africa	2.2	1.9	0.3
16	Finland	2.2	1.9	0.3
17	Iran	2.2	1.2	1.0

TABLE 73.7
The Most Prolific Scopus Subject Categories in the Lignocellulosic Biomass-based Bioethanol Fuels

No.	Scopus Subject Categories	Sample Papers (%)	Population Papers (%)	Surplus (%)
1	Chemical Engineering	57.1	45.9	11.2
2	Environmental Science	48.4	38.0	10.4
3	Energy	43.4	36.2	7.2
4	Biochemistry, Genetics and Molecular Biology	41.2	35.4	5.8
5	Immunology and Microbiology	31.3	27.8	3.5
6	Agricultural and Biological Sciences	9.3	13.4	−4.1
7	Chemistry	8.8	13.3	−4.5
8	Engineering	8.2	11.3	−3.1
9	Materials Science	3.8	5.8	−2.0

Energy, Biochemistry, Genetics and Molecular Biology, and Immunology and Microbiology with 48%, 43%, 41%, and 31% of the sample papers, respectively. It is notable that Social Sciences including Economics and Business account for 1% and 2% of the sample and population studies, respectively.

On the other hand, the Scopus subject categories with the most citation impact are the Chemical Engineering and Environmental Science with 11% and 10% surplus, respectively. The other influential subject areas are Energy and Biochemistry, Genetics and Molecular Biology with 7% and 6% surplus, respectively. Similarly, the least influential subject categories are Chemistry, Agricultural and Biological Sciences, and Engineering with 5%, 4%, and 3% deficit, respectively.

73.3.9 The Most Prolific Keywords in the Lignocellulosic Biomass-based Bioethanol Fuels

Information about the Scopus keywords used with at least 6.6% or 4.1% of the sample or population papers, respectively, is given in Table 73.8. For this purpose, keywords related to the keyword set given in the appendix are selected from a list of the most prolific keyword set provided by Scopus database.

These keywords are grouped under the five headings: Lignocellulosic biomass, pretreatments, fermentation, hydrolysis and hydrolysates, and products.

The most prolific keyword related to the biomass and biomass constituents is lignin with 82% of the sample papers, followed by cellulose, lignocellulose, and biomass with 56%–68% of the sample papers. The other prolific keywords are lignocellulosic biomass, hemicellulose, and carbohydrate with 14%–35% of the sample papers each.

Further, the prolific keyword related to the pretreatments is pre-treatment with 31% of the sample papers. The other prolific keywords are the enzymes, pretreatments, and degradation with 16%–24% of the sample papers each.

The most prolific keyword related to the fermentation is fermentation with 37% of the sample papers. The other prolific keywords are fungi and saccharomyces with 16% and 12% of the sample papers, respectively.

Further, the most prolific keyword related to the hydrolysis and hydrolysates is hydrolysis with 57% of the sample papers. The other prolific keywords are enzyme activity, enzymatic hydrolysis, cellulases, and sugars with 26%–29% of the sample papers each.

Finally, the most prolific keyword related to the products is biofuels with 41% of the sample papers, followed by ethanol with 39% of the sample papers. The other prolific keywords are bioethanol, biorefinery, and ethanol production with 12%–17% of the sample papers each.

TABLE 73.8
The Most Prolific Keywords in the Lignocellulosic Biomass-based Bioethanol Fuels

No.	Keywords	Sample Papers (%)	Population Papers (%)	Surplus (%)
1	**Lignocellulosic Biomass**			
	Lignin	80.2	57.6	22.6
	Cellulose	68.7	48.3	20.4
	Lignocellulose	65.9	48.3	17.6
	Biomass	56.0	41.9	14.1
	Lignocellulosic biomass	35.2	30.3	4.9
	Hemicellulose	21.4	8.4	13.0
	Carbohydrate	14.3	8.0	6.3
	Zea	12.1	6.4	5.7
	Lignocellulosic material	9.3	5.9	3.4
	wood	7.7	5.4	2.3
	Straw	3.8	6.0	–2.2
	Triticum	3.8	4.5	–0.7
	Bagasse		4.5	–4.5
2	**Pretreatments**			
	Pre-treatment	30.2	13.7	16.5
	Enzymes	24.2	20.9	3.3
	Pretreatment	20.9	12.1	8.8
	Degradation	15.9	13.9	2.0
	Ionic liquids	12.6	8.9	3.7
	Delignification	8.2	5.4	2.8
	Temperature	7.7	5.4	2.3
	Ammonia	7.1		7.1
	Pretreatment technology	7.1		7.1
	Enzyme inhibition	6.6		6.6
	pH		6.3	–6.3
3	**Fermentation**			
	Fermentation	37.4	25.9	11.5
	Fungi	15.9	16.3	–0.4
	Saccharomyces	12.1	6.8	5.3
	Detoxification	8.2	4.0	4.2
	Yeast	7.7	7.4	0.3
	Furfural	5.5	5.2	0.3
	Acetic acids	4.4	4.4	0.0
4	**Hydrolysis**			
	Hydrolysis	52.7	30.4	22.3
	Enzyme activity	29.1	22.5	6.6
	Enzymatic hydrolysis	29.1	19.1	10.0
	Cellulases	26.9	19.9	7.0
	Sugar	26.4	16.3	10.1
	Saccharification	14.8	15.9	–1.1

(Continued)

TABLE 73.8 (*Continued*)
The Most Prolific Keywords in the Lignocellulosic Biomass-based Bioethanol Fuels

No.	Keywords	Sample Papers (%)	Population Papers (%)	Surplus (%)
	Glucose	13.2	13.8	–0.6
	Xylose	7.1	6.8	0.3
	Enzymolysis	4.9	4.1	0.8
5	**Products**			
	Biofuels	41.2	24.6	16.6
	Ethanol	38.5	25.7	12.8
	Bioethanol	16.5	15.5	1.0
	Biorefinery	12.1	9.5	2.6
	Ethanol production	11.5	4.1	7.4
	Biofuel production	10.4	3.9	6.5
	Cellulosic ethanol	9.3	7.7	1.6
	Bio-ethanol production	4.9	5.6	–0.7

On the other hand, the most prolific keywords across all of the research fronts are lignin, cellulose, lignocellulose, biomass, and hydrolysis with 53%–80% of the sample papers each. The other prolific keywords are biofuels, ethanol, fermentation, lignocellulosic biomass, and pre-treatment with 30%–41% of the sample papers each.

Similarly, the most influential keywords are lignin, hydrolysis, cellulose, lignocellulose, biofuels, pre-treatment, biomass, hemicellulose, ethanol, fermentation, sugar, enzymatic hydrolysis with 10%–23% surplus each.

73.3.10 The Most Prolific Research Fronts in the Lignocellulosic Biomass-based Bioethanol Fuels

Information about the thematic research fronts for the sample papers in the lignocellulosic biomass-based bioethanol fuels is given in Table 73.9. As this table shows, the most prolific research front is the pretreatments of the lignocellulosic feedstocks with 70% of the sample papers. The other prolific research fronts are the hydrolysis of the lignocellulosic feedstocks, hydrolysate fermentation, and production and evaluation of the bioethanol fuels with 34%, 19%, 15%, and 7% of the sample papers, respectively.

TABLE 73.9
The Most Prolific Thematic Research Fronts for the Lignocellulosic Biomass-based Bioethanol Fuels

No.	Research Fronts	N Paper % Sample
1	Biomass pretreatments	70.3
2	Biomass hydrolysis	33.5
3	Hydrolysate fermentation	19.2
4	Bioethanol production	15.4
5	Bioethanol fuel evaluation	7.1

N paper (%) sample: The number of papers in the population sample of 182 papers.

73.4 DISCUSSION

73.4.1 Introduction

The crude oil-based gasoline fuels have been widely used in the transportation sector since the 1920s. However, there have been great public concerns over the adverse environmental and human impact of these fuels. Hence, biomass-based bioethanol fuels have increasingly been used in blending gasoline fuels, in the fuel cells, and in the biochemical production in a biorefinery context.

However, it is necessary to pretreat the biomass to enhance the yield of the bioethanol prior to the bioethanol production through the hydrolysis and fermentation. One of the most-studied feedstocks for the bioethanol fuels has been the lignocellulosic biomass at large. The research in the field of the lignocellulosic biomass-based bioethanol fuels has intensified in this context in the key research fronts of the pretreatment and hydrolysis of the lignocellulosic biomass, fermentation of the lignocellulosic biomass-based hydrolysates, and production and evaluation of the lignocellulosic biomass-based bioethanol fuels. Thus, it emerges as a distinctive research field, complementing the research on the second generation bioethanol fuels from the agricultural residues, food, industrial, urban, and forestry wastes among others.

However, it is essential to develop efficient incentive structures for the primary stakeholders to enhance the research in this field. This is especially important to maintain energy security in the cases of supply shocks such as oil price shocks, war-related chocks as in the case of Russian invasion of Ukraine, or COVID-19 shocks.

The scientometric analysis has been used in this context to inform the primary stakeholders about the current state of the research in a selected research field. As there has been no scientometric study in this field, this book chapter presents a scientometric study of the research in the lignocellulosic biomass-based bioethanol fuels. It examines the scientometric characteristics of both the sample and population data presenting scientometric characteristics of these both datasets in the order of documents, authors, publication years, institutions, funding bodies, source titles, countries, Scopus subject categories, Scopus keywords, and research fronts.

As a first step for the search of the relevant literature, the keywords were selected using the first most-cited 400 papers, The selected keyword list was then optimized to obtain a representative sample of papers for the searched research field, A copy of this extended keyword list was provided in the appendix for future replicative studies. Further, a selected list of the keywords was presented in Table 73.8.

As a second step, two sets of data were used for this study. First, a population sample of 3,644 papers was used to examine the scientometric characteristics of the population data. Secondly, a sample of 182 most-cited papers, corresponding to 5% of the population dataset, was used to examine the scientometric characteristics of these citation classics.

The scientometric characteristics of these sample and population datasets were presented in the order of documents, authors, publication years, institutions, funding bodies, source titles, countries, Scopus subject categories, Scopus keywords, and research fronts.

Lastly, the key scientometric findings for both datasets were discussed to highlight the research landscape for lignocellulosic biomass-based bioethanol fuels. Additionally, a number of brief conclusions were drawn and a number of relevant recommendations were made to enhance the future research landscape.

73.4.2 The Most Prolific Documents in the Lignocellulosic Biomass-based Bioethanol Fuels

Articles (together with conference papers) dominate both the sample (51%) and population (79%) papers with 28% deficit (Table 73.1). Further, review papers have a surplus (35%) and the representation of the reviews in the sample papers is quite extraordinary (49%).

Scopus differs from the Web of Science database in differentiating and showing articles (48%) and conference papers (3%) published in the journals separately. However, it should be noted that these conference papers are also published in journals as articles, compared to those published only in the conference proceedings. Hence, the total number of articles and review papers in the sample dataset is 51% and 49%, respectively.

It is observed during the search process that there has been inconsistency in the classification of the documents in Scopus as well as in other databases such as Web of Science. This is especially relevant for the classification of papers as reviews or articles as the papers not involving a literature review may be erroneously classified as a review paper. There is also a case of review papers being classified as articles. For example, the total number of the reviews in the sample dataset was manually found as nearly 59% compared to 49% as indexed by Scopus, decreasing the number of articles and conference papers to 41% for the sample dataset.

In this context, it would be helpful to provide a classification note for the published papers in the books and journals at the first instance. It would also be helpful to use the document types listed in Table 73.1 for this purpose. Book chapters may also be classified as articles or reviews as an additional classification to differentiate review chapters from the experimental chapters as it is done by the Web of Science. It would be further helpful to additionally classify the conference papers as articles or review papers as well as it is done in the Web of Science database.

73.4.3 The Most Prolific Authors in the Lignocellulosic Biomass-based Bioethanol Fuels

There have been most prolific 35 authors with at least 1.6% of the sample papers each as given in Table 73.2. These authors have shaped the development of the research in this field.

The most prolific authors are Bruce E. Dale, John N. Saddler, Leif J. Jonsson, and to a lesser extent Arthur J. Ragauskas, Venkatesh Balan, Barbel Hahn-Hagerdal, and Shishir P. S. Chundawat. Further, the most influential authors Bruce E. Dale, John N. Saddler, Leif J. Jonsson and to a lesser extent Barbel Hahn-Hagerdal, Shishir P. S. Chundawat, Venkatesh Balan, Arthur J. Ragauskas, and Nils-Olof Nilvebrant.

It is important to note the inconsistencies in indexing of the author names in Scopus and other databases. It is especially an issue for the names with more than two components such as 'Blake Sam de Hyun Dale'. The probable outcomes are 'Dale, B.S.D.H.', 'de Hyun Dale, B.S.', or 'Hyun Dale, B.S.D.'. The first choice is the gold standard of the publishing sector as the last word in the name is taken as the last name. In most of the academic databases such as PUBMED and EBSCO databases, this version is used predominantly. The second choice is a strong alternative while the last choice is an undesired outcome as two last words are taken as the last name. It is good practice to combine the words of the last name by a hyphen: 'Hyun-Dale, B.S.D.'. It is notable that inconsistent indexing of the author names may cause substantial inefficiencies in the search process for the papers as well as allocating credit to the authors as there are different author entries for each outcome in the databases.

There are also inconsistencies in the shortening Chinese names. For example, 'YangYing Zhao' is often shortened as 'Zhao, Y.', 'Zhao, Y.-Y.', and 'Zhao, Y.Y.' as it is done in the Web of Science database as well. However, the gold standard in this case is 'Zhao, Y' where the last word is taken as the last name and the first word is taken as a single forename. In most of the academic databases such as PUBMED and EBSCO, this first version is used predominantly. However, it makes sense to use the third option to differentiate Chinese names, efficiently: 'Zhao, Y.Y.'. Therefore, there have been difficulties in locating papers for the Chinese authors. In such cases, the use of the unique author codes provided for each author by the Scopus database has been helpful.

There is also a difficulty in allowing credit for the authors especially for the authors with common names such as 'Zhao, X.' in conducting scientometric studies. These difficulties strongly influence the efficiency of the scientometric studies as well as allocating credit to the authors as there are the same author entries for different authors with the same name, for example, 'Zhao, X.' in the databases.

In this context, the coding of authors in Scopus database is a welcome innovation compared to the other databases such as Web of Science. In this process, Scopus allocates a unique number to each author in the database (Aman, 2018). However, there might still be substantial inefficiencies in this coding system especially for common names. For example, some of the papers for a certain author maybe allocated to another researcher with a different author code. It is possible that Scopus uses a number of software programs to differentiate the author names and the program may not be false-proof (Kim, 2018).

In this context, it does not help that author names are not given in full in some journals and books. This makes difficult to differentiate authors with common names and makes the scientometric studies further difficult in the author domain. Therefore, the author names should be given in all books and journals at the first instance. There is also a cultural issue where some authors do not use their full names in their papers. Instead, they use initials for their forenames: 'Dale, H.J.', 'Dale, H.', or 'Dale, J.' instead of 'Dale, Hyun Jae'.

There are also inconsistencies in naming of the authors with more than two components by the authors themselves in journal papers and book chapters. For example, 'Dale, A.P.C.' might be given as 'Dale, A.' or 'Dale, A.C.' or 'Dale, A.P.' or 'Dale, C.' in the journals and books. This also makes the scientometric studies difficult in the author domain. Hence, contributing authors should use their name consistently in their publications.

The other critical issue regarding the author names is the inconsistencies in the spelling of the author names in the national spellings (e.g., Özğümüş, Şençöl) rather than in the English spellings (e.g., Ozgumus, Sencol) in Scopus database. Scopus differs from the Web of Science database and many other databases in this respect where the author names are given only in the English spellings. It is observed that national spellings of the author names do not help much in conducting scientometric studies as well in allocating credits to the authors as sometimes there are the different author entries for the English and National spellings in the Scopus database.

The most prolific institutions for the sample dataset are Lund University, Imperial College, Lawrence Berkeley National Laboratory and to a lesser extent NREL, Purdue University, and Tsinghua University. Further, the most prolific countries for the sample dataset are the USA and to a lesser extent Sweden, UK, China, and Denmark. These findings confirm the dominance of the USA, Europe, and China in this field.

On the other hand, the primary research fronts are the pretreatments and hydrolysis of the lignocellulosic biomass while the other research fronts are fermentation of the lignocellulosic biomass-based hydrolysates and production and evaluation of bioethanol fuels from the lignocellulosic biomass.

It is also notable that there is significant gender deficit for the sample dataset as surprisingly with a representation rate of 11%. This finding is the most thought-provoking with strong public policy implications. Hence, institutions, funding bodies, and policy makers should take efficient measures to reduce the gender deficit in this field as well as other scientific fields with strong gender deficit. In this context, it is worth to note the level of representation of the researchers from the minority groups in science on the basis of race, sexuality, age, and disability, besides the gender (Blankenship, 1993; Dirth and Branscombe, 2017; Konur, 2000, 2002a–c, 2006a,b, 2007a,b).

73.4.4 The Most Prolific Research Output by Years in the Lignocellulosic Biomass-based Bioethanol Fuels

The research output observed between 1970 and 2022 is illustrated in Figure 73.1. This figure clearly shows that the bulk of the research papers in the population dataset were published primarily in the 2010s and early 2020s. Similarly, the bulk of the research papers in the sample dataset were published in the 2010s and 2000s.

Further, the number of publications for the population papers increased between 2009 and 2017 and thereafter it steadied around 8% of the population papers for each year. However, there was a sharp rise in the research output for the population papers in 2020 and 2021 possibly due to the supply shocks.

These findings suggest that the most prolific sample and population papers were primarily published in the 2010s. Further, a significant portion of the sample and population papers were published in the early 2020s and 2000s, respectively.

These are the thought-provoking findings as there has been significant research boom in since 2010 and 2007 for the population and sample papers, respectively. In this context, the increasing public concerns about climate change (Change, 2007), greenhouse gas emissions (Carlson et al., 2017), and global warming (Kerr, 2007) have been certainly behind the boom in the research in this field since 2007. Furthermore, the recent supply shocks experiences due to the COVID-19 pandemics and the Ukrainian war might also be behind the research boom in this field since 2019.

Based on these findings, the size of the population papers likely to more than double in the current decade, provided that the public concerns about climate change, greenhouse gas emissions, and global warming, as well as the supply shocks are translated efficiently to the research funding in this field.

73.4.5 The Most Prolific Institutions in the Lignocellulosic Biomass-based Bioethanol Fuels

The most prolific 29 institutions publishing papers on the lignocellulosic biomass-based bioethanol fuels with at least 1.6% of the sample papers each given in Table 73.3 have shaped the development of the research in this field.

The most prolific institutions are the Lund University, NREL, and to a lesser extent Michigan State University, University of British Columbia, Chinese Academy of Sciences, Technical University of Denmark, Oak Ridge National Laboratory, and USDA Forest Service. Similarly, the top countries for these most prolific institutions are the USA, and to a lesser extent Sweden, China, and Denmark. In total, ten countries house these top institutions.

On the other hand, the institutions with the most citation impact are the Lund University, NREL, and to a lesser extent University of British Columbia, USDA Forest Service, Imperial College, Oak Ridge National Laboratory, and Technical University of Denmark. These findings confirm the dominance of the institutions from the USA, Europe, and to a lesser extent China and Canada.

73.4.6 The Most Prolific Funding Bodies in the Lignocellulosic Biomass-based Bioethanol Fuels

The most prolific 18 funding bodies funding at least 1.6% of the sample papers each is given in Table 73.4. It is notable that only 38% and 43% of the sample and population papers were funded, respectively.

The most prolific funding bodies are the US DOE, the European Commission, and to a lesser extent National Natural Science Foundation of China, Office of Science, and National Science Foundation. On the other hand, the most prolific countries for these top funding bodies are China, Japan, Sweden, the USA, and to a lesser extent India and the EU. In total, only seven countries and the EU house these top funding bodies.

The funding bodies with the most citation impact are the European Commission, the US DOE, and to a lesser extent Knut and Alice Wallenberg Foundation, Swedish National Board for Industrial and Technical Development, Government of West Bengal, and Swedish Research Council.

These findings on the funding of the research in this field suggest that the level of the funding, mostly since 2010 has been largely instrumental in enhancing the research in this field (Ebadi and Schiffauerova, 2016) in light of North's institutional framework (North, 1991). It is also notable that the funding rate in this field is relatively modest compared to those in the other research fronts of the bioethanol fuels such as algal bioethanol fuels. Further, it is expected that this high funding rate would improve in light of the recent supply shocks. Further, it emerges that China, Japan, Europe, the USA, and to a lesser extent India have heavily funded the research on the lignocellulosic biomass-based bioethanol fuels.

73.4.7 The Most Prolific Source Titles in the Lignocellulosic Biomass-based Bioethanol Fuels

The most prolific 15 source titles publishing at least 1.6% of the sample papers each in the lignocellulosic biomass-based bioethanol fuels have shaped the development of the research in this field (Table 73.5).

The most prolific source titles are the Bioresource Technology and to a lesser extent Green Chemistry, Biotechnology for Biofuels, Biotechnology and Bioengineering, and Biotechnology Advances. On the other hand, the source titles with the most citation impact are the Bioresource Technology and to a lesser extent by Green Chemistry, Biotechnology Advances, Renewable and Sustainable Energy Reviews, Biotechnology and Bioengineering, Enzyme and Microbial Technology, and Current Opinion in Biotechnology.

It is notable that these top source titles are primarily related to the bioresources, biotechnology, and to a lesser extent energy and microbial technology. This finding suggests that Bioresource Technology, and the other prolific journals in these fields have significantly shaped the development of the research in this field as they focus primarily on the lignocellulosic biomass-based bioethanol fuels with a high yield. In this context, the influence of the top journal is quite extraordinary.

73.4.8 The Most Prolific Countries in the Lignocellulosic Biomass-based Bioethanol Fuels

The most prolific 17 countries publishing at least 2.2% of the sample papers each have significantly shaped the development of the research in this field (Table 73.6).

The most prolific countries are the USA, Sweden, China, and to a lesser extent India, Denmark, Canada, and the UK. It is also notable that China is the largest producer of the population papers with a 21.3% publication rate. Further, seven European countries listed in Table 73.6 produce 39% and 20% of the sample and population papers, respectively, with a 19% surplus.

On the other hand, the countries with the most citation impact are the USA, Sweden, and to a lesser extent Denmark, Belgium, Canada, and the UK. Similarly, the countries with the least impact are China and to a lesser extent India and Brazil.

The close examination of these findings suggests that the USA, Europe, China, India, and to a lesser extent Canada, Japan, and S. Korea are the major producers of the research in this field. It is a fact that the USA has been a major player in science (Leydesdorff and Wagner, 2009). The USA has further developed a strong research infrastructure to support its corn- and grass-based bioethanol industry (Gillon, 2010).

However, China has been a rising mega star in scientific research in competition with the USA and Europe (Leydesdorff and Zhou, 2005). China is also a major player in this field as a major producer of bioethanol (Fang et al., 2010).

Next, Europe has been a persistent player in the scientific research in competition with both the USA and China (Leydesdorff, 2000). Europe has also been a persistent producer of bioethanol along with the USA and Brazil (Gnansounou, 2010).

Further, Japan (Negishi et al., 2004), S. Korea (Leydesdorff and Zhou, 2005), Canada (Tahmooresnejad et al., 2015), and India (Basu and Kumar, 2000) are the other countries with substantial research activities in bioethanol fuels.

73.4.9 The Most Prolific Scopus Subject Categories in the Lignocellulosic Biomass-based Bioethanol Fuels

The most prolific nine Scopus subject categories indexing at least 4% of the sample papers each, given in Table 73.7 have shaped the development of the research in this field.

The most prolific Scopus subject categories in the lignocellulosic biomass-based bioethanol fuels are Chemical Engineering, Environmental Science, Energy, and to a lesser extent Biochemistry, Genetics and Molecular Biology, and Immunology and Microbiology. It is also notable that Social Sciences including Economics and Business have a minimal presence in both sample and population studies.

On the other hand, the Scopus subject categories with the most citation impact are Chemical Engineering, Environmental Science, Energy and Biochemistry, Genetics and Molecular Biology. Similarly, the least influential subject categories are Chemistry, Agricultural and Biological Sciences, and Engineering.

These findings are thought-provoking, suggesting that the primary subject categories are related to energy, chemical engineering, environmental sciences, and to a lesser extent genetics and microbiology as the core of the research in this field concerns with the production and utilization of the lignocellulosic biomass-based bioethanol fuels. The other finding is that social sciences are not well represented in both the sample and population papers as in line with the most fields in bioethanol fuels. The social, environmental, and economics studies account for the field of social sciences.

73.4.10 The Most Prolific Keywords in the Lignocellulosic Biomass-based Bioethanol Fuels

A limited number of keywords have shaped the development of the research in this field as shown in Table 73.8 and the appendix. These keywords are grouped under the five headings: lignocellulosic biomass, pretreatments, fermentation, hydrolysis and hydrolysates, and products.

The most prolific keywords across all of the research fronts are lignin, cellulose, lignocellulose, biomass, hydrolysis, and to a lesser extent biofuels, ethanol, fermentation, lignocellulosic biomass, and pre-treatment. Similarly, the most influential keywords are lignin, hydrolysis, cellulose, lignocellulose, biofuels, pre-treatment, biomass, hemicellulose, ethanol, fermentation, sugar, enzymatic hydrolysis.

These findings suggest that it is necessary to determine the keyword set carefully to locate the relevant research in each of these research fronts. Additionally, the size of the samples for each keyword highlights the intensity of the research in the relevant research areas for both sample and population datasets. These findings also highlight different spelling of some strategic keywords such as pretreatment v. pre-treatment and ethanol v. bio-ethanol, etc. However, there is tendency toward the use of the connected keywords without using a hyphen.

73.4.11 The Most Prolific Research Fronts in the Lignocellulosic Biomass-based Bioethanol Fuels

Information about the thematic research fronts for the sample papers in the lignocellulosic biomass-based bioethanol fuels is given in Table 73.9. As this table shows, the most prolific research front is the pretreatments of the lignocellulosic feedstocks. The other prolific research fronts are the hydrolysis of the lignocellulosic feedstocks, hydrolysate fermentation, and production and evaluation of the bioethanol fuels.

These findings are thought-provoking in seeking ways to increase lignocellulosic biomass feedstock-based bioethanol yield at the global scale. It is clear that all of these research fronts have public importance and merit substantial funding and other incentives. Further, it is notable that lignocellulosic biomass-based bioethanol fuels have become a core unit of the bioethanol research to make it more competitive with the crude oil-based gasoline and petrodiesel fuels, especially for the USA, Europe, and China. It also solves the tremendous waste treatment (Morrissey and Browne, 2004; Wilson, 2007) problem of the huge amounts of the lignocellulosic biomass.

In comparison with the other feedstock-based research fronts, it is notable that the pretreatment of the lignocellulosic biomass emerges as a primary research front for this field. However, the research fronts of the hydrolysis of the lignocellulosic biomass, fermentation of the lignocellulosic biomass-based hydrolysates and the bioethanol production from the lignocellulosic biomass-based hydrolysates are also important.

Further, the field of the evaluation and utilization of bioethanol fuels is a neglected area. This suggests that the primary stakeholders have been primarily interested in these key processes of the bioethanol production. It is also notable that evaluation of the lignocellulosic biomass-based bioethanol fuels such as technoeconomics, life cycle, economics, social, land use, labor, and environment-related studies emerges as a case study for the bioethanol fuels. Similarly, the utilization of these biofuels in the gasoline or petrodiesel engines is also an important research field from a societal perspective. In this context, the USA and Brazil have been the global leaders in the production and use of the corn- and sugarcane-based bioethanol fuels since the 1970s in the aftermath of the global crude oil crisis in the early 1970s.

It is further notable that the research on the lignocellulosic biomass-based bioethanol fuels complements the research on the first generation bioethanol fuel research from sugar and starch feedstocks, extracting further value from these primary feedstocks. It also emerges as a distinctive research field complementing the research on the bioethanol fuels from the residual sugar and starch feedstocks, food, industrial, urban, and forestry wastes among others.

In the end, these most-cited papers in this field hint that the production of lignocellulosic biomass-based bioethanol fuels could be optimized using the structure, processing, and property relationships of these lignocellulosic biomass in the fronts of the feedstock pretreatment and hydrolysis, and hydrolysate fermentation (Formela et al., 2016; Konur, 2018a, 2020b, 2021a–d; Konur and Matthews, 1989).

73.5 CONCLUSION AND FUTURE RESEARCH

The research on the lignocellulosic biomass-based bioethanol fuels has been mapped through a scientometric study of both sample (182 papers) and population (3,644 papers) datasets.

The critical issue in this study has been to obtain a representative sample of the research as in any other scientometric study. Therefore, the keyword set has been carefully devised and optimized after a number of runs in the Scopus database. It is a representative sample of the wider population studies. This keyword set was provided in the appendix, and the relevant keywords are presented in Table 73.8. However, it should be noted that it has been very difficult to compile a representative keyword set since this research field has been connected closely with many other fields. Therefore, it has been necessary to compile a keyword list to exclude papers concerned with the other research fields.

The other issue has been the selection of a multidisciplinary database to carry out the scientometric study of the research in this field. For this purpose. Scopus database has been selected. The journal coverage of this database has been notably wider than that of Web of Science and other multisubject databases.

The key scientometric properties of the research in this field have been determined and discussed in this book chapter. It is evident that a limited number of documents, authors, institutions, publication years, institutions, funding bodies, source titles, countries, Scopus subject categories, Scopus keywords, and research fronts have shaped the development of the research in this field.

There is ample scope to increase the efficiency of the scientometric studies in this field in the author and document domains by developing consistent policies and practices in both domains across all the academic databases. In this respect, it seems that authors, journals, and academic databases have a lot to do. Furthermore, the significant gender deficit as in most scientific fields emerges as a public policy issue. The potential deficits on the basis of age, race, disability, and sexuality need also to be explored in this field as in other scientific fields.

The research in this field has boomed since 2010 and 2007 for the population and sample papers, respectively, possibly promoted by the public concerns on global warming, greenhouse gas emissions, and climate change. Furthermore, the recent COVID-19 pandemics and Russian invasion of Ukraine have resulted in the global supply shocks shifting the recent focus of the stakeholders from the crude oil-based fuels to biomass-based fuels such as bioethanol fuels.

It is expected that there would be further incentives for the key stakeholders to carry out the research for the lignocellulosic biomass-based bioethanol fuels to increase the ethanol yield and to make it more competitive with the crude oil-based gasoline and petrodiesel fuels. This might be truer for the crude oil- and foreign exchange-deficient countries to maintain the energy and food security at the face of the global supply shocks. It also solves the tremendous waste treatment problem of the huge amounts of the lignocellulosic biomass.

The relatively modest funding rate of 38% and 43% for the sample and population papers, respectively, suggests that funding in this field significantly enhanced the research in this field primarily since 2010, possibly more than doubling in the current decade. However, it is evident that there is ample room for more funding and other incentives to enhance the research in this field further.

The institutions from the USA, and to a lesser extent Sweden, China, and Denmark have mostly shaped the research in this field. Further, the USA, Europe, China, and to a lesser extent India and Canada have been the major producers of the research in this field as the major producers and users of bioethanol fuels. It is evident that these countries have well-developed research infrastructure in bioethanol fuels and their derivatives.

It emerges that ethanol is more popular than bioethanol as a keyword with strong implications for the search strategy. In other words, the search strategy using only bioethanol keyword would not be much helpful. On the other hand, the Scopus keywords are grouped under the five headings: lignocellulosic biomass, pretreatments, fermentation, hydrolysis and hydrolysates, and products.

Further, Table 73.9 shows that most prolific research fronts are the pretreatments of the lignocellulosic biomass and to a lesser extent hydrolysis of the lignocellulosic biomass, fermentation of the lignocellulosic biomass-based hydrolysates, and bioethanol production. The other minor research front is the bioethanol fuel evaluation. The first four research fronts dominate the research in this field while the field of the utilization and evaluation of the lignocellulosic biomass-based bioethanol fuels is relatively a neglected research field. In this context, it is notable that there is ample room for the improvement of the research on social and humanitarian aspects of the research on the bioethanol fuels from the lignocellulosic biomass such as scientometric and user studies.

These findings are thought-provoking in seeking ways to increase lignocellulosic biomass feedstock-based bioethanol yield at the global scale. It is clear that all of these research fronts have public importance and merit substantial funding and other incentives. Further, it is notable that lignocellulosic biomass-based bioethanol fuels have become a core unit of the bioethanol research to make it more competitive with the crude oil-based gasoline and petrodiesel fuels, especially for the USA, Europe, Brazil, and China. It is further notable that the research on the lignocellulosic biomass-based bioethanol emerges as a distinctive research field, complementing the research on the second generation bioethanol fuels from the agricultural residues, food, industrial, urban, and forestry wastes among others.

Thus, the scientometric analysis has a great potential to gain valuable insights into the evolution of the research in this field as in other scientific fields especially in the aftermath of the significant global supply shocks such as COVID-19 pandemics and the Russian invasion of Ukraine.

It is recommended that further scientometric studies are carried out for the primary research fronts. It is further recommended that reviews of the most-cited papers are carried out for each primary research front to complement these scientometric studies. Next, the scientometric studies of the hot papers in these primary fields are carried out.

ACKNOWLEDGMENTS

The contribution of the highly cited researchers in the field of the lignocellulosic biomass-based bioethanol fuels has been gratefully acknowledged.

APPENDIX: THE KEYWORD SET FOR LIGNOCELLULOSIC BIOMASS-BASED BIOETHANOL FUELS

(((TITLE (lignocellulos* OR lignocellulolytic*) OR SRCTITLE (lignocellulos*)) AND TITLE (ethanol* OR bioethanol OR hydroly* OR pretreat* OR "pre treat*" OR saccharif* OR ferment* OR ssf OR fractionat* OR detoxif* OR *cellulase OR delignif* OR "consolidated bioprocessing" OR recalcitrance OR "ionic liquid*" OR decompos* OR degrad* OR dissolution OR *degradation OR solvents OR xylanase* OR deconstruct* OR enzyme* OR termite* OR xylose OR sugar OR glucosidase OR "steam treat*" OR "cellulose accessibility" OR aspergillus OR glucanase* OR "hydrothermal treat*" OR digestibility OR breakdown)) AND NOT (TITLE ("thermal degrad*" OR "thermal pretreat*" OR biogas OR anti* OR anaerobic OR pyrolysis OR *hydrogen OR *diesel OR polyols OR feeds OR *methane OR aerogels OR carbonization OR *sorption OR *composites OR silica OR *fiber OR *composting OR adhesive* OR "thermal decompos*" OR *chars OR immun* OR levulinic OR butanediol OR h2 OR *fibre OR lactic OR lipid OR nanocry* OR wetland OR chitin OR xylitol OR gasif* OR "activated carbons" OR aviation OR jet OR succinic OR *butyrate OR furans OR *oligosaccharides) OR SUBJAREA (medi OR phar OR vete OR heal) OR SRCTITLE (polymer* OR macromol* OR hydrogen OR materials OR food* OR pyrolysis OR organic OR plant))) OR (TITLE (peroxide AND lignocellulos*)) AND (LIMIT-TO (DOCTYPE, "ar") OR LIMIT-TO (DOCTYPE, "re") OR LIMIT-TO (DOCTYPE, "cp") OR LIMIT-TO (DOCTYPE, "ch") OR LIMIT-TO (DOCTYPE, "ed") OR LIMIT-TO (DOCTYPE, "sh") OR LIMIT-TO (DOCTYPE, "le") OR LIMIT-TO (DOCTYPE, "no") OR LIMIT-TO (DOCTYPE, "bk")) AND (LIMIT-TO (LANGUAGE, "English")) AND (LIMIT-TO (SRCTYPE, "j") OR LIMIT-TO (SRCTYPE, "b") OR LIMIT-TO (SRCTYPE, "k")).

REFERENCES

Aman, V. 2018. Does the Scopus author ID suffice to track scientific international mobility? A case study based on Leibniz laureates. *Scientometrics* 117:705–720.

Angelici, C., B. M. Weckhuysen and P. C. A. Bruijnincx. 2013. Chemocatalytic conversion of ethanol into butadiene and other bulk chemicals. *ChemSusChem* 6:1595–1614.

Antolini, E. 2007. Catalysts for direct ethanol fuel cells. *Journal of Power Sources* 170:1–12.

Antolini, E. 2009. Palladium in fuel cell catalysis. *Energy and Environmental Science* 2:915–931.

Basu, A. and B. V. Kumar. 2000. International collaboration in Indian scientific papers. *Scientometrics* 48:381–402.

Beaudry, C. and V. Lariviere. 2016. Which gender gap? Factors affecting researchers' scientific impact in science and medicine. *Research Policy* 45:1790–1817.

Blankenship, K. M. 1993. Bringing gender and race in: US employment discrimination policy. *Gender & Society* 7:204–226.

Burnham, J. F. 2006. Scopus database: A review. *Biomedical Digital Libraries* 3:1–8.

Carlson, K. M., J. S. Gerber and D. Mueller, et al. 2017. Greenhouse gas emissions intensity of global croplands. *Nature Climate Change* 7:63–68.
Change, C. 2007. Climate change impacts, adaptation and vulnerability. *Science of the Total Environment* 326:95–112.
Dirth, T. P. and N. R. Branscombe. 2017. Disability models affect disability policy support through awareness of structural discrimination. *Journal of Social Issues* 73:413–442.
Ebadi, A. and A. Schiffauerova. 2016. How to boost scientific production? A statistical analysis of research funding and other influencing factors. *Scientometrics* 106:1093–1116.
Fang, X., Y. Shen, J. Zhao, X. Bao and Y. Qu. 2010. Status and prospect of lignocellulosic bioethanol production in China. *Bioresource Technology* 101:4814–4819.
Fauci, A. S., H. C. Lane and R. R. Redfield. 2020. Covid-19-navigating the uncharted. *New England Journal of Medicine* 382:1268–1269.
Fernando, S., S. Adhikari, C. Chandrapal and M. Murali. 2006. Biorefineries: Current status, challenges, and future direction. *Energy & Fuels* 20:1727–1737.
Formela, K., A. Hejna, L. Piszczyk, M. R. Saeb and X. Colom. 2016. Processing and structure-property relationships of natural rubber/wheat bran biocomposites. *Cellulose* 23:3157–3175.
Garfield, E. 1955. Citation indexes for science. *Science* 122:108–111.
Gillon, S. 2010. Fields of dreams: Negotiating an ethanol agenda in the Midwest United States. *Journal of Peasant Studies* 37:723–748.
Gnansounou, E. 2010. Production and use of lignocellulosic bioethanol in Europe: Current situation and perspectives. *Bioresource Technology* 101:4842–4850.
Hahn-Hagerdal, B., M. Galbe, M. F. Gorwa-Grauslund, G. Liden and G. Zacchi. 2006. Bio-ethanol: The fuel of tomorrow from the residues of today. *Trends in Biotechnology* 24:549–556.
Hamelinck, C. N., G. van Hooijdonk and A. P. C. Faaij. 2005. Ethanol from lignocellulosic biomass: Techno-economic performance in short-, middle- and long-term. *Biomass and Bioenergy* 28:384–410.
Hamilton, J. D. 1983. Oil and the macroeconomy since World War II. *Journal of Political Economy* 91:228–248.
Hamilton, J. D. 2003. What is an oil shock? *Journal of Econometrics* 113:363–398.
Hendriks, A. T. W. M. and G. Zeeman. 2009. Pretreatments to enhance the digestibility of lignocellulosic biomass. *Bioresource Technology* 100:10–18.
Hill, J., E. Nelson, D. Tilman, S. Polasky and D. Tiffany. 2006. Environmental, economic, and energetic costs and benefits of biodiesel and ethanol biofuels. *Proceedings of the National Academy of Sciences of the United States of America* 103:11206–11210.
Hill, J., S. Polasky and E. Nelson, et al. 2009. Climate change and health costs of air emissions from biofuels and gasoline. *Proceedings of the National Academy of Sciences of the United States of America* 106:2077–2082.
Hsieh, W. D., R. H. Chen, T. L. Wu and T. H. Lin. 2002. Engine performance and pollutant emission of an SI engine using ethanol-gasoline blended fuels. *Atmospheric Environment* 36:403–410.
Huang, H. J., S. Ramaswamy, U. W. Tschirner and B. V. Ramarao. 2008. A review of separation technologies in current and future biorefineries. *Separation and Purification Technology* 62:1–21.
Jones, T. C. 2012. America, oil, and war in the Middle East. *Journal of American History* 99:208–218.
Jorgensen, H., J. B. Kristensen and C. Felby. 2007. Enzymatic conversion of lignocellulose into fermentable sugars: Challenges and opportunities. *Biofuels, Bioproducts and Biorefining* 1:119–134.
Kerr, R. A. 2007. Global warming is changing the world. *Science* 316:188–190.
Kilian, L. 2008. Exogenous oil supply shocks: How big are they and how much do they matter for the US economy? *Review of Economics and Statistics* 90:216–240.
Kilian, L. 2009. Not all oil price shocks are alike: Disentangling demand and supply shocks in the crude oil market. *American Economic Review* 99:1053–1069.
Kim, J. 2018. Evaluating author name disambiguation for digital libraries: A case of DBLP. *Scientometrics* 116:1867–1886.
Konur, O. 2000. Creating enforceable civil rights for disabled students in higher education: An institutional theory perspective. *Disability & Society* 15:1041–1063.
Konur, O. 2002a. Access to nursing education by disabled students: Rights and duties of nursing programs. *Nurse Education Today* 22:364–374.
Konur, O. 2002b. Assessment of disabled students in higher education: Current public policy issues. *Assessment and Evaluation in Higher Education* 27:131–152.
Konur, O. 2002c. Access to employment by disabled people in the UK: Is the Disability Discrimination Act working? *International Journal of Discrimination and the Law* 5:247–279.

Konur, O. 2006a. Participation of children with dyslexia in compulsory education: Current public policy issues. *Dyslexia* 12:51–67.

Konur, O. 2006b. Teaching disabled students in higher education. *Teaching in Higher Education* 11:351–363.

Konur, O. 2007a. A judicial outcome analysis of the Disability Discrimination Act: A windfall for the employers? *Disability & Society* 22:187–204.

Konur, O. 2007b. Computer-assisted teaching and assessment of disabled students in higher education: The interface between academic standards and disability rights. *Journal of Computer Assisted Learning* 23:207–219.

Konur, O. 2011. The scientometric evaluation of the research on the algae and bio-energy. *Applied Energy* 88:3532–3540.

Konur, O. 2012a. The evaluation of the biogas research: A scientometric approach. *Energy Education Science and Technology Part A: Energy Science and Research* 29:1277–1292.

Konur, O. 2012b. The evaluation of the educational research: A scientometric approach. *Energy Education Science and Technology Part B: Social and Educational Studies* 4:1935–1948.

Konur, O. 2012c. The evaluation of the global energy and fuels research: A scientometric approach. *Energy Education Science and Technology Part A: Energy Science and Research* 30:613–628.

Konur, O. 2012d. The evaluation of the research on the biodiesel: A scientometric approach. *Energy Education Science and Technology Part A: Energy Science and Research* 28:1003–1014.

Konur, O. 2012e. The evaluation of the research on the bioethanol: A scientometric approach. *Energy Education Science and Technology Part A: Energy Science and Research* 28:1051–1064.

Konur, O. 2012f. The evaluation of the research on the biofuels: A scientometric approach. *Energy Education Science and Technology Part A: Energy Science and Research* 28:903–916.

Konur, O. 2012g. The evaluation of the research on the biohydrogen: A scientometric approach. *Energy Education Science and Technology Part A: Energy Science and Research* 29:323–338.

Konur, O. 2012h. The evaluation of the research on the microbial fuel cells: A scientometric approach. *Energy Education Science and Technology Part A: Energy Science and Research* 29:309–322.

Konur, O. 2012i. The scientometric evaluation of the research on the production of bioenergy from biomass. *Biomass and Bioenergy* 47:504–515.

Konur, O. 2015. Current state of research on algal bioethanol. In *Marine Bioenergy: Trends and Developments*, Eds. S. K. Kim and C. G. Lee, pp. 217–244. Boca Raton, FL: CRC Press.

Konur, O., Ed. 2018a. *Bioenergy and Biofuels*. Boca Raton, FL: CRC Press.

Konur, O. 2018b. Bioenergy and biofuels science and technology: Scientometric overview and citation classics. In *Bioenergy and Biofuels*, Ed. O. Konur, pp. 3–63. Boca Raton: CRC Press.

Konur, O. 2019. Cyanobacterial bioenergy and biofuels science and technology: A scientometric overview. In *Cyanobacteria: From Basic Science to Applications*, Eds. A. K. Mishra, D. N. Tiwari and A. N. Rai, pp. 419–442. Amsterdam: Elsevier.

Konur, O. 2020a. The scientometric analysis of the research on the bioethanol production from green macroalgae. In *Handbook of Algal Science, Technology and Medicine,* Ed. O. Konur, pp. 385–401. London: Academic Press.

Konur, O., Ed. 2020b. *Handbook of Algal Science, Technology and Medicine*. London: Academic Press.

Konur, O., Ed. 2021a. *Handbook of Biodiesel and Petrodiesel Fuels: Science, Technology, Health, and Environment*. Boca Raton, FL: CRC Press.

Konur, O., Ed. 2021b. *Handbook of Biodiesel and Petrodiesel Fuels: Science, Technology, Health, and Environment. Volume 1. Biodiesel Fuels: Science, Technology, Health, and Environment.* Boca Raton, FL: CRC Press.

Konur, O., Ed. 2021c. *Handbook of Biodiesel and Petrodiesel Fuels: Science, Technology, Health, and Environment. Volume 2. Biodiesel Fuels based on the Edible and Nonedible Feedstocks, Wastes, and Algae: Science, Technology, Health, and Environment.* Boca Raton, FL: CRC Press.

Konur, O., Ed. 2021d. *Handbook of Biodiesel and Petrodiesel Fuels: Science, Technology, Health, and Environment. Volume 3. Petrodiesel Fuels: Science, Technology, Health, and Environment.* Boca Raton, FL: CRC Press.

Konur, O. and F. L. Matthews. 1989. Effect of the properties of the constituents on the fatigue performance of composites: A review. *Composites* 20:317–328.

Kruyt, B., D. P. van Vuuren, H. J. de Vries and H. Groenenberg. 2009. Indicators for energy security. *Energy Policy* 37:2166–2181.

Leydesdorff, L. 2000. Is the European Union becoming a single publication system? *Scientometrics* 47:265–280.

Leydesdorff, L. and C. Wagner. 2009. Is the United States losing ground in science? A global perspective on the world science system. *Scientometrics* 78:23–36.

Leydesdorff, L. and P. Zhou. 2005. Are the contributions of China and Korea upsetting the world system of science? *Scientometrics* 63:617–630.

Li, H., S. M. Liu, X. H. Yu, S. L. Tang and C. K. Tang. 2020. Coronavirus disease 2019 (COVID-19): Current status and future perspectives. *International Journal of Antimicrobial Agents* 55:105951.

Limayem, A. and S. C. Ricke. 2012. Lignocellulosic biomass for bioethanol production: Current perspectives, potential issues and future prospects. *Progress in Energy and Combustion Science* 38:449–467.

Lin, Y. and S. Tanaka. 2006. Ethanol fermentation from biomass resources: Current state and prospects. *Applied Microbiology and Biotechnology* 69:627–642.

Ma, X., L. Sun and C. Song. 2002. A new approach to deep desulfurization of gasoline, diesel fuel and jet fuel by selective adsorption for ultra-clean fuels and for fuel cell applications. *Catalysis Today* 77:107–116.

Morrissey, A. J. and J. Browne. 2004. Waste management models and their application to sustainable waste management. *Waste Management* 24:297–308.

Morschbacker, A. 2009. Bio-ethanol based ethylene. *Polymer Reviews* 49:79–84.

Mosier, N., C. Wyman and B. Dale, et al. 2005. Features of promising technologies for pretreatment of lignocellulosic biomass. *Bioresource Technology* 96:673–686.

Najafi, G., B. Ghobadian and T. Tavakoli, et al. 2009. Performance and exhaust emissions of a gasoline engine with ethanol blended gasoline fuels using artificial neural network. *Applied Energy* 86:630–639.

Negishi, M., Y. Sun and K. Shigi. 2004. Citation database for Japanese papers: A new bibliometric tool for Japanese academic society. *Scientometrics* 60:333–351.

Newman, P. W. G. and J. R. Kenworthy. 1989. Gasoline consumption and cities: A comparison of U.S. cities with a global survey. *Journal of the American Planning Association* 55:24–37.

North, D. C. 1991. Institutions. *Journal of Economic Perspectives* 5:97–112.

Olsson, L. and B. Hahn-Hagerdal. 1996. Fermentation of lignocellulosic hydrolysates for ethanol production. *Enzyme and Microbial Technology* 18:312–331.

Palmqvist, E. and B. Hahn-Hagerdal. 2000a. Fermentation of lignocellulosic hydrolysates. I: Inhibition and detoxification. *Bioresource Technology* 74:17–24.

Palmqvist, E. and B. Hahn-Hagerdal. 2000b. Fermentation of lignocellulosic hydrolysates. II: Inhibitors and mechanisms of inhibition. *Bioresource Technology* 74:25–33.

Reeves, S. 2014. To Russia with love: How moral arguments for a humanitarian intervention in Syria opened the door for an invasion of the Ukraine. *Michigan State University International Law Review* 23:199.

Sanchez, O. J. and C. A. Cardona. 2008. Trends in biotechnological production of fuel ethanol from different feedstocks. *Bioresource Technology* 99:5270–5295.

Sassner, P., M. Galbe and G. Zacchi. 2008. Techno-economic evaluation of bioethanol production from three different lignocellulosic materials. *Biomass and Bioenergy* 32:422–430.

Sun, Y. and J. Cheng. 2002. Hydrolysis of lignocellulosic materials for ethanol production: A review. *Bioresource Technology* 83:1–11.

Taherzadeh, M. J. and K. Karimi. 2007. Enzyme-based hydrolysis processes for ethanol from lignocellulosic materials: A review. *Bioresources* 2:707–738.

Taherzadeh, M. J. and K. Karimi. 2008. Pretreatment of lignocellulosic wastes to improve ethanol and biogas production: A review. *International Journal of Molecular Sciences* 9:1621–1651.

Tahmooresnejad, L., C. Beaudry and A. Schiffauerova. 2015. The role of public funding in nanotechnology scientific production: Where Canada stands in comparison to the United States. *Scientometrics* 102:753–787.

Wilson, D. C. 2007. Development drivers for waste management. *Waste Management & Research* 25:198–207.

Winzer, C. 2012. Conceptualizing energy security. *Energy Policy* 46:36–48.

Yang, B. and C. E. Wyman. 2008. Pretreatment: The key to unlocking low-cost cellulosic ethanol. *Biofuels, Bioproducts and Biorefining* 2:26–40.

Zaldivar, J., J. Nielsen and L. Olsson. 2001. Fuel ethanol production from lignocellulose: A challenge for metabolic engineering and process integration. *Applied Microbiology and Biotechnology* 56:17–34.

74 Lignocellulosic Biomass-based Bioethanol Fuels
Review

Ozcan Konur
(Formerly) Ankara Yildirim Beyazit University

74.1 INTRODUCTION

Crude oil-based gasoline fuels (Ma et al., 2002; Newman and Kenworthy, 1989) have been widely used in the transportation sector since the 1920s. However, there have been great public concerns over the adverse environmental and human impact of these fuels (Hill et al., 2006, 2009). Hence, biomass-based bioethanol fuels (Hill et al., 2006; Konur, 2012, 2015, 2019, 2020) have increasingly been used in blending gasoline fuels (Hsieh et al., 2002; Najafi et al., 2009), in fuel cells (Antolini, 2007, 2009), and in the biochemical production (Angelici et al., 2013; Morschbacker, 2009) in a biorefinery context (Fernando et al., 2006; Huang et al., 2008).

However, it is necessary to pretreat the biomass (Alvira et al., 2010; Taherzadeh and Karimi, 2008) to enhance the yield of bioethanol (Hahn-Hagerdal et al., 2006; Sanchez and Cardona, 2008) prior to the bioethanol fuel production from feedstocks through hydrolysis (Sun and Cheng, 2002; Taherzadeh and Karimi, 2007) and fermentation (Lin and Tanaka, 2006; Olsson and Hahn-Hagerdal, 1996) of the biomass and hydrolysates, respectively.

One of the most studied feedstocks for bioethanol fuels has been lignocellulosic biomass. Research in the field of lignocellulosic biomass-based bioethanol fuels has intensified in this context in the key research fronts of pretreatment (Eriksson et al., 2002; Zhang et al., 2007) and hydrolysis (Kristensen et al., 2009; Li et al., 2008) of the lignocellulosic biomass, fermentation (Larsson et al., 2009; Martinez et al., 2001) of the lignocellulosic biomass-based hydrolysates, and production (Ballesteros et al., 2004; Sukumaran et al., 2009) and evaluation (Hamelinck et al., 2005; Wyman, 1994) of lignocellulosic biomass-based bioethanol fuels. Thus, it emerges as a distinctive research field, complementing primarily the research on second generation bioethanol fuels from agricultural residues as well as food, industrial, urban, and forestry wastes among others.

However, it is essential to develop efficient incentive structures (North, 1991) for the primary stakeholders to enhance the research in this field (Konur, 2000, 2002a–c, 2006a,b, 2007a,b). Although there have been a number of review papers on lignocellulosic biomass-based bioethanol fuels (Hendriks and Zeeman, 2009; Mosier et al., 2005; Sun and Cheng, 2002), there has been no review of the 25 most cited papers in this field.

Thus, this book chapter presents a review of the 25 most cited articles in the field of lignocellulosic biomass-based bioethanol fuels. Then, it discusses the key findings of these highly influential papers and comments on future research priorities in this field.

74.2 MATERIALS AND METHODS

The search for this study was carried out using the Scopus database (Burnham, 2006) in October 2022.

 DOI: 10.1201/9781003226550-101

As the first step for the search of the relevant literature, the keywords were selected using the first 400 most cited population papers. The selected keyword list was then optimized to obtain a representative sample of papers for the searched research field. This final keyword set was provided in the appendix of Konur (2023) for future replication studies.

As the second step, a sample dataset was used for this study. The first 25 articles with at least 276 citations each were selected for the review study. Key findings from each paper were taken from the abstracts of these papers and were discussed. Additionally, a number of brief conclusions were drawn and a number of relevant recommendations were made to enhance the future research landscape.

74.3 RESULTS

The brief information about 25 most cited papers with at least 276 citations each on lignocellulosic biomass-based bioethanol fuels is given below. The primary research fronts are hydrolysis of the lignocellulosic biomass and production and evaluation of lignocellulosic biomass-based bioethanol fuels with 14 and 11 highly cited papers (HCPs), respectively.

74.3.1 Lignocellulosic Biomass Hydrolysis

The brief information about 14 most cited papers on the hydrolysis of lignocellulosic biomass with at least 277 citations each is given in Table 74.1. These papers also cover the research on the pretreatment of the lignocellulosic biomass.

Eriksson et al. (2002) explored the mechanism of surfactant effect in enzymatic hydrolysis of lignocellulosic biomass in a paper with 751 citations. They screened a number of surfactants for their ability to improve enzymatic hydrolysis of steam-pretreated spruce. They found that non-ionic surfactants were the most effective and both anionic and non-ionic surfactants reduced enzyme adsorption to the lignocellulose substrate. The approximate reduction of enzyme adsorption was from 90% adsorbed enzyme to 80% with surfactant addition. However, surfactants had only a weak effect on cellulase temperature stability. They explained the improved conversion of lignocellulose with the surfactant by the reduction of the unproductive enzyme adsorption to the lignin part of the substrate. This was due to hydrophobic interaction of the surfactant with lignin on the lignocellulose surface, which released un-specifically bound enzyme.

Zhang et al. (2007) fractionated lignocellulosic biomass to cellulose, hemicellulose, lignin, and acetic acid at modest reaction conditions (50°C and atmospheric pressure) in a paper with 496 citations. They used a non-volatile cellulose solvent (concentrated phosphoric acid), a highly volatile organic solvent (acetone), and water. They attributed the highest sugar yields after enzymatic hydrolysis to no sugar degradation during the fractionation and the highest enzymatic cellulose digestibility (~97% in 24 h) during the hydrolysis step at the enzyme loading of 15 filter paper units (FPU) of cellulase and 60 international units (IU) of β-glucosidase per gram of glucan. Further, isolation of high-value lignocellulose components (lignin, acetic acid, and hemicellulose) would greatly increase potential revenues of a lignocellulose biorefinery.

Kristensen et al. (2009) explored the enzymatic hydrolysis of lignocellulosic biomass at high solid concentrations in a paper with 477 citations. They found that the decreasing enzymatic conversion of lignocellulosic biomass at increasing solids concentrations was a generic or intrinsic effect, describing a linear correlation from 5% to 30% initial total solids content (w/w). Neither lignin content nor hemicellulose-derived inhibitors were responsible for the decrease in yields. Product inhibition by glucose and in particular cellobiose at the increased concentrations at high solids loading played a role, but could not completely account for the decreasing conversion. Adsorption of cellulases decreased at increasing solids concentrations. Hence, there was a strong correlation between the decreasing cellulase adsorption and enzymatic conversion of biomass, indicating that the inhibition of cellulase adsorption to cellulose was causing the decrease in yield. In conclusion,

TABLE 74.1
The Hydrolysis of Lignocellulosic Biomass

No.	Papers	Wastes	Prts.	Parameters	Keywords	Lead Authors	Affil.	Cits.
2	Eriksson et al. (2002)	Lignocellulosic biomass spruce	Surfactants, enzymes, steam	Enzymatic hydrolysis, surfactant effect and mechanism, enzyme adsorption	Lignocellulose, hydrolysis	Tjerneld, Folke 7006446969	Lund Univ. Sweden	751
3	Zhang et al. (2007)	Lignocellulosic biomass	Solvents, enzymes	Enzymatic hydrolysis, solvents, fractionation, enzymatic digestibility	Lignocellulose, fractionating	Zhang, Yi-Heng P. 34876090400	Tianjin Inst. Ind. Biotechnol. China	496
4	Kristensen et al. (2009)	Lignocellulosic biomass	Enzymes	Enzymatic hydrolysis, solids content, conversion yield, cellulase adsorption inhibition	Lignocellulose, hydrolysis	Jorgensen, Henning 7202554496	Univ. Copenhagen Denmark	477
7	Li et al. (2008)	Lignocellulosic biomass corn stalk, rice straw, pine wood, bagasse	Acids	Acid hydrolysis, pretreatments, ILs, sugar yield, biomass	Lignocellulose, hydrolysis	Zhao, Zongbao K. 56972812400	Chinese Acad. Sci. China	424
9	Chundawat et al. (2011)	Lignocellulosic biomass	Ammonia	Enzymatic hydrolysis, pretreatment mechanisms, enzyme accessibility, morphology, pore surface area	Lignocellulosic, pretreatment	Chundawat, Shishir P. S. 12803763300	Rutgers, State Univ. N. J. USA	405
12	Brandt et al. (2011)	Lignocellulosic biomass miscanthus, pine, willow	IL, enzymes	Enzymatic hydrolysis, IL pretreatment, ILs, biomass, cellulose digestibility	Lignocellulosic, pretreatment	Welton, Tom 7003503272	Imperial Coll. UK	383
15	Kim et al. (2011)	Lignocellulosic biomass	Enzymes, acids	Enzymatic hydrolysis, hydrolysis inhibitors, biomass detoxification, pretreatments, hydrolysis rate, enzyme inhibition	Lignocellulosic, biomass	Ladisch, Michael R. 7005670397	Purdue Univ. USA	351
16	Hu et al. (2011)	Lignocellulosic biomass	Enzymes, steam	Enzymatic hydrolysis, xylanase-enzyme interactions, protein loading, cellulose accessibility, hydrolysis rate, cellulase loading	Lignocellulosic, hydrolysis	Saddler, John N. 7005297559	University of British Columbia Canada	334

(Continued)

TABLE 74.1 (*Continued*)
The Hydrolysis of Lignocellulosic Biomass

No.	Papers	Wastes	Prts.	Parameters	Keywords	Lead Authors	Affil.	Cits.
17	Berlin et al. (2007)	Lignocellulosic biomass corn stover	Enzymes	Enzymatic hydrolysis, accessory enzymes, cellulose hydrolysis, protein loading	Lignocellulose hydrolysis	Berlin, Alex 8639650700	Novozymes Biotech Inc. USA	333
18	Lenihan et al. (2010)	Lignocellulosic biomass potato peels	Acids	Acid hydrolysis, temperature, acid concentration, optimization, sugar yield	Lignocellulosic biomass, hydrolysis	Walker, Gavin M 35583992400	Univ. Limerick Ireland	310
20	Kristensen et al. (2007)	Lignocellulosic biomass wheat straw	Surfactants, acid, steam, enzymes, ammonia H_2O_2	Enzymatic hydrolysis, enzyme loading, pretreatments, surfactants, cellulase adsorption, lignin-surfactant reaction	Lignocellulose hydrolysis	Kristensen, Jan B 35322181100	Novozymes Inc. Denmark	306
21	Rollin et al. (2011)	Lignocellulosic biomass switchgrass	Solvents, enzymes, ammonia, BSA	Enzymatic hydrolysis, cellulose accessibility, lignin removal, CAC	Lignocellulose, fractionation	Zhang, Yi-Heng P. 34876090400	Tianjin Inst. Ind. Biotechnol. China	289
22	Viikari et al. (2007)	Lignocellulosic biomass spruce, corn stover	Enzymes	Enzymatic hydrolysis, accessory enzymes, protein and enzyme loading, hydrolysis rate	Lignocellulose hydrolysis	Viikari, Liisa* 7006720604	Univ. Helsinki inland	286
25	Zhang et al. (2016)	Lignocellulosic biomass corncob	Solvents	Enzymatic hydrolysis DESs, delignification, hydrolysis efficiency, acid amount and strength	lignocellulosic biomass, pretreatment	Xia, Shu-Qian* 7202893257	Tianjin Univ. China	277

Prt., Biomass pretreatments; Cits., Number of citations received for each paper; *, Female.

the inhibition of enzyme adsorption by hydrolysis products was the main cause of the decreasing yields at increasing substrate concentrations in the enzymatic decomposition of cellulosic biomass.

Li et al. (2008) used acid in ionic liquid (IL) for the hydrolysis of lignocellulosic biomass in a paper with 424 citations. They showed that this pretreatment was as an efficient system for hydrolysis of lignocellulosic materials with improved total reducing sugar (TRS) yield under mild conditions. TRS yields were up to 66%, 74%, 81%, and 68% for hydrolysis of corn stalk, rice straw, pine wood, and bagasse, respectively, in [C_4mim]Cl in the presence of 7 wt% hydrogen chloride (HCl) at 100°C under atmospheric pressure within 60 min. Different combinations between ILs, such as [C_6mim]Cl, [C_4mim]Br, [Amim]Cl, [C_4mim]HSO_4, and [Sbmim]HSO_4, and acids, including sulfuric acid, nitric acid, phosphoric acid, as well as maleic acid, afforded similar results though a longer reaction time was generally required compared with the combination of [C_4mim]Cl and HCl. Further, the modification of lignin occurred during sulfuric acid-catalyzed hydrolysis.

Chundawat et al. (2011) characterized corn stover cell walls to elucidate the mechanism of ammonia fiber expansion (AFEX) pretreatment in a paper with 405 citations. They observed that AFEX first dissolved, then extracted, and, as the ammonia evaporated, redeposited cell wall decomposition products (e.g., amides, arabinoxylan oligomers, lignin-based phenolics) on outer cell wall surfaces. As a result, nanoporous tunnel-like networks were formed within the cell walls. They proposed that this highly porous structure greatly enhanced enzyme accessibility to embedded cellulosic microfibrils. The shape, size (10–1,000 nm), and spatial distribution of the pores depended on their location within the cell wall and the pretreatment conditions used. Exposed pore surface area per unit AFEX-pretreated cell wall volume ranged between 0.005 and 0.05 nm^2 per nm^3. AFEX thus resulted in ultrastructural and physicochemical modifications within the cell wall that enhanced the enzymatic hydrolysis yield by four- to fivefold over that of untreated cell walls.

Brandt et al. (2011) performed the IL pretreatment of lignocellulosic biomass with IL–water mixtures in a paper with 383 citations. They pretreated miscanthus, pine, and willow with IL–water mixtures of 1-butyl-3-methylimidazolium methyl sulfate ([Bmim]$MeSO_4$) and 1-butyl-3-methylimidazolium hydrogen sulfate [Bmim]HSO_4). They found that up to 90% of the glucose and 25% of the hemicellulose contained in the original biomass were released by the combined IL pretreatment and the enzymatic hydrolysis. After the pretreatment, the IL liquor contained the majority of the lignin and the hemicellulose. The lignin portion was partially precipitated from the liquor upon dilution with water. They found that IL liquors containing methyl sulfate, hydrogen sulfate, and methanesulfonate anions were most effective in terms of lignin/cellulose fractionation and enhancement of cellulose digestibility.

Kim et al. (2011) detoxified the lignocellulosic biomass in a paper with 351 citations. They noticed the effect of the soluble inhibitors on the enzymatic hydrolysis when an increase in the concentration of pretreated biomass in a hydrolysis slurry resulted in decreased cellulose conversion, even though the ratio of enzyme to cellulose was kept constant. They used lignin-free cellulose, Solka Floc, combined with mixtures of soluble components released during pretreatment of wood, to prove that the decrease in the rate and extent of cellulose hydrolysis was due to a combination of enzyme inhibition and deactivation. They extracted the causative agents from wood pretreatment liquid using the polyethylene glycol (PEG) surfactant, activated charcoal, or ethyl acetate and then desorbed, recovered, and added back to a mixture of enzyme and cellulose. At enzyme loadings of either 1 or 25 mg protein/g glucan, they found that the most-inhibitory components, phenolics, decreased the rate and extent of cellulose hydrolysis by half due to both inhibition and precipitation of the enzymes. However, full enzyme activity occurred when the phenols were removed. Hence, detoxification of pretreated woods through phenol removal reduced enzyme loadings and therefore reduced enzyme costs, for a given level of cellulose conversion.

Hu et al. (2011) enhanced the enzymatic hydrolysis of lignocellulosic substrates by the addition of accessory xylanase enzymes partially replacing cellulase enzymes in a paper with 334 citations. For this reason, they assessed the interaction between cellulase and xylanase enzymes and their potential to improve the hydrolysis efficiency of various pretreated lignocellulosic substrates when added at low protein loadings. When xylanases were added to the minimum amount of cellulase enzymes required to achieve 70% cellulose hydrolysis of steam-pretreated corn stover (SPCS), or used to partially replace the equivalent cellulase dose, they observed that both approaches resulted in enhanced enzymatic hydrolysis. However, the xylanase supplementation approach increased the total protein loading required to achieve significant improvements in hydrolysis (an additive effect), whereas the partial replacement of cellulases with xylanase resulted in similar improvements in hydrolysis without increasing enzyme loading (a synergistic effect). The enhancement resulting from xylanase-aided synergism was higher when enzymes were added simultaneously at the beginning of hydrolysis. This cohydrolysis of the xylan also influenced the gross fiber characteristics, resulting in increased accessibility of the cellulose to the cellulase enzymes. These apparent increases in accessibility enhanced the SPCS digestibility, resulting in three times faster cellulose and xylan hydrolysis, a sevenfold decrease in cellulase loading, and a significant increase in the hydrolysis performance of the optimized enzyme mixture. In conclusion, the blocking effect of xylan was one of the major mechanisms that limited the accessibility of the cellulase enzymes to the cellulose. However, the synergistic interaction of the xylanase and cellulase enzymes significantly improved cellulose accessibility through increasing fiber swelling and fiber porosity and also played a major role in enhancing enzymatic accessibility.

Berlin et al. (2007) enhanced the enzymatic hydrolysis of lignocellulosic substrates by the addition of accessory xylanase, pectinase, and β-glucosidase enzymes, complementing cellulase enzymes, in a paper with 333 citations. They observed that the ability of a commercial *Trichoderma reesei* cellulase preparation (Celluclast 1.5L) to hydrolyze the cellulose and xylan components of pretreated corn stover was significantly improved by supplementation with three types of crude commercial enzyme preparations nominally enriched in xylanase, pectinase, and β-glucosidase activity. Although the product inhibition by β-glucosidase contributed to the observed improvement in cellulase performance, they attributed significant benefits to enzymes components that hydrolyze non-cellulosic polysaccharides. They suggested that these accessory enzymes such as xylanase and pectinase stimulated cellulose hydrolysis by removing non-cellulosic polysaccharides such as glucan and xylan that coated cellulose fibers. They produced an optimally supplemented enzyme mixture. This mixture allowed for a nearly twofold reduction in the total protein required to reach glucan-to-glucose and xylan-to-xylose hydrolysis targets (99% and 88% conversion, respectively), thereby validating this approach toward enzyme improvement and process cost reduction for lignocellulose hydrolysis.

Lenihan et al. (2010) optimized the dilute acid hydrolysis of lignocellulosic biomass in a paper with 310 citations. They performed the hydrolysis in a 1-L high-pressure pilot batch reactor using dilute phosphoric acid. Process parameters included reactor temperature (from 135°C to 200°C) and acid concentration (from 2.5% (w/w) to 10% (w/w)). They observed that the high conversion of cellulose to glucose was apparent while arabinose conversion was quite low due to thermal instability. However, they obtained an overall sugar yield of 82.5% at 135°C and 10% (w/w) acid concentration. They produced 55.2 g sugar/100 g dry potato peel after a time of 8 min.

Kristensen et al. (2007) enhanced the enzymatic hydrolysis of lignocellulosic substrates by the addition of surface active additives in a paper with 306 citations. They observed that the addition of non-ionic surfactants to enzymatic hydrolysis of lignocellulosic substrates of wheat straw increased the conversion of cellulose into soluble, fermentable sugars up to 70%. This provided an opportunity of decreasing enzyme loading while retaining the same degree of hydrolysis. Further, surfactants had a more pronounced effect on acid- and steam-pretreated straw than, for example, ammonia- and hydrogen peroxide-pretreated straw. Thus, lignin content was not directly proportional to the potential surfactant effect. The main mechanism behind the surfactant effect was prevention of unspecific adsorption of enzyme on the substrate lignin due to hydrophobic interaction between lignin and the surfactant, causing steric repulsion of enzyme from the lignin surface.

Rollin et al. (2011) showed that improving the surface area accessible to cellulase was a more important factor for achieving a high sugar yield compared to the lignin removal in a paper with 289 citations. They compared the pretreatment of switchgrass by cellulose solvent- and organic solvent-based lignocellulose fractionation (COSLIF) and soaking in aqueous ammonia (SAA). Following these pretreatments, they performed enzymatic hydrolysis at two cellulase loadings, 15 and 3 FPU/g glucan, with and without BSA blocking of lignin absorption sites. They observed that the lignin remaining after SAA had a significant negative effect on cellulase performance, despite the high level of delignification achieved with this pretreatment. However, there was no negative effect due to lignin for the COSLIF-treated substrate. Further, COSLIF fully disrupted the cell wall structure, resulting in a 16-fold increase in cellulose accessibility to cellulase (CAC), while SAA caused a 1.4-fold CAC increase.

Viikari et al. (2007) enhanced the enzymatic hydrolysis of lignocellulosic substrates by the addition of accessory enzymes in a paper with 286 citations. For this purpose, they cloned thermostable cellulases, cellobiohydrolase, endoglucanase, and β-glucosidase, produced them in *T. reesei*, and mixed them to compose a mixture of thermostable cellulases. They evaluated this new optimized thermostable enzyme mixtures in high temperature hydrolysis experiments on steam-pretreated spruce and corn stover. They obtained about 90% of theoretical hydrolysis with the thermostable enzymes at 60°C as with the commercial enzymes at 45°C. They thus obtained more efficient hydrolysis per assayed FPU or per amount of cellobiohydrolase I protein used. The maximum FPU activity of the novel enzyme mixture was about 25% higher at the optimum temperature at 65°C, as compared with the highest activity of the commercial reference enzyme at 65°C. These products could have high temperature stability in process conditions in the range of 55°C–60°C and clearly improved specific activity, essentially decreasing the protein dosage required for an efficient hydrolysis of lignocellulosic substrates.

Zhang et al. (2016) explored the roles of various deep eutectic solvent (DES) combinations, pretreatment temperature, and time to better utilize the DESs in the pretreatment of lignocellulosic biomass in a paper with 277 citations. They prepared three kinds of DESs, monocarboxylic acid/choline chloride, dicarboxylic acid/choline chloride, and polyalcohol/choline chloride, and used them in the pretreatment of corncob. They found that the enhanced delignification and subsequent enzymatic hydrolysis efficiency were related to the acid amount, acid strength, and the nature of hydrogen bond acceptors. Further, the structures of corncob were disrupted by the removal of lignin and hemicellulose in the pretreatment process. In addition, the optimal pretreatment temperature and time were 90°C and 24 h, respectively.

74.3.2 Lignocellulosic Biomass-based Bioethanol Production and Evaluation

There are 11 HCPs for the production and evaluation of the lignocellulosic biomass-based bioethanol fuels with at least 276 citations each (Table 74.2). Further, there are six and five HCPs for the production and evaluation of the bioethanol fuels, respectively. As the pretreatment and hydrolysis of the lignocellulosic biomass are the fundamental parts of bioethanol production, these narrated papers often cover these processes too.

74.3.2.1 Lignocellulosic Biomass-based Bioethanol Production

There are six HCPs for the production of lignocellulosic biomass-based bioethanol fuels with at least 286 citations each (Table 74.2). These papers also cover the fermentation of the hydrolysates of the lignocellulosic biomass. As the pretreatment and hydrolysis are the fundamental parts of bioethanol production, these narrated papers often cover these processes too.

Larsson et al. (1999) compared the different methods for the detoxification of lignocellulose hydrolysates of spruce to improve both cell growth and ethanol production in a paper with 430 citations. They used a dilute acid hydrolysate of spruce with *Saccharomyces cerevisiae* strains. They determined

TABLE 74.2
The Production and Evaluation of Lignocellulosic Biomass-based Bioethanol Fuels

No.	Papers	Wastes	Res. Fronts	Prts.	Yeasts	Parameters	Keywords	Lead Authors	Affil.	Cits.
1	Hamelinck et al. (2005)	Lignocellulosic biomass	Evaluation	Enzymes	Yeasts	Techno-economics, bioethanol production efficiency and costs, investment costs, hydrolysis, fermentation, biomass costs	Lignocellulosic, biomass, ethanol	Hamelinck, Carlo N. 6603008025	Utrecht Univ. Netherlands	1,214
5	Larsson et al. (1999)	Lignocellulosic biomass spruce	Production	Acids, enzymes	*S. cerevisiae*	Fermentation, hydrolysate detoxification methods, fermentation inhibitors, sugars	Lignocellulose, hydrolysates	Jonsson, Leif J. 7102349315	Ume Univ. Sweden	430
6	Jorgensen et al. (2007)	Lignocellulosic biomass straw	Production	Enzymes	*S. cerevisiae*	Enzymatic hydrolysis, solids content, glucose content, ethanol yield	Lignocellulose, liquefaction	Jorgensen, Henning 7202554496	Univ. Copenhagen Denmark	429
8	Delgenes et al. (1996)	Lignocellulosic biomass	Production	Pretreatment	*S. cerevisiae, Z. mobilis, P. stipitis, C. shehatae*	Hydrolysate fermentation, fermentation inhibitors, yeast strains	Lignocellulose, ethanol, fermentation	Delgenes, Jean P. 7005849678	Univ. Montpellier France	418
10	Ballesteros et al. (2004)	Lignocellulosic biomass poplar, eucalyptus, sorghum bagasse, wheat straw, brassica residue	Production	Enzymes	*K. marxianus*	Bioethanol production, pretreatment, hydrolysis, fermentation, SSF, ethanol yield	Lignocellulosic materials, ethanol	Manzanares, Paloma* 55779406300	CIEMAT Spain	389
11	Wyman (1994)	Lignocellulosic biomass	Evaluation	Enzymes	Yeasts	Techno-economics, bioethanol production, pretreatment, SSF, ethanol selling price	Lignocellulosic biomass, ethanol	Wyman, Charles E. 7004396809	Univ. Calif. Riverside USA	384

(Continued)

TABLE 74.2 (*Continued*)
The Production and Evaluation of Lignocellulosic Biomass-based Bioethanol Fuels

No.	Papers	Wastes	Res. Fronts	Prts.	Yeasts	Parameters	Keywords	Lead Authors	Affil.	Cits.
13	Sukumaran et al. (2009)	Lignocellulosic biomass sugarcane bagasse, rice straw, and water hyacinth	Production	Enzymes	*S. cerevisiae*	Bioethanol production, enzymatic hydrolysis, fermentation, sugar and ethanol yield	Lignocellulose, bio-ethanol	Pandey, Ashok 7201771319	CSIR India	365
14	Sassner et al. (2008)	Lignocellulosic biomass spruce, salix, corn stover	Evaluation	Steam	Yeasts	Techno-economics, bioethanol production, production costs, energy demand, SSF, capital cost, coproducts	Lignocellulosic materials, ethanol	Sassner, Per 8610538900	Lund Univ. Sweden	361
19	Soccol et al. (2010)	Lignocellulosic biomass	Evaluation	Na	Na	Brazilian bioethanol experience, ethanol and land use, ethanol yield	Lignocellulose, bioethanol	Soccol, Carlos R. 7004252959	Fed. Univ. Parana Brazil	305
23	Martinez et al. (2001)	Lignocellulosic biomass	Production	Pretreatment	Na	Hydrolysate detoxification, lime content, fermentation inhibitors	Lignocellulose, hydrolysates	Ingram, Lonnie O. 7102962097	Univ. Florida USA	286
24	Klein-Marcuschamer et al. (2011)	Lignocellulosic biomass	Evaluation	IL	Na	Techno-economics, bioethanol production, production costs, energy demand, SSF, capital cost, coproducts	Lignocellulosic, ethanol	Blanch, Harvey W. 7006259341	Univ. Calif. Riverside USA	276

Prt., Biomass pretreatments; Na, Non available; Cits., Number of citations received for each paper; *, Female.

the changes in the concentrations of fermentable sugars and three groups of fermentation inhibitors, aliphatic acids, furan derivatives, and phenolic compound as well as the fermentability of the detoxified hydrolysate. Further, the applied detoxification methods included treatment with alkali (sodium hydroxide (NaOH) or lime ($Ca(OH)_2$), treatment with sulfite (0.1% [w/v] or 1% [w/v] at pH 5.5 or 10), evaporation of 10% or 90% of the initial volume, anion exchange (at pH 5.5 or 10), enzymatic detoxification with the phenoloxidase laccase, and detoxification with the *T. reesei*. They found that an ion exchange at pH 5.5 or 10, treatment with laccase, treatment with lime, and treatment with *T. reesei* were the most efficient detoxification methods. Evaporation of 10% of the initial volume and treatment with 0.1% sulfite were the least efficient detoxification methods. Treatment with laccase was the only detoxification method that specifically removed only phenolic compounds. Anion exchange at pH 10 was the most efficient method for removing all three major groups of inhibitory compounds. However, it also resulted in loss of fermentable sugars.

Jorgensen et al. (2007) produced bioethanol through the enzymatic hydrolysis and fermentation of the lignocellulosic biomass at high solids concentrations in a paper with 429 citations. They performed the enzymatic liquefaction and saccharification of pretreated wheat straw with up to 40% (w/w) initial dry matter (DM). In <10 h, they observed that the structure of the biomass was changed from intact straw particles (length 1–5 cm) into a paste/liquid that could be pumped. There was no significant effect of mixing speed in the range 3.3–11.5 rpm on the glucose conversion after 24 h and ethanol yield after subsequent fermentation for 48 h. Liquefaction and saccharification for 96 h using an enzyme loading of 7 FPU/g·DM and 40% DM resulted in a glucose concentration of 86 g/kg. Experiments conducted at 2%–40% (w/w) initial DM revealed that cellulose and hemicellulose conversion decreased almost linearly with increasing DM. Performing the experiments as simultaneous saccharification and fermentation (SSF) also revealed a decrease in ethanol yield at increasing initial DM. *S. cerevisiae* was capable of fermenting hydrolysates up to 40% DM. They obtained the highest ethanol concentration, 48 g/kg, using 35% (w/w) DM.

Delgenes et al. (1996) studied the effects of six fermentation inhibitors on ethanol fermentations of glucose and xylose by four yeast strains in a paper with 418 citations. They used *S. cerevisiae* and *Zymomonas mobilis* for the glucose fermentation and *Pichia stipitis* and *Candida shehatae* for the xylose fermentation in batch cultures. They added the inhibitors with varying concentrations and determined the subsequent inhibitions on growth and ethanol production. They found that vanillin was a strong inhibitor of both growth and ethanol production by xylose-fermenting yeasts and *S. cerevisiae* when it was added to the culture media at a concentration of 1 g/L. Further, the fermentative activities of *Z. mobilis* were greatly sensitive to the presence of hydroxybenzaldehyde (0.5 g/L). However, some of the inhibitors, particularly vanillin and furaldehyde, could be assimilated by these strains, which resulted in the partial recovery in both growth and ethanol production processes on prolonged incubation.

Ballesteros et al. (2004) produced ethanol from lignocellulosic materials in a paper with 389 citations. They used woody (poplar and eucalyptus) and herbaceous (*Sorghum* sp. bagasse, wheat straw and *Brassica carinata* residue) materials as feedstocks and *Kluyveromyces marxianus* CECT 10875 strain. They first pretreated biomass samples in a steam explosion pilot plant and performed the SSF experiments under laboratory conditions at 42°C, 10% (w/v) substrate concentration, and 15 FPU/g substrate of commercial cellulase. They reached SSF yields in the range of 50%–72% of the maximum theoretical SSF yield, based on the glucose available in pretreated materials, in 72–82 h. Further, maximum ethanol contents from 16 to 19 g/L were obtained in fermentation media, depending on the feedstocks tested.

Sukumaran et al. (2009) produced bioethanol from lignocellulosic biomass in a paper with 365 citations. They produced cellulolytic enzymes for biomass hydrolysis using solid-state fermentation on wheat bran as substrate with *T. reesei* RUT C30 and *Aspergillus niger* MTCC 7956, respectively. They hydrolyzed sugarcane bagasse, rice straw, and water hyacinth biomass using 50 FPU of cellulase and 10 U of β-glucosidase per gram of pretreated biomass. They obtained highest yield of reducing sugars (26.3 g/L) from rice straw followed by sugarcane bagasse (17.79 g/L). They finally used the enzymatic hydrolysate of rice straw as substrate for ethanol production by *S. cerevisiae*, and they obtained the yield of ethanol of 0.093 g per gram of pretreated rice straw.

Martinez et al. (2001) detoxified dilute acid hydrolysates of lignocellulosic biomass with lime in a paper with 286 citations. Using *Escherichia coli* LY01 as the biocatalyst, they observed that the optimal lime addition for detoxification varied and depended on the concentration of mineral acids and organic acids in each hydrolysate. They predicted this optimum on the basis of the titration of hydrolysate with 2 N NaOH at ambient temperature to either pH 7.0 or pH 11.0. The average composition of 15 hydrolysates prior to treatment was as follows (per L): 95.24 g sugar, 5.3 g acetic acid, 1.305 g total furans (furfural and hydroxymethylfurfural – HMF), and 2.86 g phenolic compounds. Optimal overliming resulted in a 51% reduction in total furans, a 41% reduction in phenolic compounds, and a 8.7% decline in sugar. Acetic acid levels were unchanged.

74.3.2.2 Lignocellulosic Biomass-based Bioethanol Evaluation

There are five HCPs for the evaluation of the lignocellulosic biomass-based bioethanol fuels with at least 276 citations each (Table 74.2).

Hamelinck et al. (2005) evaluated ethanol production costs from lignocellulosic biomass in a paper with 1,214 citations. They found that the technology available as of the early 2000s, which was based on dilute acid hydrolysis, had about 35% efficiency (higher heating value, HHV) from biomass to ethanol. The overall efficiency, with bioelectricity co-produced from the lignin, was about 60%. They foresaw that the improvements in pretreatment and advances in biotechnology, especially through process combinations, could bring the ethanol efficiency to 48% and the overall process efficiency to 68%. They estimated investment costs as of the early 2000s at 2.1 k€/kWHHV (at 400 MWHHV input, i.e., a nominal 2,000 tonne dry/day input). However, a future technology in a five times larger plant (2 GWHHV) could have investments of 900 k€/kWHHV. They further found that a combined effect of higher hydrolysis and fermentation efficiency, lower specific capital investments, increase of scale, and cheaper biomass feedstock costs (from 3 to 2 €/GJHHV) could bring the ethanol production costs from 22 €/GJHHV in the next 5 years, to 13 €/GJ over the 10–15 year time scale, and down to 8.7 €/GJ in 20 or more years.

Wyman (1994) evaluated the impact of the advancements in the pretreatment and SSF processes for the conversion of cellulose and hemicellulose from lignocellulosic biomass on the production costs of the bioethanol fuels in a paper with 284 citations. He found that these advancements reduced the projected gate price of ethanol from about US$0.95/L (US$3.60/gallon, US$113/barrel) in 1980 to only about US$0.32/L (US$1.22/gallon, US$38) in 1994. He further located the technical targets to bring the selling price down to about US$0.18/L (US$0.67/gallon, US$21), a level that is competitive when oil prices exceed US$25/barrel. Finally, he estimated that the projected costs of bioethanol from lignocellulosic biomass could be competitive with bioethanol from corn, particularly if lower-cost feedstocks or other niche markets were used.

Sassner et al. (2008) performed the technoeconomic evaluation of bioethanol production from three different lignocellulosic biomass in a paper with 361 citations. They considered spruce, Salix, and corn stover. They compared production cost and energy demand using a process concept based on SO_2-catalyzed steam explosion pretreatment followed by SSF. They showed the importance of a high ethanol yield and the necessity of utilizing the pentose fraction for ethanol production to obtain good process economy, especially when using salix or corn stover. Furthermore, a less-energy-demanding process, mainly achieved by increasing the DM content in SSF, reduced the capital cost and resulted in higher coproduct credit, and therefore has a significant effect on the overall process economy.

Soccol et al. (2010) narrated the Brazilian experience of the production and use of bioethanol fuels from sugarcane and its bagasse as of 2010 in a paper with 305 citations. The National Alcohol Program led to less dependency on fossil fuels since its conception in 1975. The addition of 25% ethanol to gasoline (E25) reduced the import of 550 million barrels oil and also reduced the CO_2 emissions by 110 million tons. Nearly 44% of the Brazilian energy matrix was renewable and 13.5%

was derived from sugarcane as of 2010. Brazil has a land area of 851 million hectares, of which 54% are preserved, including the Amazon forest (350 million hectares). From the land available for agriculture (340 million hectares), only 0.9% was occupied by sugarcane as energy crop, showing a great expansion potential. In the coming years, ethanol yield per hectare of sugarcane, which was 6,000 L/ha as of 2010, could reach 10,000 L/ha, if 50% of the produced bagasse could be converted to ethanol.

Klein-Marcuschamer et al. (2011) performed the technoeconomic analysis of a lignocellulosic ethanol biorefinery with IL pretreatment in a paper with 276 citations. They identified the most significant areas in terms of cost savings/revenue generation that must be addressed before IL pretreatment could compete with other, more established, pretreatment technologies. They thus evaluated this pretreatment technology through the perspective of a virtual operating biorefinery and concluded that although there were significant challenges that must be addressed, there was a potential that could enable commercialization of this pretreatment.

74.4 DISCUSSION

74.4.1 INTRODUCTION

Crude oil-based gasoline fuels have been widely used in the transportation sector since the 1920s. However, there have been great public concerns over the adverse environmental and human impact of these fuels. Hence, biomass-based bioethanol fuels have increasingly been used in blending gasoline and petrodiesel fuels, in fuel cells, and in the biochemical production in a biorefinery context.

However, it is necessary to pretreat the biomass to enhance the yield of bioethanol prior to the bioethanol fuel production from feedstocks through hydrolysis and fermentation of the biomass and hydrolysates, respectively.

One of the most studied feedstocks for the bioethanol fuels has been the lignocellulosic biomass at large. Research in the field of lignocellulosic biomass-based bioethanol fuels has intensified in this context in the key research fronts of the pretreatment and hydrolysis of the lignocellulosic biomass, fermentation of the lignocellulosic biomass-based hydrolysates, and production and evaluation of the lignocellulosic biomass-based bioethanol fuels. Thus, it emerges as a distinctive research field, complementing primarily the research on the second generation bioethanol fuels from the agricultural residues as well as food, industrial, urban, and forestry wastes among others.

However, it is essential to develop efficient incentive structures for the primary stakeholders to enhance the research in this field. Although there have been a number of review papers for this field, there has been no review of the 25 most cited articles in this field.

Thus, this book chapter presents a review of the 25 most cited articles on the bioethanol fuel production and evaluation from the lignocellulosic biomass. Then, it discusses the key findings of these highly influential papers and comments on the future research priorities in this field.

As the first step for the search of the relevant literature, the keywords were selected using the first 400 most cited population papers. The selected keyword list was then optimized to obtain a representative sample of papers for the searched research field. This keyword list was provided in the appendix of Konur (2023) for future replicative studies.

As the second step, a sample dataset was used for this study. The first 25 articles with at least 276 citations each were selected for the review study. Key findings from each paper were taken from the abstracts of these papers and were discussed. Additionally, a number of brief conclusions were drawn and a number of relevant recommendations were made to enhance the future research landscape.

Information about the thematic research fronts for the sample papers in the lignocellulosic biomass-based bioethanol fuels is given in Table 74.3. As this table shows, the most prolific research fronts for this field are the pretreatment and hydrolysis of the lignocellulosic biomass with 96% and 88% of the HCPs, respectively. The other prolific research fronts are the hydrolysate fermentation

TABLE 74.3
The Most Prolific Thematic Research Fronts for the Lignocellulosic Biomass-based Bioethanol Fuels

No.	Research Fronts	N Paper % Review	N Paper 2% Sample	Surplus (%)
1	Biomass pretreatments	96.0	70.3	25.7
2	Biomass hydrolysis	88.0	33.5	54.5
3	Hydrolysate fermentation	36.0	19.2	16.8
4	Bioethanol production	36.0	15.4	20.6
5	Bioethanol fuel evaluation	20.0	7.1	12.9

N Paper (%) review: The number of papers in the sample of 25 reviewed papers. N paper (%) sample: The number of papers in the population sample of 182 papers.

and bioethanol production with 36% of the sample papers each. The other research front is the evaluation of bioethanol fuels with 20% of the HCPs.

Further, the most influential research fronts are pretreatment and hydrolysis of the lignocellulosic biomass with 26% and 55% surplus, respectively. Similarly, the other influential research fronts are hydrolysate fermentation, and production and evaluation of bioethanol fuels with 17%, 21%, and 13% surplus, respectively.

74.4.2 Lignocellulosic Biomass Hydrolysis

The brief information about 14 most cited papers on the hydrolysis of lignocellulosic biomass with at least 277 citations each is given in Table 74.1. These papers also cover the research on the pretreatment of the lignocellulosic biomass. It is notable that as Table 74.3 shows, 96% and 88% of these HCPs are related to the pretreatments and hydrolysis of the lignocellulosic biomass, respectively. These findings show that both pretreatments and hydrolysis of the lignocellulosic biomass are the fundamental processes for the bioethanol production from the lignocellulosic biomass.

These narrated studies highlight the importance of the pretreatment and hydrolysis processes for the production of the bioethanol fuels from the lignocellulosic biomass with a high ethanol yield. These pretreatments, primarily enzymatic and chemical pretreatments, fractionate the lignocellulosic biomass and enhance the enzymatic digestibility of the biomass.

Eriksson et al. (2002) explored the mechanism of surfactant effect in enzymatic hydrolysis of lignocellulosic biomass and found that non-ionic surfactants were the most effective and both anionic and non-ionic surfactants reduced enzyme adsorption to the lignocellulosic biomass. Further, Zhang et al. (2007) fractionated lignocellulosic biomass to cellulose, hemicellulose, lignin, and acetic acid and attributed the highest sugar yields after enzymatic hydrolysis to no sugar degradation during the fractionation and the highest enzymatic cellulose digestibility.

Kristensen et al. (2009) explored the enzymatic hydrolysis of lignocellulosic biomass at high solid concentrations and found that the decreasing enzymatic conversion of lignocellulosic biomass at increasing solid concentrations was a generic or intrinsic effect. Further, Li et al. (2008) used acid in IL for the hydrolysis of lignocellulosic biomass and showed that this IL pretreatment was as an efficient system for hydrolysis of lignocellulosic biomass with improved TRS yield under mild conditions.

Chundawat et al. (2011) characterized corn stover cell walls to elucidate the mechanism of AFEX pretreatment and observed that AFEX first dissolved, then extracted, and, as the ammonia evaporated, redeposited cell wall decomposition products on outer cell wall surfaces. Further, Brandt et al. (2011) performed the IL pretreatment of lignocellulosic biomass with IL–water mixtures and found that up to 90% of the glucose and 25% of the hemicellulose contained in the original biomass were released by the combined IL pretreatment and the enzymatic hydrolysis.

Kim et al. (2011) detoxified the lignocellulosic biomass and noticed the effect of the soluble inhibitors on the enzymatic hydrolysis when an increase in the concentration of pretreated biomass in a hydrolysis slurry resulted in decreased cellulose conversion. Further, Hu et al. (2011) enhanced the enzymatic hydrolysis of lignocellulosic biomass by the addition of accessory xylanase enzymes partially replacing cellulase enzymes.

Berlin et al. (2007) enhanced the enzymatic hydrolysis of lignocellulosic biomass by the addition of accessory enzymes. Further, Lenihan et al. (2010) optimized the dilute acid hydrolysis of lignocellulosic biomass and observed that the high conversion of cellulose to glucose was apparent while arabinose conversion was quite low due to thermal instability.

Kristensen et al. (2007) enhanced the enzymatic hydrolysis of lignocellulosic biomass by the addition of surface active additives. Further, Rollin et al. (2011) showed that improving the surface area accessible to cellulase was a more important factor for achieving a high sugar yield compared to the lignin removal.

Viikari et al. (2007) enhanced the enzymatic hydrolysis of lignocellulosic biomass by the addition of accessory enzymes. Finally, Zhang et al. (2016) explored the roles of various DES combinations, pretreatment temperature, and time to better utilize the DESs in the pretreatment of lignocellulosic biomass and found that the enhanced delignification and subsequent enzymatic hydrolysis efficiency were related to the acid amount, acid strength, and the nature of hydrogen bond acceptors.

74.4.3 Lignocellulosic Biomass-based Bioethanol Production

There are 11 HCPs for the production and evaluation of lignocellulosic biomass-based bioethanol fuels with at least 276 citations each (Table 74.2). Further, there are six and five HCPs for the production and evaluation of the bioethanol fuels, respectively. As the pretreatment and hydrolysis of the lignocellulosic biomass are the fundamental parts of the bioethanol production, these narrated papers often cover these processes too. It is notable that as Table 74.3 shows, 36% and 20% of these HCPs are related to the production and evaluation of bioethanol fuels from the lignocellulosic biomass, respectively. However, there are no HCPs on the utilization of biofuels in diesel or gasoline engines, partially displacing petrodiesel or gasoline fuels.

74.4.3.1 Lignocellulosic Biomass-based Bioethanol Production

There are six HCPs for the production of the lignocellulosic biomass-based bioethanol fuels with at least 286 citations each (Table 74.2). These papers also cover the fermentation of the hydrolysates of the lignocellulosic biomass. As the pretreatment and hydrolysis are the fundamental parts of bioethanol production, these narrated papers often cover these processes too. It is notable that as Table 74.3 shows, 36% of these HCPs are related to the production of bioethanol fuels from the lignocellulosic biomass.

These narrated studies highlight the importance of the pretreatment (primarily chemical, enzymatic, or hydrothermal) and hydrolysis (primarily enzymatic or acid) processes as well as of the fermentation processes (SSF or SHF) on the production of bioethanol fuels from the lignocellulosic biomass with a high ethanol yield. Further, some fermentation studies focus on the detoxification of the lignocellulosic hydrolysates to improve the ethanol yield.

Larsson et al. (1999) compared the different methods for the detoxification of lignocellulosic hydrolysates of spruce to improve both cell growth and ethanol production and found that an ion exchange treatment with laccase, treatment with lime, and treatment with *T. reesei* were the most efficient detoxification methods. Further, Jorgensen et al. (2007) produced bioethanol fuels through the enzymatic hydrolysis and fermentation of the lignocellulosic biomass at high solid concentrations and observed a decrease in ethanol yield at increasing initial DM.

Delgenes et al. (1996) studied the effects of six fermentation inhibitors on ethanol fermentations of glucose and xylose by four yeasts strains and found that vanillin was a strong inhibitor of both growth and ethanol production by xylose-fermenting yeasts and *S. cerevisiae*. Further, Ballesteros et al. (2004) produced ethanol from lignocellulosic biomass and reached SSF yields in the range of 50%–72% of the maximum theoretical SSF yield, based on the glucose available in pretreated materials.

Sukumaran et al. (2009) produced bioethanol from lignocellulosic biomass and obtained the yield of ethanol of 0.093 g per gram of pretreated rice straw. Further, Martinez et al. (2001) detoxified dilute acid hydrolysates of lignocellulosic biomass with lime and observed that the optimal lime addition for detoxification varied and depended on the concentration of mineral acids and organic acids in each hydrolysate.

74.4.3.2 Lignocellulosic Biomass-based Bioethanol Evaluation

There are five HCPs for the evaluation of lignocellulosic biomass-based bioethanol fuels with at least 276 citations each (Table 74.2).

These narrated studies often focus the technoeconomics and environmental impact of the bioethanol fuels from the lignocellulosic biomass. These technoeconomic studies show that the lignocellulosic ethanol fuels are cost competitive in relation to the crude oil-based gasoline and petrodiesel fuels, which ethanol fuels partially replace as the ethanol price reached $150 per barrel in 2022 following the invasion of Ukraine by Russia.

Hamelinck et al. (2005) evaluated ethanol production costs from lignocellulosic biomass and found that a combined effect of higher hydrolysis and fermentation efficiency, lower specific capital investments, increase of scale, and cheaper biomass feedstock costs could bring the ethanol production costs from 22 €/GJHHV in the next 5 years, to 13 €/GJ over the 10–15 year time scale, and down to 8.7 €/GJ in 20 or more years. Further, Wyman (1994) evaluated the impact of the advancements in the pretreatment and SSF processes for the conversion of cellulose and hemicellulose from lignocellulosic biomass on the production costs of the bioethanol fuels and found that these advancements reduced the projected gate price of ethanol from about US$0.95/L (US$3.60/gallon, US$113/barrel) in 1980 to only about US$0.32/L (US$1.22/gallon, US$38/barrel) in 1994, which is highly competitive.

Sassner et al. (2008) performed the technoeconomic evaluation of bioethanol production from three different lignocellulosic biomass and showed the importance of a high ethanol yield and the necessity of utilizing the pentose fraction for ethanol production to obtain good process economy, especially when using salix or corn stover. Further, Soccol et al. (2010) narrated the Brazilian experience of the production and use of the bioethanol fuels from sugarcane and its bagasse as of 2010 and found that in the coming years, ethanol yield per hectare of sugarcane, which was 6,000 L/ha as of 2010, could reach 10,000 L/ha, if 50% of the produced bagasse could be converted to ethanol. Finally, Klein-Marcuschamer et al. (2011) performed the technoeconomic analysis of a lignocellulosic ethanol biorefinery with IL pretreatment and concluded that although there were significant challenges that must be addressed, there was a potential that could enable commercialization of this pretreatment.

74.5 CONCLUSION AND FUTURE RESEARCH

The brief information about the key research fronts covered by the 25 most cited papers with at least 276 citations each is given under two primary headings: The hydrolysis of the lignocellulosic biomass and production and evaluation of the bioethanol fuels.

The usual characteristics of these HCPs are that the pretreatments and hydrolysis of the lignocellulosic biomass and fermentation of the resulting hydrolysates are the primary processes for the bioethanol fuel production from lignocellulosic biomass to improve the ethanol yield as the lignocellulosic biomass is one of the most studied feedstocks at large for the bioethanol production, especially for the countries with the large farmlands, forests, and crude oil deficiency.

The key findings on these research fronts should be read in light of the increasing public concerns about climate change, GHG emissions, and global warming as these concerns have been certainly behind the boom in the research on lignocellulosic biomass-based bioethanol fuels as an alternative to crude oil-based gasoline and petrodiesel fuels in the last decades. It is also a sustainable alternative to first generation food crop-based bioethanol fuels such as corn grain- or sugarcane-based bioethanol fuels. The recent supply shocks caused by the COVID-19 pandemics and the Russian invasion of Ukraine also highlight the importance of the production and utilization of the bioethanol fuels from the lignocellulosic biomass as an alternative to the crude oil-based gasoline and petrodiesel fuels.

As Table 74.3 shows, the most prolific thematic research fronts for this field are the pretreatment and hydrolysis of the lignocellulosic biomass and to a lesser extent the hydrolysate fermentation and bioethanol production and evaluation.

These studies emphasize the importance of proper incentive structures for the efficient production of lignocellulosic biomass-based bioethanol fuels in light of North's institutional framework (North, 1991). In this context, the major producers and users of bioethanol fuels such as the USA and Brazil with vast forests and farmlands have developed strong incentive structures for efficient lignocellulosic biomass-based bioethanol fuels.

In light of the recent supply shocks caused primarily by the COVID-19 pandemic and Russian invasion of Ukraine, it is expected that the incentive structures such as public funding would be enhanced to increase the share of bioethanol fuels from the lignocellulosic biomass in the global fuel portfolio as a strong alternative to crude oil-based gasoline and petrodiesel fuels.

In this context, it is expected that the most prolific researchers, institutions, countries, funding bodies, and journals in this field would have a first-mover advantage to benefit from such potential incentives. This is especially true for the US stakeholders as the USA has become the global leader in both the production and utilization of second generation bioethanol fuels from the lignocellulosic biomass.

It is recommended that such review studies are performed for the primary research fronts of the lignocellulosic biomass-based bioethanol fuels.

ACKNOWLEDGMENTS

The contribution of the highly cited researchers in the field of lignocellulosic biomass-based bioethanol fuels has been gratefully acknowledged.

REFERENCES

Alvira, P., E. Tomas-Pejo, M. Ballesteros and M. J. Negro. 2010. Pretreatment technologies for an efficient bioethanol production process based on enzymatic hydrolysis: A review. *Bioresource Technology* 101:4851–4861.

Angelici, C., B. M. Weckhuysen and P. C. A. Bruijnincx. 2013. Chemocatalytic conversion of ethanol into butadiene and other bulk chemicals. *ChemSusChem* 6:1595–1614.

Antolini, E. 2007. Catalysts for direct ethanol fuel cells. *Journal of Power Sources* 170:1–12.

Antolini, E. 2009. Palladium in fuel cell catalysis. *Energy and Environmental Science* 2:915–931.

Ballesteros, M., J. M. Oliva, M. J. Negro, P. Manzanares and I. Ballesteros. 2004. Ethanol from lignocellulosic materials by a simultaneous saccharification and fermentation process (SFS) with *Kluyveromyces marxianus* CECT 10875. *Process Biochemistry* 39:1843–1848.

Berlin, A., V. Maximenko, N. Gilkes and J. Saddler. 2007. Optimization of enzyme complexes for lignocellulose hydrolysis. *Biotechnology and Bioengineering* 97:287–296.

Brandt, A., M. J. Ray and T. Q. To, et al. 2011. Ionic liquid pretreatment of lignocellulosic biomass with ionic liquid-water mixtures. *Green Chemistry* 13:2489–2499.

Burnham, J. F. 2006. Scopus database: A review. *Biomedical Digital Libraries* 3:1–8.

Chundawat, S. P. S., B. S. Donohoe and L. da Costa Sousa, et al. 2011. Multi-scale visualization and characterization of lignocellulosic plant cell wall deconstruction during thermochemical pretreatment. *Energy and Environmental Science* 4:973–984.

Delgenes, J. P., R. Moletta and J. M. Navarro. 1996. Effects of lignocellulose degradation products on ethanol fermentations of glucose and xylose by *Saccharomyces cerevisiae*, *Zymomonas mobilis*, *Pichia stipitis*, and *Candida shehatae*. *Enzyme and Microbial Technology* 19:220–225.

Eriksson, T., J. Borjesson and F. Tjerneld. 2002. Mechanism of surfactant effect in enzymatic hydrolysis of lignocellulose. *Enzyme and Microbial Technology* 31:353–364.

Fernando, S., S. Adhikari, C. Chandrapal and M. Murali. 2006. Biorefineries: Current status, challenges, and future direction. *Energy & Fuels* 20:1727–1737.

Hahn-Hagerdal, B., M. Galbe, M. F. Gorwa-Grauslund, G. Liden and G. Zacchi. 2006. Bio-ethanol: The fuel of tomorrow from the residues of today. *Trends in Biotechnology* 24:549–556.

Hamelinck, C. N., G. van Hooijdonk and A. P. C. Faaij. 2005. Ethanol from lignocellulosic biomass: Techno-economic performance in short-, middle- and long-term. *Biomass and Bioenergy* 28:384–410.

Hendriks, A. T. W. M. and G. Zeeman. 2009. Pretreatments to enhance the digestibility of lignocellulosic biomass. *Bioresource Technology* 100:10–18.

Hill, J., E. Nelson, D. Tilman, S. Polasky and D. Tiffany. 2006. Environmental, economic, and energetic costs and benefits of biodiesel and ethanol biofuels. *Proceedings of the National Academy of Sciences of the United States of America* 103:11206–11210.

Hill, J., S. Polasky and E. Nelson, et al. 2009. Climate change and health costs of air emissions from biofuels and gasoline. Proceedings of the National Academy of Sciences of the United States of America 106:2077–2082.

Hsieh, W. D., R. H. Chen, T. L. Wu and T. H. Lin. 2002. Engine performance and pollutant emission of an SI engine using ethanol-gasoline blended fuels. *Atmospheric Environment* 36:403–410.

Hu, J., V. Arantes and J. N. Saddler. 2011. The enhancement of enzymatic hydrolysis of lignocellulosic substrates by the addition of accessory enzymes such as xylanase: Is it an additive or synergistic effect? *Biotechnology for Biofuels* 4:36.

Huang, H. J., S. Ramaswamy, U. W. Tschirner and B. V. Ramarao. 2008. A review of separation technologies in current and future biorefineries. *Separation and Purification Technology* 62:1–21.

Jorgensen, H., J. Vibe-Pedersen, J. Larsen and C. Felby. 2007. Liquefaction of lignocellulose at high-solids concentrations. *Biotechnology and Bioengineering* 96:862–870.

Kim, Y., E. Ximenes, N. S. Mosier and M. R. Ladisch. 2011. Soluble inhibitors/deactivators of cellulase enzymes from lignocellulosic biomass. *Enzyme and Microbial Technology* 48:408–415.

Klein-Marcuschamer, D., B. A. Simmons and H. W. Blanch. 2011. Techno-economic analysis of a lignocellulosic ethanol biorefinery with ionic liquid pre-treatment. *Biofuels, Bioproducts and Biorefining* 5:562–569.

Konur, O. 2000. Creating enforceable civil rights for disabled students in higher education: An institutional theory perspective. *Disability & Society* 15:1041–1063.

Konur, O. 2002a. Access to nursing education by disabled students: Rights and duties of nursing programs. *Nurse Education Today* 22:364–374.

Konur, O. 2002b. Assessment of disabled students in higher education: Current public policy issues. *Assessment and Evaluation in Higher Education* 27:131–152.

Konur, O. 2002c. Access to employment by disabled people in the UK: Is the Disability Discrimination Act working? *International Journal of Discrimination and the Law* 5:247–279.

Konur, O. 2006a. Participation of children with dyslexia in compulsory education: Current public policy issues. *Dyslexia* 12:51–67.

Konur, O. 2006b. Teaching disabled students in higher education. *Teaching in Higher Education* 11:351–363.

Konur, O. 2007a. A judicial outcome analysis of the Disability Discrimination Act: A windfall for the employers? *Disability & Society* 22:187–204.

Konur, O. 2007b. Computer-assisted teaching and assessment of disabled students in higher education: The interface between academic standards and disability rights. *Journal of Computer Assisted Learning* 23:207–219.

Konur, O. 2012. The evaluation of the research on the bioethanol: A scientometric approach. *Energy Education Science and Technology Part A: Energy Science and Research* 28:1051–1064.

Konur, O. 2015. Current state of research on algal bioethanol. In *Marine Bioenergy: Trends and Developments*, Eds. S. K. Kim and C. G. Lee, pp. 217–244. Boca Raton, FL: CRC Press.

Konur, O. 2019. Cyanobacterial bioenergy and biofuels science and technology: A scientometric overview. In Cyanobacteria: From Basic Science to Applications, Eds. A. K. Mishra, D. N. Tiwari and A. N. Rai, pp. 419–442. Amsterdam: Elsevier.

Konur, O. 2020. The scientometric analysis of the research on the bioethanol production from green macroalgae. In *Handbook of Algal Science, Technology and Medicine,* Ed. O. Konur, pp. 385–401. London: Academic Press.

Konur, O. 2023. Lignocellulosic biomass-based bioethanol fuels: Scientometric study. In *Feedstock-based Bioethanol Fuels. II. Waste Feedstocks: Agricultural, Food, Industrial, Urban, Forestry, and Lignocellulosic Waste-based Bioethanol Fuels. Handbook of Bioethanol Fuels Volume 4*, Ed. O. Konur, pp. 325–349. Boca Raton, FL: CRC Press.

Kristensen, J. B., C. Felby and H. Jorgensen. 2009. Yield-determining factors in high-solids enzymatic hydrolysis of lignocellulose. *Biotechnology for Biofuels* 2:11.

Kristensen, J. B., J. Borjesson, J., M. H. Bruun, F. Tjerneld and H. Jorgensen. 2007. Use of surface active additives in enzymatic hydrolysis of wheat straw lignocellulose. *Enzyme and Microbial Technology* 40:888–895.

Larsson, S., A. Reimann, N. O. Nilvebrant and L. J. Jonsson. 1999. Comparison of different methods for the detoxification of lignocellulose hydrolyzates of spruce. *Applied Biochemistry and Biotechnology, Part A: Enzyme Engineering and Biotechnology* 77–79:91–103.

Lenihan, P., A. Orozco and E. O'Neill, et al. 2010. Dilute acid hydrolysis of lignocellulosic biomass. *Chemical Engineering Journal* 156:395–403.

Li, C., Q. Wang and Z. K. Zhao. 2008. Acid in ionic liquid: An efficient system for hydrolysis of lignocellulose. *Green Chemistry* 10:177–182.

Lin, Y. and S. Tanaka. 2006. Ethanol fermentation from biomass resources: Current state and prospects. *Applied Microbiology and Biotechnology* 69:627–642.

Ma, X., L. Sun and C. Song. 2002. A new approach to deep desulfurization of gasoline, diesel fuel and jet fuel by selective adsorption for ultra-clean fuels and for fuel cell applications. *Catalysis Today* 77:107–116.

Martinez, A., M. E. Rodriguez and M. L. Wells, et al. 2001. Detoxification of dilute acid hydrolysates of lignocellulose with lime. *Biotechnology Progress* 17:287–293.

Morschbacker, A. 2009. Bio-ethanol based ethylene. *Polymer Reviews* 49:79–84.

Mosier, N., C. Wyman and B. Dale, et al. 2005. Features of promising technologies for pretreatment of lignocellulosic biomass. *Bioresource Technology* 96:673–686.

Najafi, G., B. Ghobadian and T. Tavakoli, et al. 2009. Performance and exhaust emissions of a gasoline engine with ethanol blended gasoline fuels using artificial neural network. *Applied Energy* 86:630–639.

Newman, P. W. G. and J. R. Kenworthy. 1989. Gasoline consumption and cities: A comparison of U.S. cities with a global survey. *Journal of the American Planning Association* 55:24–37.

North, D. C. 1991. Institutions. *Journal of Economic Perspectives* 5:97–112.

Olsson, L. and B. Hahn-Hagerdal. 1996. Fermentation of lignocellulosic hydrolysates for ethanol production. *Enzyme and Microbial Technology* 18:312–331.

Rollin, J. A., Z. Zhu, N. Sathitsuksanoh and Y. H. P. Zhang. 2011. Increasing cellulose accessibility is more important than removing lignin: A comparison of cellulose solvent-based lignocellulose fractionation and soaking in aqueous ammonia. *Biotechnology and Bioengineering* 108:22–30.

Sanchez, O. J. and C. A. Cardona. 2008. Trends in biotechnological production of fuel ethanol from different feedstocks. *Bioresource Technology* 99:5270–5295.

Sassner, P., M. Galbe and G. Zacchi. 2008. Techno-economic evaluation of bioethanol production from three different lignocellulosic materials. *Biomass and Bioenergy* 32:422–430.

Soccol, C. R., de Souza Vandenberghe, L. P. and A. B.P. Medeiros, et al. 2010. Bioethanol from lignocelluloses: Status and perspectives in Brazil. *Bioresource Technology* 101:4820–4825.

Sukumaran, R. K., R. R. Singhania, G. M. Mathew and A. Pandey. 2009. Cellulase production using biomass feed stock and its application in lignocellulose saccharification for bio-ethanol production. *Renewable Energy* 34:421–424.

Sun, Y. and J. Cheng. 2002. Hydrolysis of lignocellulosic materials for ethanol production: A review. *Bioresource Technology* 83:1–11.

Taherzadeh, M. J. and K. Karimi. 2007. Enzyme-based hydrolysis processes for ethanol from lignocellulosic materials: A review. *Bioresources* 2:707–738.

Taherzadeh, M. J. and K. Karimi. 2008. Pretreatment of lignocellulosic wastes to improve ethanol and biogas production: A review. *International Journal of Molecular Sciences* 9:1621–1651.

Viikari, L., M. Alapuranen, T. Puranen, J. Vehmaanpera and M. Siika-Aho. 2007. Thermostable enzymes in lignocellulose hydrolysis. *Advances in Biochemical Engineering/Biotechnology* 108:121–145.

Wyman, C. E. 1994. Ethanol from lignocellulosic biomass: Technology, economics, and opportunities. *Bioresource Technology* 50:3–15.

Zhang, C. W., S. Q. Xia and P. S. Ma. 2016. Facile pretreatment of lignocellulosic biomass using deep eutectic solvents. *Bioresource Technology* 219:1–5.

Zhang, Y. H. P., S. Y. Ding and J. R. Mielenz, et al. 2007. Fractionating recalcitrant lignocellulose at modest reaction conditions. *Biotechnology and Bioengineering* 97:214–223.

75 Production and Uses of Bioethanol in an Integrated Biorefinery from Agro- and Forest-Industrial Waste

Carolina M. Mendieta, Julia Kruyeniski, María E. Vallejos, and María C. Area
National University of Misiones

75.1 2G GLOBAL BIOETHANOL FUEL PRODUCTION AND CONSUMPTION

The renewable biomass contains a high carbohydrates fraction (60%–80%) in the form of storage carbohydrates (starch, inulin, and sucrose) or structural polysaccharides (cellulose, hemicelluloses, and chitin). In the lignocellulosic biomass (LCB), the polysaccharides are the main components of the cell walls. That is why pretreatment and enzymatic depolymerization are required to obtain fermentable sugars (Sheldon, 2021). First generation (1G) bioethanol is produced by fermentation of sugars from carbohydrates of food crops that contain starch (corn and wheat) or sugars (sugarcane and sugar beet). It is a well-established technology, which has been commercially applied for many years. However, the controversial focus is due to its limited reduction of greenhouse emissions, and those raw materials compete with food production (Sheldon, 2021; Susmozas et al., 2020).

Second generation (2G) bioethanol fuels have emerged in recent years to counter these critics. They are produced from sugars obtained from (enzymatic or acid) hydrolysis of cellulose and hemicelluloses followed by fermentation of these sugars to ethanol. Cellulose and hemicellulose are obtained by pretreatment or fractionation (Kruyeniski, 2017).

An advantage of LCB as raw materials is their availability in high amounts, low cost, and sustainability. Besides, they do not compete with food production. However, they are more complex to process than starch/juice. Some examples are wood, agricultural waste, straw, grasses, and different agro- and forest-industrial residues. This technology is still in development, in its initial commercialization phase (Area et al., 2019). Ethanol reduces greenhouse gas (GHG) emissions per unit of energy produced. Compared to gasoline, in a modeling study of the US Department of Energy (DOE), corn ethanol reduces GHG emissions by 18%–28%, while the reduction of cellulosic ethanol is 87% (Nguyen and Bowyer, 2017).

Global production of bioethanol was 105 billion liters in 2020 (REN21, 2020). Its use as fuel represents 80%–90% of the consumption. Nevertheless, governments promote bioethanol production by subsidies to compete with oil prices (Clauser et al., 2018).

The leading producers of 1G bioethanol are the USA and Brazil, accounting for 84% of global production from corn starch and sugarcane sucrose, respectively (Sheldon, 2021). The USA is the chief producer with 58% of participation in the market. Most of its ethanol remains at a national level, considering that it represents half of worldwide consumption. The following producer is Brazil, with 26% of the participation. Brazil's industry is becoming dominated by foreign capital that invests in modern facilities at a large scale. Other producers are the Europe and China, which represent 16% of the market. In terms of demand, the values are 23% for Brazil, 9% for China, and 8% for Europe.

DOI: 10.1201/9781003226550-102

The capacity of 2G bioethanol represents <1% of the total bioethanol produced worldwide (3 billons of liters in 2018) (Morales et al., 2021), and its production is even lower, between 25% and 35%. In 2007, the Renewable Fuel Standard was modified to include 2G bioethanol requirements (EPA, 2007). The installed capacity in 2015 was 35%, 24%, and 22% in the USA, China, and Canada, respectively (UNCTAD, 2016). Most commercial plants are in North America. In 2015, the USA had 35% of the commercial installed capacity worldwide for 2G bioethanol production (Nguyen and Bowyer, 2017). As the industry advances, many demonstration plants were built, and in 2014 some were scaled up at the commercial scale, for example, Abengoa (Kansas) and POET-DSM Advanced Fuels (EIA, 2014).

In China, Henan Tianguan Group began producing cellulosic ethanol in two facilities from corn stover and wheat straw in 2009 and 2011, respectively. Starting in 2012, Shangdong Longlive Biotechnology is currently the chief 2G bioethanol producer (ETIP Bioenergy, 2016). The Tianguan Group projected 100 2G bioethanol plants by 2020 (Zhang et al., 2016).

75.2 LIGNOCELLULOSIC BIOETHANOL PRODUCTION

Biofuel feedstocks should not compete with food and feed production. 1G ethanol production from sugarcane and sugar beet juices is known worldwide. However, the processing waste should be exploited to produce 2G bioethanol because it is a cheaper input that does not compete with food (Mendieta et al., 2021a). In crop residues, the lignocellulosic matrix needs destructuration to reach the sugars that are the fermentation platform to obtain ethanol, whereas the crop juice is ready for saccharification (Mendieta et al., 2021b).

The 2G bioethanol production-processing stages are LCB pretreatment, polysaccharide fraction hydrolysis, and sugars fermentation (Figure 75.1). The chosen pretreatment depends on the raw material and influences the subsequent steps (Kruyeniski, 2017). Pretreatment aims to break the LCB matrix to obtain the structural carbohydrates (cellulose and hemicelluloses) and lignin.

Enzymatic hydrolysis is a catalytic process in which enzymes act synergistically to produce glucose monomers by bond cleavage between polysaccharides under mild conditions (generally, pH 4.5–5.0 and temperatures between 40°C and 50°C) (Robak and Balcerek, 2018). Following this stage, the use of fermentation strategies allows obtaining the target product.

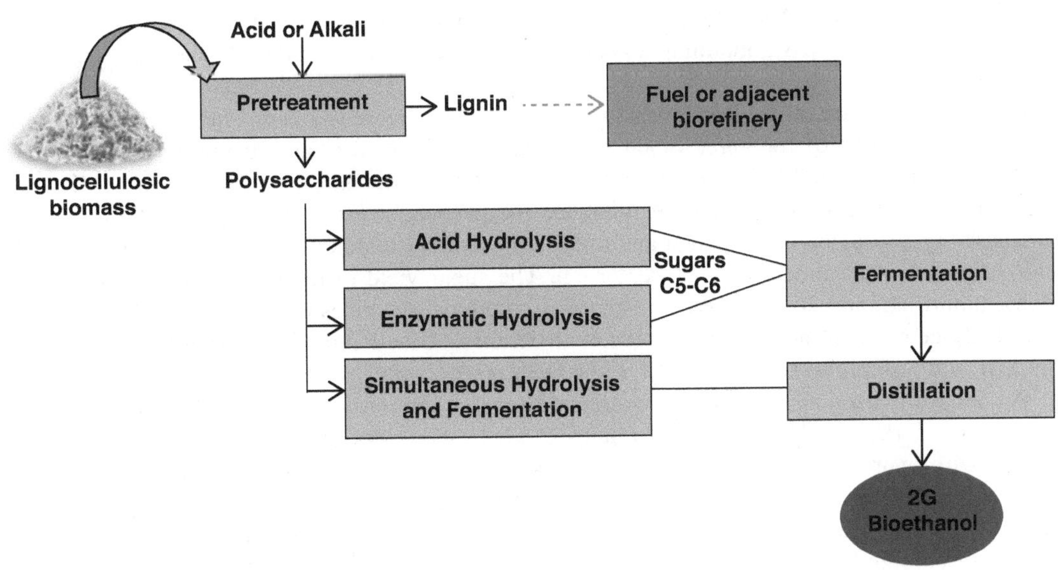

FIGURE 75.1 General scheme to obtain 2G bioethanol from LCB.

Alkaline pretreatment is one of the most studied for lignocellulosic material delignification. It facilitates biomass fractionation and improves its enzymatic conversion (Sannigrahi et al., 2010), obtaining high-purity cellulose since almost all lignin and hemicelluloses are dissolved in the process (Stoffel et al., 2017). Mendieta et al. (2021b) have obtained 2G bioethanol from pine sawdust, using an alkaline organosolv delignification (soda-ethanol) under the following conditions: liquor-to-wood ratio (L:W) = 5.44:1, maximum temperature 170°C, time to maximum temperature (60 min), and EtOH/H_2O (35/65% v/v), 23.3 NaOH (% w/w), 140, employing separate hydrolysis and fermentation (SHF) and simultaneous saccharification and fermentation (SSF) strategies and using Cellic® CTec2 commercial enzymes and *Saccharomyces cerevisiae* IMR 1181 yeast. Bioethanol yields were 89.3% (in 13 h) for SHF and 100% (in 72 h) for SSF.

Kruyeniski (2017) studied different pretreatments to obtain bioethanol using an SHF strategy. Despite relevant differences in enzymatic hydrolysis yields (ranging from 20% to 80%), bioethanol yields varied between 88% and 93% in all cases in 48 h. The studied pretreatments and their conditions were: (i) alkaline pretreatment plus steam explosion (90°C, 60 min, 5% over-dry wood NaOH + 200°C, 5 min, 3% over-dry material H_2SO_4); (ii) steam explosion followed by alkaline washing (190°C, 7.5 min, 0.75% over-dry material H_2SO_4 plus 60°C, 60 min, 0.4% over-dry material NaOH); (iii) Soda-anthraquinone (AQ)-oxygen treatment (170°C, 170 min, 25 g/L NaOH, LSR (liquid-to-solid ratio): 10/1 plus 120°C, 60 min, 10% consistency, 6 kg/cm^2); (iv) Kraft-AQ (170°C, 170 min, 25 g/L NaOH, LSR: 10/1) and (v) soda-AQ treatment (170°C, 140 min, 55.17 g/L NaOH, LSR: 5/1).

Although their lignin content could be similar, the reactivity of softwoods and hardwoods is different due to lignin structure and distribution and lignin–carbohydrate complex. Dagnino et al. (2013) found for *Prosopis nigra* (commonly known as black carbon tree) subjected to optimum conditions of dilute acid pretreatment (1.2% of sulfuric acid and 10.2 min) an enzymatic hydrolysis yield of 70%, much higher than non-pretreated sawdust, that gives 23.3%.

Another hardwood example is eucalyptus, which gives a 42% ethanol yield applying pretreatment with dilute acid and a steam explosion followed by a presaccharification SSF (pSSF) at the pilot scale (Vallejos et al., 2017a). Romani et al. (2012) evaluated the optimization of 2G bioethanol production from autohydrolyzed *E. globulus* wood, at a proportion of 8 g of water/g of oven-dry wood, temperature profile until reaching the range 210°C–230°C, operating at high substrate loadings using the SSF strategy. The enzymes used in this study were Celluclast 1.5L cellulases from *Trichoderma reesei* and NS50010 β-glucosidase from *Aspergillus niger* and *Saccharomyces cerevisiae* CECT-1170 as yeast, obtaining a good ethanol conversion (up to 86.4%) and a concentration (26.7 g ethanol/L).

Currently, there are other alternatives to produce 2G bioethanol from agro-industrial raw materials. Pabon et al. (2020) compared bioethanol production from rice husks using the SSF and SHF strategies, using Cellic® CTec2 commercial enzymes and *Saccharomyces cerevisiae* IMR 1181 yeast, and, in addition, evaluated the influence the particle size on the enzymatic hydrolysis of rice husk. Types of rice husk, milled and unmilled, were pretreated with 3% w/V NaOH and 10% w/V biomass in an autoclave at 121°C for 1 h. The sugar yield in the enzymatic hydrolysis was 35.5% (unmilled) and 36.6% (milled). The maximum glucose values obtained for both milled and unmilled rice husk did not present significant differences. Bioethanol yields for milled rice husk in the SHF and SSF processes were 35.3% and 43.9%, respectively.

A commonly used agroresidue is sugarcane bagasse, composed of 40%–60% fiber and 20%–30% pith (Clarke and Edye, 1996). Vallejos et al. (2017a), working with bagasse sequentially treated with hydrothermal and organosolv steps and only organosolv delignification, found that both lignin and xylan removal contributed to an increase in the accessibility of cellulases to cellulose. Enzymatic yields ranged between 18.1% and 83.4% for the bagasse treated only by organosolv delignification and between 61.8% and 78.9% for a two-step treatment. High conversions were only achieved with organosolv delignification when the residual material was highly delignified (severe conditions).

Neves et al. (2016), in a study of bagasse pretreated with steam explosion prepared by autohydrolysis, found that SHF gave ethanol yields higher than those obtained using SSF and pSSF. However, considering total processing time, pSSF provided the best overall ethanol volumetric productivity.

Wong and Sanggari (2014) studied 2G bioethanol production from sugarcane bagasse (10 g of sugarcane, 200 mL of distilled water, and 0.5 mL of NaOH to adjust the pH) and determined the effect of pH and temperature on bioethanol yield. Enzymes such as α-amylase and glucoamylase were used to break down the cellulose in sugarcane bagasse and *Saccharomyces cerevisiae* as yeast for the fermentation process, obtaining the highest ethanol content in water, which is 14.8% at pH 4.5 and 13.7% at 35°C.

75.3 UTILIZATION OF 2G BIOETHANOL FUELS

Almost 80%–90% of bioethanol fuels are used as biofuel in blends with gasoline (5%–10%), which represents a saving of 5%–27% petroleum resources. Other uses of bioethanol are as a solvent for flavoring and coloring products, perfumes, medicines, dyes, additives in the paint industry, agricultural chemicals, and odor agents (Mika et al., 2018; Rana and Parikh, 2020). The bioethanol production for beverage applications is not included in this chapter.

The production cost varies considerably according to variations in crude oil prices. The tendency indicates that the changes in oil prices prevent 1G and 2G biofuels to achieve a cost-competitive level comparable with other alternative energies (Araujo, 2016). It is projected that the increase in the yields of the bioethanol production processes and the reduction in the production costs through the optimization of the processes favor its use as a chemical intermediary of high-valued chemical products. Bioethanol as raw material for the manufacturing of chemical products has accomplished the development of the market, identified as one of the main bioproducts for the chemical industry, as can be seen in Figure 75.2.

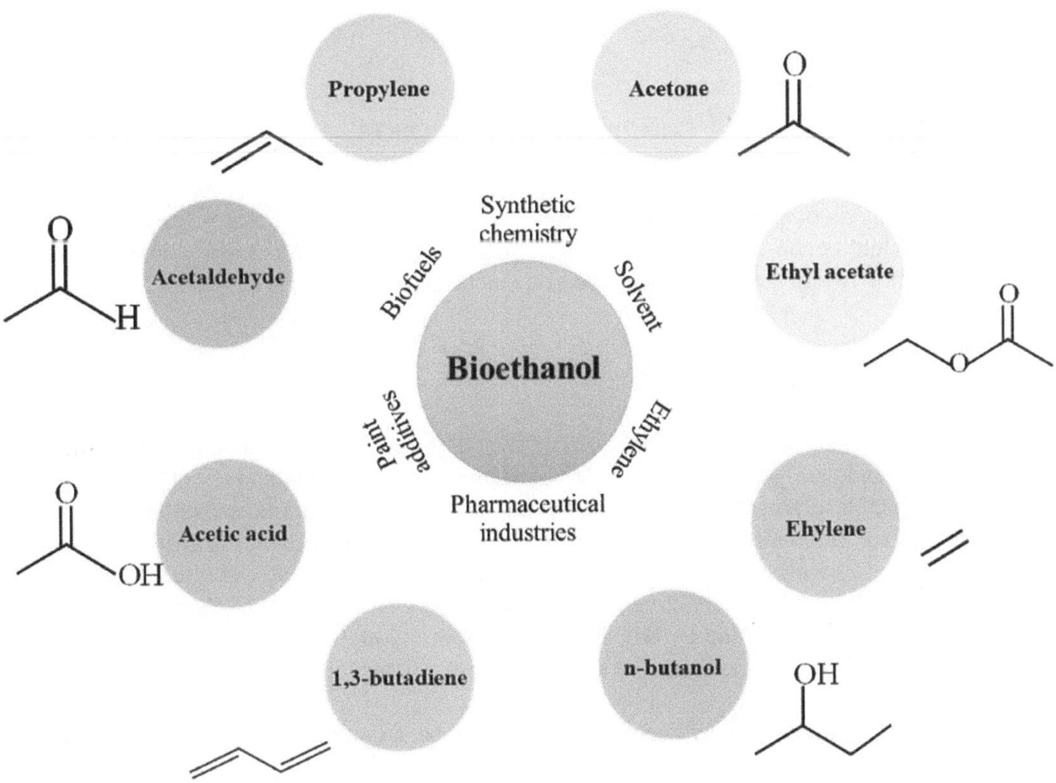

FIGURE 75.2 Bioethanol derivatives by catalytic conversion.

Bioethanol could become a primary intermediate in industrial organic chemistry based on renewables. Heterogeneous catalysis is fundamental in developing and improving processes to produce renewable products from bioethanol (Garbarino et al., 2020). There have been technological advances in the catalytic conversion of bioethanol to relevant biochemicals such as ethylene, propylene, 1,3-butadiene, iso-butylene, hydrogen, acetaldehyde, ethylene oxide, n-butanol, acetic acid, ethyl acetate, acetone, and dimethyl ether, which could be more cost-effective concerning its use as a fuel additive. Some of them are currently derived from fossil resources. The conversion processes could involve dehydration, dehydrogenation, oxidation, reforming, gasification, decomposition, coupling, among others. Many developments are still in the initial stages of design (laboratory or pilot scale).

75.3.1 Conversion of 2G Bioethanol Fuels to Oxygenated Compounds

The oxidation of ethanol is a green catalytic process because the secondary product is water. However, it is necessary to consider that the O_2-based oxidation processes present high flammability and explosion risk. Bioethanol can be the source of bio-based acetaldehyde, acetic acid, ethyl acetate, and other oxygenated compounds. The metal catalysts show advantages for industrial application because of the easier catalyst separation and cost-effectiveness (Xiang, et al., 2017).

Acetic acid, ethyl acetate, acetic anhydride, butanol, ethyl acetate, among others, can be produced from acetaldehyde. It can be produced from ethane, acetylene, ethylene, and ethanol, by reactions such as oxidation, hydration, oxidation, oxidative dehydrogenation, and dehydrogenation. It is commercially produced from ethylene by hydration to ethanol and subsequent dehydrogenation or partial oxidation to acetaldehyde (Rana and Parikh, 2020). A renewable alternative to acetaldehyde production is its conversion from bioethanol. Bioethanol can be oxidized to form acetaldehyde, and the resulting aldehyde can then undergo further oxidation to acetic acid.

$$CH_3CH_2OH_{(g)} + \frac{1}{2}O_2 \rightarrow CH_3CHO + H_2O$$

$$CH_3CHO + \frac{1}{2}O_2 \rightarrow CH_3COOH$$

As dehydrogenation competes with dehydration reactions at high temperatures, the production of acetaldehyde could occur. Therefore, dehydrogenation to acetaldehyde is favored above 327°C. It is an endothermic reaction efficiently catalyzed by metal-based catalysts (Ob-eye et al., 2019). The reaction is:

$$CH_3CH_2OH_{(g)} \leftrightarrow CH_3CHO_{(g)} + H_{2(g)}$$

Acetaldehyde production by dehydrogenation is affected by equilibrium limitations, moderate reaction selectivity, and catalysis deactivation due to carbon deposition. Oxidative dehydrogenation is considered a suitable alternative to overcome these issues. The main parameters studied in the ethanol conversion are temperature (250°C–350°C), pressure (1–5 atm), ethanol in feed (10%–100%), time (4–36 h), and catalysts (Ag/CeO_2, Ag/ZrO_2, Au/SiO_2, $Au/MgCuCr_2O$, Cu/AlMgO-P, $Cu/ZnAl_2O_4$, others). Ethanol conversion reported in these studies was between 24% and 100%, with a 31%–100% selectivity (Rana and Parikh, 2020). High conversion (92%) and selectivity (90%) were achieved using 100% ethanol as feedstock and 0.5 wt% of $Au/MgCuCr_2O$ as a catalyst at 250°C for 10 h (Liu et al., 2015).

Acetic acid is a simple carboxylic acid used as a food preservative and food additive, as a vinyl acetate monomer, in ester production (ethyl acetate, propyl acetate, and isobutyl acetate), as a acetic anhydride (cellulose acetate production), as a solvent (paints, coatings, and inks), and in medical applications (otitis, others). It is another product that could be obtained from bioethanol, which is currently produced by carbonylation of methanol. This process involves an oxidation stage of ethanol to acetaldehyde and a sequential oxidation stage to acetic acid. Several catalysts have been studied in the gas phase ($Mo_{0.61}V_{0.31}Nb_{0.08}O_x/TiO_2$, V_2O_5/TiO_2, Mo–CeO_x/SnO_2, others) and also in the liquid phase ($Au/MgAl_2O_4$, $Au/Ni_{0.95}Cu_{0.05}O_x$, RuO_x/CeO_2, $Ru(OH)_x/CeO_2$, others) using O_2 as an oxidant (pure or air) (Xiang et al., 2017). Higher ethanol conversion (63%) and selectivity to acetic acid (81%) were reported using $Au/CuFe_2O_4$ as a catalyst at 140°C and 3 MPa (Hu et al., 2018). Acetic acid can also be produced from ethanol–water solutions using the Cu/ZnO–ZrO_2–Al_2O_3 catalyst at 250°C–320°C and atmospheric pressure with 80%–90% selectivity and 60%–80% ethanol conversion. In this process, hydrogen is produced as a by-product (almost 2 moles H_2/L mole of acetic acid) (Brei et al., 2013; Galvita et al., 2001).

Partial oxidation with oxygen produces hydrogen from ethanol in a reaction catalyzed by noble metals.

$$CH_3CH_2OH_{(g)} + \frac{1}{2}O_2 \rightarrow 2CO_2 + 3H_2$$

Steam reforming is a way to produce syngas composed of CO_2 and hydrogen from a renewable source. The chief reaction is:

$$CH_3CH_2OH_{(g)} + H_2O_{(g)} \leftrightarrow 2CO_{2(g)} + 4\ H_{2(g)}$$

Ethyl acetate can be produced by reactions with various reagents such as acid anhydrides and chlorides, amides, ethers, aldehydes, alcohols, etc. Currently, ethyl acetate production in the chemical industry is from ethanol and acetic acid by the Fischer esterification reaction. It is an industrial solvent, and its low toxicity and biodegradability characterize it. However, this process produces problems due to high corrosiveness and toxicity. Oxidation and dehydrogenation are two more environmental and economic processes to produce ethyl acetate, which generate water and hydrogen as by-products, respectively (Piotrowski and Kubica, 2021). Bi-functional zeolite-supported Pd nanoparticle catalysts allowed high yields by aerobic oxidation of ethanol to ethyl acetate under mild conditions (selectivity and ethanol conversion of 94.7% and 98.6%, respectively) at a low temperature of 150°C for 73 h (Chen et al., 2016).

Acetone is an industrial solvent used to synthesize various chemicals (methyl methacrylate and meta-acrylic acid monomers, bisphenol A, methyl isobutyl ketone isopropanol, among others), drugs, and polymers (polycarbonate). It is industrially produced from benzene and propylene as fossil raw material by the cumene process (phenol as a by-product). However, acetone can also be produced from bioethanol without the generation of phenol as a by-product:

$$2CH_3CH_2OH + H_2O \leftrightarrow CH_3COCH_3 + CO_2 + 4H_2$$

Different catalytic systems were studied (Fe_2O_3–ZnO, Fe_2O_3–CaO, and Fe_2O_3–Mn, oxides based on Fe–Zn and Zn–Ca, Cu/CeO_2, Pd/CeO_2 and CeO_2, others) (Rodrigues et al., 2017). High ethanol conversion (100%) and selectivity to acetone (94%) were reported using Fe–Zn oxide as a catalyst in the presence of water at 440°C (Nakajima et al., 1994).

75.3.2 Conversion of 2G Bioethanol Fuels to Hydrocarbon Compounds

Ethylene is a chemical building block employed in the petrochemical sector. It is traditionally obtained by steam cracking the hydrocarbons from petroleum and recovered from the refinery cracked gas. Its chemical structure is widely used to produce polymers (polyethylene, polyvinylchloride, and polystyrene), ethylene glycol, ethylene oxide, among other chemicals. In recent years, sustainable alternatives to ethylene were researched to decrease greenhouse emissions and reduce the dependency on fossil fuels (Mendieta et al., 2021a). Bioethylene can be produced from dehydration of bioethanol at a temperature range of 200°C–450°C, generating diethyl ether as a by-product):

$$CH_3CH_2OH \leftrightarrow C_2H_4 + H_2O$$

Butanol is a biofuel used as an additive in gasoline and industrial solvent. It is commercially produced by hydroformylation of propylene to generate butyraldehyde and subsequent hydrogenation of butyraldehyde to butanol. It is used to produce acrylic esters and acrylic acid (Wu et al., 2018). This process is expensive due to the problems for product recovery, catalyst, and environmental impacts. Ethanol conversion to butanol involves a direct dimerization of two molecules of ethanol, followed by a multistep dehydrogenation to acetaldehyde, subsequent formation of crotonaldehyde by self-aldol condensation, and finally hydrogenation of crotonaldehyde to butanol (Guerbet process).

Fermentatively produced biobutanol is an attractive alternative to bioethanol as fuel for transport, and there is a growing demand from the automotive industry. One key opportunity is to replace ethanol as fuel. Its advantages are a higher energy content and its capacity to blend with oil in higher concentrations, and it does not require motors modifications. Besides, it prevents moisture absorbers, reduces motors corrosion, and generates lower carbon emissions (Area et al., 2019). Biobutanol is obtained by oxo synthesis or the ABE (acetone–butanol-ethanol) fermentative process, with yields of 0.19–0.40t diatomaceous earth (DE)-pretreated biobutanol/t DE-pretreated biomass. Existing ethanol plants can be adapted to produce biobutanol. The improvements in the fermentation stage and lower prices of raw material (2G sugars) can help biobutanol competition in production costs with fossil-based butanol.

The main participants of the bioisobutanol market are Gevo and Butamax Advanced Biofuels and the reproducer of n-biobutanol Green Biologics. Another relevant market participant is Green Biologics. It is a renewable chemical products biotechnological company in the United Kingdom, producing n-butanol from corncob and corn craps as raw material (Area et al., 2019).

Diethyl ether's chief applications are fuel and fuel additives, propellants, solvents, chemical intermediates, and extractive mediums. It is mainly used in the automotive, plastics, pharmaceutical, and fragrance industries. Currently, one of the major factors driving the market is its demand for industrial and laboratory solvents (Mordor Intelligence, 2020). The following reaction represents ethanol dehydration to diethyl ether (DDE, or ethyl ether):

$$CH_3CH_2OH_{(g)} \leftrightarrow \frac{1}{2}(CH_3CH_2)_2\,O_{(g)} + \frac{1}{2}H_2O_{(g)}$$

This reaction competes with its dehydration to ethylene. For this reason, the possibility to achieve a very high yield of DDE is practically impossible in a conventional flow reactor. The ethylene production can occur by two alternative formal reactions: the cracking (dehydration) of ethanol and/or cracking of DDE.

Propylene is a commodity from petroleum used to produce polypropylene, polyacrylonitrile, acrolein, acrylic acid, and propylene oxide. It is industrially produced from methanol as a mixture of olefins using HZSM-5 as a catalyst. Bioethanol can be used to produce a mixture of ethylene, C_3–C_4 olefins, and other long-chain hydrocarbons. This process is based on bioethanol dehydration to ethylene and sequential oligomerization–cracking of ethylene to propylene. The oligomerization stage has a low yield of propylene (20%–30%) because other by-products are generated as ethylene, butenes, and aromatic hydrocarbons (Song et al., 2009; Xue et al., 2017).

The ethanol to propylene (ETP) process is particularly promising to produce biopropylene. The studies regarding ethanol conversion to propylene focus on zeolite, mainly ZSM-5 and metal oxides as catalysts. Reaction determining factors involve reaction temperature, pressure, co-feeding water, and contact time. Phung et al. (2021) conclude that too high reaction temperatures favor unwanted successive polymerization reactions to form higher hydrocarbons and coke, but, at the too low reaction temperature, ethylene is the main product in the ethanol dehydration reaction. Generally, a temperature around 500°C and atmospheric pressure are considered. Lehmann and Seidel-Morgenstern (2014) analyzed the thermodynamic of the gas-phase ethanol to propylene and obtained a maximum propylene yield of 42% at 1 bar and 600°C.

Water coexists in ethanol solution derived from lignocellulosic fermentation; however, only a few studies focused on this case. Water can poison the active sites of catalysts during the reaction. It has also been reported that co-feeding water can reduce ethanol conversion and aromatics selectivity. However, in the case of ZSM-5 modified with Fe and P, the selectivity of propylene and C_{3+} olefins increased with co-feeding water. In the case of zeolites, the primary product of ethanol conversion is ethylene, which later produces propylene through intermediate hydrocarbons. When catalysts have a higher density of acid sites, ethanol dehydration to ethylene and later lo intermediates and propylene is faster.

1,3-Butadiene is used to produce a variety of synthetic rubbers, resin, elastomers, and other polymeric materials. It can also be produced from ethanol by a multistep reaction (dehydrogenation, condensation, and dehydration). However, the reaction complexity makes it necessary to use efficient catalysts to achieve high yields (Pomalaza et al., 2016).

Kyriienko et al. (2021) achieved 1,3-butadiene formation selectivities higher than 60% using 80 vol% ethanol in the conversion of the ethanol–water mixture in the presence of $ZnO/MgO–SiO_2$ catalysts with ratios $MgO:SiO_2$ of (1:1) and (3:1).

In ethanol conversion to 1,3-butadiene (ETB), Wang et al. (2022) analyzed a series of bi-functional ZnCe@SBA-15 catalysts with different Zn/Ce ratios, reaching a 45.2% selectivity of 1,3-butadiene under relatively mild reaction conditions (375°C, 101.325 kPa, and 0.54 g/catalyst h).

75.4 CASE STUDIES OF 2G BIOETHANOL FUELS INTEGRATED INTO A BIOREFINERY

2G bioethanol is a sustainable alternative to obtain high-value products in a pine sawdust biorefinery (Vallejos et al., 2017b) due to its benefit and potential demand in the market (Huang et al., 2011; Perez et al., 2020). One of these products is bioethylene, obtained through the catalytic dehydration of 2G bioethanol (Mendieta et al., 2021a). Currently, bioethylene is produced exclusively from 1G ethanol based on crops (sugarcane, corn, and sugar beet) as raw material, using editable sugars. However, the primary use of these renewable sources is food. On the contrary, the bioethylene production route from 2G bioethanol does not compete with food production and tries to maximize the use of lignocellulosic residues that are neither exploited nor used for industrial purposes.

Bioethylene from 2G bioethanol requires a previous treatment of the LCB (Stoffel, 2016). The ideal reagents are inexpensive and easily recoverable with low energy consumption, adapted to various substrates to achieve efficient delignification (Kruyeniski, 2017; Vallejos et al., 2019). Cellulose is the main polysaccharide obtained from pretreatment. It is the source of sugars for conversion to ethanol by depolymerization into glucose monomers through a hydrolysis process (Rodriguez et al., 2017). Although acid-catalyzed hydrolysis is faster and more effective than enzymatic, the latter is preferred for bioethanol production because it offers a bioconversion process under milder operating conditions (Hou et al., 2019; Kruyeniski et al., 2019), generally between pH (4.5 and 5.0) and temperature between 40°C and 50°C (Chang and Holtzapple, 2000). It follows a fermentation stage, where glucose is transformed into ethanol using yeast. The traditional one is *Saccharomyces cerevisiae*, the microorganism commonly used in industrial production due to its efficient ability to ferment glucose and other hexoses into bioethanol. It presents tolerance to ethanol and other inhibitors in lignocellulosic hydrolysates and works in an acidic pH range (Azhar et al., 2017). Its optimal development is usually between 25°C and 35°C (Cunha et al., 2020), it can withstand temperatures of up to 38°C, but its performance progressively decreases (Azhar et al., 2017).

The main strategies for bioethanol production are SHF and SSF (Balat, 2011). SHF has several advantages. First, both hydrolysis and fermentation can be carried out under optimal process conditions. Since it works at optimal temperature, the saccharification process requires a lower enzymatic load than the simultaneous strategy (Mendieta et al., 2021b). Besides, the fermentation in liquid broth facilitates the mass transfer and yeast recycling after fermentation (filtration or centrifugation) (Galbe et al., 2011). Its disadvantages are the generation of inhibition products, glucose for hydrolysis, and ethanol for fermentation (Araque et al., 2008). This process is expensive (Neves et al., 2016) because of its limited capacity to produce high concentrations of lignocellulosic ethanol.

The generated sugars restrain the cellulase activity and the hydrolysis rate (Robak and Balcerek, 2018). However, the simultaneous process seems to be the most feasible and profitable alternative to produce bioethanol due to the advantages in reducing inhibitor products and a single reactor use throughout the process, limiting investment costs (Olofsson et al., 2008). During saccharification, the released sugar monomers are immediately fermented by microorganisms, reducing the risk of microbial contamination (Robak and Balcerek, 2018), so glucose is instantly fermented in bioethanol (Wyman et al., 1992), regardless of the optimal temperature and pH parameters for hydrolysis and fermentation (Mendes et al., 2020).

The SSF bioprocess strategy for ethanol production was further improved by incorporating a short pSSF. This alternative process has provided benefits because it supports high solid loads, rapidly reducing the initial viscosity of the substrate that leads to increased production of ethanol and its yield (Manfredi et al., 2018). First, this process involves substrate incubation with hydrolytic enzymes in a short period, usually between 8 and 24h. Then, the microorganism is inoculated for the saccharification process. Adjusting the different optimal temperatures between the enzymes (50°C) and the traditional yeasts fermentation (30°C) allows the process optimization (Mcintosh et al., 2016).

The dehydration of 2G bioethanol requires a previous purification because the ethanol comes from a fermentation broth that contains nutrients, microorganisms, and reaction by-products. Impurities must be eliminated to reach bioethanol purities higher than 95% in weight. Usually, recovery and purification processes involve distillation, adsorption, extraction (Ramis et al., 2017), membrane pervaporation separation, and extractive or heteroazeotropic distillation (Feng and Huang, 1997). Following this step, the production of ethylene involves a process of energy adsorption (endothermic reaction) using high temperatures (higher than 573 K) and catalysts capable of achieving high conversions of ethanol and selectivity of ethylene as the main product (Mohsenzadeh et al., 2017).

The most used are alumina, silica, zeolites, clays, and phosphoric acid, among others. Bioethanol conversion yields are between 95.5% and 99.5%, with an ethylene selectivity of 95.0%–99.0% (Mendieta et al., 2021a; Morschbacker, 2009). Bioethylene from pine sawdust could compete with those based on fossil fuels if the production stages are optimized to reduce costs as raw material is abundant and not currently used for industrial purposes in some countries: Consumption of enzymes, reagents, energy, and water, efficient catalysts, purification techniques, integration of mass and energy, among others (Mendieta et al., 2021b).

Bioethylene is a widely used precursor in the production of bioplastics. For example, 2G biopolyethylene (BioPE) from the dehydration of bioethanol produced from pine sawdust is a polymer with a long chain of carbon atoms linked with two hydrogen atoms (a repeat of CH_2 units). It is currently a non-biodegradable plastic of great interest due to its usefulness in society, low raw material cost, versatility, and chemical stability (Mendieta et al., 2019).

There are varieties of PE with unique characteristics and applications. The main types are high-density PE (HDPE), linear low-density PE (LLDPE), and very-low-density polyethylene (VLDPE). HDPE production is carried out at relatively low temperatures and pressures in the presence of a catalyst (Ziegler-Natta, Phillips, or metallocene processes), and low-density polyethylenes are obtained by free radical polymerization mechanisms using high pressure and temperature (LDPE and LLDPE) (Sidek et al., 2019).

BioPE is produced from sugarcane (Kikuchi et al., 2013). It is converted to LDPE in an autoclave or high-pressure tubular reactors, HDPE in low-pressure reactors using suspension or gas phase polymerization, or LLDPE at relatively low pressures and temperatures by polymerization in the solution or gas phase (Tsiropoulos et al., 2015). The chemical composition of BioPE is the same as that of PE of fossil origin. The chemical composition of BioPE is the same as that of PE of fossil origin. Therefore, it can be used in rigid and flexible packaging or other applications such as containers, toys, shampoo bottles, cosmetics, beverages, food, retailing, and household items (Chamas et al., 2020; Hong et al., 2021; Sujuthi and Liew, 2016).

Ehman et al. (2020) evaluated the potential advantages of manufacturing biocomposites for 3D printing filaments using sugarcane-derived BioPE, sugarcane bagasse pulp, and fossil and bio-based compatibilizers. The evaluation involved water absorption, mechanical properties, thermal stability, decomposition temperature (thermo-gravimetric analysis (TGA)). After the testing of filaments, the fracture area was evaluated by scanning electron microscopy (SEM). Finally, some shapes were 3D-printed for demonstrations with the filaments.

75.5 CONCLUSION

The LCB based on the agro- and forest-industrial wastes could supply the growing energy demands worldwide. A sustainable alternative is the production of 2G bioethanol fuels employing the LCB as a sustainable alternative to the 1G bioethanol fuels from corn or sugarcane. It is also possible to produce biofuels and biochemicals from the bioethanol fuels produced from the LCB.

REFERENCES

Araque, E., C. Parra and J. Freer, et al. 2008. Evaluation of organosolv pretreatment for the conversion of *Pinus radiata* D. Don to ethanol. *Enzyme and Microbial Technology* 43:214–219.

Araujo, W. A. 2016. Ethanol industry: Surpassing uncertainties and looking forward. In *Global Bioethanol*, Eds. S. L. Monteiro Salles-Filho, L. A. Barbosa Cortez and M. G. Derengowski Fonseca, pp. 1–33. Chennai: Academic Press.

Area, M. C., M. E. Vallejos, N. M., Clauser, L. G. Covinich and G. Gonzalez. 2019. *Consultoria Para el Desarrollo de Nuevos Productos Foresto-Industriales en la Argentina. La Sostenibilidad Como un Instrumento de Desarrollo de Sectores Productivos Estrategicos [Consulting for the Ddevelopment of New Forest-Industrial Products in Argentina. Sustainability as an Instrument for the Development of Strategic Productive Sectors].* Washington, DC: Inter-American Development Bank.

Azhar, S. H. M., R. Abdulla and S. A. Jambo, et al. 2017. Yeasts in sustainable bioethanol production: A review. *Biochemistry and Biophysics Reports* 10:52–61.

Balat, M. 2011. Production of bioethanol from lignocellulosic materials via the biochemical pathway : A review. *Energy Conversion and Management* 52:858–875.

Brei, V. V., M. E. Sharanda, S. V. Prudius and E. A. Bondarenko. 2013. Synthesis of acetic acid from ethanol-water mixture over Cu/ZnO-ZrO_2-Al_2O_3 Catalyst. *Applied Catalysis A: General* 458:196–200.

Chamas, A., H. Moon and J. Zheng, et al. 2020. Degradation rates of plastics in the environment. *ACS Sustainable Chemistry and Engineering* 8:3494–3511.

Chang, V. S. and M. T. Holtzapple. 2000. Fundamental factors affecting biomass enzymatic reactivity. *Applied Biochemistry and Biotechnology* 84–86:5–38.

Chen, H., Y. Dai, X. Jia, H. Yu and Y. Yang. 2016. Highly selective gas-phase oxidation of ethanol to ethyl acetate over bi-functional Pd/zeolite catalysts. *Green Chemistry* 18:3048–3056.

Clarke, M. A. and L. A. Edye. 1996. Sugar beet and sugarcane as renewable resources. *ACS Symposium Series* 647:229–247.

Clauser, N. M., S. Gutierrez, M. C. Area, F. E. Felissia and M. E. Vallejos. 2018. Techno-economic assessment of carboxylic acids, furfural, and pellets production in a pine sawdust biorefinery. *Biofuels, Bioproducts & Biorefining* 12:997–1012.

Cunha, J. T., P. O. Soares, S. L Baptista, C. E. Costa and L. Domingues. 2020. Engineered *saccharomyces cerevisiae* for lignocellulosic valorization : A review and perspectives on bioethanol production and perspectives on bioethanol production. *Bioengineered* 11:883–903.

Dagnino, E. P., E. R. Chamorro, S. D. Romano, F. E. Felissia and M. C. Area. 2013. Optimization of the pretreatment of *Prosopis nigra* sawdust for the production of fermentable sugars. *BioResources* 8:499–514.

Ehman, N. V., D. Ita-nagy and F. E. Felissia, et al. 2020. Hydrothermal-Alkaline sugarcane bagasse pulp and coupled with a bio-based compatibilizer. *Molecules* 25:1–16.

EIA. 2014. *Commercial-Scale Cellulosic Ethanol Plant Opens*. Washington, DC: U.S. Energy Information Administration. Retrieved October 21, 2021. https://www.eia.gov/todayinenergy/detail.php?id=17851.

EPA. 2007. *Overview for Renewable Fuel Standard*. Washington, DC: U. S. Environmental Protection Agency. Retrieved October 21, 2021. (https://www.epa.gov/renewable-fuel-standard-program/overview-renewable-fuel-standard).

ETIP Bioenergy. 2016. *Cellulosic Ethanol (CE). An Introduction to Cellulosic Ethanol Technology*. European Technology and Innovation Platform Bioenergy. Retrieved October 21, 2021. https://www.etipbioenergy.eu/?option=com_content&view=article&id=273.

Feng, X. and R. Y. M. Huang. 1997. Liquid separation by membrane pervaporation: A review. *Industrial and Engineering Chemistry Research* 36:1048–1066.

Galbe, M., O. Wallberg and G. Zacchi. 2011. Techno-Economic aspects of ethanol production from lignocellulosic agricultural crops and residues. In *Comprehensive Biotechnology*, Ed. M. Moo-Young, pp. 615–628. London: Elsevier.

Galvita, V. V., G. L. Semin and V. D. Belyaev, et al. 2001. Synthesis gas production by steam reforming of ethanol. *Applied Catalysis A: General*, 220:123–127.

Garbarino, G., G. Pampararo, T. K. Phung, P. Rianiand and G. Busca. 2020. Heterogeneous catalysis in (bio) ethanol conversion to chemicals and fuels: Thermodynamics, catalysis, reaction paths, mechanisms and product selectivities. *Energies* 13:1–20.

Hong, L. G., N. Y. Yuhana and E. Z. E. Zawawi. 2021. Review of bioplastics as food packaging materials. *AIMS Materials Science* 8:166–184.

Hou, J., J. Tang and J. Chen, et al. 2019. Evaluation of inhibition of lignocellulose-derived by-products on bioethanol production by using the QSAR method and mechanism study. *Biochemical Engineering Journal* 147:153–162.

Hu, W., D. Li and Y. Yang, et al. 2018. Copper ferrite supported gold nanoparticles as efficient and recyclable catalyst for liquid-phase ethanol oxidation. *Journal of Catalysis* 357:108–117.

Huang, R., R. Su, W. Qi and Z. He. 2011. Bioconversion of lignocellulose into bioethanol: Process intensification and mechanism research. *Bioenergy Research* 4:225–245.

Kikuchi, Y, M. Hirao and K. Narita, et al. 2013. Environmental performance of biomass-derived chemical production : A case study on sugarcane-derived polyethylene. *Journal of Chemical Engineering of Japan* 46:319–25.

Kruyeniski, J. 2017. *Influencia del Pretratamiento de Residuos Forestoindustriales Sobre la Producción de Bioetanol [Influence of Pretreatment of Forestry-Industrial Waste on Bioethanol Production]*, Ph.D. Thesis. Posadas: National University of Misiones.

Kruyeniski, J., P. J. T. Ferreira and M. da G. V. Sousa Carvalho, et al. 2019. Physical and chemical characteristics of pretreated pine sawdust and its enzymatic hydrolysis. *Industrial Crops & Products journal* 130:528–536.

Kyriienko, P. V., O. V. Larina and D. Y. Balakin, et al. 2021. 1,3-Butadiene production from aqueous ethanol over $ZnO/MgO-SiO_2$ catalysts: Insight into H_2O effect on catalytic performance. *Applied Catalyst A, General* 616:118081.

Lehmann, T. and A. Seidel-Morgenstern. 2014. Thermodynamic appraisal of the gas phase conversion of ethylene or ethanol to propylene. *Chemical Engineering Journal* 242:422–432.

Liu, P., X. Zhu, S. Yang, T. Li and E. J. M. Hensen. 2015. On the metal-support synergy for selective gas-phase ethanol oxidation over $MgCuCr_2O_4$ supported metal nanoparticle catalysts. *Journal of Catalysis* 331:138–146.

Manfredi, A. P., I. Ballesteros and F. Saez, et al. 2018. Bioresource technology integral process assessment of sugarcane agricultural crop residues conversion to ethanol. *Bioresource Technology* 260:241–247.

Mcintosh, S., Z. Zhang and J. Palmer, et al. 2016. Pilot-scale cellulosic ethanol production using eucalyptus biomass pre-treated by dilute acid and steam. *Biofuel, Bioproducts & Biorefining* 4:77–93.

Mendes, C. V. T., P. Vergara and J. M. Carbajo, et al. 2020. Bioconversion of pine stumps to ethanol: Pretreatment and simultaneous saccharification and fermentation. *Holzforschung* 74:212–216.

Mendieta, C. M., F. E. Felissia, A. M. Arismendy, J. Kruyeniski and M. C. Area. 2021b. Enzymatic hydrolysis and fermentation strategies for the biorefining of pine sawdust. *BioResources* 16:7474–7491.

Mendieta, C. M., M. E. Vallejos, F. E. Felissia, G. Chinga-Carrasco and M. C. Area. 2019. Review: Bio-polyethylene from wood wastes. *Journal of Polymers and the Environment* 28:1–16.

Mendieta, C. M., R. E. Cardozo and F. E. Felissia, et al. 2021a. Bioconversion of wood waste to bio-ethylene: A review. *BioResources* 16:1–27.

Mika, L. T., E. Csefalvay and A. Nemeth. 2018. Catalytic conversion of carbohydrates to initial platform chemicals: Chemistry and sustainability. *Chemical Reviews* 118:505–613.

Mohsenzadeh, A., A. Zamani and M. J. Taherzadeh. 2017. Bioethylene production from ethanol: A review and techno-economical evaluation. *ChemBioEng Reviews* 4:75–91.

Morales, M., A. Arvesen and F. Cherubini. 2021. Integrated process simulation for bioethanol production: Effects of varying lignocellulosic feedstocks on technical performance. *Bioresource Technology* 328:124833.

Mordor Intelligence. 2020. *Diethyl Ether Market - Growth, Trends, Covid-19 Impact, and Forecasts (202-2026).* Telangana: Mordor. https://www.mordorintelligence.com/industry-reports/dimethyl-ether-market.

Morschbacker, A. 2009. Bio-ethanol based ethylene. *Polymer Reviews* 49:79–84.

Nakajima, T., H. Nameta, S. Mishima, I. Matsuzaki and K. Tanabe. 1994. A highly active and highly selective oxide catalyst for the conversion of ethanol to acetone in the presence of water vapour. *Journal of Materials Chemistry* 4:853.

Neves, P. V, A. P. Pitarelo and L. P. Ramos. 2016. Production of cellulosic ethanol from sugarcane bagasse by steam explosion: Effect of extractives content, acid catalysis and different fermentation technologies. *Bioresource Technology* 208:184–194.

Nguyen, Q. and J. Bowyer. 2017. *Global Production of Second Generation Biofuels: Trends and Influences.* Report. Minneapolis: Dovetail Partners, Inc. https://dovetailinc.org/upload/tmp/1579558792.pdf.

Ob-eye, J., P. Praserthdam and B. Jongsomjit. 2019. Dehydrogenation of ethanol to acetaldehyde over different metals supported on carbon catalysts. *Catalysts* 9:66.

Olofsson, K., M. Bertilsson and G. Liden. 2008. A short review on SSF: An interesting process option for ethanol production from lignocellulosic feedstocks. *Biotechnology for Biofuels* 1:1–14.

Pabon, A. M. A., F. E. Felissia, C. M. Mendieta, E. R. Chamorro and M. C. Area. 2020. Improvement of bioethanol production from rice husks. *Cellulose Chemistry and Technology* 54:689–698.

Perez, V., A. Pascual and A. Rodrigo, et al. 2020. Integrated innovative biorefinery for the transformation of municipal solid waste into biobased products. In *Waste Biorefinery,* Eds. T. Bhaskar, A. Pandey, E. R. Rene and D. C. W. Tsang, pp. 41–80. Amsterdam: Elsevier.

Phung, T. K., T. L. M. Pham, K. B. Vu and G. Busca. 2021. (Bio)propylene production processes: A critical review. *Journal of Environmental Chemical Engineering* 9:105673.

Piotrowski, W. and R. Kubica. 2021. Integration of the process for production of ethyl acetate by an enhanced extraction process. *Processes* 9:1425.

Pomalaza, G., M. Capron, V. Ordomsky and F. Dumeignil. 2016. Recent breakthroughs in the conversion of ethanol to butadiene. *Catalysts* 6:203.

Ramis, G., I. Rossetti, A. Trizpodi and M. Compagnoni, 2017. Diluted bioethanol solutions for the production of hydrogen and ethylene. *Chemical Engineering Transactions* 57:1663–1668.

Rana, P. H. and A. Parikh. 2020. Catalytic transformation of ethanol to industrially relevant fine chemicals. In *Biorefinery of Alternative Resources: Targeting Green Fuels and Platform Chemicals*, Eds. S. Nanda, D. V. N. Vo and S. K. Sarangi, pp. 49–74. Singapore: Springer.

REN21. 2020. *Renewables Global Status Report.* Market and Industry Trends. https://www.ren21.net/gsr-2021/chapters/chapter_03/chapter_03/#target_62_1.

Robak, K. and M. Balcerek. 2018. Review of second generation bioethanol production from residual biomass. *Food Technology and Biotechnology* 56:174–187.

Rodrigues, C. P., P. da Costa Zonetti and L. G. Appel. 2017. Chemicals from ethanol: The acetone synthesis from ethanol employing $Ce_{0.75}Zr_{0.25}O_2$, ZrO_2 and $Cu/ZnO/Al_2O_3$. *Chemistry Central Journal* 11:30.

Rodriguez, M. D., M. L. Castrillo and J. E. Velazquez, et al. 2017. Obtencion de azucares fermentables a partir de aserrin de pino pretratado secuencialmente con acido-base [Fermentable sugars obtained from pine sawdust pretreated sequentially with acid-base]. *Revista Internacional de Contaminacion Ambiental* 33:317–324.

Romani, A., G. Garrote and J. C. Parajo. 2012. Bioethanol production from autohydrolyzed *Eucalyptus globulus* by simultaneous saccharification and fermentation operating at high solids loading. *Fuel* 94:305–312.

Sannigrahi, P., S. J. Miller and A. J. Ragauskas. 2010. Effects of organosolv pretreatment and enzymatic hydrolysis on cellulose structure and crystallinity in Loblolly pine. *Carbohydrate Research* 345:965–970.

Sheldon, R. A. 2021. Biomass processing with biocatalysis. In *Biomass Valorization: Sustainable Methods for the Production of Chemicals*, Eds. D. Ravelli and C. Samor, pp. 113–46. Weinheim: Wiley-VCH.

Sidek, I. S., S. F. S. Draman, S. R. S. Abdullah and N. Anuar. 2019. Current development on bioplastics and its future prospects: An introductory review. *I Tech Mag* 2019:3–8.

Song, Z., A. Takahashi, N. Mimura and T. Fujitani. 2009. Production of propylene from ethanol over ZSM-5 zeolites. *Catalysis Letters* 131:364–369.

Stoffel, R. B. 2016. *Fraccionamiento de Aserrin de Pino Destinado a una Biorrefineria Forestal* [*Pine Sawdust Fractionation for a Forest Biorefinery*]. Ph.D. Thesis. Posadas: National University of Misiones.

Stoffel, R. B., P. V. Neves and F. E. Felissia, et al. 2017. Hemicellulose extraction from slash pine sawdust by steam explosion with sulfuric acid. *Biomass and Bioenergy* 107:93–101.

Sujuthi, R. A. F. M. and K. C. Liew. 2016. Properties of bioplastic sheets made from different types of starch incorporated with recycled newspaper pulp. *Transactions on Science and Technology* 3:257–264.

Susmozas, A., R. Martin-Sampedro and D. Ibarra, et al. 2020. Process strategies for the transition of 1G to advanced bioethanol production. *Processes* 8:1310.

Tsiropoulos, I., A. P. C. Faaij and L. Lundquist, et al. 2015. Life cycle impact assessment of bio-based plastics from sugarcane ethanol. *Journal of Cleaner Production* 90:114–127.

UNCTAD. 2016. *Second Generation Biofuel Markets: State of Play, Trade and Developing Country Perspectives*. Geneva: UCNTAD. https://unctad.org/system/files/official-document/ditcted2015d8_en.pdf.

Vallejos, M. E., F. E. Felissia and M. C. Area. 2017a. Hydrothermal treatments applied to agro- and forest-industrial waste to produce high added-value compounds. *BioResources* 12:2058–2080.

Vallejos, M. E., J. Kruyeniski and M. C. Area. 2017b. Second-generation bioethanol from industrial wood waste of South American species. *Biofuel Research Journal* 4:654–667.

Vallejos, M. E., M. D. Zambon, M. C. Area and A. A. S. Curvelo. 2019. Influence of the chemical composition on the enzymatic hydrolysis of hot water and organosolv pretreated sugarcane bagasse. *Waste and Biomass Valorization* 11:3337–3344.

Wang, Z., S. Li, S. Wang, J. Liu, Y. Zhao and X. Ma. 2022. Coupling effect of bifunctional ZnCe@SBA-15 catalyst in 1,3-butadiene production from bioethanol. *Chinese Journal of Chemical Engineering* 22:162–170.

Wong, Y. C. and V. Sanggari. 2014. Bioethanol production from sugarcane bagasse using fermentation process. *Oriental Journal of Chemistry* 30:507–513.

Wu, X., G. Fang and Y. Tong, et al. 2018. Catalytic upgrading of ethanol to n-butanol: Progress in catalyst development. *ChemSusChem* 11:71–85.

Wyman, C. E., D. D. Spindler and K. Grohmann. 1992. Simultaneous and fermentation of several lignocellulosic feedstocks to fuel ethanol. *Biomass and Bioenergy* 3:301–307.

Xiang, N., P. Xu, N. Ran and T. Ye. 2017. Production of acetic acid from ethanol over CuCr catalysts via dehydrogenation-(aldehyde-water shift) reaction. *RSC Advances* 7:38586–88593.

Xue, F., C. Miao, Y. Yue, W. Hua and Z. Gao. 2017. Direct conversion of bio-ethanol to propylene in high yield over the composite of In_2O_3 and zeolite beta. *Green Chemistry* 19:5582–5590.

Zhang, Z., S. Hu, D. Chen and B. Zhu. 2016. An analysis of an ethanol-based, whole-crop refinery system in China. *Chinese Journal of Chemical Engineering* 24:1609–1618.

Index

FSC
www.fsc.org
MIX
Papier | Fördert gute Waldnutzung
FSC® C083411

Kolibri 360 GmbH
Ferdinand-Jühlke-Straße 7
99095 Erfurt, Deutschland
produktsicherheit@kolibri360.de